Ceramic Materials

Ceramic Materials

Processes, Properties and Applications

Edited by
Philippe Boch
Jean-Claude Nièpce

First published in France by Hermès Science Publications in two volumes entitled "Matériaux et processus céramiques" and " Propriétés et applications des céramiques". First published in Great Britain and the United States in 2007 by ISTE Ltd

ISTE Ltd
6 Fitzroy Square
London W1T 5DX
UK

ISTE USA
4308 Patrice Road
Newport Beach, CA 92663
USA

www.iste.co.uk

© ISTE Ltd.
© HERMES Science Europe Ltd.

First South Asian Edition 2007

The rights of Philippe Boch and Jean-Claude Niepce to be identified as the authors of this work have been asserted by them in accordance with the Copyright, Designs and Patents Act 1988.

This edition is for sale in India, Pakistan, Bangladesh, Sri Lanka and Nepal only. Not for export elsewhere.

Library of Congress Cataloging-in-Publication Data

[Materiaux et processus céramiques]
Ceramic materials: processes, properties and applications edited by Philippe Boch, Jean-Claude Niepce.
   p. cm.
First published in France by Hermes Science Publications in two volumes entitled "Matériaux et processus céramiques" and "Propriétés et applications des céramiques".
Includes bibliographical references and index.
ISBN 10: 1-905209-23-1
ISBN 13: 978-1-905209-23-1
1. Ceramic materials. I. Boch, Philippe. II. Niepce, Jean-Claude.
TA430.M38713   2006
620.1'4—dc22

2006028746

British Library Cataloguing-in-Publication Data
A CIP record for this book is available from the British Library
ISBN 10: 1-905209-23-1
ISBN 13: 978-1-905209-23-1

Printed by Brijbasi Art Press Ltd., I-72, Sector-9, Noida, U.P. India.

Table of Contents

Preface xv

Part I: Ceramics: Materials and Processes 1

Chapter 1. Ceramic Compounds: Ceramic Materials. 3
Philippe BOCH and Jean-François BAUMARD

1.1. Ceramics 3
    1.1.1. Ceramics and terra cotta 3
    1.1.2. Ceramics: physics, chemistry and materials engineering 4
    1.1.3. Powders; sintering 6
    1.1.4. A few definitions 8
1.2. Ceramic compounds. 9
    1.2.1. Chemistry of ceramics 9
    1.2.2. Silicate ceramics and non-silicate ceramics. 11
1.3. Silicate ceramics. 12
1.4. Non-silicate ceramics 14
    1.4.1. Structural ceramics 15
    1.4.2. Functional ceramics 16
1.5. Ceramic structures and microstructures 18
    1.5.1. Ceramic structures 18
    1.5.2. Polymorphism: crystals and glasses 23
    1.5.3. Ceramic microstructures. 25
1.6. Specificity of ceramics 26
1.7. Bibliography 27

Chapter 2. History of Ceramics   29
Anne BOUQUILLON

2.1. Ceramics and clays   29
2.2. The first ceramics: sporadic occurrences as early as the end
of the Paleolithic   30
2.3. The Neolithic era: the true beginning.   31
    2.3.1. Forming and firing   34
    2.3.2. Decorations   35
2.4. Chinese stoneware and porcelains: millenniums ahead   43
    2.4.1. Stoneware   44
    2.4.2. Porcelains   45
2.5. In the quest for porcelains in the East and in the West   46
    2.5.1. Siliceous pastes and glass frit pastes.   47
    2.5.2. Faience fine   48
    2.5.3. The first veritable porcelains in Europe   49
2.6. Conclusion: the beginnings of industrialization   50
2.7. Bibliography   50

Chapter 3. Sintering and Microstructure of Ceramics   55
Philippe BOCH and Anne LERICHE

3.1. Sintering and microstructure of ceramics   55
3.2. Thermodynamics and kinetics: experimental aspects of sintering. . . .   56
    3.2.1. Thermodynamics of sintering.   56
    3.2.2. Matter transport   58
    3.2.3. Experimental aspects of sintering   58
3.3. Interface effects   61
3.4. Matter transport   65
    3.4.1. Viscous flow of vitreous phases   66
    3.4.2. Atomic diffusion in crystallized phases   67
    3.4.3. Grain size distribution: scale effects   68
3.5. Solid phase sintering   70
    3.5.1. The three stages of sintering.   70
    3.5.2. Grain growth.   72
    3.5.3. Competition between consolidation and grain growth   73
    3.5.4. Normal grain growth   74
    3.5.5. Abnormal grain growth   77
3.6. Sintering with liquid phase: vitrification.   78
    3.6.1. Parameters of the liquid phase   78
    3.6.2. The stages in liquid phase sintering   80
3.7. Sintering additives: sintering maps   82
3.8. Pressure sintering and hot isostatic pressing   85
    3.8.1. Applying a pressure during sintering   85

3.8.2. Pressure sintering   86
3.8.3. Hot isostatic pressingn (HIP)   88
3.8.4. Densification/conformity of shapes in HIP   90
3.9. Bibliography   92

Chapter 4. Silicate Ceramics   95
Jean-Pierre BONNET and Jean-Marie GAILLARD

4.1. Introduction.   95
4.2. General information   96
4.3. The main raw materials   98
    4.3.1. Introduction   98
    4.3.2. Clays   98
    4.3.3. Kaolinite   99
    4.3.4. Feldspars   101
    4.3.5. Silica   102
4.4. Enamel and decorations   103
    4.4.1. Nature of enamel   103
    4.4.2. Enamel/shard combination   104
    4.4.3. Optical properties of enamel   104
    4.4.4. Decorations   105
4.5. The products   105
    4.5.1. Classification   105
    4.5.2. Terra cotta products.   106
    4.5.3. Earthenwares.   108
    4.5.4. Stonewares   110
    4.5.5. Porcelains   114
    4.5.6. Vitreous china   117
4.6. Evolution of processes: the example of crockery.   118
    4.6.1. Introduction . . . . —   118
    4.6.2. The case of crockery   118
4.7. Conclusion   119
4.8. Bibliography   120

Chapter 5. Ceramic Forming Processes   123
Thierry CHARTIER

5.1. Introduction.   123
5.2. Ceramic powders   125
    5.2.1. Chemical characteristics   125
    5.2.2. Physical properties   126
    5.2.3. Particle packing   128
    5.2.4. Influence of powders on the rheology of mixtures   131
    5.2.5. Modifications in the characteristics of a powder.   132

5.3. Ceramic particle suspensions   133
    5.3.1. Introduction   133
    5.3.2. Surface charge of oxides in water   133
    5.3.3. Oxide/solution interface   135
    5.3.4. The zeta potential (ξ).   136
    5.3.5. Measurement of the zeta potential (ξ)   137
    5.3.6. Electrostatic stabilization   138
    5.3.7. Steric and electrosteric stabilizations   141
    5.3.8. Rheology of ceramic systems   142
5.4. Casting   147
    5.4.1. Slip casting.   147
    5.4.2. Pressure casting   152
    5.4.3. Tape casting   153
    5.4.4. Consolidation of concentrated suspensions   158
5.5. Pressing   159
    5.5.1. Introduction   159
    5.5.2. Granulation.   160
    5.5.3. Uniaxial pressing   164
    5.5.4. Isostatic pressing   168
    5.5.5. Semi-isostatic pressing.   168
    5.5.6. Roller compressing   169
5.6. Extrusion-injection molding   170
    5.6.1. Introduction   170
    5.6.2. The choice of components.   171
    5.6.3. Mixing   172
    5.6.4. Shaping the mixture   176
    5.6.5. Common flaws   182
    5.6.6. Other plastic shaping techniques of traditional ceramics   183
5.7. Extraction of organic shaping additives   184
    5.7.1. Introduction   184
    5.7.2. Thermal debinding   184
    5.7.3. Other debinding techniques   187
5.8. Deposition techniques.   191
    5.8.1. Vacuum deposition [RIC 94]   191
    5.8.2. Thermal spray deposition [FAU 00]   193
5.9. Bibliography   194

Chapter 6. Alumina, Mullite and Spinel, Zirconia   199
Philippe BOCH and Thierry CHARTIER

6.1. Alumina, silica and mullite, magnesia and spinel, zirconia.   199
6.2. Alumina   200
    6.2.1. Alpha alumina   200
    6.2.2. Alumina and its numerous varieties: the Bayer process   202
6.3. Alumina ceramics   205
    6.3.1. Structural applications of alumina   205
    6.3.2. Functional applications of alumina   209
6.4. Sintering of dense alumina   211
6.5. Point defects and diffusion in alumina   213
6.6. Al_2O_3-SiO_2 system mullite   213
6.7. Al_2O_3-MgO system: magnesia and spinel   216
    6.7.1. Magnesia   216
    6.7.2. Spinel   217
6.8. Zirconia   219
    6.8.1. Polymorphism of zirconia   219
    6.8.2. Ceramic steel?   220
    6.8.3. Transformation toughening   222
    6.8.4. ZrO_2 based materials: treatments and microstructures   225
6.9. Other oxides   227
6.10. Bibliography   228

Chapter 7. Non-oxide Ceramics   231
Paul GOURSAT and Sylvie FOUCAUD

7.1. Introduction   231
7.2. Synthesis of non-oxides   232
    7.2.1. Powders   232
    7.2.2. Fibers   235
    7.2.3. Monocrystals   238
    7.2.4. Depositions-coatings   239
7.3. Sintering and microstructure   243
    7.3.1. Silicon nitride   244
    7.3.2. Aluminum nitride   246
    7.3.3. Silicon carbide   246
    7.3.4. Boron carbide   247
7.4. Chemical stability and behavior at high temperature   248
    7.4.1. Oxygen oxidation of Si_3N_4 and SiC   249
    7.4.2. Water vapor oxidation   252
    7.4.3. Corrosion by molten salts   252
7.5. Properties and applications   253
7.6. Bibliography   259

Part II. Properties and Applications of Ceramics   261

Chapter 8. Mechanical Properties of Ceramics   263
Tanguy ROUXEL

8.1. Brittleness and ductility   264
    8.1.1. Brittle behavior   264
    8.1.2. The R-curve effect   271
    8.1.3. Statistical approach to fracture   275
    8.1.4. Brittle/ductile transition   277
    8.1.5. Experimental techniques   279
8.2. Friction, wear and abrasion   283
    8.2.1. Friction and wear   283
    8.2.2. Abrasion   286
8.3. Deformation   287
    8.3.1. Elasticity   287
    8.3.2. Irreversible deformation (by heat)   297
8.4. Damage and in-service behavior   306
    8.4.1. Elastic behavior and slow growth of flaws   307
    8.4.2. Viscoplastic behavior and creep damage   310
    8.4.3. Thermal shocks   313
8.5. Conclusion   318
8.6. Acknowledgements   318
8.7. Bibliography   318

Chapter 9. Materials for Cutting, Drilling and Tribology   325
Henri PASTOR

9.1. Introduction   325
9.2. Materials for cutting   326
    9.2.1. Machining and properties of tools   326
    9.2.2. Processing of materials   327
    9.2.3. Material properties   329
    9.2.4. High speed steels   329
    9.2.5. Sintered hard metals   333
    9.2.6. Cermets   336
    9.2.7. Ceramics   337
    9.2.8. Superhard materials   339
    9.2.9. Conclusion on cutting materials   340
9.3. Materials for drilling (petroleum, natural gas), mining, buildings
and public works   340
    9.3.1. Sintered hard metals   340
    9.3.2. Diamonds and PCD   344

9.4. Materials for tools and components resisting wear, fatigue
and corrosion 346
    9.4.1. Cold working tools 347
    9.4.2. Hot working tools 347
    9.4.3. Components 348
9.5. Conclusion 349
9.6. Bibliography 355

Chapter 10. Refractory Materials 357
Jacques POIRIER

10.1. Introduction 357
10.2. Characteristics and service properties of refractory materials 358
    10.2.1. Definition of a refractory product 358
    10.2.2. Role and function of refractory products 359
    10.2.3 Classification of refractory products 360
    10.2.4. Design and components of electrofused refractory products . . 361
    10.2.5. Design and components of "cohesive particles"
    refractory products 363
    10.2.6. Manufacturing principles 372
    10.2.7. Service properties of refractory products 374
10.3. Wear and degradation factors 375
    10.3.1. Thermochemical behavior of refractories 376
    10.3.2. Thermomechanical behavior of refractories 381
10.4. Conclusion 386
10.5. Bibliography 386

Chapter 11. Ceramics for Electronics 389
Pierre ABÉLARD

11.1. Conductors and insulators 389
    11.1.1. Insulators 391
    11.1.2. Semi-conductors 392
    11.1.3. Metal conductors 401
    11.1.4. Transition metal oxides 402
    11.1.5. Localization or delocalization? 404
    11.1.6. Contribution of ions to conductivity 407
11.2. Dielectrics 414
    11.2.1. Basic concepts 414
    11.2.2. Ferroelectric materials 423
    11.2.3. Relaxors 431
11.3. Magnetic materials 433
    11.3.1. Paramagnetism 434
    11.3.2. Antiferromagnetism 435

11.3.3. Ferrimagnetism   435
11.4. Electronic properties of surfaces and interfaces in
semi-conductor ceramic materials   439
    11.4.1. Surface   440
    11.4.2. Metal semi-conductor interface   444
    11.4.3. Polycrystalline materials   445
11.5. Influence of microstructure on electrical properties   447
    11.5.1. Modeling of apparent conductivity   447
    11.5.2. Composite materials   453
    11.5.3. Complex impedance measurements: a technique for studying
    heterogenous systems   457
11.6. Ceramic components in electronics   459
    11.6.1. Substrates, interconnection circuits and hybrid circuits   459
    11.6.2. Capacitors.   463
    11.6.3. Electromechanical transducers   469
    11.6.4. Resonators   473
    11.6.5. Heating elements   475
    11.6.6. Temperature sensors   475
    11.6.7. Varistors   483
    11.6.8. Ceramic magnets   485
    11.6.9. Gas sensors and fuel batteries   487
11.7. Bibliography   490

Chapter 12. Bioceramics   493
Christèle COMBES and Christian REY

12.1. Introduction and history.   493
12.2. Biomedical ceramics and their field of use   494
    12.2.1. Usage properties of biomedical ceramics   494
    12.2.2. Multipurpose ceramics   494
    12.2.3. Ceramics for specific uses   497
12.3. Biological properties.   503
    12.3.1. Ceramic-tissue interactions   503
    12.3.2. Cell-ceramic interactions.   505
    12.3.3. Biodegradation.   508
    12.3.4. Standards and biological tests   512
12.4. Processing of bioceramics   513
    12.4.1. Massive ceramics   513
    12.4.2. Thin coatings.   514
    12.4.3. Cements.   515
    12.4.4. Composites   516
12.5. Bibliography   517

Chapter 13. Nuclear Ceramics: Fuels, Absorbers and Inert Matrices . . .   523
Clément LEMAIGNAN and Jean-Claude NIEPCE

13.1. Introduction   523
13.2. Fuel element   524
    13.2.1. Fuel fabrication (UO_2 or MOX).   525
    13.2.2. Behavior of nuclear fuel under irradiation.   528
13.3. Absorptive ceramics   534
    13.3.1. Fabrication of ceramic absorber materials.   536
    13.3.2. Behavior of B_4C in reactor.   536
13.4. "Inert matrix" ceramics of the fuel and other nuclear ceramics   537
13.5. Bibliography   538

Chapter 14. Sol-gel Methods and Optical Properties   539
Jean-Pierre BOILOT and Jacques MUGNIER

14.1. Physico-chemistry of gels   539
    14.1.1. Sols and silica gels   539
    14.1.2. Growth and structure of sol-gel polymers   542
    14.1.3. From gel to materials   548
14.2. Sol-gel process for multi-elemental oxides   550
    14.2.1. Metallo-organic molecular precursors   550
    14.2.2. Different methods of synthesis   552
    14.2.3. Crystallization problem.   553
14.3. Optical properties of sol-gel thin layers   554
    14.3.1. Sol-gel thin layers   554
    14.3.2. Planar optical waveguides   555
    14.3.3. Development and method of analysis of planar sol-gel
    wave guides   556
    14.3.4. Other applications of thin sol-gel coatings   559
14.4. Conclusions and future prospects   560
14.5. Bibliography   561

List of Authors   565

Index   567

Preface

It would be a misstatement to claim that this title provides an "exhaustiveness" coverage of the world of ceramic materials, if only because this subject covers such a wide area that only one book would not be able to do it justice – even if it is supplemented by a book dedicated to ceramic composites and by several volumes devoted to the cousins of ceramics, namely glasses, cements and concretes, and geomaterials. However, we believe that this problem can also be an advantage, because it resulted in the production of a concise work that provides the essence of the subject. In fact, initially this book was written for non-ceramists – students and engineers – interested in an introduction to the knowledge of this vast family of materials, in other words, for readers whose interests are not limited only to ceramics but for those who are aware that ceramics can act as a support (often) or as a rival (sometimes) to other materials.

The necessity of concision has led to some restrictions. We have eliminated recapitulations, for example, thermodynamics or crystallography; we have not discussed materials that seemed far from the interests of our potential readers, such as monocrystals (jewelry and technical uses), ceramics for chemistry (catalysis and filtration), or superconductive oxides; finally, we have not dealt with "black ceramics" (graphite and carbons) or with deposits of diamond or hard carbon.

As regards the division of this book into two parts, our choice was to devote the first part to the fundamentals of *materials* and *processes* and to devote the second to *properties* and *applications*. An essential difference between metals and ceramics lies in the fact that, for the former, there is generally a separation between the industry that produces the *material* (for example, in the form of sheets) and the industry that manufactures the *part* (for example, the car body structure), while, for the latter, it is the same ceramist who is in charge of manufacturing, almost simultaneously, both *the material and the part*.

This explains why, in Part 1, we have given an important place to *production processes*: raw materials, processing of powders, forming and, finally, sintering (Chapters 1, 3 and 5).

Being materials older than metals and materials that had experienced unequalled aesthetic and technical successes before metals (the porcelains of China), ceramics become more interesting, even in their latest forms, when observed from a viewpoint that does not exclude history. We have therefore presented ancient ceramics (Chapter 2), as well as silicate ceramics – ceramics that are wrongly described as "traditional", a term which could conceal the permanent innovations that they undergo (Chapter 4).

"Technical ceramics" (to use another commonly used term) are mainly oxides (Chapter 6), but non-oxides (chapter 7) have recently experienced a number of developments. The presentation of the main compounds – oxides and non-oxides – throws light on the characteristics and therefore the potential uses of the corresponding materials, which opens the way to the second part devoted to properties and applications.

In Part 2, Chapter 8 describes the mechanical properties of ceramics by emphasizing the specificities of these materials. The fields of abrasion, cutting and tribology highlight the importance of mechanical properties, *cermets* (Chapter 9) here bridging the gap between the world of CERamics and the world of METals. Refractories (Chapter 10) widen the requirements by combining mechanical stresses, the effects of high temperature and severe chemical aggressions.

It is not just structural materials which require satisfactory mechanical properties; functional materials are also demanding: broken spectacles no longer correct vision and henceforth manufacturers of spectacles work mainly to obtain better resistances to scratches rather than fine-tuning optical performances! We have therefore given functional ceramics their due, and particularly ceramics for electronics (Chapter 11) which represent the bulk of "technical ceramics".

Bioceramics (Chapter 12) also illustrate the complementary nature of structural performances and functional performances: high mechanical properties are required but these should not enter into conflict with biocompatibility. If necessary, these must even disappear from the pedestal of bone reconstruction; this richness and complexity in behavior could not be ignored.

France is one of the largest electronuclear countries and the nuclear world is the perfect example for the interlacing of questions and answers that a ceramist could encounter: this forms the subject matter of Chapter 13.

Lastly, chemistry forms a basic discipline for ceramists, particularly when it requires making materials as demanding as materials in optical applications: Chapter 14 describes the contribution of "soft chemistry" with the help of sol-gel methods.

Ceramics show that diversity and unity are not contradictory. The compounds are multiple, the applications are varied, the properties brought into play are different, even contradictory – ceramic oxides range from superconductors to the best electrical insulators. However, the chemistry of the systems brought into play, the nature of the interatomic bonds and, from the engineer's view point, the production processes serve as the connecting point. Naysayers could suggest that there is another common point (and which, worse, is a weak point): brittleness. It is true (except at high temperatures where it is creep that poses problem), but I believe our readers are informed enough to understand that the term "brittle" (in the meaning of "non-ductile") is a bad quality only for those who do not understand its profound significance. As Chapter 8 shows, brittleness is a hydra that can be controlled and, moreover, it is the price to be paid for high modulii of elasticity and high hardness. The horizontal vision (*materials science and materials engineering*) of the Anglo-Saxons offers the advantage of including within a single vast landscape metals, polymers and ceramics – a term then understood in its broadest meaning of "non-metallic inorganic solids". In this vision, the varied characteristics of materials are less seen as "strong points" or "weak points" than as givens, admittedly often exclusive of each other, but the comparison of usage specifications makes it possible to select them as effectively as possible.

Having begun my scientific career with the study of metals and continuing to cooperate with metallurgists, surrounded by polymerist colleagues, then having become a ceramist and now particularly interested in cementing materials, filled with wonder for minerals – they are beautiful and efficient and show all that we can achieve if we had enough time and if we could work more easily under high pressures – I am persuaded that it is by comparing materials with one another that we can best understand them. I hope that this book on ceramics becomes the tree that does not hide the forest.

Philippe Boch

At the time where the French version of this work would have to become the English version, Philippe had to leave us on the bad side of the forest. I became in charge of the tree.

Philippe, the English version of your book on ceramics is now ready; I hope it is faithful to that you wished and that the tree is now on the right side of the forest.

Jean-Claude Nièpce

PART I

Ceramics: Materials and Processes

Chapter 1

Ceramic Compounds:
Ceramic Materials

1.1. Ceramics

Defining what the term "ceramic" means is not simple, as there is no single definition on which everyone agrees; there are in fact various definitions depending on the point of view adopted. We can thus consider the points of view of a historian, a scientist (physicist, chemist, etc.), an engineer or a manufacturer.

1.1.1. *Ceramics and terra cotta*

The concept of ceramics is historically related to the concept of terra cotta and pottery, from which the Greek term κεραμοσ derives. This vision refers to the soils, the crushed rocks, that is, the geological materials and it also highlights firing: ceramic art is the art of fire, even if the ceramist has his feet in clay...

The potter chooses suitable soils, primarily clay soils, which in the wet state offer the plasticity required to model them in the desired form: cup; vase or statuette. Then the piece is dried and water loss makes it lose its plasticity, but rehydration would restore the clay's initial properties. In fact, dried clay is not yet a ceramic, although it is utilized in the production of rudimentary bricks used in very dry countries – like Saharan Africa. It is in fact the firing that causes the irreversible

Chapter written by Philippe BOCH and Jean-François BAUMARD.

physicochemical transformations resulting in a material that has lost its plasticity and is no longer capable of rehydration: ceramic. Terra cotta bricks or flower pots are examples of these products, whose visual appearance is not very different from dried clays, but whose mechanical resistance is much higher and, for which water-insensitivity constitutes an essential property.

The identification between "ceramic" and "terra cotta" gathers together the basic concepts that we will continue to encounter throughout this book: powdery mineral raw materials [RIN 96], the shaping which is made possible by the plasticity of wet clay, the heat treatments which start by drying (reversible dehydration) and continue with firing (irreversible dehydration and permanent physicochemical modifications). We have not yet mentioned in the description a major characteristic that conditions the preparation techniques as well as the uses of ceramics: brittleness. The flower pot is hard (it can scratch a metal sheet) but is vulnerable to impact. This brittleness is a hydra with many heads, as it implies:

i) lack of ductility and plasticity;

ii) low toughness and therefore great sensitivity to the notch effect;

iii) poor mechanical impact resistance, i.e. poor resilience;

iv) vulnerability to differential expansions, therefore vulnerability to thermal shocks;

v) mechanical tensile strength significantly lower than compression strength;

vi) significant dispersion of the mechanical strengths in samples believed to be identical.

But the hydra can be tamed and these drawbacks come with a set of qualities that make ceramics irreplaceable for innumerable applications.

We must also point out that artists, historians or museologists often use the term "ceramic" as a noun rather than as an adjective: "a Greek ceramic..." meaning "a Greek ceramic vase...".

1.1.2. Ceramics: physics, chemistry, science and materials engineering

The definitions depend on the point of view adopted:

– solid state physicists are particularly interested in the electronic structure of solids and their conduction properties [KIT 98]. It is therefore natural for them to classify solids – as they seldom speak about materials – in a ternary division: i) insulators ii) semiconductors and iii) conductors and superconductors. The term ceramic is rarely used, but when it is, it refers to oxides, whether they are insulators or conductors, even superconductors like cuprates, which earned Bednorz and Müller the Nobel Prize [BED 86];

– solid state chemists attach particular importance to the nature of the bonding forces, by distinguishing three types of strong bonds (metallic bond, ionic bond and covalent bond) and various variants of weak bonds (like the Van der Waals bond) [JAF 88, WEL 84, WES 90]. Ceramics are therefore solids with essential bonds. However, solid state chemists usually reserve the term ceramics for polycrystalline materials, as opposed to crystals (monocrystalline) and glasses. Thus, a crystallized polycrystalline silica (SiO_2) – for example, in the form of quartz – obtained by sintering (we will explain the meaning of this term a little further on) [GER 86] is regarded as a ceramic, but not the corresponding monocrystal (it is a crystal), or quartz glass – whose amorphous structure makes it unworthy of being qualified as a ceramic;

– materials science adopts a ternary classification that distinguishes i) organic materials, ii) inorganic and metallic materials and iii) inorganic and non-metallic materials. The concept of ceramic compounds is extremely broad here, as it is synonymous with the third category of inorganic and non-metallic materials. Therefore, ceramic compounds include most minerals and rocks, i.e. almost the entire crust of our planet. It should be noted that this definition of ceramics shifts the problem to other definitions, including the exact meaning that must be given to the term metallic. Tungsten carbide (WC), for example, can be classified among ceramics, but it is an electronic conductor that physicists classify within the group of metallic solids;

– materials engineering is an engineers' science, and thus considers not only the chemical composition, structure and properties of solids, but also their method of preparation. It is therefore appropriate to split in two each of the three categories put forward by materials science and separate natural materials from synthetic materials. Organic materials are then differentiated into natural products (like wood) and synthetic products (like most polymers) – but under what can we classify plywood? Metallic materials distinguish the rare native metals (gold, sometimes copper) from all metals and alloys that are derived from industrial processes. Finally, inorganic and non-metallic materials, classify on one side minerals and on the other rocks and ceramics or, more exactly, the triplet: ceramics + glasses + hydraulic binders (cement and plaster are examples of hydraulic binders).

All these definitions are useful and they are not mutually exclusive. It is advantageous, depending on the subject treated, to adopt a particular point of view – the vision of the physicist being undoubtedly preferable when we consider the electronic conduction properties of ceramics and the vision of the materials engineers being undoubtedly more relevant when we have industrial concerns. Now we will discuss the most widely used manufacturing techniques, as they help us to locate and define what we will mean by ceramics in this book..

1.1.3. *Powders; sintering*

We have said that brittleness is a predominant feature of ceramics – at least at normal temperatures, because at very high temperatures (typically above 1,000°C), a certain plasticity can be observed. Brittleness causes limitations in the uses of ceramics (it is advisable to avoid bulls in a china shop…), but it also induces restrictions in the production techniques applicable to ceramics. This opposition is most evident between metals and ceramics:

– a metallic object is typically obtained in two stages: production of the material in the form of a semi-finished product (bars, sheets, wires, etc.), and then production of the object (body of a car or a bolt, etc.). The processing methods take advantage of the fact that most metals are sufficiently ductile and malleable to laminate them, to deform them plastically, to stretch them, etc. and that they are sufficiently soft to cut them, turn them, mill them, bore them, etc.;

– the production of a ceramic object is not dissociated from the processing of the material: the potter does not cut his vase in an already consolidated ceramic preform! The firing of the wet clay rough shape solidifies the object in its final form while simultaneously allowing physicochemical reactions transforming the raw materials into ceramic phases. In most cases, the ceramist creates the object whilst working on the material that constitutes it.

It is true that some metal parts are produced by foundry, a process where the processing of the material and preparation of the object are concomitant. But foundry is seldom applied to ceramics because of the high temperatures that would be required – as the melting points of ceramics are generally higher than those of common metals – and also because of its brittleness, which makes it necessary to avoid thermal shocks and differential expansions. All these limitations on the possible manufacturing processes explain why the basic technique for the preparation of ceramic parts is sintering, i.e. the transformation, using the mechanisms of atomic diffusion, of a powdery substance – a non-cohesive granular medium made up of loosely agglomerated particles, therefore without any marked mechanical properties – into a consolidated substance – a cohesive granular medium whose grains are strongly linked to one another, hence with strong mechanical properties of a solid and not the poor performance of a powder. We can give the example of a snowball, a non-cohesive material that we can throw at someone's face without any risk of hurting that person, but which can be transformed into a block of ice, a cohesive material, which can hurt when thrown like a stone. The movements of matter (atomic diffusion) that allow sintering are activated thermally and it is only when the temperature is sufficiently high (typically 0.5 T_f to 0.8 T_f, where T_f is the melting point expressed in Kelvin) that sintering occurs at a usable speed. Snow is transformed easily into ice because, even in very cold areas, temperatures remain high in relation to melting, at 273 K. But for ceramics, where T_f generally exceeds

1,300 K and can exceed 2,500 K, sintering requires high temperature firing: that is why ceramics is an art of fire. We may note that sintering is the basic process for the preparation of ceramic parts, but it is also increasingly used for the manufacture of metallic objects (powder metallurgy).

What is true for ceramics can be transposed, partly, to glasses and hydraulic binders: these three types of materials are interrelated and we can say that all three bring into play ceramic compounds. But if we consider that a ceramic object is made of ceramic compounds processed by *ceramic* techniques, we can differentiate these materials by the order in which the three fundamental steps of the process take place: powders (P), forming of the object (F) and heat treatments of drying and sintering (HT):

– P→F→HT: the manufacture of a ceramic component starts from a powdery medium (P), continues with its forming (F), and then ends with the heat treatments (HT). The consolidation of the material is done during sintering, therefore during the high temperature treatment;

– P→HT→F: the manufacture of a glass component also starts with the powdery medium (P), but this is followed by heat treatments (HT), which must result in melting, whereas the consolidation occurs at the end of the process, at the time of the solidification of the magma on cooling, therefore at the same time as the forming of the object (F);

– HT→P→F: the use of hydraulic binders starts with heat treatments (HT) – for example the firing, at about 1,450°C, of a mixture of limestone and clay for the preparation of cement clinker – then the reactive powders (P) thus produced are formed (F) and consolidated at normal temperatures, due to chemical reactions.

This differentiation of ceramics, glasses and hydraulic binders based on the order of operations P/F/HT is rather simplistic, as it refers only to the most common cases. In fact, some ceramics are prepared by other sequences than the P→F→HT trio, some glasses are fashioned by sintering, and there are other exceptions, some of which will be discussed in detail in this book. However, the classification suggested is sufficiently relevant to be retained here, which implies that this volume on ceramics does not consider either glasses [ZAR 82] or hydraulic binders [BAR 96, TAY 97], these materials being studied in separate books. A logic emphasizing "the scientific aspects" rather than "the technical and industrial aspects" could have, on the other hand, justified combining the three categories of materials.

The distinction made here between industrial ceramics and glasses lies in the method of preparation, not in the difference between crystalline solids and amorphous solids. In fact, some ceramics are mainly made up of vitreous (amorphous) phases, whereas vitroceramics are obtained by crystallizing a glass (devitrification).

1.1.4. *A few definitions*

In the reference work on ceramics [KIN 76], the authors place "the art and the science of ceramics in the production and use of objects formed of solids, whose essential components are inorganic and non-metallic materials". This definition includes not only potteries, porcelains, refractory materials, terra cotta products, abrasives, sheet enamels, cements and glasses, but also magnetic non-metallic materials, ferroelectric materials, synthetic monocrystals and vitroceramics, not to mention "a variety of other products that did not exist a few years ago and the many others that do not yet exist...". Kingery *et al.* stress that this definition largely exceeds terra cotta products consolidated by firing, to which the Greek term κεραμοσ refers, just as it exceeds the definitions given by most dictionaries – but our *Petit Larousse* is apt, defining the adjective "ceramic" as "relating to the manufacture of potteries and other terra cotta objects (including earthenware, stoneware, porcelain)" but also "ceramic material or ceramic: manufactured material that is neither a metal nor an organic material". Kingery *et al.* adopt a scientific approach, which considers compounds rather than materials and thus includes glasses and cements among ceramics, but they do not make the jump to natural materials, since they mention only synthetic monocrystals, and yet the quartz that man has synthesized to cut out small piezoelectric resonators that form the heart of modern watches is a twin of the natural rock crystal.

In the *Dictionary of Ceramic Science and Engineering* [OBA 84], we find a rather restrictive definition of ceramics: "Any inorganic and non-metallic product prepared by treatment at temperatures higher than 540°C (1,000°F) or used under conditions implying these temperatures, which includes metallic oxides and borides, carbides, nitrides and mixtures of these compounds".

We will see that *kaolinite* $Al_2(Si_2O_5)(OH)_4$ – which is the main mineral of clays, hence the name *kaolin* given to some clays rich in kaolinite – undergoes irreversible reactions that transform it into metakaolinite when it is fired above approximately 500°C, which could justify the temperature of 540°C selected in the definition used to make the distinction between dried earth and terra cotta.

In the *Concise Encyclopedia of Advanced Ceramic Materials* [BRO 91], which presents the common viewpoint held in Europe, we can distinguish materials and processes:

– ceramic materials are based on inorganic non-metallic compounds, primarily oxides, but also nitrides, carbides, silicides; they must contain at least 30% of crystallized phases in volume; they exhibit a fragile behavior, with a stress-strain curve that obeys Hooke's law of linear elasticity;

– ceramic processes bring sintering primarily into play, at temperatures higher than 800°C.

We will now pause the study on the definitions of the word "ceramic", which was necessary to bring to light some of the keywords that we will find throughout this book (non-metallic inorganic compounds, mineral powders, heat treatments and sintering, brittleness, etc.) and also to justify the differences that arise from the variety of possible points of view. If we propose our own definition, it could be the following: *ceramic materials are synthetic materials, mainly composed of iono-covalent inorganic phases, not fully amorphous, and generally consolidated by the sintering at high temperatures of a powdery "compact" formed into the shape of the desired object, the starting powders being frequently prepared from crushed rocks.* We do not think it necessary to define a threshold for the sintering temperature; it is important, with the development of composites, to suggest that besides inorganic phases, minority organic phases can occur; the existence of interesting ceramic materials with electronic conduction means that the reference to the iono-covalent character of the bonds does not exclude a partially metallic nature; lastly, it is necessary to insist on the closeness between the world of ceramics and that of minerals and rocks. If we were to expand this idea, we could affirm that "ceramics are synthetic rocks".

1.2. Ceramic compounds

Defining a ceramic compound as a *non-metallic inorganic compound* is a paradoxical choice, because it supposes that the concept of metal is sufficiently unambiguous to serve as basis for clarifying its antonym. However, confusions are frequent here, and are aggravated by the fact that ceramic compounds are metallic-non-metallic compounds, which we shall now study in detail.

1.2.1. *Chemistry of ceramics*

Metallic elements form a majority among the elements of the periodic table. We know that these elements are located on the left of this table, and are therefore electropositive elements – which tend to lose electrons to yield positively charged cations. Non-metallic elements, sometimes denoted by their old name metalloids, are located on the right of the table: they are electronegative and tend to capture electrons. The ionic bond is illustrated by the attractions that develop between a metal, sodium, and a non-metal, chlorine, to create sodium chloride $NaCl$ – also written as Na^+Cl^-.

Oxygen, a non-metal, is the most abundant element in the 45 km-thick Earth's crust (approximately 47% in mass) [EMS 95]; then comes an element at the border between metals and non-metals, silicon ($\approx$ 28%), and then metallic elements, particularly aluminum ($\approx$ 8%) and iron ($\approx$ 4%).

Aluminum and iron are the most widespread metals – if we reason in terms of elements, but in a world rich in oxygen, almost all metals tend to oxidize, so that only noble metals remain in the native state, more stable than their oxides: we can find in the soil a few gold nuggets and a little native copper, but never blocks of aluminum or iron; in other words, we do not find in an isolated state materials whose constitutive elements are so abundant. Let us remind ourselves that we left the Stone Age only a few thousand years ago to reach the Bronze Age and then the Iron Age; that it was only in 1824 that Berzelius isolated silicon and in 1825 that Oersted isolated aluminum. In short, it is obvious that the reduction towards the constitutive metal of very stable metallic oxides – their enthalpies of formation are several hundred kilojoules per mole – is a difficult process.

The essentially iono-covalent, non-metallic compounds that constitute ceramics are compounds formed between metals and non-metals. The opposition – the word is not too strong – between a metallic material and a metallic oxide can be illustrated by the comparison between a metal, aluminum, and its oxide, Al_2O_3 (alumina in English, where oxides are named by adding the ending "a" to the name of the metal: aluminum/alumina, silicon/silica, magnesium/magnesia, uranium/urania, etc.):

– aluminum melts at low temperatures (660°C); it is an excellent conductor of electricity and heat, opaque to visible light, soft and very ductile, its modulus of elasticity is low (one-third that of steel) and it is vulnerable to the aggressions of multiple chemical reactants;

– alumina melts at high temperatures (2,050°C); it is one of best known electrical insulators and a poor heat conductor; it is transparent to visible light, very hard but brittle, its modulus of elasticity is high (double that of steel) and it resists most aggressions of chemical reactant: in particular, it is perfectly stable in an oxidizing medium, as it "is already oxidized"

This comparison brings to light the fact that metals and metallic oxides exhibit different, even opposite, characteristics.

Regarded as compounds formed between metals and non-metals, ceramics can therefore be classified with reference to the non-metal that is involved in the bond. The abundance of oxygen is such that the largest share belongs to oxides, while within these oxides, the abundance of silicon privileges silicates – which combine oxygen and silicon. Besides oxides, we find carbides and nitrides (the non-metals being respectively carbon and nitrogen), as well as borides, silicides (not to be confused with silicates), even halides (chlorides, fluorides, iodides, etc.), sulphides, etc. However, another separation distinguishes silicate ceramics and non-silicate ceramics [CER 99].

1.2.2. *Silicate ceramics and non-silicate ceramics*

The abundance of oxygen, silicon and aluminum implies that the Earth's crust consists predominantly (more than 97%) of silicate minerals, particularly aluminosilicates. Granite, for example, is a rock made up of quartz (the usual crystallized form of silica SiO_2), feldspars (for example potassic feldspar $KAlSi_3O_8$) and various micas-aluminosilicates containing iron, magnesium or sodium; clay is a rock that always contains a high proportion of the mineral kaolinite $Al_2(Si_2O_5)(OH)_4$, etc. We can mention here a practice of ceramists that can lead to confusion: they write chemical formulas of oxide compounds as if they were elementary mixtures of oxides and not compounds: $KAlSi_3O_8$ is written as $K_2O\text{-}Al_2O_3\text{-}6SiO_2$ and $Al_2(Si_2O_5)(OH)_4$ is written as $Al_2O_3\text{-}2SiO_2\text{-}2H_2O$. This does not obviously mean that potassic feldspar is an agglomerate of the oxide of potassium, aluminum and silicon, and that kaolinite is a mixture of alumina, silica and water. A common error is therefore to think that the high proportion of aluminum in the soil, which is the reason behind the frequent occurrence of "alumina" in the writing of the constitution of minerals, makes this alumina a common mineral: this is not the case, as can be testified by the fortunate but rare owners of the main gem whose chemical composition is Al_2O_3: sapphire!

Since the Earth's crust is primarily made up of silicates, we can understand why all ceramics that come close, by near or by far, to terra cotta are silicate materials. Silicate ceramics form, in tonnage, the majority of the world of ceramics. They are often described as traditional ceramics, a term which we do not endorse because it may be understood as opposed to progress and technical improvements – whereas many silicate ceramics are sophisticated materials – but which is justified by history: it was only at the end of the 19th century that non-silicate ceramics came to the scene, with specific uses that explain their other name, technical ceramics. Our choice here is to use "silicate ceramics" and "non-silicate ceramics", rather than opposing tradition and advanced technology. We may note that almost all industrial glasses and cements are also silicate compounds.

1.3. Silicate ceramics

If we consider industrial products, and forget for the moment glasses and cements, the world of silicate ceramics (see Chapter 4) consists of the usual products:

– terra cotta products, the most important of these being bricks and tiles, sometimes known as red products because of the color that they owe to the iron oxides they contain;

– ceramic tiles;

– sanitary ceramics;

– tableware: colored, opaque and waterproof sandstone, often colored, opaque and porous earthenware, but generally covered with a waterproofing enamel – sometimes well, sometimes badly when it is cracked, but the esthetic quality is then enhanced – white, translucent and waterproof porcelain; white, opaque and waterproof vitreous objects;

– technical ceramics, for example porcelains for electric insulation or porcelains for resistance to chemical attacks;

– bioceramics (dental prostheses, etc.);

– certain refractory materials, which serve as protection or insulation elements within devices that must function at high temperatures (lining of chimney hearths, etc.);

– enamel sheets, to cover and protect steel sheets that are part of our refrigerators and washing machines. These enamel sheets are too often forgotten in works dealing with ceramics and glasses – but as these enamels are primarily amorphous and prepared by melting, it is better to classify them among glasses than among ceramics;

– silica products – but it is common practice to classify them among technical ceramics, therefore paradoxically, to exclude them from silicate ceramics.

Though non-exhaustive, this list shows the variety of the products involved. As regards the composition on the one hand, and the raw materials brought into play on the other, a silicate ceramics can be located primarily in a ternary diagram:

– by adopting the conventional writing in the form of "mixtures" of elementary oxides, the composition of most silicate ceramics is in the pseudo-ternary diagram SiO_2-Al_2O_3-M_xO_y, where M_xO_y is an oxide like Fe_2O_3, K_2O, Na_2O, MgO, TiO_2, etc. The natural abundance of the element iron explains, as we have seen, the red, brown and yellow coloring of a number of silicate ceramics, where the degree of oxidation of iron (Fe^{3+}: ferric-iron or Fe^{2+}: ferrous-iron) modifies the coloring, as it is well known to producers of tiles and bricks who modify the atmospheres – oxidizing or reducing – of the kilns;

– by considering now the raw materials, we can find a ternary composition, because the three components of silicate ceramics are: i) clays, ii) sand and iii) fluxes – i.e. compounds contributing to the firing thanks to the development of phases with low melting points. As kaolinite clay can be written: Al_2O_3-$2SiO_2$-$2H_2O$, quartz sand: SiO_2, and potassic feldspar, which is frequently used as a flux: K_2O-Al_2O_3-$6SiO_2$, we again find the ternary SiO_2-Al_2O_3-M_xO_y (if $M_xO_y = K_2O$). Figure 1.1 shows the equilibrium diagram Al_2O_3-SiO_2-MgO and locates some of the main compounds that come under it [KEI 52, KIN 76]. Magnesia MgO is useful for refractory materials in iron metallurgy; mullite $3Al_2O_3$-$2SiO_2$, a unique compound defined in the binary diagram Al_2O_3-SiO_2, is a crystallized phase present in many ceramics; cordierite $2MgO$-$2Al_2O_3$-$5SiO_2$ is characterized by very poor thermal expansion: it is used for example as catalyst support in exhaust pipes, etc.

Hydroxyls OH^- are present in many hydrated raw materials and water H_2O allows the plasticity of clays, but because the ions and the corresponding molecules are eliminated in the heat treatments (this is called ignition loss), they are not taken into account in the composition of ceramics after firing.

It is important to distinguish between impregnated water (which occurs as a mixture with rock particles and whose reversible departure is caused by simple drying, with possibility of rehydration in a wet environment) and combined water (which corresponds to the hydroxyls of the hydrated phases, for example to the four OH^- in the formula of kaolinite $Al_2(Si_2O_5)(OH)_4$). The departure of this "water" is accompanied by the disturbance of the crystallographic structure, hence the irreversible transformation at the end of firing beyond approximately 500°C.

Silicate ceramics make the most of the versatility of silica (see section 1.5.2), which can exist in crystallized form (particularly quartz) or in amorphous form (silica glass) and, as a result, contain both crystallized phases and vitreous phases.

The interatomic bonds brought into play in silicate ceramics are typically iono-covalent (SiO_2 exhibiting a fine compromise, because its bonds are regarded as 50% ionic and 50% covalent), therefore these ceramics are almost always electrical insulators. The accentuation of the ionic nature yields hydrolysable compounds: halides can be regarded as ceramic compounds, but the salt-marsh workers are not classified among the producers of ceramic powders!

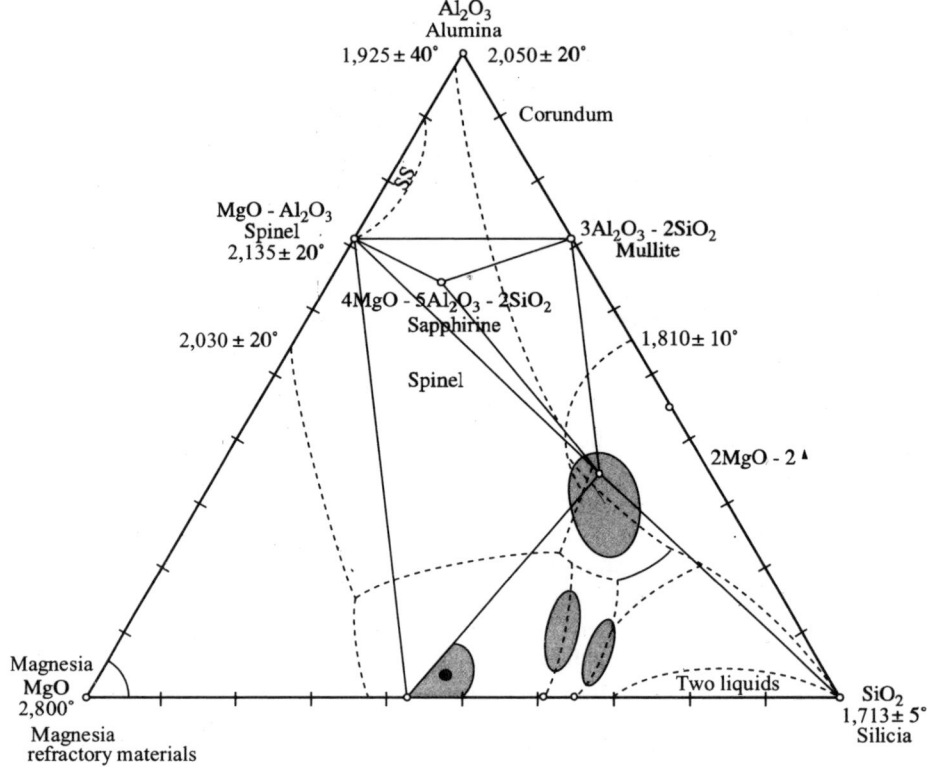

Figure 1.1. *Main compounds in the diagram Al$_2$O$_3$-SiO$_2$-MgO [KIN 76]*

1.4. Non-silicate ceramics

To classify any material, the user can consider two main categories: i) structural materials, whose operating performances are essentially mechanical, even thermal, in nature, and ii) functional materials, whose operating performances are primarily electrical, magnetic, optical, etc. We have said "primarily", because we must underline, especially in a book of this nature, that no application can be exempt from mechanical properties. For example, the glasses in our spectacles are functional materials, designed in such a manner that their optical characteristics correct the defects in our vision, but their impact resistance or their scratch resistance are variables that are more difficult to improve than the optical properties. In addition to their functionalities, functional materials must in general exhibit a sufficient level of mechanical properties.

1.4.1. *Structural ceramics*

The uses of these ceramics vary according to their characteristics:

– for ceramics with high mechanical performances, the established markets include abrasives, cutting tools and tribological applications: wear resistance and friction resistance (see Chapters 6, 7, 8 and 9);

– for ceramics used at high temperatures, the established markets include refractory materials, essential for equipments of iron and steel, glass, cement or incineration industries (see Chapter 10);

– for ceramics that must combine high mechanical performances and high temperatures the markets are more recent, but are growing rapidly. Only these ceramics are sometimes referred to as structural ceramics, but it is better to call them thermomechanical ceramics [CHE 89] (see Chapter 7). Thermostructural composites form the vanguard of thermomechanical ceramics: we will not discuss these composites here.

The first two subdivisions that we have classified here among structural ceramics (abrasives, cutting tools and wear parts on the one hand, industrial refractory materials on the other) are often classified outside the field of ceramics. This is logical for abrasives, because if abrasive grains are ceramic compounds (primarily alumina Al_2O_3 or silicon carbide SiC), abrasives are themselves multi-material systems, for example, grinding stones whose matrix can be a glass or a ceramic, but frequently also a polymeric resin or a metal, or fabrics and papers (sandpaper) whose base is organic. The most widely used cutting tools and wear parts are made of tungsten carbide (WC) grains bonded together by a metal matrix, typically of cobalt. These cemented carbides fall under the category of cermets (for "ceramic-metal"), which are materials prepared by powder metallurgy, and this explains why they are claimed by ceramists and metallurgists. Our choice has been to include cermets among structural ceramics and to cover them in Chapter 9.

Finally, as regards industrial refractory materials, their importance sometimes justifies their being regarded as a distinct category when we speak of "ceramics and refractory materials". Here again, our choice has been to include refractory materials among ceramics (see Chapter 10), which is currently the commonly accepted view, but this does not however imply that refractory materials are always classified among structural ceramics. These remarks are essential to decipher economic data: spread across the three categories that we are considering here, structural ceramics represent a larger market than that of functional ceramics about which we will speak later on, but if reduced to thermomechanical ceramics and thermostructural composites, they represent only a small market in comparison with functional ceramics.

1.4.2. *Functional ceramics*

Functional ceramics are characterized by their:

– electrical properties: insulators (very often), semiconductors (often), conductors (less frequently) and superconductors (a scientifically exciting field, but whose industrial applications are yet to be explored at the time of writing);

– magnetic properties: hard magnets (permanent magnets) or soft magnets (winding cores); the field of magnetic recording is among one of the most spectacular scientific and technical advances with enormous industrial stakes;

– optical properties;

– chemical properties: catalysis, sensors;

– "nuclear" properties: fuels, moderators;

– biological properties: biomaterials and prostheses;

– monocrystals for varied uses, for example for ionizing radiation detectors.

Unlike silicate ceramics, raw materials used for the preparation of non-silicate ceramics are generally synthetic powders and not mixtures of crushed rocks. But these synthetic powders can result from natural products, which the English terminology makes easy to understand by distinguishing between "starting materials" (for example, alumina powders) and "raw materials" (bauxite rocks, in this case, whose treatment by the Bayer process yields the alumina powders) [CAS 90].

Ceramic compositions offer in general a simple chemistry, but microstructural parameters are complex.

1.4.2.1. *Oxide ceramics*

Alumina Al_2O_3 is by far the foremost basic compound for "technical ceramics", because alumina exhibits exceptional versatility: abrasion, cut, friction and wear, refractory uses, electricity and electronics, optics, biomedical, jewelry and the list can go on and on [CAS 90].

Silica SiO_2 is also a basic compound both for ceramists and glassmakers; the alumina-silica diagram has for ceramists the same importance as the iron-carbon diagram has for metallurgists.

Magnesia (MgO) and spinel $(MgAl_2O_4)$ are primarily used as refractory materials in the iron and steel industry.

Zirconia ZrO_2 (not to be confused with zirconium silicate, called zircon, $ZrSiO_4$) is used in the ceramic colors, but also for ionic conduction, mechanical purposes or in jewelry.

Uranium oxide UO_2 is the basic constituent of nuclear fuels, if necessary, as a mixture with a little plutonium oxide PuO_2 (the mixture gives MOXs, or "mixed oxide nuclear fuels").

Barium titanate $BaTiO_3$ is dielectric or a semiconductor, depending on its doping and its stoichiometry. It is the basic material in the industry of ceramic capacitors and it is also used for the manufacture of various types of probes and sensors.

Soft ferrites and hard ferrites or hexaferrites are important materials for magnetic uses. "Soft" ferrites are crystals with a spinel structure whose reference is magnetite Fe_3O_4; "hard" hexaferrites are crystals with a hexagonal structure, whose type is $BaFe_{12}O_{19}$.

Almost all metallic oxides have uses in ceramics, for example yttrium oxide Y_2O_3, beryllium oxide BeO, zinc oxide ZnO, tin oxide SnO_2, superconductive cuprates like $YBa_2Cu_3O_7$, and others.

Most ceramic oxides are electrical insulators, whose electronic conduction is very weak (major exception: superconductors), but whose ionic conduction can be remarkable (for example, zirconia); those oxides that are semiconductors are frequently extrinsic semiconductors, whose performances vary considerably with the nature of the doping agents and their concentration.

1.4.2.2. Non-oxide ceramics

Carbides form the main category of non-oxides [MCC 83], the most important of which are silicon carbide SiC, which is a semiconductor, but whose chemical is essentially covalent, and tungsten carbides, whose name comes from a typically metallic band structure, which therefore exhibits high electronic conductivity: tungsten carbide WC is the main industrial material in this class, which includes many other compounds, for instance, titanium carbide TiC.

Nitrides primarily include silicon nitride Si_3N_4 and aluminum and silicon oxynitrides, also called sialons, aluminum nitride AlN, and various metallic nitrides, including TiN.

Some borides have industrial applications, for example titanium diboride TiB_2 or lanthanum hexaboride LaB_6, and some boron compounds are conventionally classified among borides, including boron carbide B_4C and boron nitride BN, a material which has three polymorphs, including two isostructural carbon polymorphs: graphite and diamond.

Silicides are numerous, but only one of them presents great industrial interest: molybdenum disilicide $MoSi_2$ (not to be confused with molybdenum disulphide MoS_2), which is used for the manufacture of the heating elements of very high temperature electric ovens (1,750°C), in air.

Halides, finally, are more model materials in the chemistry of solids than usable ceramics, even if some of them are used for their optical properties; some chalcogenides could also enter the field of ceramics.

This list omits a class of materials that has not yet been mentioned, in spite of its importance: carbonaceous materials – diamond, graphite, and more or less crystallized carbons that are obtained by heat treatments of tar and pitch, not to mention carbon fullerenes and nanotubes, which have not yet actually reached the stage of industrial products. Although some carbonaceous materials are prepared from organic raw materials, the trend is to classify the materials themselves under inorganic products: we endorse the term black ceramics [LEN 92].

Whereas oxides are mainly electrical insulators, non-oxides equally include insulators (for example, Si_3N_4 and AlN), semiconductors (for example, SiC) and conductors (for example, "metallic" carbides and borides and carbon products other than diamond, of which graphite is the most important).

1.5. Ceramic structures and microstructures

1.5.1. *Ceramic structures*

This discussion on the crystalline structure of ceramics presupposes that the reader is familiar with the basics of crystallography [BUR 90, GIA 85, HAH 89].

Most oxides and silicates have crystalline structures obeying Pauling's rules for ionic crystals, where the ions of small size (generally cations) enter the interstices of big ions (generally anions). The three main rules relate to the coordination number of cations, the coordination number of anions and the coordination number of polyhedra.

The cation coordination number (N_c): the geometry of the polyhedron of anions around a cation depends on the ratio $R = r_{cation}/r_{anion}$. The cation must be in contact with the anions. The ionic radius varies, for a given element, with the charge of the ion and its coordination number.

The anion coordination number (N_a): the geometry of the polyhedron of cations around an anion is such that the sum of the electrostatic attractions resulting in the anion is equal to the charge (p) of this anion.

For $M^{q+} X^{p-}$: s = electrostatic attraction of the link = q/N_a $\Sigma s = p$

The force of the link is obtained by dividing the charge of the cation by its coordination number:

EXAMPLE 1.– NaCl: $Na^{VI}Cl^{VI}$: each Na^+ at the center of an octahedron of 6 Cl^- contributes $+ 1/VI = 1/6$, therefore $N_a(Cl^-) = 6$.

EXAMPLE 2.– SiO_2: $Si^{IV}O^{II}_2$: each Si^{4+} at the center of a tetrahedron of 4 O^{2-} contributes $+ 4/IV = 1$, therefore $N_a (O^{2-}) = 2$.

The linking of polyhedra: the stability of the crystal decreases if the cations are too close, it decreases; therefore a linking of the coordination polyhedra at the corners is more favorable than at their edges and, even more so than at their faces.

Many ceramic compounds have structures that bring into play an appreciably compact stacking of anions with cations in tetrahedral (four neighbors) or octahedral (six neighbors) coordination (see Table 1.1).

Formula	Cation: anion coordination	Type and number of interstices occupied	Compact cubic stacking	Compact hexagonal stacking
MX	6:6	All the octa.	NaCl, FeO, MnS, TiC	NiAs, FeS, NiS
	4:4	Half the tetra.	ZnS blende, CuCl, AgI-γ	ZnS würtzite, AgI-β
MX$_2$	8:4	All the tetra.	CaF$_2$ fluorine, ThO$_2$, ZrO$_2$, UO$_2$	
	6:3	Half the octa. Alternating layers with all the occupied sites	CdCl$_2$	CdI$_2$, TiS$_2$
MX$_3$	6:2	1/3 of the octa. Alternating pairs of layers with 2/3 of the octa. occupied		BiI$_3$, FeCl$_3$, TiCl$_3$, VCl$_3$
M$_2$X$_3$	6:4	2/3 of the octa.		Al$_2$O$_3$ corundum, Fe$_2$O$_3$, V$_2$O$_3$, Ti$_2$O$_3$, Cr$_2$O$_3$
ABO$_3$		2/3 of the octa.		FeTiO$_3$ ilmenite
AB$_2$O$_4$		1/8 of the tetra. and 1/2 of the octa.	MgAl$_2$O$_4$ spinel, MgFe$_2$O$_4$ inverse spinel	Mg$_2$SiO$_4$ olivine

Table 1.1. *Coordination and stacking in a few typical structures*

The structures are varied and we will mention only five of the most important ones (MgO, ZrO$_2$, BaTiO$_3$, Al$_2$O$_3$ and diamond), before discussing the rudiments of the structure of silicates:

– MgO is the example of oxides with NaCl structure (space group Fm $\overline{3}$ m) with Mg in site 4a (0, 0, 0) and O in 4b (1/2, 1/2, 1/2);

– CaF_2 (fluorine) and K_2O (antifluorine) also crystallize in the space group $Fm\overline{3}m$, with Ca in 4a and F in 8c ± (1/4, 1/4, 1/4); zirconia ZrO_2 and urania UO_2 adopt this type of structure;

– $BaTiO_3$ adopts a perovskite structure, with the oxygen octahedra at the center of which are titaniums, linked at their corners and surrounding a perovskite cage occupied by the large barium. A "beads on rods" representation of this structure places titanium at the eight corners of the cube, oxygen at the twelve centers of the edges and barium at the center of the cube (or barium at the eight corners of the cube, oxygen at the six centers of the faces and titanium at the center of the cube). Cuprate superconductors frequently have structures based on the perovskite structure;

– alumina defines the corundum structure where oxygens form a compact stacking with the hexagonal aluminum ions placed in two-thirds of the octahedral sites, which decreases the overall symmetry towards the rhombohedric space group $R\overline{3}c$;

– if it is true that most ceramics have iono-covalent bonds which lead to structures that reasonably obey Pauling's rules, others are markedly covalent. This is the case with silicon carbide, whose structure is similar to that of diamond (or silicon). We can think of a giant covalent molecule, extended to the scale of a crystal: the network is cubic, face centered and the pattern is composed of two carbon atoms, one located at 0, 0, 0 and the other located at 1/4, 1/4, 1/4;

– as regards the various silicates, the description of the structure depends on the manner in which the Si-O bond is modeled. The ionic model predicts a compact stacking of O^{2-}, with Si^{4+} and the other cations that occur in the various interstices. However, most silicates do not have a compact stacking of O^{2-} and the coordination numbers observed often violate the rules deduced from the r_{cation}/r_{anion} ratio: the ionic model is imperfect. The covalent model describes the Si-O bonds by bonding orbitals, which explains the tetrahedral coordination of silicon and the angles between the bonds are close to the theoretical value of 109.5°. But the covalent model stumbles on some hurdles and explains less well than the ionic model the chemical formulas of most silicates and the substitution of silicon by aluminum, which correspond to formal charges: Si^{4+}, Al^{3+}, O^{2-}, etc. In fact, the Si-O bond is 50% ionic and 50% covalent, the structure of silicates having been described based on tetrahedra $[SiO_4]^{4-}$ linked such that: i) the tetrahedra are linked at the corners, ii) a bridging oxygen is common only to two tetrahedra and iii) the formal charges of the ions are Si^{4+} and O^{2-}. The sequencing of the tetrahedra makes it possible to classify the various silicates under six categories, based on an increasing degree of polymerization [PUT 92]:

1) tetrahedra isolated from one another, without bridging oxygens, Si/O ratio = 1/4, (for example, olivine Mg_2SiO_4);

2) two tetrahedra forming a dimer, with oxygens bridging two tetrahedra, each tetrahedron having one bridging and three non-bridging oxygens: Si/O ratio = 1/3.5; charge of the dimer: $[Si_2O_7]^{6-}$, (for example, rankinite $Ca_3Si_2O_7$);

3) single chain silicates, each tetrahedron having two bridging and two non-bridging oxygens: Si/O ratio = 1/3; a chain with N links has a charge $[SiO_3]_n^{2n-}$ (for example, enstatite $MgSiO_3$);

4) double chain silicates, half of the tetrahedra with two bridging and two non-bridging oxygens (Si/O = 1/3) and other half three bridging and one non-bridging (Si/O = 1/(2.5): in total Si/O = 2/5.5 and the charge is $[Si_4O_{11}]_n^{6n-}$ (for example, anthophyllite $Mg_7Si_8O_{22}(OH)_2$, the OHs being independent of the tetrahedra);

5) silicates forming two dimensional layers, each tetrahedron with three bridging and one non-bridging oxygens: Si/O = 1/(2.5); charge of a layer $[Si_2O_5]_n^{2n-}$ (for example, minerals of clays and micas or talc: $Mg_6Si_8O_{20}(OH)_4$, the OHs being here again independent of the tetrahedra);

6) lastly, silicates where the tetrahedra are linked at all their corners: four bridging oxygens per tetrahedron, Si/O = 1/2 (for example, quartz SiO_2). Quartz is part, like diamond, of a covalent description where the molecule extends to the scale of the entire crystal, regularly in the three-dimensional space.

In addition to this classification, we can observe that:

– when Al substitutes Si in the tetrahedron, we must consider the (Al+Si)/O ratio: for example, plagioclase feldspars, which range from albite $NaAlSi_3O_8$ to anorthite $CaAl_2Si_2O_8$, the (Al+Si)/O ratio always being 1/2;

– Al is generally in a tetrahedral site, instead of Si, but can be in an octahedral site: for example, muscovite mica $K_2Al_4^{octa}[Si_6Al_2O_{20}](OH)_4$, where tetrahedral coordination group is the one located between brackets [ ];

– an important point for the structure of hydrous silicates is the fact that O^{2-} and OH⁻ have the same ionic radius: 1.40 Å.

We must not be misled by the examples of MgO, $BaTiO_3$ or diamond: most ceramics do not crystallize in a cubic group and this implies that the many physical properties that are described by a second order tensor are not isotropic [NYE 87]. Thermal expansion, optical index, electric and thermal conductivities, permittivity and permeability are at first view anisotropic, which can misinform some metallurgists, because the most common metals (iron, aluminum, copper) are cubic. An effect of the anisotropy of thermal expansion is to create residual stress at the grain boundaries of the polycrystals. Beyond the properties described by a second order tensor, low symmetries combine with the properties of iono-covalent bonds to make dislocations rare and relatively immobile, which explains the lack of ductility and the impossibility of plastic deformation.

1.5.2. Polymorphism: crystals and glasses

Many compounds of ceramic interest can exist in various varieties. We can mention the polymorphism of zirconia ZrO_2 – cubic at high temperatures, then quadratic (tetragonal) and finally monoclinic at decreasing temperatures – a polymorphism that was regarded for a long time as a disadvantage, then understood as an advantage when the possibilities that it offered for the development of high mechanical performance ceramics were discovered [HEU 81] (see Chapter 6). However, it is the polymorphism of silica SiO_2 that is most frequently used in ceramic and glass industry.

Sand is primarily made up of silica, often highly pure (more than 98%), SiO_2 having been crystallized in the form of quartz α, also known as low quartz (point group 32, without center of symmetry). When heated to 573°C, quartz α transforms to quartz β (high quartz, point group 622, centrosymmetric). Higher treatment temperatures help distinguish two types of behavior, depending on whether thermodynamically stable phases are achieved or whether kinetic effects favor metastable phases. These effects depend on the relative ease with which the transformations occur: displacive transformations – which require only small atomic movements to change the structure of a phase and modify its symmetries – are easier than reconstructive transformations – which require the structure to be destroyed and then recomposed. Figure 1.2 schematizes the evolutions between the various possible phases: the transformations indicated by vertical arrows are fast and always happen; those indicated by horizontal arrows are slow and often require, in order to occur, the addition of impurities that play the role of mineralizers. Thus, the transformation of quartz into tridymite is generally not achieved in the temperature range in which it is predicted by the equilibrium diagram, because what is formed is cristobalite, which is metastable.

High cristobalite melts at 1,723°C to produce an extremely viscous liquid (4 MPa.s). On cooling, this high viscosity generally prohibits crystallization, from which it maintains a super-molten liquid and then, below the glass transition temperature [ZAR 82], a silica glass (molten silica, often called, incorrectly, molten quartz or, even worse, quartz). Heated at a sufficiently high temperature to allow sufficient atomic mobility (for example, about 1,100°C), silica glass tends to devitrify to produce cristobalite. The crystallized varieties of silica have properties that are very different from silica glass: the former exhibits anisotropic characteristics in general, whereas glass is isotropic and they have remarkable expansion coefficients (in the order of $10^{-5}K^{-1}$), whereas silica glass has exceptionally poor thermal expansion (about $0.5.10^{-6}K^{-1}$).

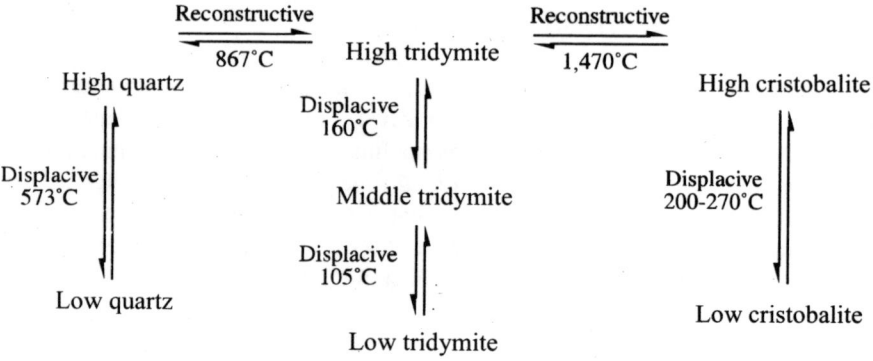

Figure 1.2. *Main crystallized varieties of silica*

Figure 1.3 illustrates the difference between crystallized quartz, where the tetrahedra $[SiO_4]^{4-}$ are linked at their four corners to form an architecture regular in its angles and its ranges (crystal = triperiodicity, i.e. long-range order), and silica glass, where a short-range order continues to exist, significantly similar to the one that exists in the crystal, but with dangling bonds and distortions in angles and variations in length that disorganize the structure.

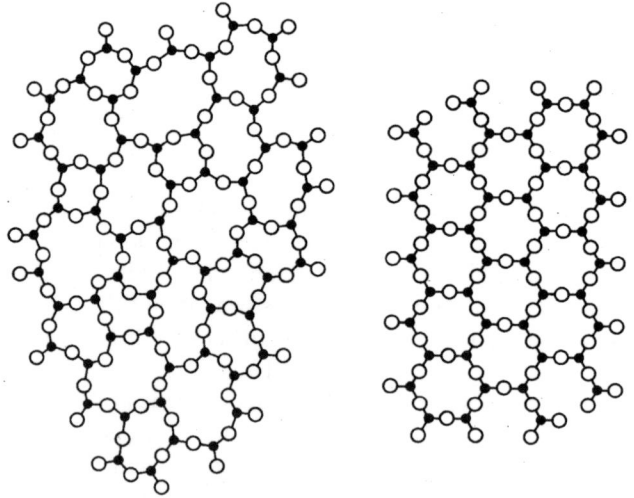

Figure 1.3. *Illustration of the structure of quartz (on the right) and silica glass (on the left); this two-dimensional diagram must be imagined in three dimensions, the silicon atom (full circle) being at the center of a tetrahedron of four oxygen atoms (hollow circle), of which only three are represented here*

1.5.3. *Ceramic microstructures*

The microstructural aspects are discussed in detail in Chapter 3, but we must underline the decisive role that the ceramic microstructure plays in relation to their properties, particularly sensitive properties like mechanical resistance or electric conductivity. The performances of the other categories of materials also depend on their microstructure, but seldom to the same extent as in the case of ceramics. There are several reasons for this sensitivity of ceramics to microstructural parameters:

– the material is processed whilst the object is manufactured, therefore the causes for any disparity in the material are multiplied by the disparity of the processes;

– sintering is accompanied by considerable dilatometric effects, with strong variations in porosity: pores and defects due to differential expansions are inherent to ceramic microstructures; they are rarer in metals prepared by plastic deformation or machining;

– the granular properties of ceramics are more often anisotropic than in the case of other categories of materials;

– ceramics have poor toughness, with critical stress intensity factors (K_c) generally lower than 5 MPa.m$^{1/2}$, i.e. an order of magnitude lower than that of most metals. However, as mechanical resistance to brittle fracture (σ_f) is proportional to K_c and inversely proportional to the square root of the equivalent size of the critical defect (a_c), the defects must be, for a given value of σ_f, 100 times smaller in ceramics than they can be in metals. This is all the more difficult to control since the size of the grains of ceramics is often lower than that of metals: it is much more difficult to avoid 50 μm defects in a ceramic whose grains are 5 μm than notches of a centimeter in a metal whose grains are 100 μm. The absence of plasticity does not allow the relaxation of excessive stresses;

– the iono-covalent bond is less receptive to impurities than the metallic bond, and therefore segregations are more frequent in ceramics than in metals; given that electric conductivity can vary by more than 20 orders of magnitude between a conductor and an insulator, we understand that the presence of a insulating film at the grain boundaries of a material expected to be a conductor, or that of a conducting phase interlinked in a matrix expected to be insulating, can destroy the expected functionalities;

– the frequency of the polymorphism of ceramic phases and the ability of a number of silicate phases to be crystallized or vitreous introduce additional variables into the complexity of ceramic microstructures.

Despite the small number of really useful ceramic compounds, the variety of microstructures makes a very large number of different applications possible. We will limit ourselves to two examples: a ceramic prosthesis in alumina must be dense and fine-grained – to optimize mechanical resistance and tribological properties –

whereas an alumina refractory material must be porous and coarse-grained – to optimize resistance to thermal shocks and creep strength. Controlling the properties of ceramics requires controlling their microstructures.

1.6. Specificity of ceramics

The variety of ceramics is such that they do not exhibit uniform characteristics, but there are common features that give them an undeniable specificity. Most of these common features have already been mentioned, but it is useful to recapitulate the overall physiognomy:

– the iono-covalent bonds confer properties of electric insulation and transparency in the visible range, even if some semiconducting compounds and those that have a partially metallic nature are an exception to this rule;

– the thermal conductivity of ceramics is often poor, because electrons do not, or hardly take part in it, but conduction by network vibrations (phonons) can be considerable: it is diamond, a non-metal, that is the best thermal conductor at ambient temperature and certain ceramics (AlN, BeO, SiC) perform better than copper in this respect;

– strong and directed, subject to electrostatic restrictions, the iono-covalent bonds of ceramics do not allow the movement of the dislocations, i.e. linear defects in atomic stacking that are the reason for the plasticity of metals; hence their brittleness. On the other hand, ceramics offer a high degree of hardness and high moduli of elasticity; their mechanical resistance can be remarkable; they are light; their melting point is generally high;

– most ceramics exhibit good resistance to chemical aggressions. As regards oxidation, we must distinguish oxides (stable in an oxidizing atmosphere, which helps us to take advantage of their refractoriness) from non-oxides, which can oxidize at relatively low temperatures and whose use at high temperatures requires protective, neutral or reducing atmospheres. Graphites and carbons are thus ultrarefractory (sublimation beyond 3,500°C), but they can be used only in protective gas. Among non-oxides, silicon compounds (silicon carbide SiC, silicon nitride Si_3N_4, molybdenum disilicide $MoSi_2$) exhibit the remarkable ability of self-protection from oxidation thanks to a tight and overlapping silica layer SiO_2 (until about 1,800°C for $MoSi_2$). It is erroneous to identify refractoriness with a high melting point because a high melting point is a necessary but insufficient condition: chemical compatibility with the environment and sufficient thermomechanical performances (sufficiently low creep rate, in particular) are also necessary.

1.7. Bibliography

[BAR 96] BARON J. (ed.), *Les bétons; bases et données pour leur formulation*, Eyrolles, 1996.

[BED 86] BEDNORZ J.G. and MÜLLER K.A., "Possible high-Tc superconductivity in the Ba-La-Cu-o systems", *Z. Phys., UB* 64, p. 189-193, 1986.

[BRO 91] BROOK R.J. (ed.), *Concise Encyclopedia of Advanced Ceramic Materials*, Pergamon Press, 1991.

[BUR 90] BURNS G. and GLAZER A.M., *Space Groups for Solid State Scientists*, Academic Press, 1990.

[CAS 90] CASTEL A., *Les alumines et leurs applications*, Nathan, 1990.

[CHE 89] CHERMANT J.L., *Les céramiques thermomécaniques*, Presses du CNRS, 1989.

[COL 99] COLLECTIVE WORK, *Ceramics Monographs, Handbook of Ceramics*, updated by *Interceram*, Verlag Schmidt (since 1982), 1999.

[EMS 95] EMSLEY J., *The Elements*, Clarendon Press, 1995.

[GER 96] GERMAN R.M., *Sintering Theory and Practice*, John Wiley, 1996.

[GIA 85] GIACOVAZZO C. (ed.), *Fundamentals of Crystallography*, Oxford Science Publications, 1985.

[HAH 89] HAHN T. (ed.), *International Tables for Crystallography*, Vol. A, Kluwer Academic Publ., 1989.

[HEU 81] HEUER A.H. and HOBBS L.W. (eds), *Advances in Ceramics*, vol. 3, American Ceramic Society, 1981.

[JAF 88] JAFFE H.W., *Introduction to Crystal Chemistry*, Cambridge University Press, 1988.

[KIN 76] KINGERY W.D., BOWEN H.K. and UHLMANN D.R., *Introduction to Ceramics*, 2nd ed., John Wiley & Sons, 1976.

[KIN 84] KINGERY W.D. (ed.), *Structure and Properties of MgO and Al_2O_3 Ceramics*, American Ceramic Society, 1984.

[KIT 98] KITTEL C., *Physique de l'état solide*, Dunod, 1998.

[LEG 92] LEGENDRE A., *Le matériau carbone*, Eyrolles, 1992.

[MCC 83] MCCOLM I.J., *Ceramic Science for Materials Technologists*, Leonard Hill, 1983.

[NYE 87] NYE J.F., *Physical Properties of Crystals*, Oxford Science Publications, 1987.

[OBA 84] O'BANNON L.S., *Dictionary of Ceramic Science and Engineering*, Plenum Press, 1984.

[PUT 92] PUTNIS A., *Introduction to Mineral Sciences*, Cambridge University Press, 1992.

[RIN 96] RING T., *Fundamentals of Ceramic Powder Processing and Synthesis*, Academic Press, 1996.

[TAY 97] TAYLOR H.F.W., *Cement Chemistry*, Academic Press, 1997.

[WEL 84] WELLS A.F., *Structural Inorganic Chemistry*, Oxford University Press, 1984.

[WES 90] WEST A.R., *Solid State Chemistry and its Applications*, John Wiley, 1990.

[ZAR 82] ZARZYCKI J., *Les verres et l'état vitreux*, Masson, 1982.

Chapter 2

History of Ceramics

2.1. Ceramics and clays

Chapter 1 showed that ceramic, or rather ceramics, are varied and complex materials; the study of the structural transformations and physicochemical reorganizations at all stages of their manufacture constitutes a vast field of research that is still vibrant and largely open, since many new ceramic products continue to be developed for the needs of the most advanced technology. But to trace the history of ceramics, we need to reiterate some basic concepts.

Brongniart observes in the introduction to his famous *Traité des arts céramiques* that "clay is undoubtedly the most widespread raw material on the surface of the earth, the easiest to work with immediately and to transform, but also the one that allows the most utilitarian and artistic productions". This universality and this ease explain why as early as the end of the Stone Age, ceramics gradually became what we can truly call a ubiquitous invention, insofar as it emerged in many human settlements, on all the continents, at extremely different eras.

However, when a terra cotta artifact, whether prehistoric, antique and often even more recent, is found, it is always through a comparative analysis of the material and forms associated with a rigorous dating and an exhaustive study of the archaeological context that we can affirm whether this artifact is the creation of a local craftsmanship that emerged and developed *in situ*, or the result of a foreign

Chapter written by Anne BOUQUILLON.

know-how, established there in the wake of migrations and conquests, or whether it arrived fortuitously through exchanges or expeditions.

The location of the birth, evolution and progress of ceramic art in time and space, particularly in the first few millennia of its existence would require the extensive break up of the subject in order to get a sufficiently clear picture. The scope of this chapter would not allow it, and therefore we will limit ourselves to the presentation of some of the most outstanding facts in the history of this material.

But first, what is clay? It is a rock resulting either from the disintegration of pre-existent crystalline rocks (granites, gneiss, etc.), or from a formation inside large sedimentary basins. Clays are made up of fine pseudo-hexagonal particles often a few micrometers in size. We can distinguish, from a ceramic point of view, at least two types of deposits: primary clays are found on the site of formation; they are coarse, mixed with residues of original rocks (quartz, flint, micas, feldspars, etc.); and secondary clays, which have generally undergone a long transportation that induced a natural decantation. These deposits contain clays that are much finer, more homogenous.

In mineralogical and chemical terms, clays are diverse and as a result have particular properties depending on the family considered: the refractory nature of kaolinite, swelling properties of smectites, propensity to vitrification of illite, high absorptive capacity of vermiculites and palygorskites. However, almost all clays have a layered morphology (phyllo-silicates) and high water content (15% H_2O), hence a plasticity that allows easy forming. The addition of water at the time of shaping (approximately 25%) enables the potter to create a shape and to preserve it during drying. To make it permanent, firing is required and the irreversible chemical and crystallographic modifications resulting from the rise in temperature will make it possible to keep the ceramics. These ceramics are fragile and cannot be re-used once they are broken, but the material hardly deteriorates when used or buried; this explains why they are frequently unearthed during excavations. They must be considered as "fossils" embodying great civilizations and their evolution.

Primitive man learnt very quickly to take advantage of the plasticity of raw clay; observation, deduction, and the fortuitous contact between this material and heat sources would put them on the track to this vital essential discovery: ceramics.

2.2. The first ceramics: sporadic occurrences as early as the end of the Paleolithic

The first ceramics appeared at Dolni Vestonice (in the former Czechoslovakia), as early as 26,000 BC. They were both anthropomorphic figurines and diverse artifacts whose shards were discovered in the thousands. The technology was rudimentary: use of raw local loesses, shaping and firing in open "horseshoe shaped" kilns at temperatures not exceeding 900°C [VAN 90].

This manifestation at the end of the Stone Age apparently remained very isolated and it would be several thousands of years later that the Japanese Jomon ceramics (more than 12,000 years old and discovered in the Fukui caves, near Nagasaki) marked the beginning of a production that has subsisted until now. As early as this period, there was a more complex use of argillaceous earth, which was prepared by adding, probably voluntarily, organic fibers and mica [HAR 97].

It is interesting to note that the first ceramics appeared sporadically in hunter-gatherer, semi-nomadic societies, in which an elaborate social structure appeared. For the moment, in the current state of discoveries, these very early appearances of terra cotta in the form of statuettes are rare and do not seem to last (except in the case of Jomons). They are reported in Siberia towards 12,000 BC, then in China in the Jiangxi province 1,000 years later. In Mesopotamia, on the Mureybet site, ceramic fragments dating from circa 8,000 BC have been found; they come from small coarsely modeled artifacts [MAR 91]. They are very few, not well fired, but nonetheless constitute a proof of the early appearance of terra cotta in the Middle East. They are referred to as intermittent ceramics [MAR 91] and they all belong to a phase that may also be called "pre-ceramic" (without potteries) of the Neolithic era.

The other civilizations used, particularly as containers, stone, fired clay, sometimes dried in the sun or possibly rubefied holes fired in the heat of a hearth. More interesting for the history of pyrotechnology, and therefore of ceramics, are these "white wares", these floor and wall tiles and these lime and gypsum sculptures found on sites in Palestine first (Beidha, 8,300–7,600 BC) then in Iraq, Syria, Anatolia and Jericho [KIN 92].

2.3. The Neolithic era: the true beginning

The explosion of the use of this material would be observed in Asia as well as in the Near and Middle East and in Europe circa 7,000–6,000 BC. It resulted from an important change in lifestyles, needs and beliefs in the wake of sedentarization, cultivation and cattle-raising. This is generally referred to by the term neolithization [RIC 90]. Some, however, believe that the first ceramics had a more symbolic than utilitarian role (Figure 2.1) [PER 94].

From a technical point of view, all ceramics must be classified under soft pastes, i.e. pastes with high porosity, fired at a temperature lower than 1,000°C. In the first few millennia that followed the invention of ceramics, pastes were often coarse, and potters understood very quickly the need to modify the intrinsic properties of raw earth to obtain better resistance to drying and firing. For this purpose, they added to the raw material non-plastic particles (tempers), which to some extent constitute the "skeleton" of the artifact. These tempers are of various types: mineral, organic, natural or anthropic. Their precise characterization is essential in archeometric studies to specify the function of a ceramic, determine the know-how, define a culture or establish an origin.

Figure 2.1. *Anthropomorphic vase of the final Neolithic*
(© National Museum of History, Sofia)

A few examples:

– to determine the function: in the Neolithic era, the uses of ceramic vases were manifold and each of them largely determined the technique; the so-called "provision" vases were large earthenware jars made of porous paste, with coarse tempers, intended to protect grains, fruits, etc. from insects, rodents or moisture; other less porous vases could hold thick liquids, dairy products, oils, etc. The vases with more meticulous surface treatment could contain drinks and, finally, others undoubtedly show specific residues from culinary preparations or materials such as birch pitch used as an adhesive [REG in press], [REG 03]. Researches undertaken on the Clairvaux sites in the Jura or Chalain in the Savoy have brought to light a fine diversity of these ceramics in the limited context of villages. The addition of plant matter to the paste was not insignificant; in fact, these materials do not resist heating, except very rarely, and they leave large pores when they are burnt. Ceramics with vegetal temper in our regions often identify artifacts used as coolers, or that had to undergo violent and repeated thermal shocks which were mitigated by the pores [RYE 89];

– to determine the know-how: the use of Grog as temper (fragment of crushed fired ceramic) is the most logical (Figure 2.2); in fact, these non-plastic particles exhibit the best properties: perfect compatibility with clay, optimum expansion/shrinkage coefficient (the earth has already been fired) and reuse of wastes and wasters. This is found very early, in the Neolithic era, on the megalithic site of Bougon (Deux-Sèvres) for example;

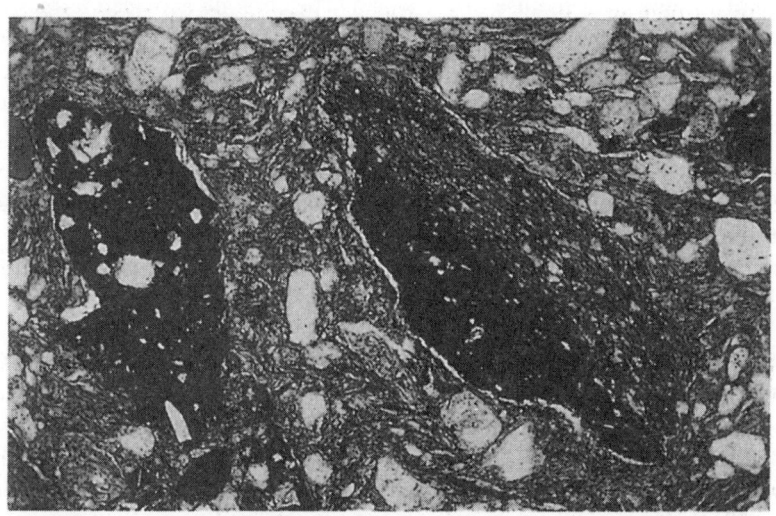

Figure 2.2. *Grog temper of a ceramic in Northern France – petrographic microscopy on thin section (© C2RMF – photo: A. Leclaire)*

– to define a culture: in the Neolithic era, ceramic pastes of the so-called "Cerny" culture are characterized by a temper made of splinters of bone [BRI 90]. It has been sometimes said that this temper had a ritual significance;

– to establish an origin: in Brittany, the campaniform (bell-shaped) culture developed during the Chalcolithic age; characterized by bell-shaped vases, this culture was found all along the Atlantic coastline, in Spain, Portugal, etc. The presence on the sites of the Southern Finistere of some shards containing fragments of volcanic rocks among ceramics tempered with fragments of granitic rocks typical of the region has revealed probable imports from the Rhine areas in this period [CON 98].

2.3.1. *Forming and firing*

2.3.1.1. *Forming*

If the composition of pastes varies from one region to another, techniques of shaping vases follow an identical evolution: besides simple modeling, primitive traditional methods include coiling (wads of clay are shaped then placed on a base depending on the desired shape), assembly by juxtaposition of clay plates shaped beforehand and the forming by mining or digging – by pinching or drawing a ball or a thick plate of clay. These techniques are still used today. The greatest revolution in the field of forming was the use of the wheel. The ceramic is placed in the center of a table revolving at 50-150 rotations/minute, caused by the potter himself. This technique, which probably appeared in Asia around the fourth millennium BC [ROU 98], a little later in the Middle East and was apparently unknown in America until the arrival of the conquistadors, changed ceramic techniques considerably. In fact, turning ceramics implies a very special know-how and a mastery that can be acquired only after a long period of training: the table must be turned with regularity to thin the walls symmetrically. The generalization of the use of the wheel necessarily required a modification of the pastes: these must be markedly finer and more homogenous.

Turned ceramics are fine, the shapes are regular and more elaborate and, especially for a skillful craftsman, it takes less time to produce pottery. The time was ripe to enter the era of mass production. Most of the time, modeled and turned wares continued to be manufactured simultaneously, sometimes on the same sites and it is not always easy to affirm that a particular ceramic was manufactured using the wheel; in fact, the typical traces left on the ground by this method are sometimes ambiguous, and we must not mistake them for the marks of a hand or a tool intended to smooth and refine the surface of the potteries made by coiling or by another technique described above, but worked with a turntable, the predecessor of the wheel. Later, other methods would come to enrich this ancient technology of terra cotta: so, for example, molding using earthen or plaster molds would become frequent, as can be seen particularly in Gallo-Roman terra sigillata wares.

We note here that these sketches of the history of ceramics before the 19th century focus especially on pottery itself. To review all the techniques of manufacture, assembly and shaping that were born out of terra cotta arts, we must also take into account all the other types of productions: floor tiles, architectural elements, more important statuettes and sculptures, various artifacts, etc. The scope of this book does not allow us to delve into this subject in detail.

2.3.1.2. *Kilns*

Since the most primitive kilns, comprising a hole in the ground with pots and hearths in a single structure covered with branches or earth, the design of kilns has changed significantly, making it possible to have a better control of the temperatures and atmospheres, to fire a greater quantity of objects in a single batch and a more efficient use of fuel. These kilns were also reusable after one firing. As early as the 4th century BC, Greek kilns were very sophisticated, easily enabling a change from an oxidizing atmosphere to a reducing atmosphere.

On certain sites of the major Gallo-Roman workshops (Gueugnon, Lezoux, Graufesenque, etc.), up to 12 different types of kilns have been discovered, each one dedicated to a particular type of production: from the half-buried single chamber in which the pots to be fired were placed directly on two hearths especially used to fire the dark common potteries of Gallic tradition, to the more sophisticated kiln dedicated to the famous terra sigillata, built in such a manner that the vases to be fired were protected from the products of combustion of the hearth, which were located under a hearth. Heat was distributed by a system of pipes present both in the floor and the wall nearest to the kiln chamber. The most widespread kilns were, however, simpler in design with a half-buried "laboratory" separated from the fuel by a pierced floor. An opening, often made on the top of the kiln, helped control the temperature and, more importantly, the atmosphere of the kiln [DUF 96].

2.3.2. *Decorations*

This know-how and these techniques would unleash creative imagination through the millennia, to the four corners of the world: a universe burgeoning with shapes and decors. The shapes, from the simple useful object to the prestigious work of art, are innumerable, so varied from one culture to another. As for decorations, for reasons of accuracy, we will now present the techniques of the most characteristic and most universal decorations.

2.3.2.1. *The very first decorations*

As early as the first potteries, decoration was immediately an essential element, a symbolic system with which a whole culture identified itself. The simplest shapes are incisions, nail marks, scratches, etc. An example taken from the old Neolithic era [MOH 98] will illustrate the importance of the study of decoration in the knowledge of cultures. Along the Mediterranean, around 6,000 BC, ceramics with cardial decorations flourished: we can trace the expansion of the culture between 5,000 and 4,500 BC towards the Massif Central and the Atlantic coasts by discovering the presence of ceramics in the excavated sites.

At the same time, in Central Europe, around 4,750 BC, another culture developed, that of ceramics with linear decoration, where the decorations are ribbons, volutes, horseshoes, etc. covering the Danube valley and the Rhine valley.

Later, in the West, the civilization of megaliths would develop, characterized, as regards ceramics, by shapes that are not decorated or with a few gripping buttons.

Painted ceramics

Painted ceramics also became very quickly a developed means of expression, as early as the first ceramic periods. Paintings, generally of mineral origin, slip (very watered-down fine clay) of red metallic oxides (hematite, goethite or red clays like ochres), black metallic oxides (manganese), kaolinite or calcite for white, are executed either before firing on dried ceramics or after firing. This technique would produce remarkable works in various sites on all the continents. Speaking about antique ceramics, we can cite among many others the ceramics of Susa, pre-Indus ceramics (fourth and third millennia BC) at the Nausharo site (Figure 2.3), those of the Banshan culture in the third and second millennia BC, the art of the Cyclades, Cretan ceramics, vases with geometrical decorations of the first millennium BC in Greece and painted potteries of the Iron Age discovered in Champagne.

Figure 2.3. *Painted vase of Mehrgarh (Indus-Pakistan valley),*
pre-Indus period (© C. Jarrige)

Added decorations

Here again, the diversity is remarkable: the addition of small clay buttons like on the Carn ceramics, the addition of fine clots on certain Jomon potteries from a more recent period or on some large earthenware jars of Cnossos.

Subsequently, finer decorations would be obtained by molding (see Figure 2.4) or stamping on finely engraved molds or carved punches.

This type of decorations is found on Gallo-Roman terra sigillata ceramics. "Scenes" so finely represented on the bodies of vases are obtained by molding. An artist first creates a hollow matrix. The finer the details are, the finer the paste must be, obtained by decanting a basic argillaceous earth a number of times. These decanting procedures were developed by making the clay pass through a series of basins.

Figure 2.4. *Tanagra statuette – inv. 556 – Louvre Museum – Department of Greek, Etruscan and Roman Antiquities (© C2RMF – photo: G. Koatz)*

2.3.2.2. *Ceramic of classical antiquity*

Decorative effects could also be achieved by the play of the oxidations/reductions of certain metallic compounds. But for this, firing had to be controlled. As early as the third millennium BC, artisans already knew how to fire or smoke potteries in a reducing atmosphere by adding grass or green wood, stopping all the air intakes or, on the contrary, in an oxidizing atmosphere. The Egyptians had first fired two-tone ceramics with a red base and blackened neck, called *black-top*, by designing half-buried kilns; the ceramic was half-buried in the sand, the protected part fired red, the other part, directly in contact with the atmosphere of the kiln, fired black [NOB 88]. But it was undoubtedly with Greek potteries featuring black Attic decorations that we can best see the remarkable mastery of the potters of the 6th century BC in Greece (Figure 2.5): this technique relied on a precise preparation of the clay of the paste and the decoration as well as on a perfect mastery of the various phases of firing.

Figure 2.5. *Attic crater inv. MNE938 – Department of Greek, Etruscan and Roman Antiquities PAS 00 (© C2RMF photo: D. Vigears)*

Attic clay naturally contains iron, therefore fired dark red in the normal oxidizing conditions of the kilns of the time. The potters shaped the ceramic on the wheel, allowed it to dry in the sun and then painted the decoration scenes with a slip that consisted of a water suspension of the same clay used for the body but very decanted. This slip was applied on all the areas that had to be black. Firing was done only once but had three phases:

– phase 1: oxidizing firing at 900°C: the entire ceramic is red; iron is in the form of Fe_2O_3;

– phase 2: the kiln is closed; the addition of green wood as fuel creates a reducing atmosphere; the entire ceramic becomes black under the effect of the reduction of iron in the form of FeO or Fe_3O_4;

– phase 3: the kiln is opened again; oxygen is reintroduced: only those areas of the paste that are still porous will be able to change to red again; the areas covered with slip are vitrified (effect of the potassium in the clays that acts as flux on very fine particles); the process of the reoxidation of iron cannot take place. Thus the famous Attic "black varnish" was created; the technique was rediscovered only in 1948 by Schumann [SCH 48].

The red varnishes of terra sigillata ceramics or Italic ceramics were achieved in the same way but in a single phase of oxidizing firing.

2.3.2.3. *Glazes*

These black and red varnishes were often compared with glazes, but incorrectly; in fact, glaze, discovered approximately 7,000 years ago simultaneously in Egypt, in Mesopotamia and in the Indus valley is a glass made up of a mixture of sand, fluxes (vegetal ashes, natron, natural sodium carbonate or lead compounds) and coloring or opacifying oxides. Such a vitreous glaze has a two-fold advantage: waterproofing as well as coloring and decorating a porous terra cotta.

2.3.2.3.1. First alkaline glazes: the middle of the second millennium BC

The first glazes were found on stones (quartz or steatite) and it was much later, about the second millennium BC, that the first glazed ceramics appeared. This delay is explained by the great technical difficulties encountered in the production of a glazed ceramic: the mixture had to be free from impurities, have a composition such that the melting point was compatible with the kilns of the time and its thermal behavior on heating and cooling (expansion/shrinkage) had to be compatible with that of the underlying paste to avoid frequent and permanent firing accidents (crazing, bursting, etc.) [DAY 85]. Recent research has revealed that the appearance of the first glazed ceramics dates back to 1,600–1,500 BC in Northern Iraq, in Alalakh in particular, and Northeast Syria [HED 82]. This technique would be used very quickly in the Middle East for architectural decoration, statues and vases. It is interesting to note that in Egypt it was not until the Islamic era in the 8th century AD that the first glazes appeared. In the beginning, glazes were alkaline, often monochromic and blue or blue-green, colored by copper oxides. Polychromy developed later, in the first half of the first millennium BC.

2.3.2.3.2. Appearance of lead-glazes: the Roman era

1,500 years later, in the 1st century AD, the first glazes appeared in the Roman world, in England and in Asia Minor. Their main flux was lead oxide [HAT 94]. It was a timid appearance: only a few specimens have been found. The technical features are good, low point melting, good adherence, iridescent colors, etc. However, this type of glaze would be abundantly used by the Byzantine artisans. In Europe, after the fall of the Roman Empire, ceramics became less sophisticated, often modeled, little decorated, without or almost without glazes until the 6th or 7th century. [ENC 98]. Lead potteries would make a comeback around the 9th century and they would have a long history.

2.3.2.3.3. The glossy decorations of the Islamic world: lusters

This technique, directly derived from the traditions of goldsmiths, made it possible to apply metallic salts (gold, silver, copper, etc.) on the vitreous support of an opaque glaze. The first examples were produced at the court of the Abbasid sovereigns in Baghdad, but it was primarily Egyptian artisans, as early as the 9th century AD under the reign of the Fatimides, who honed this technique to perfection. In the wake of the migrations of artisans or recipes, luster was introduced into the entire Islamic world as far as Spain, where it flourished during the entire Hispanic-Moorish period [DAR 05].

How was a luster produced? The body of the pottery, made up of a siliceous or a clayey paste, was covered with a transparent or an opaque alkaline-lead glaze. A painting containing metallic salts was placed on the glaze to execute the decoration. This painting was very complex and contained two types of principal components: metallic salts, of course, and a non-reactive "binder" that helped to apply the painting in a regular way. Metallic salts were mixed according to the ancient recipes of Abul Qasim [ALL 73] with vinegar to form acetates. The whole was fired again at low temperatures (about 240°C) in a reducing atmosphere kiln. A two-fold phenomenon occurred: there was a slight diffusion of the metal in the glaze and a reduction and precipitation of this metal on the surface. A quick polishing after firing highlighted the metallic aspect by the play of the refractions/diffusions of light. The colors of the luster varied considerably and in general depend on the size, the concentration of the particles, the nature of the metal used and the control of the last firing [KIN 86].

2.3.2.3.4. The opacification of the glazes by the addition of tin: an innovation of the Islamic artisans

Opacifying a glaze was an important milestone for the artisans to cross. An opaque glaze has many advantages including the obvious one of hiding a not very esthetic paste color and allowing greater freedom and greater possibilities of decoration. There are several ways to do this: the presence of gas bubbles in large quantities diffuses the light and gives an "opalescent" appearance; the persistence within the vitreous matrix of large-sized grains, generally non-molten quartz or feldspar grains, yield an opaque glaze. Finally, the growth of secondary crystals at the interface between the paste and the glaze also gives an opaque aspect to the glaze [MAS 97]. We should point out that the Egyptian antique white or antique yellow, opaque glasses or glazes were obtained by adding calcium antimoniate [KAC 83].

With the discovery of the properties of the cassiterite (tin oxide), opacification would become simpler and would especially give an esthetic aspect that the above-mentioned methods could not offer. The first tin opacification "tests" date back to the 9th century AD in Syria and then in Egypt. Tin was added to a glaze often using a mixture of lead and tin called calcine. On firing, lead and tin dissociated and tin oxidized to produce tiny cassiterite grains. 10% of cassiterite was sufficient to opacify a glaze and to give it a perfect white color.

2.3.2.3.5. Western faience: emergence in the 13th century

The tradition of ceramics with stanniferous glaze developed first in the entire Mediterranean Basin and thereafter across the Western world. In the 13th century, there was a substantial production of ceramics decorated with tin oxide in Spain. They would be exported in large quantities to the entire Mediterranean Basin, particularly to Italy and Southern France. It was at this time that the first centers for the production of faience were set up in Italy and shortly thereafter in Marseilles [COL 95]. Expansion was also favored by the recruitment of Spanish artists to work on royal building sites where they made use of their know-how by adapting it to local materials. Artisans from Saragossa, Jehan de Valence and Jehan-le-Voleur are known on the royal building sites of Mehun sur Yèvre, where the first faience tiles have been found [BON 90].

During this period, the production of majolica began in Italy and many large centers were established: Faenza of course, but also Urbino. Each of these centers developed an iconography and a specific type of decoration, but all shared the same technique, described by Piccolpasso in his work *Three Books of the Potter's Art* published in 1548. The body of the ceramic is worked with fine marly earth; it is fired first at about 950°C; thereafter several stages are necessary for the production of the glaze: preparation of the calcine (a mixture of lead and tin); preparation of a sand frit, calcine, wine dreg (KNO_3); firing and crushing of the frit which is added to water and addition, if necessary, of coloring metallic pigments. This mixture is applied on the piece to be decorated and the whole is reheated again at about 950°C. Later, there could even be more than three firings when colors other than the high fire colors (cobalt blue, copper green, manganese purple, antimony yellow, etc.) are used, i.e. low fire colors (pink, green, red, etc.).

2.3.2.3.6. Productions of the Renaissance

Towards the 15th century, it was in Spain with luster and in Italy with majolicas that earthenware experienced their most spectacular growth. The greatest Italian centers were Faenza, Deruta, Gubbio, Urbino [PAD 03] and Casteldurante [GIA 35], but also Florence, particularly with the productions of Della Robbia. This family of sculptors and ceramists used glazed terra cotta as a new material for the sculpture of busts, retables, decorative tiles, vases, etc. They succeeded in achieving such a degree of perfection (see Figure 2.6) with respect to both sculpture and colors that these productions immediately sparked off a great interest everywhere in Europe. It is even said that Palissy, on seeing the works of Della Robbia, relentlessly sought to unveil the secret of their so perfect marmoreal white.

Figure 2.6. *Bust of young man attributed to Della Robbia – inv. OA1932 Department of sculptures – Louvre Museum (© C2RMF – photo: D. Bagault)*

We cannot speak about Renaissance ceramics without mentioning Palissy and his achievements, both in the domains of marbled earth and earthenware with rubble. According to his writings and also all the material found in his workshop, he was the first ceramist to experiment so much in order to achieve the desired color and effect.

Some have even attributed to him a prestigious production (Figure 3.7), known as Henry II faiences, or Saint Porchaire wares, which constitutes the beginnings of hard pastes. Less than 60 specimens in the whole world are known and the esthetic quality, the great technical mastery both in shaping and in the production of the decorations have given rise to varied interpretations, put forward even recently [COL 97]: was this a production of Palissy himself? Is it a Parisian production reserved for the king and nobles? Is this a production of Saintonge?

Figure 2.7. *St Porchaire ewer – inv. Ec83 – National Museum of the Renaissance Ecouen (© C2RMF – photo: D. Bagault)*

The skills of the Italian artists in particular, in the wake of the travels of artisans and the disclosure of trade secrets, inspired the creation in the 16th century in France of the greatest earthenware makers still operating today. First in Lyon, then especially in Nevers, Rouen, Strasbourg, etc.; it is in the 16th and 17th centuries that stanniferous earthenware would evolve, as a high quality production was demanded. However, from the French Revolution onwards, a period of durable recession would follow, consecutive to a fall in demand, of course but, more importantly, along with the rise in prices of wood and raw materials, tin, lead and the arrival on the market of a remarkable English production: fine earthenware. Only a few great centers would resist and succeed in the industrial, technical and stylistic changes necessary for their survival: for example, the use of coal as fuel, discoveries of certain processes (pouncing patterns, etc.) that accelerated the various phases of decoration, the use of a more complete pallet of colors, thanks to the development of so-called "low fire" colors [ROS 91] and the diversification of productions. However, this type of ceramic would soon fall into disuse.

2.4. Chinese stoneware and porcelains: millenniums ahead

We have talked so far about relatively porous ceramics, fired at about 1,000–1,050°C maximum. With stoneware and porcelains, porosity decreases (less than 5%

for stoneware, less than 1% for porcelain) and vitrification becomes increasingly significant. Firing temperatures exceed 1,200°C for stoneware and 1,300°C for porcelain. These characteristics have two corollaries: firstly the need for specific clays or mixtures that allow melting and more importantly kilns for reaching such temperatures. These two conditions were met in China as early as 1,000 BC for stoneware under the Shang dynasty in South China and towards 600 AD for porcelains in North China.

2.4.1. *Stoneware*

Stoneware clays have particular characteristics. They are in general very siliceous, aluminous and contain rather significant proportions of potassium, which acts as flux. These clays have the property of vitrifying gradually with the rise in temperature without becoming deformed; they yield an opaque material, often brown in color, variable according to the impurities contained in the initial mixture.

Archeologists have discovered in the Harappan sites of the Indus valley, dating back to the third millennium BC [VID 90] a specific production of "stoneware" bracelets. They were made up of very fine clays, fired according to very sophisticated processes, in a reducing atmosphere and inside special containers. This is not exactly stoneware in the meaning that this term has today, but it needs to be mentioned in this context. As we mentioned above, the first attested stoneware are Chinese artifacts dating from the Shang dynasty (1,500–1,050 BC). This stoneware is covered with a glaze very rich in calcium, composed of a mixture of sand and vegetal ashes. Later, ashes would be replaced by a substantial addition of mineral carbonates [WOO 99]. The stoneware tradition would last a long time in China and masterpieces would be created during the following centuries (celadons, Yue, etc.). Stoneware would appear 200 years later in Japan and Korea, but the first stoneware would be manufactured in Europe only in the 10th century.

In France, the important regions of stoneware production relied on the presence of Sparnacian clays which were so plastic that a high content of quartz had to be added to avoid very high shrinkage. The glazes, fired at the same time as the paste, were of three types: sea salt glazes, which give a beautiful varnish (see Figure 2.8), blast-kiln slag glazes as in Puisaye and ash glazes. Right from its first appearance, stoneware became a huge success and its production has never ceased.

Figure 2.8. *Salt glazed stoneware vase. Martainvill (© O. Leconte)*

2.4.2. *Porcelains*

Porcelains were developed in China. As we mentioned earlier, the concurrence of several factors led the Chinese to develop these products: specific raw materials, mastery of firing conditions and the possibility of firing at high temperatures (Figure 2.9).

China has numerous kaolin deposits, which were exploited very early on. These fireclays fire white. Depending on the geographical area in question, Northern or Southern China, the composition of these kaolins is a little different. In the North, clays were associated with coal deposits: they were rich in alumina (approximately 30%) and low in flux elements (alkaline, alkaline-earths) and iron. It was therefore necessary, in order to fire ceramics, to reach temperatures estimated at 1,200–1,350°C [HAR 98]. In the South, on the other hand, kaolins resulted from the deterioration of igneous rocks and as a result they were enriched with flux elements; they could be fired at about 1,200°C.

As early as the end of the Neolithic era, Chinese kilns were very sophisticated. The ovoid kilns of Jingdezhen are often cited. The sizes of these kilns, their firing chamber being in the form of an egg, made it possible to reach more than 1,350°C everywhere in the kiln. Temperature control, essential for performing the firing, was done by an ingenious system of windows. The fuel used was made up of small branches and pinewood [HUL 97]. The ceramics were placed in saggers, a kind of small refractory terra cotta boxes which insulated them and which also allowed better heat distribution.

Figure 2.9. *White-blue ewer of Yuan era (1335) – inv. MA 5657*
Guimet Museum (© C2RMF – photo: D. Bagault)

In the North, the kilns were dug directly into the mountains, on the hillside, sometimes at more than 100 m [WOO 99] with a slope of about 15 to 20°. These "dragon kilns" were already extremely sophisticated, as early as the Song period. Firing started at the base of the kiln. The upper part then served as a pre-heating chamber for the ceramics that were placed there inside saggers. When the firing temperature was achieved in the lower zone, the chimney of the following zone was blocked using branches, in order for the heat to be propagated in this zone, and so on until it reached the top.

It is obvious that this system resulted in many wasters, but it also made it possible to fire thousands of pieces in a single batch.

The porcelains thus obtained are characterized by a vitrified paste which contains generally high mullite concentrations in microcrystals, mullite being derived from the high temperature treatment of kaolin. All these components (glass, microcrystals, bubbles) gave the much desired translucidity and hardness.

2.5. The quest for porcelains in the East and the West

The arrival of Chinese porcelains of the Yuan period, first on the Islamic markets in the 9th century, then later on the European markets in the wake of the voyages of

Marco Polo, triggered an unrestrained quest to uncover the secrets of this matter to which all virtues were attributed, even that of detecting poisoned substances [ROS 95].

Even if, at least initially, it was the esthetic qualities of porcelains that people sought the most: whiteness, translucidity, etc., soon their properties of hardness, resistance to thermal impact and also savings in terms of firing time gave an impetus to the research.

Those who battled with the problem explored two essential directions: glass frit pastes and faience fine.

2.5.1. *Siliceous pastes and glass frit pastes*

By refining the recipe of "archaeological earthenware" already known in the fourth millennium BC in Egypt and Mesopotamia, the artisans of the Islamic period added to variable quantities of plastic clays, quartz and glass frits a synthetic material made up primarily of sand and fluxes. This paste had a two-fold advantage: firstly an esthetic one, since it was very white and slightly translucent and then a technical one, since it expanded highly on heating just like the alkaline glazes that decorated it. These pastes were abundant as early as the 9th century AD and widespread in the entire Islamic and Hispano-Moorish world as support for luster or wall tiles.

In the West, the first successes are attributed to the Italians in the time of Francesco de' Medici. Under the impetus of the Renaissance artists and with the protection of the Grand Duke, artisans developed around 1570 a white paste, fired at 1,100°C, whose recipe was an ingenious mixture of Islamic siliceous pastes and Italian majolica traditions. In fact, the paste was made up of a "frit" (*marzacotta*) prepared with silica, wine dregs and various salts, crushed and added to a white clay enriched with quartz. The resulting paste was very white, porous and exhibited an important vitreous phase. A lead glaze covered a decoration drawn in cobalt blue. This glaze, slightly under-fired, produced an artificial effect of translucidity thanks to the combined effect of thousands of microbubbles, incompletely molten grains of quartz, feldspars and calcium phosphates. Only a small number of artifacts exist today and the production ceased after a few years.

Soft-paste porcelain is one of the most beautiful achievements of the 18th century, especially in France and England. Designed on the same principle as the Medici porcelain, soft-paste porcelain combines translucidity and intricacy of shapes made possible by the smoothness and plasticity of the paste. A transparent lead glaze made the subtle combinations of colors possible, but it was fragile and easily scratched; moreover, this porcelain was not very resistant to thermal shocks. Several workshops manufactured it, initially Rouen in 1673, then Saint-Cloud, Chantilly, etc.

The Vincennes production [PRE 91], representative of the compositions of the French soft-paste porcelains, reveals the complexity of the pastes. A frit was prepared from saltpeter (KNO_3), salt, alum, soda, gypsum and sand. After firing and fine crushing, chalk and marl from Argenteuil, a very plastic illitic clay, were added to it. The paste thus obtained was turned, molded or sculpted directly and then fired in oxidizing atmosphere in a kiln at approximately 1,100°C. The glaze was transparent, made up of lead, alkaline, sand and calcined flints. The decoration was very delicate to execute and the pallet of colors evolved progressively with discoveries of flux compositions and the development of continuous special kilns.

Other types of soft-paste porcelains developed (Figure 2.10), particularly in England where the most famous were porcelains containing bone ashes, kaolin and Cornish stone (feldspathic rock used as flux).

Figure 2.10. *Bone porcelain of Minton 1872 (© O. Leconte)*

2.5.2. Faience fine

In this case, it was not translucidity, obtained by adding glass, that people were seeking. Through the working of the paste, they were looking for brilliancy, density and sonority, whiteness, as well as ease of mass productions. It was under the influence of the English (the most famous being Wedgwood) that this new type of production, faience fine, would be born, and then spread out primarily in the wake of the Industrial Revolution. Faience fine was characterized by an opaque paste made up of very fine plastic clay, mixed with flint, fine quartz and grog. The clay was often kaolin, used as white firing element associated with one or more plastic clays. A few feldspars or limestone played the role of fluxing and tempering agents. The glaze was transparent. The classification of soft-pastes was in fact based on the various compositions of the pastes [MUN 54]. In France, the main production centers were in Northern (Douai) and Eastern France (Lunéville) and then the Paris area (Montereau, Creil) where clay deposits were abundant and accessible [GIA 35]. It is important to stress that these products truly marked the beginning of mechanical processes in the preparation of the paste and the decorations and that the establishment of the first manufactures at least was contingent on the proximity of large deposits of coal or raw material in order to overcome constraints of economic profitability.

This faience fine would be appreciated by the middle classes, in whose homes they would eventually replace traditional stanniferous faience. It would soon compete with porcelains.

2.5.3. *The first veritable porcelains in Europe*

In 1709 Böttger, working in the Meissen manufacture in Germany, revealed that he had for the first time succeeded in recreating genuine porcelain, although he achieved this by using a recipe very different from the ones normally used which were based on kaolin. With this new method, these ceramics were fired at high temperatures; they were siliceous and aluminous of course, but contained, in the first stages of development, large quantities of calcium (nearly 5%) which conferred on the matter great resistance to thermal shocks. Later, when the recipes evolved after the death of Böttger, lime would be replaced by feldspars.

In France, it was in Sèvres, under the aegis of the Academy of Sciences and under the impetus given by Macquer and de Dufour, that the first and the finest successes of hard-paste porcelain would be created. A. d'Albis recently published a work on the "conquest of porcelain in Sèvres" [ALB 99]. He describes in detail the atmosphere of competition, betrayals, and the ceaseless research of the chemists. Three great stages marked this research: the discovery of substantial kaolin deposits at St Yrieix, in 1767; a new model of cylindrical kiln with two superimposed chambers which were capable, in particular, of reaching an extreme phase of reduction necessary for the perfect whiteness and translucidity of the paste; and especially the development, from 1778, of a recipe of impeccable paste and glaze by adding feldspar to them in the form of pegmatite. This recipe has given the Sèvres manufacture an uncontested supremacy until today (Figure 2.11).

Figure 2.11. *Vase of Sèvres – 19ᵗʰ century – Museum of Amiens*
(© C2RMF – photo: O. Leconte)

2.6. Conclusion: the beginnings of industrialization

The Industrial Revolution would introduce into the field of ceramics radical and ceaseless changes during the entire 20ᵗʰ century, in the modes of preparation, manufacture, decorations, coloring and firing and especially with the use of the electric power, advances in chemistry and material sciences. New applications that take advantage of the resistance of the material to thermal shocks (insulators, heat shields, etc.) as well as the inalterability and harmlessness of bioceramics would be implemented.

However, we can affirm that all the great traditions of ceramic art are still alive! A material that is so close, so flexible, so faithful will always express the innermost, the most immediate and the most elated preoccupations of man; thousands of examples spread across the entire history of humanity testify to this in such a striking manner.

2.7. Bibliography

[ALB 99] D'ALBIS A., "Sèvres 1756-1783 – La conquête de la porcelaine dure", *Dossiers de l'Art,* no. 54, 1999.

[ALL 73] ALLAN J.W., "Abu'l Qasim's treatise on ceramics", *Iran*, vol. 11, 1973.

[BON 92] BON P., *Les premiers bleus de France – Carreaux de faience au décor peint fabriqués pour le duc de Berry en 1384,* Picard, 1992.

[BRI 89] BRIARD J., *Poterie et civilisations,* Errance, 1989.

[BRO 77] BRONGNIART A., *Traité des Arts Céramiques ou des Poteries, fac-similé de l'édition de 1877,* Editions Dessain et Tolra, 1977.

[COL 95] COLLECTIF, *Le vert et le brun-De Kairouan à Avignon, céramiques du Xe au XVe siècle,* Editions Musées de Marseille, Réunion des Musées Nationaux, 1995.

[COL 97] COLLECTIF, *Une orfèvrerie de terre – Bernard Palissy et la céramique de Saint Porchaire,* Editions Musée d'Ecouen, Réunion des Musées Nationaux, 1997.

[CON 98] CONVERTINI F. and QUERRE G., "Apports des études céramologiques en laboratoire à la connaissance du Campaniforme: résultats, bilan et perspectives", *Bulletin de la Société Préhistorique Française,* vol. 95, no. 3, 1998.

[COU 95] COURTY M.A. and ROUX V., "Identification of wheel throwing on the basis of ceramic surface features and microfabrics", *Journal of Archaeological Science,* vol. 22, 1995.

[DAR 05] DARQUE-CERETTI E., HELARY D., BOUQUILLON A., AUCOUTURIER M., "Gold-like lustre: nanometric surface treatment for decoration of glazed ceramics in ancient Islam, Moresque Spain and Renaissance Italy", *Surface Engineering,* vol. 21, no. 5, p. 1-7, 2005.

[DAY 85] DAYTON J., *Minerals metals glazing and man or who was Sesostris I,* Harrap, 1985.

[DUF 96] DUFAY B., "Les ateliers – organisation, localisation structures de commercialisation", *Dossiers d'archéologie,* no. 215, 1996.

[ENC 98] *Encyclopaedia Universalis,* CD Rom version 4.0 France, 1998.

[GIA 35] GIACOMOTTI J., *La céramique: la faience fine, la porcelaine tendre et la porcelaine dure,* vol. III, Editions les Arts décoratifs, 1934.

[GIA 34] GIACOMOTTI J., *La céramique: la faience en Europe du moyen Age au XVIIIe siècle,* vol. II, Editions les Arts décoratifs, 1934.

[HAR 97] HARRIS V., "Jomon Pottery in Ancient Japan", *Pottery in the making, world ceramic traditions,* Freestone I, Gaimster D. (ed.), British Museum Press, 1997.

[HAR 97] HARRISON HALL J., "Chinese porcelain from Jingdezhen", *Pottery in the making, world ceramic traditions,* Freestone I, Gaimster D. (ed.), British Museum Press, 1997.

[HUL 97] HU J.Q. and LI H.T., "The Jingdezhen egg-shaped kiln", *The Prehistory and History of Ceramic Kilns, Ceramics and Civilization,* vol. VII, American Ceramic Society, 1997.

[KAC 83] KACZMARCZYK A. and HEDGES R.E.M., *Ancient Egyptian Faience – An Analytical Survey of Egyptian Faience from Predynastic to Roman Times,* Aris & Phillips, 1983.

[KIN 86] KINGERY W.D. (ed.), "The development of European porcelain", *Ceramics and Civilization*, vol. III, Kingery and Lense (ed.), American Ceramic Society, p. 153-180, 1986.

[KIN 86] KINGERY W.D. and VANDIVER P.B., *Ceramic Masterpieces – Art, Structure and Technology*, The Free Press Editions, 1986.

[KIN 91] KINGERY W.D., "Attic pottery gloss technology", *Archeomaterials*, vol. 5, p. 47-54, 1991.

[KIN 92] KINGERY W.D., VANDIVER P.B., NOY T., "An 8500-year-old sculpted plaster head from Jericho (Israel)", *MRS Bulletin*, vol. XVII, 1992.

[KLE 86] KLEINMANN B., "History and development of early islamic pottery glazes", *Proceedings of the 24th International Archaeometry Symposium*, Smithsonian Institution p. 73-84, 1986.

[MAR 91] MARGUERON J.C., *Les mésopotamiens*, Armand Colin, 1991.

[MAS 97] MASON R.B. and TITE M.S., "The beginning of tin-opacification of pottery glazes", *Archaeometry*, vol. 39, no. 1, p. 41-58, 1997.

[MOH 99] MOHEN J.P. and TABORIN Y., *Les sociétés de la préhistoire*, Hachette supérieur, 1999.

[MUN 57] MUNIER P., *Technologie des earthenwares*, Editions Gauthier-Villars, 1957.

[NOB 88] NOBLE J.V., *The Techniques of Painted Attic Pottery*, Thames & Hudson, 1988.

[PAD 03] PADELETTI G., FERMO P., "How the masters in Umbria, Italy, generated and used nanoparticules in art fabrication during the Renaissance period", *Applied Physics A*, vol. 76, p. 515-525, 2003.

[PAS 00] PASQUIER A., "Un cratère-rafraichissoir au musée du Louvre: du vin frais pour un banquet de luxe", *Monuments Piot*, vol. 78, p. 5-51, 2000.

[PER 94] PERLES C. and VITELLI K.D., "Technologie et fonction des premières productions céramiques de Grèce", *XIVe rencontres internationales d'Archéologie et d'Histoire d'Antibes,* APDCA eds, Juan les Pins, p. 225-242, 1994.

[PIC 78] PICOLPASSO C., *The Three Books of the Potter's Art: A Facsimile of the Manuscript in the Victoria and Albert Museum,* Lightborn R. and Caiger Smith A. (ed.), 2 vol., Scolar Press, 1980.

[PRE 91] PREAUD T. and D'ALBIS A., *La porcelaine de Vincennes*, Adam Biro, 1991.

[REG in press] REGERT M., "Elucidating pottery function using a multi-step analytical methodology combining infrared spectroscopy", *Mass Spectrometry and Chromatographic Procedures*, British Archaeological Reports, in press.

[REG 03] REGERT M., VACHER S., MOULHERAT C., DECAVALLAS O., "Study of adhesive production and pottery function during Iron Age at the site of Grand Aunay (Sarrthe, France)", *Archaeometry*, vol. 48, p. 101-120, 2003.

[RIC 87] RICE P.M., *Pottery Analysis – A Sourcebook*, The University of Chicago Press, 1987.

[ROS 95] ROSEN J., *La earthenware en France du XIVe au XIXe siècle*, Errance, 1995.

[RYE 81] RYE O.S., *Pottery Technology – Principles and Reconstruction*, Manuals on archeology 4, Taraxacum, 1981.

[SCH 42] SCHUMANN, *Berichte der Deutschen Keramischen Gesselschaft*, vol. 23, p. 408-426, 1942.

[VAN 90] VANDIVER P.B., SOFFER O., KLIMA B., SVOBODA J., "Venuses and wolverines: the origins of ceramic technology, ca. 26000 B.P.", in *The Changing Roles of Ceramics in Society: 26000 B.P. to the Present*, Kingery W.D. (ed.), American Ceramic Society, p. 13-81, 1990.

[WOO 99] WOOD N., *Chinese glazes*, Ed A & C Black, 1999.

Chapter 3

Sintering and Microstructure of Ceramics

3.1. Sintering and microstructure of ceramics

We saw in Chapter 1 that sintering is at the heart of ceramic processes. However, as sintering takes place only in the last of the three main stages of the process (powders → forming → heat treatments), one might be surprised to see that the place devoted to it in written works is much greater than that devoted to powder preparation and forming stages. This is perhaps because sintering involves scientific considerations more directly, whereas the other two stages often stress more technical observations – in the best possible meaning of the term, but with manufacturing secrets and industrial property aspects that are not compatible with the dissemination of knowledge. However, there is more: being the last of the three stages – even though it may be followed by various finishing treatments (rectification, decoration, deposit of surfacing coatings, etc.) – sintering often reveals defects caused during the preceding stages, which are generally optimized with respect to sintering, which perfects them – for example, the granularity of the powders directly impacts on the densification and grain growth, so therefore the success of the powder treatment is validated by the performances of the sintered part.

Sintering allows the consolidation – the non-cohesive granular medium becomes a cohesive material – whilst organizing the microstructure (size and shape of the grains, rate and nature of the porosity, etc.). However, the microstructure determines to a large extent the performances of the material: all the more reason why sintering

---

Chapter written by Philippe BOCH and Anne LERICHE.

deserves a thorough attention, and the reason for which this chapter interlaces "sintering" and "microstructures". We will now describe the overall landscape and the various chapters in this volume will present, on a case-by-case basis, the specificities of the sintering of the materials they deal with.

Sintering is the basic technique for the processing of ceramics, but other materials can also use it: metals, carbides bound by a metallic phase and other cermets, as well as natural materials, primarily snow and ice.

Among the reference works on sintering, we recommend above all [BER 93] and [GER 96]; the latter refers to more than 6,000 articles and deals with both ceramics and metals. We also recommend [LEE 94], which discusses ceramic microstructures and [RIN 96], which focuses on powders.

3.2. Thermodynamics and kinetics: experimental aspects of sintering

3.2.1. *Thermodynamics of sintering*

Sintering is the consolidation, under the effect of temperature, of a powdery agglomerate, a non-cohesive granular material (often called compact, even though its porosity is typically 40% and therefore its compactness is only 60%), with the particles of the starting powder "welding" with one another to create a mechanically cohesive solid, generally a polycrystal.

The surface of a solid has a surplus energy (energy per unit area: γ_{SV}, where $_S$ is for "solid" and $_V$ is for "vapor") due to the fact that the atoms here do not have the normal environment of the solid which would minimize the free enthalpy. In a polycrystal, the grains are separated by grain boundaries whose surplus energy (denoted γ_{SS}, or γ_{GB}, where $_{SS}$ is for "solid-solid" and $_{GB}$ for "grain boundary") is due to the structural disorder of the boundary. In general, $\gamma_{SS} < \gamma_{SV}$, so a powder lowers its energy when it is sintered to yield a polycrystal: the thermodynamic engine of sintering is the reduction of system's interfacial energies.

Mechanical energy is the reduction of the system's free enthalpy:

$$\Delta G_T = \Delta G_{VOL} + \Delta G_{GB} + \Delta G_S$$

where ΔG_T is the total variation of G and where $_{VOL}$, $_{GB}$ and $_S$ correspond to the variation of the terms associated respectively with the volume, the grain boundaries and the surface.

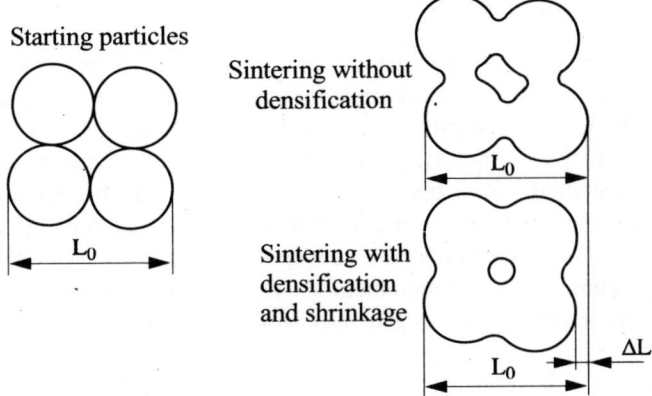

Figure 3.1. *Sintering of four powder particles. In general, we want sintering to be "densifying", in which case the reduction of porosity implies a shrinkage: $L_{final} = L_0 - \Delta L$. Some mechanisms are non-densifying and allow only grain growth. This diagram shows a two-dimensional system but the powder is a three-dimensional system. We could consider an octahedral configuration where the interstice between the four particles is closed below and above by a fifth and a sixth particle [KIN 76]*

The interfacial energy has the form $G = \gamma A$, where γ is the specific interface energy and A its surface area. The lowering of energy can therefore be achieved in three ways: i) by reducing the value of γ, ii) by reducing the interface area A, and iii) by combining these effects. The replacement of the solid-vapor surfaces by grain boundaries decreases γ, when γ_{SS} is lower than γ_{SV}. The reduction of A is achieved by grain growth: for example, the coalescence of n small spheres with surface s and volume v results in a large sphere with volume $V = nv$ but with surface $S < ns$ (this coalescence can be easily observed in water-oil emulsions). In fact, the term sintering includes four phenomena, which take place simultaneously and often compete with each other:

– consolidation: development of necks that "weld" the particles to one another;

– densification: reduction of the porosity, therefore overall contraction of the part (sintering shrinkage);

– grain coarsening: coarsening of the particles and the grains;

– physicochemical reactions: in the powder, then in the material under consolidation.

3.2.2. *Matter transport*

Sintering is possible only if the atoms can diffuse to form the necks that weld the particles with one another. The transport of matter can occur in vapor phase, in a liquid, by diffusion in a crystal, or through the viscous flow of a glass. Most mechanisms are activated thermally because the action of temperature is necessary to overcome the potential barrier between the initial state of higher energy (compacted powder) and the final state of lower energy (consolidated material). Atomic diffusion in ceramics is sufficiently rapid only at temperatures higher than 0.6-0.8 T_F, where T_F is the melting point (in K). For alumina, for example, which melts at around 2,320 K the sintering temperature chosen is generally around 1,900 K.

3.2.3. *Experimental aspects of sintering*

The parameters available to us to regulate sintering and control the development of the microstructure are primarily the composition of the starting system and the sintering conditions:

– composition of the system: i) chemical composition of the starting powders, ii) size and shape of the particles, and iii) compactness rate of the pressed powder;

– sintering conditions: i) treatment temperature, ii) treatment duration, iii) treatment atmosphere and, as the case may be, iv) pressure during the heat treatment (for pressure sintering).

Pressureless sintering and pressure sintering

In general, sintering is achieved solely by heat treatment at high temperature, but in difficult cases it can be assisted by the application of an external pressure:

– pressureless sintering: no external pressure during the heat treatment;

– pressure sintering (under uniaxial load or isostatic pressure): application of an external pressure during the heat treatment.

Pressure sintering requires a pressure device that withstands the high sintering temperatures, which is in fact a complex and expensive technique and therefore reserved for specific cases.

Sintering with or without liquid phase

Sintering excludes a complete melting of the material and can therefore occur without any liquid phase. However, it can be facilitated by the presence of a liquid phase, in a more or less abundant quantity. We can thus distinguish solid phase sintering on the one hand and sintering where a liquid phase is present; the latter case can be either liquid phase sintering or vitrification, depending on the quantity of liquid (see Figure 3.2):

– for solid phase sintering, the quantity of liquid is zero or is at least too low to be detected. Consolidation and elimination of the porosity require a disruption of the granular architecture: after the sintering, the grains of the polycrystal are generally much larger than the particles of the starting powder and their morphologies are also different. Solid phase sintering requires very fine particles (micrometric) and high treatment temperatures; it is reserved for demanding uses, for example, transparent alumina for public lamps;

– for liquid phase sintering, the quantity of liquid formed is too low (a few vol.%) to fill the inter-particle porosities. However, the liquid contributes to the movements of matter, particularly thanks to phenomena of dissolution followed by reprecipitation. The partial dissolution of the particles modifies their morphology and can lead to the development of new phases. A number of technical ceramics (refractory materials, alumina for insulators, $BaTiO_3$-based dielectrics) are sintered in liquid phase;

– lastly, for vitrification, there is an abundant liquid phase (for example, 20 vol.%), resulting from the melting of some of the starting components or from products of the reaction between these components. This liquid fills the spaces between the non-molten particles and consolidation occurs primarily by the penetration of the liquid into the interstices due to capillary forces, then solidification during cooling, to give crystallized phases or amorphous glass. This type of sintering is the rule for silicate ceramics, for example, porcelains. However, the quantity of liquid must not be excessive, and its viscosity must not be too low, otherwise the object would collapse under its own weight and would lose the shape given to it.

Sintering with and without reaction

We can speak of reactive sintering for traditional ceramics, where the starting raw materials are mixtures of crushed minerals that react with one another during sintering. The presence of a liquid phase often favors the chemical reactions between the liquid and the solid grains. However, for solid phase sintering, reactive sintering is generally avoided: either we have the powders of the desired compound already, or sintering is preceded by calcination, i.e. a high temperature treatment of the starting raw materials to allow their reaction towards the desired compound, followed by the crushing of this compound to obtain the powders that will be sintered:

– non-reactive sintering: an example is that of alumina, because the powders of this compound are available on the market;

– calcination and then sintering: an example is barium titanate ($BaTiO_3$). $BaTiO_3$ powders are expensive and some industrialists prefer to start with a less expensive mixture of barium carbonate $BaCO_3$ and titanium oxide TiO_2, the mixture being initially calcined by a high temperature treatment to form $BaTiO_3$, which is then crushed to give the powder that will be used for sintering;

– reactive sintering: an example is that of silicon nitride (Si_3N_4), for which one of the preparation methods consists of treating silicon powders in an atmosphere of nitrogen and hydrogen, so that the reaction that forms the nitride ($3Si + 2N_2 \rightarrow Si_3N_4$) is concomitant with its sintering (see Chapter 7). This technique (RBSN = reaction bonded silicon nitride) makes it possible to circumvent the difficulties of the direct sintering of Si_3N_4 and offers the advantage of minimizing dimensional variations, but the disadvantage of yielding a porous material (P > 10%). Mullite and zirconia mullite can also be prepared by reactive sintering [BOC 87 and 90].

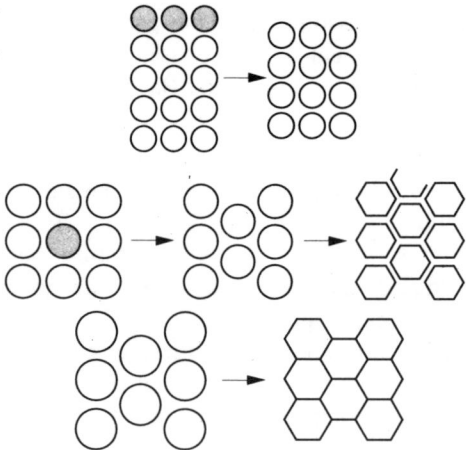

Figure 3.2. *Top, vitrification: the liquid phase is abundant enough to fill the interstices between the particles; in the middle, liquid phase sintering: the liquid is not sufficient to fill the interstices; bottom, solid phase sintering: organization and shape of the particles are extremely modified. This diagram does not show the grain coarsening: in fact, the grains of the sintered material are appreciably coarser than the starting particles [BRO 911]*

Densification: sintering shrinkage

The starting compact has a porous volume (P) of about 40% of the total volume. However, for most applications, we want relatively non-porous, even dense, ceramics (P ≈ 0%). In the absence of reactions leading to an increase in the specific volume, densification must be accompanied by an overall contraction of the part: characterized by linear withdrawal (dl/l_0), this contraction usually exceeds 10%. The control of the shrinkage is of vital importance for the industrialist: on the one hand, the shrinkage should not result in distortions of the shape and on the other hand, it must yield final dimensions as close as possible to the desired dimensions. In fact, an excessive shrinkage would make the part too small, which cannot be corrected, and an insufficient shrinkage would make the part too large; in this case machining for achieving the desired dimension must be done by rectification, often by means of diamond grinding wheel – a finishing treatment all the more expensive as the volume of matter to be abraded is large. It is difficult to control shrinkage with a relative accuracy higher than 0.5%.

Because of the phenomenon of shrinkage, dilatometry tests are widely used for the *in situ* follow-up of sintering: starting with the "green" compact to arrive at the fired product, a heating at constant speed typically comprises three stages:

i) thermal expansion, accompanied by a vaporization of the starting water and a pyrolysis of the organic binders introduced to support the pressing of the powder;

ii) a marked contraction, due to particle rearrangement, the development of sintering necks and granular changes;

iii) a resumption of the thermal expansion of the sintered product.

Many studies have sought to correlate the kinetics of shrinkage and the growth of inter-particle necks [BER 93, KUC 49].

Porosity is open as long as it is inter-connected and communicating: the material is then permeable to fluids. Porosity is closed when it is not inter-connected: even if it is not yet dense, the material can then be impermeable. The porosity level corresponding to the transformation of open pores to closed pores is about $P \approx 10\%$.

Sintering generally occurs in the absence of external pressure applied during the heat treatments (pressureless sintering); the particles of the starting powders weld with one another to form a polycrystalline material, possibly with vitreous phases; the presence or absence of a liquid phase is important. Finally, the term sintering covers four phenomena: i) consolidation, ii) densification, iii) grain coarsening and iv) physicochemical reactions. The beginning of the densification is the usual sign for the beginning of the sintering, frequently followed by dilatometry experiments.

3.3. Interface effects

From a macroscopic point of view, the driving force behind the sintering of a powder to form a polycrystalline material is the reduction of energy resulting from the reduction of solid-vapor surfaces in favor of the grain boundaries. The necessary condition for sintering is therefore that the grain boundary energy (γ_{GB}) is low compared to the energy (γ_{SV}) of the solid-vapor surfaces. But this condition is not always achieved, as shown by silicon carbide (SiC) or silicon nitride (Si_3N_4): materials where the γ_{GB}/γ_{SV} ratio is too high to allow easy sintering. The solution for sintering such materials can be i) the use of sintering additives chosen to increase

γ_{SV} or to decrease γ_{GB} or ii) the use of pressure sintering, which provides external work: $dW = - P_{external}dV$.

From a microscopic point of view, it is the differential pressure on either side of an interface that causes the matter transport making sintering possible. This pressure depends on the curvature of the surface.

Interface energy

The increase in energy (γ) at the level of the interfaces, due to the fact that the atoms do not have their normal environment, is always very insignificant: γ is typically a fraction of Joules per m^{-2}. Substances added in very small quantities can have a marked effect – this is also the case with liquids, as shown in the use of surface active agents in washing powders and detergents. The surfaces of the particles and the grain boundaries of sintered materials are frequently covered by adsorbed species, segregations or precipitations, which means that interfacial energies are in general modified by these extrinsic effects. We can give the example of silicon-based non-oxide ceramics (SiC or Si_3N_4), whose particles are covered with an oxidized skin – silica. As the specific surface of a powder grows as the inverse of the squared linear dimensions of the grain, the interfacial effects are marked in fine powders, which is generally the case with ceramic powders – the diameter of the particles measures typically from a fraction of a micrometer to a few micrometers, which corresponds to specific surfaces in the order of a few m^2g^{-1}.

The role of the curvature in the energy of an interface can be illustrated by considering a bubble blown in a soapy liquid using a straw. If we disregard the differences in density, and consequently the effects of gravity, the only obstacle for the expansion of the blown bubble under the pressure P is the increase of the energy at the interface. For a spherical bubble, the equilibrium radius r is the one for which the expansion work is equal to this increase in energy [KIN 76]:

$$\Delta Pdv = \gamma dA \qquad dv = 4\pi r^2 dr \qquad dA = 8\pi rdr$$

$$\Delta P = \gamma dA/dv = \gamma\left(\frac{8\pi rdr}{4\pi r^2 dr}\right) = 2\gamma/r$$

We note that the difference in pressure ΔP is proportional to the interfacial energy (here $\gamma = \gamma_{LV}$, liquid-vapor interface) and inversely proportional to the radius of curvature.

For a non-spherical surface with main radii of curvature r' and r

$$\Delta P = \gamma \left(\frac{1}{r'} + \frac{1}{r''} \right) \tag{3.1}$$

Likewise, the rise h of a liquid in a capillary of radius r is such that:

$$\Delta P = 2\gamma \cos\theta / r = \rho g h$$

where ρ is the density of the liquid and θ the liquid-solid wetting angle.

The relation: $\gamma = (r\rho g h)/(2\cos\theta)$ is used to assess γ by measuring θ.

These capillary effects contribute to the vitrification of silicate ceramics, because the viscous liquid formed by the molten components infiltrates into the interstices between the non-molten particles.

The difference in pressure through a curved surface implies an increase in vapor pressure and also in solubility, at a point of high curvature:

$$\Omega \Delta P = RT \ln(P/P_0) = V\gamma(1/r' + 1/r'')$$

where Ω is the molar volume, P the vapor pressure above the curved surface, and P_0 the pressure above a plane surface. Thus:

$$\ln(P/P_0) = (\Omega\gamma/RT)(1/r' + 1/r'') = (M\gamma/\rho RT)(1/r' + 1/r'') \tag{3.2}$$

where R is the ideal gas constant, T the temperature, M the molar mass, and ρ the density. Equation [3.2] is the Thomson-Kelvin equation.

For a spherical surface, this relation can be seen when we consider the transfer of a mole of the compound as a result of the vapor pressure on the surface, the work provided being equal to the product of the specific energy and the variation of surface area:

$$RT\ln P/P_0 = \gamma dA = \gamma \, 8\pi r dr$$

As the variation of volume is $dv = 4\pi r^2 dr$, the variation in radius for the transfer of a mole is $dr = \Omega/(4\pi r^2)$, consequently:

$$\ln P/P_0 = (\Omega\gamma/RT)(2/r), \text{ which is the above result when } r' = r'' = r$$

The sign convention is to consider that the radii of curvature r' and r'' to be positive for convex surfaces and negative for concave surfaces. Equation [3.1] shows that $\Delta P = 0$ for a plane surface (r' and $r'' = \infty$). A bump tends to level itself and a hole to fill itself. We can mention as an example the progressive restoration of the flatness in a skating rink surface striped by the skates, after the skaters leave the rink (it is primarily surface diffusion that makes ice get back a smooth surface).

The concept of pressure on a surface is a macroscopic concept. At the microscopic scale, atomic diffusion in a crystallized phase occurs primarily due to the movements of the vacancies. However, the equilibrium concentration of the vacancies is less under a convex surface than under a plane surface, and it is higher under a concave surface than under a plane surface. Thus, the vacancies migrate from the high concentration areas to the low concentration areas, an action which implies a movement contrary to that of the atoms. The effects are all the more obvious according to how marked the curvature ($1/r$) is and therefore the smaller particles are: sintering is facilitated by the use of fine powders (diameters of about a micrometer). However, the pressure variations and the energies brought into play by the interfacial effects still remain very low.

EXAMPLE 1.– for spherical alumina particles ($\gamma_{SV}Al_2O_3 \approx 1 Jm^{-2}$), the surplus pressure associated with particles with a diameter of 1 micrometer is 0.2%.

EXAMPLE 2.– at what size must an Al_2O_3 monocrystal be crushed to increase its energy from 500 kJmole^{-1} (500 kJ.mole^{-1} is a typical value of the energies brought into play in chemical reactions involving metallic oxides)? If $\gamma_{SV}Al_2O_3 = 1$ J m^{-2} and $\rho_0 Al_2O_3 = 4.10^3$ kg m^{-3}, the answer is: at a size less than that of the crystal cell!

EXAMPLE 3.– what is the energy variation when 1 kg of SiO_2 powder composed of beads of 2 μm diameter sinters to give a dense sphere and without internal interfaces? If $\gamma_{SV}SiO_2 \approx 0.3$ J m^{-2} and $\rho_0 SiO_2 = 2.2. 10^3$ kg m^{-3}, the answer is: 20 kJ only.

The development of a sintering neck is illustrated by the simple model of two isodiametric spherical particles (see Figure 3.3). The connection between the two particles is a neck in the shape of a horse saddle, with $r' < 0$ depending on the concavity (in the plane of the figure) and $r'' > 0$ depending on the convexity (in the plane tangent to the two spheres, perpendicular to the figure). The neck is a very curved area, which constitutes a source of matter towards which the atoms coming from the surface (the sintering is then non-densifying) or the volume (the sintering is then densifying) migrate. The movements of matter result in the progressive coarsening of the sintering neck and the consolidation of the material.

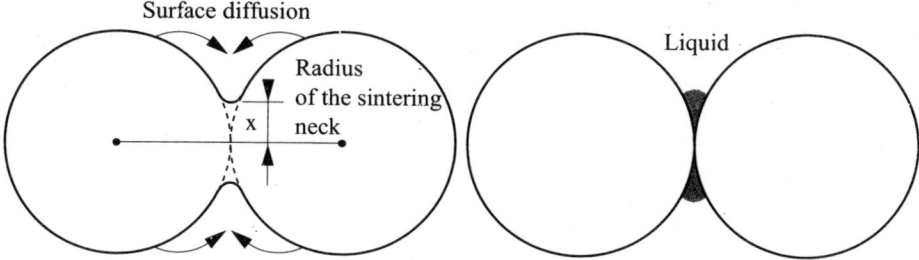

Figure 3.3. *At the beginning of the sintering, the consolidation is done by evaporation of the surfaces and condensation on the neck (on the left); this mechanism is not densifying. If there is a liquid (on the right) the capillary pressure helps the penetration of the liquid in the interstice and the dissolution-reprecipitation effects contribute to the matter transport*

3.4. Matter transport

Even if the thermodynamic condition of sintering is met ($A_{SV}\gamma_{SV} > A_{GB}\gamma_{GB}$), for the process to occur, its speed must be sufficient. However, the matter transport in a solid is very slow compared to a liquid or a gas. This matter transport can come from an overall movement (viscous flow of vitreous phases or plastic deformation of a crystal), the repetition of unit processes on an atomic scale (atomic diffusion in a crystal), from transport in vapor phase (evaporation then condensation) or in liquid phase (dissolution then reprecipitation). Speed is significant only if the temperature is sufficiently high. The diffusion (D) in a crystal or the inverse of the viscosity (η) of glass vary as $\exp(E/RT)$, where E is the apparent activation energy of the process. The usual values of E are a few hundred kilojoules per mole. The normal sintering temperatures are about 0.6 to 0.8 T_F, where T_F is the melting point of the solid in question.

The matter movement takes place from the high energy areas towards the low energy areas – primarily, the sintering neck between the particles. We must distinguish two cases depending on the location of the source of matter:

– when the source of matter is the surface, the mechanism is non-densifying, which means that the spheres take an ellipsoidal form, without their centers approaching one another. There is no macroscopic shrinkage and the porosity of the granular compact is not reduced significantly. The decrease in interfacial energy primarily comes from the grain coarsening;

– when the source of matter is inside the grains (near the boundaries, or near defects such as dislocations), the mechanism is densifying: there is shrinkage and reduction in porosity (see Table 3.1 and Figure 3.4).

For solid phase sintering, there are four ways of diffusion: i) surface diffusion, ii) volume diffusion (often called lattice diffusion), iii) vapor phase transport (evaporation-condensation), and iv) grain boundary diffusion: the boundaries are very disturbed areas, which allow "diffusion short-circuits". For liquid phase sintering, we must add dissolution-reprecipitation effects or a vitreous flow. Finally, for pressure sintering the pressure exerted allows the plastic deformation of the crystallized phases and the viscous flow of the amorphous phases.

Path in Figure 3.4	Diffusion path	Source of matter	Shaft of matter	Result obtained
1	Surface diffusion	Surface	Sintering neck	Grain coarsening
2	Volume diffusion	Surface	Sintering neck	Grain coarsening
3	Evaporation-condensation	Surface	Sintering neck	Grain coarsening
4	Grain boundary diffusion	Grain boundaries	Sintering neck	Densifying sintering
5	Volume diffusion	Grain boundaries	Sintering neck	Densifying sintering
6	Volume diffusion	Defects, like dislocations	Sintering neck	Densifying sintering

Table 3.1. *Matter transport during a solid phase sintering [ASH 75]*

3.4.1. Viscous flow of vitreous phases

The difference in pressure on either side of a curved interface causes a stress σ (a stress has the dimension of a pressure) which causes a viscous flow ε of the glass. The flow rate $d\varepsilon/dt$ is proportional to the stress and inversely proportional to the viscosity η: $d\varepsilon/dt$ proportional to σ/η.

In general, viscosity decreases exponentially when the temperature increases:

$$\eta = \eta_0 \exp(Q/RT) \rightarrow d\varepsilon/dt \text{ proportional to } \sigma/\eta\exp(Q/RT)$$

where Q is the apparent activation energy of the process.

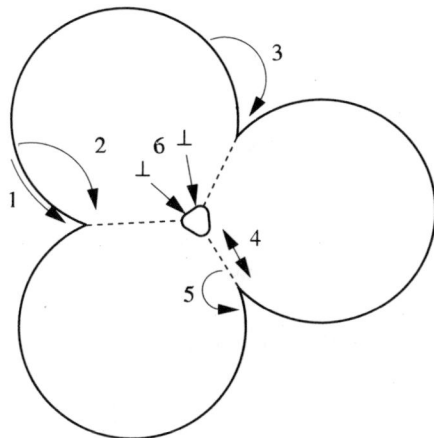

Figure 3.4. *Matter transport during a solid phase sintering; mechanisms 1, 2 and 3 are non-densifying; mechanisms 4, 5 and 6 are densifying; ⊥ schematizes a dislocation [ASH 75]*

3.4.2. *Atomic diffusion in crystallized phases*

Fick's first law

$$J = -D(\delta c/\delta x) \text{ for a unidirectional diffusion along x} \qquad [3.3]$$

J is the flow of atoms passing through a unit surface, per time unit, D the diffusion coefficient of the species that diffuses and c its concentration.

Fick's second law

$$(\delta c/\delta t) = D(\delta^2 c/\delta x^2) \qquad [3.4]$$

Nernst-Einstein's equation

The "force" that acts on the atom that diffuses is the opposite of the chemical potential gradient. The mobility of the atom i is B_i, the quotient of the speed of the atom by the driving force:

$$-B_i = v_i /[(1/N)d\mu_i/dx] \qquad [3.5]$$

$$J_i = -(1/N)(d\mu_i/x)B_i c_i$$

where N is the Avogadro number and μ_i the chemical potential of the species i.

Considering the activity equal to the unit:

$$d\mu_i = RTd(\ln c_i) \tag{3.6}$$

Substituting [3.6] in [3.5] and comparing with [3.3], we obtain:

$$J_i = -(RT/N)B_i (dc_i/dx)$$

$$D_i = kTB_i, \text{ where } k \text{ is the Boltzmann constant} \tag{3.7}$$

Therefore, the diffusion coefficient is proportional to the atomic mobility.

Besides, $d\mu/dx$ is proportional to the pressure gradient dP/dx:

$$J \approx (D/kT)(dP/dx) \tag{3.8}$$

The difference in pressure between the two sides of an interface causes a matter flow that is proportional to the pressure difference and to the diffusion coefficient of the mobile species [PHI 85].

NOTE.– $D = D_0\exp(-Q/RT)$, so despite the presence of the term kT in the denominator of [3.8], it is the exponential of the numerator that is more important: an increase of T results in a rapid increase of J.

NOTE.– the volume diffusion coefficient D_V is expressed in $m^2 s^{-1}$ (or, often, in $cm^2 s^{-1}$). As regards grain boundary diffusion (or surface diffusion), it is usual to consider the thickness of the grain boundary e_{GB} (or the thickness of the superficial area e_S), so that the diffusion term is written as $D_{GB}e_{GB}$ (or $D_S e_S$), the coefficients D_{GB} and D_S being then expressed in $m s^{-1}$ (or in $cm s^{-1}$).

3.4.3. *Grain size distribution: scale effects*

An essential objective in controlling the microstructure of a sintered material is to be able to control densification and grain growth separately. In a ceramic filter, for example, we want to preserve a notable porosity, with pores of sizes calibrated with respect to the medium to be filtered. In an optical porthole, on the contrary, we want the sintering to be accompanied by a complete densification (zero residual porosity) because the presence of residual pores would result in the diffusion of the light. However, we saw that certain matter transport mechanisms are non-densifying (like surface diffusion), while others are densifying (like grain boundary diffusion): the objective is to play on the sintering parameters in order to favor a particular mechanism. The size of the powder particles is one of the parameters at our disposal.

Although the powders in general consist of particles of irregular size and shape, the simplistic approach that considers isodiametric spherical particles makes a useful semi-quantitative analysis possible.

The laws of scale [HER 50] specify the manner in which a phenomenon associated with a "cluster" of particles must be transposed in the case of a homothetic cluster p times larger. For isodiametric spheres, the laws of scale relate to the radius of the spheres (r). Thus, the time taken to obtain a certain degree of progress in a process depends on granulometry, according to a law of scale that varies with the process brought into play. If t_1 is the time that corresponds to the small cluster and t_2 the time that corresponds to the large cluster, then $t_1/t_2 = (r_1/r_2)^n = (1/p)^n$, where the value of n depends on the process. We are interested here in matter transport processes that ensure sintering. We will deal with only two cases (flow of a vitreous phase and diffusion-reprecipitation in a liquid) and will give the results for the other mechanisms.

Viscous flow of a vitreous phase

The flow rate $d\varepsilon/dt$ is inversely proportional to the viscosity η and proportional to the stress, whose form (see equation [3.1]) is γ/u, where u is the radius of the sintering neck. The duration δt of the transport of a given quantity of matter is inversely proportional to the speed, therefore:

$$\delta t_{viscosity} \propto 1/(d\varepsilon/dt) \propto \eta u/\gamma$$

The radius of the neck u, is in a certain ratio k with the radius of the particles: $u = kr$. For a system that grows homothetically, particles p times coarser imply neck radii p times larger. Therefore, for this system that is p times larger:

$$\delta t_{viscosity} \propto p\eta u/\gamma \propto p\eta kr/\gamma \qquad\qquad [3.9]$$

This result shows that the duration is proportional to the size r of the particles and therefore that the sintering time necessary to obtain a certain degree of consolidation varies inversely to the size of the particles; for example, dividing the size of the particles by ten reduces the duration of the sintering in the same ratio.

Dissolution-reprecipitation in a liquid

We suppose that the spherical particles are covered with a thin film of liquid, with a thickness of e_L. The matter flow is:

$$J \propto (- D_{Liquid}/ kT)(\delta P/L)$$

where D_{Liquid} = transport coefficient in the liquid.

The area of the section through which the flow of diffusion passes is $A \propto e_L r$, but the pressure at the points of contact between the particles is γ/u, a term that is proportional to γ/r, therefore:

$$\delta P \propto \gamma/u \propto \gamma/r$$

The volume of matter that must diffuse to make a given level of densification possible is proportional to the cube of the linear dimensions of the system and therefore proportional to r^3. The time necessary to reach this level of densification is therefore:

$$\delta t_{liquidphase} \propto (\text{displaced volume })/(\text{speed}_{diff} \times \text{volume}_{atom})$$

$$\propto r^3/(JA\Omega) \propto r^3/[(D_{liquid}/kT)(\gamma/r^2)e_L R\Omega]$$

$$\delta t_{liquidphase} \propto [r^4 kT][D_{liquid}\, e_L \gamma\Omega], \text{ therefore } \delta t_{liquidphase} \text{ proportional to } r^4$$
[3.10]

The duration is proportional to the power of four of the size of the particles. Dividing the size of the particles by 10 helps, this time, to gain a factor of 10,000 in the sintering time.

Through a similar reasoning, we can show that the grain boundary diffusion and surface diffusion make the duration vary to the power of four of the size of the particles, the volume diffusion to the power of three, and evaporation-condensation to the power of two.

In short, liquid phase diffusion, surface diffusion and grain boundary diffusion (R^{-4} law in the three cases) are more sensitive to the reduction in size of the particles than to volume diffusion (R^{-3}), evaporation condensation (R^{-2}), and finally viscous flow (R^{-1}).

3.5. Solid phase sintering

3.5.1. *The three stages of sintering*

Solid phase sintering refers to the case where no liquid phase has been identified (but observations through electronic microscopy in transmission sometimes show the presence of a very small quantity of liquid phase, for example due to a segregation of the impurities along the grain boundaries). Solid phase sintering takes place in three successive stages:

– initial stage: the particle system is similar to a set of spheres in contact, between which the sintering necks develop. If X is the radius of the neck and R the radius of the particles, the growth of the ratio X/R in time t, for an isothermal sintering, takes the form: $(X/R)^n = Bt/D^m$, where B is a characteristic parameter of the material and the exponents n and m vary according to the process brought into play. For example, $n = 2$ and $m = 1$ for viscous flow; $n = 5$ and $m = 3$ for volume diffusion; $n = 6$ and $m = 4$ for grain boundary diffusion;

– intermediate stage: the system is schematized by a stacking of polyhedric grains intertwined at their common faces, with pores that form a canal system along the edges common to three grains, connected at the quadruple points (see Figure 3.5). The porosity is open. This diagram is valid as long as the densification does not exceed ≈ 90-92%, a threshold beyond which the interconnection of the porosity disappears;

– final stage: the porosity is closed; only isolated pores remain, often located at the quadruple points between the grains ("triple points" on a two-dimensional section) but which can be trapped in intragranular position.

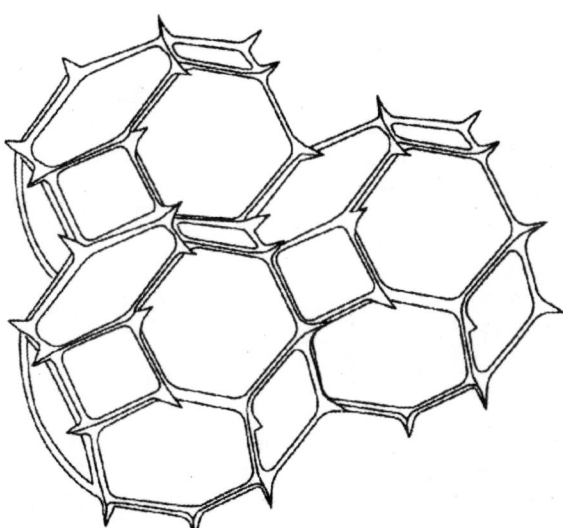

Figure 3.5. *Diagram of the porosity in the form of inter-connected canals along the edges of a polyhedron with 14 faces, typical of the intermediate stage of sintering [GER 96]*

3.5.2. *Grain growth*

As the energy of the interfaces has the form γA, where γ is the specific energy of the interface and A is the surface area of the interface, the system's energy can be reduced using two borderline cases:

– pure densification: the particles preserve their original size, but the solid-gas interfaces (γ_{SG}) are replaced by grain boundaries (γ_{SS}), with a change in the shape of the particles;

– coalescence and pure grain growth: the particles preserve their original form, but they change in size by coalescence, thus reducing the surface areas.

Pure densification has never been observed: there is always a grain growth. Owing to the difference in pressure ($\Delta P \approx \gamma/r$), the atoms diffuse from the high pressure area towards the low pressure area. In addition, a curved boundary blocked at its ends tends to reduce its length while evolving to a line segment. Because of these two causes, the boundary moves towards its center of curvature. By considering (in two dimensions) triple points with angles of 120°, the grains with less than six sides have their boundaries with the concave side turned towards the inside: the evolution towards the center of curvature makes these small grains disappear. A contrary evolution affects the grains with many sides: the small grains disappear in favor of the coarser grains, which grow (see Figure 3.6).

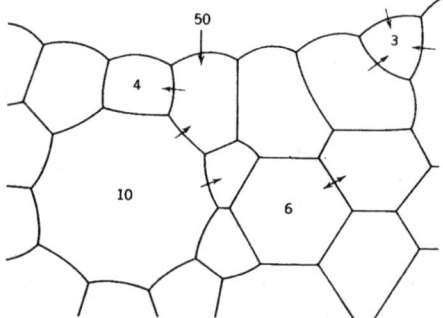

Figure 3.6. *The pressure on the curved interfaces is such that the boundaries move towards their center of curvature: the small convex grains (less than 6 sides) disappear while the coarse concave grains (more than 6 sides) grow to the detriment of the neighboring grains; the grains with rectilinear boundaries have an appreciably hexagonal form [KIN 76]*

In normal grain growth, the average grain size increases regularly, without marked modification of the relative distribution of the size; the microstructure expands homothetically. This type of grain growth is the one observed in a successful sintering.

Secondary recrystallization (or abnormal growth, or discontinuous grain growth) makes a few grains grow rapidly, to the detriment of the more moderately sized grains. The final microstructure is very heterogenous, with coexistence of very coarse grains and very small grains. This type of microstructure rarely leads to favorable properties and therefore is generally avoided.

In addition to the possibility of being homogenous or, on the contrary, heterogenous, the microstructure can be more or less isotropic. For the simple case of a mono-phased polycrystal, we can distinguish four cases:

– equiaxed microstructure and random crystalline orientation of the grains (no orientation texture): the material is isotropic by effect of average;

– equiaxed microstructure, but orientation texture: the matter loses its average isotropy and the overall anisotropy is all the more marked that the crystal in question is more anisotropic for the property concerned;

– oriented microstructure, but no orientation texture: anisotropy;

– oriented microstructure and orientation texture: maximum anisotropy. The polycrystal then offers properties close to those of the monocrystal – except for the intrinsic effect of the grain boundaries. We can cite as an example the case of graphite fibers ("carbon fibers") used for the mechanical reinforcement of composites.

The majority of ceramics are multiphased materials that comprise both crystallized and vitreous phases. Porcelain thus consists of silicate glass "reinforced" by acicular crystals of crystallized mullite, but we can also observe millimetric crystal agglomerates with a very porous microstructure (iron and steel refractory materials), or fine grained polycrystals (< 10 μm) without vitreous phases and with very low porosity (hip prosthesis in alumina or zirconia). It should be reiterated that, in addition to the chemical nature of the compound(s) in question, it is the microstructure of the material (size and shape of the grains, rate and type of porosity, distribution of the phases) that controls the properties.

3.5.3. Competition between consolidation and grain growth

Densification – and therefore the elimination of the pores – occurs effectively only if the pores remain located on the grain boundaries (intergranular position), because then the matter movements can take advantage of the grain boundary diffusion. However, a too rapid grain growth – and therefore a migration of the boundaries – leads to a separation of the pores and the boundaries: the pores are then trapped in the intragranular position, where they are difficult to eliminate, because only volume diffusion remains active. If the objective is to sinter a material to its ultimate density, and therefore eliminate all the pores, the growth of the grains must be limited.

In addition to its role in the coupling between densification and grain growth, the size of the grains (Φ) of the sintered ceramics is, together with the porosity, the essential microstructural parameter. We can give five examples:

– the brittle fracture of the ceramics is controlled by the size of the microscopic cracks, because the mechanical strength σ_f is proportional to $K_c a_c^{-1/2}$, where K_c is the toughness and a_c the length of the critical microscopic crack. However, a_c is of the same order as the size of the grain. This means that σ_f varies typically by $\Phi^{-1/2}$: ceramics with high mechanical strength (machine parts, cutting tools, hip prostheses, etc.) must be very fine-grained;

– the particle composites based on partially stabilized zirconia use mechanical reinforcement mechanisms that depend on the size of the zirconia inclusions: if they are too small ($< 0.3 \ \mu m$) they remain tetragonal and if they are too large ($1 \ \mu m$) they destabilize towards the monoclinical form, with swelling and thus micro-cracking of the surrounding matrix. The optimal effect is achieved for particles of intermediate size, metastable tetragonal, which are transformed from tetragonal to monoclinical in the stress field of a crack that propagates itself;

– the high temperature creep of refractory materials is often due to diffusion mechanisms: volume diffusion leads to the Nabarro-Herring creep and grain boundary diffusion to the Coble creep, with creep rates in Φ^{-2} and Φ^{-3}, respectively: refractory materials must therefore be coarse-grained in order to slow down the creep;

– ferroelectric or ferrimagnetic ceramics have performances sensitive to the size of the domains (size that interacts with the grain size) and to the migration of the walls (which is hampered by the grain boundaries): very fine grains are monodomain, and from them we have ferroelectric ceramics with very high dielectric constant or "hard" ferrimagnetic ceramics with very high coercitive field strength;

– finally, the transport properties (electric conduction or thermal conduction) are sensitive to the intergranular barriers due to the structural disorder of the grain boundaries or to the presence of secondary phases that are segregated there: coarse grains mean fewer grain boundaries and therefore fewer barriers.

3.5.4. Normal grain growth

In a fine-grained polycrystal heated to a sufficient temperature, the size of the grains grows and correlatively the number of grains decreases. The driving energy is the one that corresponds to the disappearance of the grain boundaries (it is of the order of a fraction of $J.m^{-2}$).

The grain growth rate is proportional to the rate of migration of the boundary; this rate (v) can be written as the product of the mobility of the grain boundary (M_{GB}) and a driving force (F):

$$v \approx d\Phi/dt \qquad \text{with: } v = M_{GB}F$$

The driving force is due to the pressure difference caused by the curvature of the boundary:

$$\Delta P = \gamma_{GB}(1/r' + 1/r'')$$

γ_{GB} is the energy of the grain boundary and r' and r'' are the curvature radii at the point in question.

When the grain growth is normal, the distribution of the grain sizes remains significantly unchanged, with a homothetic growth. Consequently:

$$(1/r' + 1/r'') \approx 1/K\Phi \quad \text{where K is a constant.}$$

Pure monophased material

A simple reasoning based on a two-dimensional microstructure (section of a polycrystal), where the equilibrium configuration of a "triple point" corresponds to angles of 120°, is that grains with less than six sides are limited by convex boundaries and therefore tend to decrease, whereas those with more than six sides are limited by concave boundaries and therefore tend to grow (see Figure 3.6). If the curvature radius of a grain is proportional to its diameter, the driving force and the growth rate are inversely proportional to its size:

$$d\Phi/dt = Cte/\Phi \quad \text{hence} \quad \Phi \approx t^{1/2} \tag{3.11}$$

The grain size must increase by the square root of the time.

Among the simplistic assumptions that have been made, we note that only the curvature of the boundary has been considered and not the crystalline anisotropy.

Obstacles to grain growth

When we express experimental results of grain growth in the form of a graph $\ln\Phi = f(\ln t)$, we obtain a straight line whose slope is, in general, less than the exponent 1/2 predicted by the parabolic law. This means that the growth is slowed down by various obstacles. Based on the interaction between a mobile grain boundary and an obstacle, we can distinguish three main cases: i) impurities in solid solution or liquid phase wetting the boundaries, ii) immobile obstacles, which block any movement of the boundary, and iii) mobile obstacles capable of migrating with the boundary.

The impurities in solid solution can slow down the movement of the boundaries because they prefer to lodge themselves close to the grain boundary and therefore the boundary can migrate either by carrying these impurities along – which slows down the movement – or by leaving them in the intragranular position – which puts them in an energetically less favorable position than before the migration of the boundary. The growth law is thus modified because of the presence of these impurities and we get:

$$d\Phi/dt \approx 1/\Phi^2 \quad \Phi \approx K\, t^{1/3} \qquad\qquad [3.12]$$

The growth law $\Phi \approx t^{1/3}$ (impure phase) is more frequently observed than the law $\Phi \approx t^{1/2}$ (pure phase).

The presence of a liquid phase that wets the boundaries tends to reduce the grain growth, by reducing the driving energy and increasing the diffusion path, since there now is a double interface. It is true that diffusion in a liquid is fast; however, the dissolution-diffusion-reprecipitation process is generally slower than the simple jump through a grain boundary. Thus, the presence of a small quantity of a molten silicate phase limits the grain growth of the sintered alumina with liquid phase. On the other hand, the presence of a liquid phase can favor chemical reactions of type $A + B \to C$ and therefore allow the growth of the grains C to the detriment of the grains A and B. This type of growth often leads to secondary recrystallization (exaggerated growth). The growth law is $\Phi \approx t^{1/3}$, as for an impure phase.

The immobile obstacles, such as precipitates and inclusions, "pin" the boundaries, reducing their energy by a quantity equal to the product of the specific pinning energy and the surface area of the inclusion. To be "undragged", the boundary must be subjected to a tearing force. As long as the migration driving force of the boundary, due to the effects of curvature, does not exceed this tearing force, the boundary remains pinned and the grain size is stable. For grains anchored by inclusions, the growth can occur only if:

– the inclusions coalesce by diffusion, to give less numerous but more voluminous inclusions (Ostwald ripening). If the coalescence takes place by volume diffusion, the radius of the inclusion (r) increases as $r^3 \approx t$, which again yields a grain growth obeying a law $\Phi \approx t^{1/3}$;

– the inclusions disappear by dissolution in the matrix: $\Phi \approx t$;

– secondary recrystallization occurs: this is the end of normal growth.

The mobile obstacles are essentially pores. If v_P and v_{GB} are the speeds of the pore and the boundary, M_P and M_{GB} are the mobilities, and F_P and F_{GB} are the corresponding "forces", we have $v_P = M_P F_P$ and $v_{GB} = M_{GB} F_{GB}$. The pore separates itself from the boundary if $v_{GB} > v_P$.

The force on the boundary F_{GB} has two components, one due to the curvature (F'_{GB}) and the other due to the pinning effect by the pores, which equals $N F_P$, if there are N pores. The condition for non-separation is therefore:

$$v_P = M_P F_P = v_{GB} = M_{GB} (F'_{GB} - N F_P)\ v_{GB} = F_{GB} (M_P M_{GB})/(N M_{GB} + M_P)$$

– if $N M_{GB} \gg M_P$, then $v_{GB} = F_{GB} (M_P/N)$: the rate of migration of the boundaries is controlled by the characteristics of the pores;

– if $N M_{GB} \ll M_P$, then $v_{GB} = F_{GB} (M_{GB})$: the rate of migration of the boundaries is controlled by the characteristics of the boundaries themselves.

Different mechanisms lead to different laws of type $\Phi \approx t^{1/n}$. The values of the exponent n depend on the mechanism and the diffusion path that control the process. For example, for control by the pores: $n = 4$ for surface diffusion, $n = 2$ for volume diffusion, and $n = 3$ for vapor phase diffusion; for control by the boundaries: $n = 2$ for a pure phase and $n = 3$ for the coalescence of a second phase by volume diffusion. The experimental studies of the grain growth consist of: i) quantifying the grain size Φ, ii) determining the exponent n of the growth law $\Phi \approx t^{1/n}$, and iii) determining the apparent activation energy E of the process. The results are semi-quantitative, because of two difficulties: i) inaccuracy of the measures of the grain size and ii) simultaneous occurrence of several processes – with different values of n and E. The law of normal grain growth that is most frequently observed is the law $\Phi \approx t^{1/3}$

3.5.5. Abnormal grain growth

Some grains develop in an exaggerated manner, the process occurring when a grain reaches a significant size with a shape limited by many concave sides: there is then a rapid growth of the coarse grain, to the detriment of fine convex grains that border it (see Figure 3.6). When the grain reaches this critical size Φ_C, much higher than the average size of the other grains in the matrix $\Phi_{average}$, the concave curvature is determined by the size of the small grains and is therefore proportional to $1/\Phi_{average}$. Hence, this apparent paradox that the use of a very fine starting powder can sometimes increase the risk of secondary recrystallization, because the presence of a few particles of size much higher than $\Phi_{average}$, is more probable there than in coarser powders where $\Phi_{average}$ is higher.

In some sintered materials, we observe very coarse grains with straight sides, whose growth cannot be explained by the surface tension on the curved boundaries. These are often materials whose grain boundary energy is very anisotropic where the growth favors the low energy facets (see Figure 3.7). This effect is observed in many rocks. They can also be materials where the impurities lead to the appearance of a small quantity of intergranular phase between the coarse grain and the matrix, which favor the growth – but a larger quantity of liquid phase would make the penetration in all the boundaries possible, limiting both normal and exaggerated growth.

Abnormal grain growth generally obeys a law $\Phi \approx t$, whereas normal growth leads to laws $\Phi \approx t^{1/3}$ or $\Phi \approx t^{1/2}$: the abnormal growth must be fought from the beginning, because, once started, its kinetics is rapid.

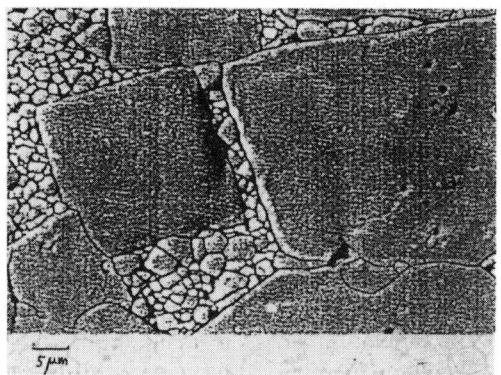

Figure 3.7. *Abnormal grain growth in In_2O_3 sintered at high temperature (1,500°C for 50 h). Some grains have grown exaggeratedly in a fine-grain matrix [NAD 97]*

3.6. Sintering with liquid phase: vitrification

3.6.1. *Parameters of the liquid phase*

In general, the presence of a liquid phase facilitates sintering. Vitrification is the rule for silicate ceramics where the reactions between the starting components form compounds melting at a rather low temperature, with the development of an abundant quantity of viscous liquid. Various technical ceramics, most metals and cermets are all sintered in the presence of a liquid phase. It is rare that sintering with liquid phase does not imply any chemical reactions, but in the simple case where these reactions do not have a marked influence, surface effects are predominant. The main parameters are therefore: i) quantity of liquid phase, ii) its viscosity, iii) its wettability with respect to the solid, and iv) the respective solubilities of the solid in the liquid and the liquid in the solid:

– quantity of liquid: as the compact stacking of isodiametric spheres leaves a porosity of approximately 26%; this value is the order of magnitude of the volume of liquid phase necessary to fill all the interstices and allow the rearrangement of the grains observed at the beginning of the vitrification. However, the presence of a small quantity of liquid (a few volumes percent) does not make it possible to fill the interstices;

– viscosity of the liquid: this decreases rapidly when the temperature increases (typically according to the Arrhenius law). Pure silica melts only at a very high temperature to produce a very viscous liquid. The presence of alkalines and alkaline earths quickly decreases the softening temperature and the viscosity of the liquid. The viscosity of the liquid should be neither too low – because then the sintered part becomes deformed in an unacceptable way – nor too high – because then the viscous flow is too limited, making grain rearrangement difficult;

– wettability: wettability is quantifiable by the experiment of the liquid drop placed on a solid, because the equilibrium shape of the drop minimizes the interfacial energies. If γ_{LV} is the liquid-vapor energy, γ_{SV} the solid-vapor energy and γ_{SL} the solid-liquid energy, the angle of contact (θ) is such that (see Figure 3.8):

$$\gamma_{LV}\cos\theta = \gamma_{SV} - \gamma_{SL} \qquad [3.13]$$

When γ_{SL} is high, the drop minimizes its interface with the solid, hence a high value of θ: $\theta > 90°$ corresponds to non-wetting (depression of the liquid in a capillary). On the contrary, when $\gamma_{SL} \ll \gamma_{SV}$, the liquid spreads on the surface of the solid: $\theta < 90°$ corresponds to wetting (rise of the liquid in a capillary); and for $\theta = 0$, the wetting is perfect.

In a granular solid that contains a liquid, the respective values of γ_{SL} and γ_{GB} (grain boundary energy) determine the value of the dihedral angle Θ:

$$2\gamma_{SL}\cos\Theta/2 = \gamma_{GB} \qquad [3.14]$$

Figure 3.9 shows the penetration of the liquid between the particles of a granular solid according to the value of Θ. For low Θ (0 to 30°), the liquid wets the boundaries; when Θ continues to grow, the occurrence of the liquid phase becomes less marked and for a high value of Θ ($\Theta > 120°$), the liquid tends to form pockets located at the "triple points" – on a two-dimensional view, but at the "quadruple points" in three-dimensional space. Based on mutual solubilities we can distinguish four cases (see Table 3.2).

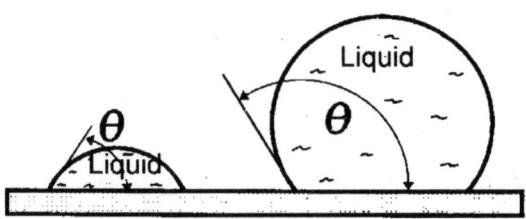

Figure 3.8. *Drop placed on a liquid; the value of θ characterizes the wettability: wetting on the left; non-wetting on the right*

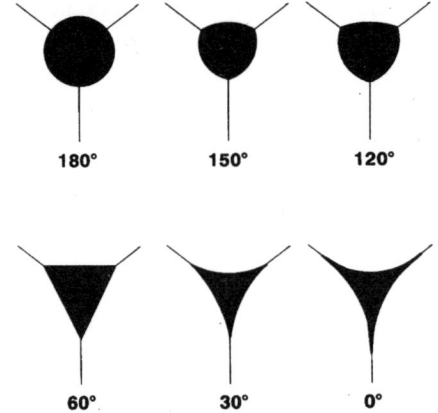

Figure 3.9. *Penetration of the liquid between the grains depending on the value of Θ [GER 96]*

	Low solubility of the solid in the liquid	High solubility of the solid in the liquid
Low solubility of the liquid in the solid	Low assistance to densification	High assistance to densification
High solubility of the liquid in the solid	Swelling, transitory liquid	Swelling, and/or densification

Table 3.2. *Effects of mutual solubilities on sintering [GER 96]*

3.6.2. *The stages in liquid phase sintering*

The shrinkage curve recorded during an isothermal treatment of liquid phase sintering shows three stages:

– viscous flow and grain rearrangement: when the liquid is formed, the limiting process consists of a viscous flow, which allows the rearrangement of the grains.

The liquid dissolves the surface asperities and also dissolves the small particles. The granular rearrangement is limited to the liquid phase sintering itself, but it can be enough to allow complete densification if the liquid phase is in sufficient quantity, as is the case in the vitrification of silicate ceramics;

– solution-reprecipitation: the solubility of the solid in the liquid increases at the inter-particle points of contact. The transfer of matter followed by reprecipitation in the low energy areas results in densification;

– development of the solid skeleton: the liquid phase is eliminated gradually by the formation of new crystals or solid solutions; we tend to approach the case of solid phase sintering and the last stage of the elimination of porosity is similar to the one observed in this case.

The disintegration of the particles attacked by the liquid results in the Ostwald ripening (coalescence of small particles to give a larger particle) and changes in the shape of the particles, with flattening of the areas of contact. As the anisotropy of crystalline growth is less hampered when a crystal grows in a liquid than when it remains in contact with solid obstacles, we sometimes observe grains whose morphology reflects these anisotropy effects: for instance, they are elongated and faceted.

The role of chemical reactions is still significant, because they bring into play energies much higher than the interfacial ones and frequently the reactions between liquid and solid result in the formation of new phases. We can thus distinguish three cases:

– weak reaction between liquid and solid: the liquid has the primary role, after cooling, of forming the matrix in which the grains that have not reacted have been glued. This is the case of abrasive materials where the grains (silicon carbide SiC or alumina Al_2O_3) are bound by a solidified vitreous phase;

– reaction between liquid and solid, solid with congruent melting: there is no appearance of new solid phases but modification of the existing ones. This is the case for silicate ceramics made of quartz sand (SiO_2) and clay (whose primary mineral is kaolinite, written as ($Al_2O_3.2SiO_2.2H_2O$), fired at rather low temperatures. The high viscosity of the silicate liquid prevents the system from reaching the equilibrium; in particular, glass of the eutectic composition does not decompose into mullite plus cristobalite, as suggested by the equilibrium diagram. Only the finest particles react; the coarsest do not dissolve. The coarse quartz grains, for example, hardly react with clay – but firing transforms them, almost completely, into cristobalite (a high temperature variety of crystallized silica);

– reaction between liquid and solid, solid with incongruent melting: an example is that of the system containing quartz (SiO_2) + kaolinite (Al_2O_3-$2SiO_2$-$2H_2O$) + potassic feldspar ($6SiO_2$-K_2O-Al_2O_3), which is the basic system of porcelains.

At about T = 1,150°C, the feldspar melts to give leucite (4SiO$_2$.Al$_2$O$_3$.K$_2$O) and a vitreous phase (with a composition close to 9SiO$_2$. Al$_2$O$_3$.K$_2$O). Leucite dissolves gradually into glass to produce a flow that is very viscous until it melts at about 1,530°C: at 1,300°C, the viscosity is equal to 10^6 poises and it decreases only slowly with temperature: at 1,400°C it is still 5.10^5 poises. Potassic feldspar is a flux (a component that, by reaction with the other components, gives rise to a phase with low melting point) which produces a liquid whose viscosity does not vary too quickly with the temperature, and which therefore does not require a very strict control of this temperature: the firing range is broad. On the contrary, certain fluxes (for example, calcic phases) have a sudden effect because they create phases with too low viscosity.

3.7. Sintering additives: sintering maps

The spectacular effect of the addition of a few hundred ppm of magnesia on the sintering of alumina is the best example of the role of sintering additives. These additives help to control the microstructure of the sintered materials; they can be classified under two categories:

– additives that react with the basic compound to give a liquid phase, for example by the appearance of an eutectic at a melting point less than the sintering temperature. We then go from the case of solid phase sintering to liquid phase sintering – even if the liquid is very insignificant. Silicon nitride Si$_3$N$_4$ ceramics are an example of where some sintering additives are selected to react with the silica layer (SiO$_2$) that covers the nitride grains, in order to produce a eutectic. Thus, magnesia MgO reacts with SiO$_2$ to form the enstatite MgSiO$_3$, from which we have a liquid phase at about 1,550°C. The liquid film wets the grain boundaries and shapes of the pockets at the triple points;

– additives that do not lead to the formation of a liquid phase and which consequently enable the sintering to take place in solid phase. This is the case of the doping of Al$_2$O$_3$ with a few hundred ppm of MgO, because the lowest temperature at which a liquid can appear in the Al$_2$O$_3$-MgO system exceeds the sintering temperature (which, for alumina, does not go beyond 1,700°C).

The explanation of the role of this second category of additives is primarily phenomenological. It considers the respective values of the diffusion coefficients and the mobility of the boundaries:

– D_L characterizes volume diffusion (L = lattice), D_b grain boundary diffusion and D_S surface diffusion;

– M_b characterizes the mobility of the grain boundaries.

The sintering maps [HAR 84] place the diameter of the grain (G) on the ordinate and densification ($\rho = d/d_0$) on the abscissa (see Figure 3.10). The two extreme cases would be i) a grain coarsening without densification (vertical trajectory) and ii) a densification with unchanged grain size (horizontal trajectory).

Experimentally, we always observe an intermediate trajectory between these two extremes because the densification is inevitably accompanied by grain growth.

In order to densify the material to 100%, the key point is to prevent the pores and the boundaries from separating because then, as we already said, the residual pores are trapped in the intragranular position, where it is practically impossible to eliminate them. The trajectory $G = f(\rho)$ must therefore be as flat as possible and must, in particular, go below the lowest point of the pore-boundary separation area (in the figure: the point ordinate G^* abscissa ρ^*). Densification cannot reach 100% if the trajectory cuts this separation area. Various ratios characterize the relationship between "contribution of the diffusion to densification" and "contribution of the diffusion to grain coarsening", with the first term in the numerator and the second term in the denominator. For example, D_L/D_L means: "densification controlled by volume diffusion" and "grain coarsening controlled by volume diffusion", whereas D_b/D_S means "densification by boundary diffusion" and "grain coarsening controlled by surface diffusion" (see Figure 3.11). The possible effect of an additive can be seen from the following observations:

– an increase in D_L flattens the trajectory without affecting the separation area: this increase of D_L is favorable to the densification;

– a decrease in M_b increases G^* and therefore shifts the separation area towards the top and slightly flattens the trajectory: this decrease in M_b also has a favorable effect on the densification;

– a decrease in D_S flattens the trajectory (which is favorable), but decreases G^* and therefore shifts the separation area to the bottom (which is unfavorable). All in all this decrease in surface diffusion – which as we said earlier leads to a non-densifying sintering – would not have a significantly useful (or harmful) effect.

The use of these sintering maps to explain the effectiveness of MgO as a sintering additive for Al_2O_3 suggests that MgO increases D_L (first favorable effect) and especially decreases M_b (second favorable effect). This phenomenological explanation does not, however, provide information on the mechanisms brought into play and in particular it does not give the reason for which MgO reduces the mobility of the boundaries. An explanation [BAE 94] would be that the traces of impurities (SiO_2 and CaO), which continue to exist even in so-called high purity alumina powders, are located along the grain boundaries, to form at the sintering temperature a thin liquid film which promotes the grain growth – "solid phase sintering" then becoming a sintering controlled by a very insignificant liquid phase. The influence of MgO would then be "to purify" the grain boundaries while reacting with SiO_2 or CaO.

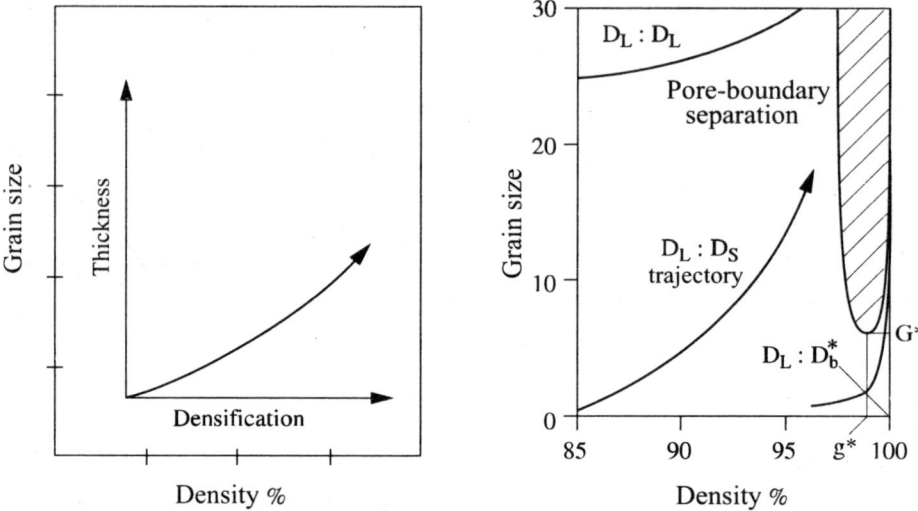

Figure 3.10. *Sintering map showing the grain size depending on the densification [HAR 84]. On the left: principle of the map; on the right: for complete densification to be possible, the sintering trajectory must not cut the hatched pore-boundary separation area*

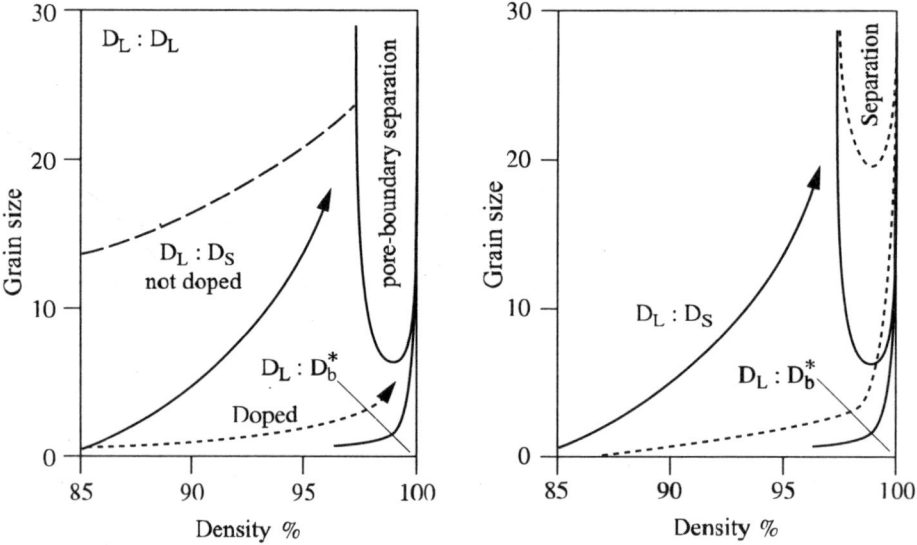

Figure 3.11. *Role of a sintering additive [HAR 84]. On the left, the effect of the doping agent is to multiply D_L by 10: the influence is favorable by the flattening of the trajectory. On the right, the effect of the doping agent is to divide M_b by 10: the influence is doubly favorable by the raising of the separation area and flatness of the trajectory*

The doping of Al_2O_3 by MgO has been transposed to various ceramic systems, for which we have determined which sintering additives limit the grain growth and make a densification close to 100% possible [NAD 97]. These studies provide answers on a case-by-case basis and there is still no general theory for the selection of the optimal additive.

The choice of the sintering temperature also plays on the relative values of the diffusion coefficients and therefore favors a densifying or a non-densifying mechanism. For example, surface diffusion has an apparent activation energy generally less than the volume diffusion. The chronothermic effect ("a long duration heat treatment at lower temperature is equivalent to a short duration heat treatment at higher temperature") therefore offers broader possibilities than those offered by the Arrhenius law with a single activation energy: low temperature sintering primarily bringing into play surface diffusion (non-densifying mechanism), and high temperature sintering volume diffusion or the grain boundary diffusion (densifying mechanisms). A high temperature treatment favors, all things being equal, high densification.

3.8. Pressure sintering and hot isostatic pressing

3.8.1. *Applying a pressure during sintering*

In most cases, ceramics are sintered by pressureless sintering and it is only for very special applications that we use "pressure sintering" or "hot pressing", which consists of applying a pressure during the heat treatment itself. The characteristic of pressure sintering is that the pressures brought into play – which are usually about 10 to 70 MPa, but can exceed 100 MPa – have considerable effects compared to capillary actions, thus offering four advantages:

i) thickening of materials whose interfacial energy balances are unfavorable;

ii) rapid densification at appreciably lower temperatures (several hundred degrees sometimes) than those demanded by pressureless sintering;

iii) possibility of reaching the theoretical density (zero porosity);

iv) possibility of limiting the grain growth.

Furthermore, it can be possible to obtain the sintered part with its exact dimensions (*net shape*), without the need for a machine finishing in applications that require high dimensional accuracy. The other side of the coin is the technical complexity of the process and the high costs incurred, as well as the limitations on the geometry of the parts, which can only have simple forms and a rather reduced size. We must have pressurization devices manufactured in materials that resist the temperatures required by sintering – and even if these temperatures are lower compared to those required by pressureless sintering, they are still high – and the chemical reactions between these materials and the environment (for example, oxidation of refractory metals), like the reactions between the mould and the ceramic powder, must be limited. One last difficulty: if the manufacture of parts with simple geometry (pellets) can be done in a piston + cylinder mould ("uniaxial pressure pressing"), obtaining more complex shapes, in particular undercut parts, cannot be done by pressure sintering. We must then apply the technique of hot isostatic pressing or "HIP", where the pressure is not transmitted by a piston but by a gas, hence the hydrostaticity (isostaticity) of the efforts, in analogy with "cold isostatic pressing" described in Chapter 5, but where the pressure transmitting fluid is a liquid and not a gas.

3.8.2. *Pressure sintering*

Graphite is the most used material for the manufacture of the mould and the piston of uniaxial pressure sintering equipments, because of its exceptional refractarity, with this originality that the mechanical strength grows when the temperature rises (until beyond 2,000°C), also taking into account its easy machinability and the generally limited speed of the reactions with the ceramic powders – often protected by a fine boron nitride deposit. But the oxidation ability of the graphite requires a reducing or neutral processing atmosphere, which is appropriate for non-oxides (primarily carbides, like HPSC, and nitrides, like HPSN; see Chapter 7), but can lead to oxygen under-stoichiometry for those oxides that are reduced easily. Refractory metals (Mo or W) and ceramics (Al_2O_3 or SiC) have also been used for the piston-cylinder couple of the mould.

The powders to be sintered are generally very fine (< 1 µm) and it is not always necessary for them to contain additives required by pressureless sintering (for example, MgO for the sintering of Al_2O_3). The justifiable applications of pressure sintering are, for example, cutting tools (ceramics or cermets) or optical parts, with the essential objectives of achieving a 100% densification and/or very fine grains – but the microstructure and the crystallographic texture can present anisotropy effects because of the uniaxiality of the pressing. Alumina for cutting tools, carbides (B_4C, for instance) or cermets are examples of materials that can benefit from pressure sintering and HIP (see further down); the same is true for metallic "superalloys" used in the hot parts of turbojets. High temperature composite materials are another example where the application of a pressure during heat treatments can be necessary to allow the impregnation of the fibrous wicks and favor the densification. Functional ceramics ($BaTiO_3$ or, especially, magnetic ferrites) can gain from very fine grains and the absence of residual porosity made possible by pressure sintering. As optical transparency is no doubt the property that is most quickly degraded by the presence of pores, even in extremely small numbers, perfectly transparent polycrystalline ceramics ($MgAl_2O_4$, Al_2O_3, Y_2O_3, etc.) are examples of materials that benefit from the use of pressure sintering.

As regards the mechanisms, pressure sintering implies: i) rearrangement of the particles, ii) lattice diffusion, iii) grain boundary diffusion, and finally iv) plastic deformation and a viscous flow. Pressureless sintering involves much less the effects i) and iv) and, as for the effects ii) and iii), the high level of the mechanical stresses (often close to and even exceeding the stresses caused by the normal operation of a part, for example a refractory part in a high temperature facility) brings them close to creep effects. This can be diffusion creep (Nabarro-Herring creep due to intragranular diffusion, Coble creep due to the grain boundary diffusion) or creep due to the movement of dislocations.

The creep equation, modified for pressure sintering, can be written as:

$$(1/\rho)(d\rho/dt) = (CD)/(kT\Phi^m)\ [\sigma^n + 2\gamma/r] \qquad [3.15]$$

where ρ is the density, C a constant, D the coefficient that controls the diffusion process, k the Boltzmann constant and T the temperature, Φ the average grain size, σ the pressure applied on the particles, γ the surface energy and r the radius of the pores. The exponents m and n characterize respectively the role of the grain size and that of the pressure applied. Table 3.3 recapitulates the relevant parameters (see Chapter 8).

Mechanism	Grain size exponent, m	Stress exponent, n	Coefficient of diffusion, D
Nabarro-Herring	2	1	Volume diff. D_V
Coble	3	1	Boundary diff. D_J
Intergranular sliding	1	1 or 2	D_J, D_V
Interface reactions	1	2	D_J, D_V
Plastic flow	0	≥ 3	D_V

Table 3.3. *Mechanisms of pressure sintering [HAR 91]*

In most cases, the use of fine grained ceramics on the one hand, and the high level of plastic flow required by iono-covalent crystals on the other, are such that the diffusion terms (Nabarro-Herring or Coble) override the plastic flow. Grain boundary diffusion dominates over volume diffusion all the more when the grains are finer and the temperature lower, because the volume exponent of the former is 3 whereas that of the latter is only 2, and the activation enthalpy of D_J is in general lower than that of D_V.

Boundary sliding is necessary in order to accommodate the variations in shape caused by the diffusion creep, which implies that the mechanisms must act sequentially and therefore that the overall kinetics is controlled by the slowest mechanism. Nonetheless, when the mechanisms can act concurrently (as is the case with diffusion creep and plastic flow), it is the fastest process that controls the overall kinetics.

An illustration [TAI 98] of pressure sintering (1 hour at 1,360°C, p = 20 MPa, graphite matrix, antiadhesive layer of BN, in vacuum) is obtaining particle composites 10%Al_2O_2-80%WC-10%Co with a mechanical strength of 1,250 MPa: pressureless sintering would not allow the densification of this type of material, whose microstructure exhibits an inter-connected matrix of WC, with precipitates of Al_2O_3 and Co_3W_3C (see Figure 3.12).

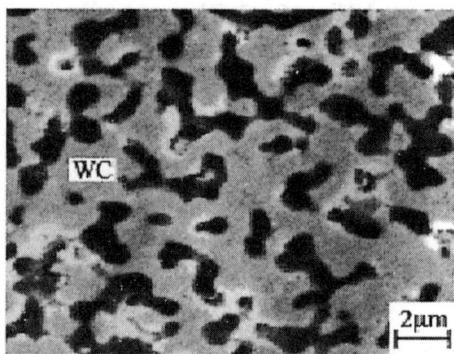

Figure 3.12. *10% Al_2O_3-80% WC-10% Co composite, sintered under pressure [TAI 98]*

3.8.3. *Hot isostatic pressing (HIP)*

Whereas for cold isostatic pressing (CIC – see Chapter 5), the pressurization fluid is a liquid, it is a gas (in general argon, but reactive atmospheres are also used, for example oxygen) that provides the pressurization in HIP. This technique was invented by the Battelle institute (USA) in the 1950s. We can imagine the risks of destructive explosions (use of a compressible fluid instead of an incompressible fluid) and the difficulties in ensuring air-tightness as well as the problems of pollution and control of thermal transfers: under a pressure of 1,000 atmospheres, a gas like argon has a density higher than that of liquid water at 20°C!

The two main methods involving HIP are direct consolidation by HIP, and HIP perfecting a pressureless sintering having preceded it (see Figure 3.13).

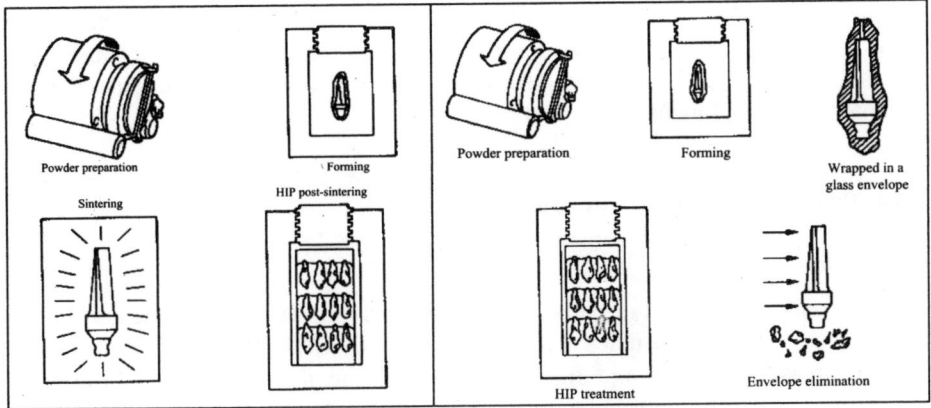

Figure 3.13. *Direct HIP (on the left) and post-sintering HIP (on the right) [DAV 91]*

Consolidation by HIP

When HIP is used directly to consolidate a powder, the "compact" must be encapsulated in an envelope in a form homothetic to that of the part to be obtained, with vacuum evacuation of gases, followed by sealing of the envelope. Soft or stainless steels can be used as envelope materials for relatively low temperature treatments (1,100–1,200°C), whereas it is necessary to use refractory metals (Ta, Mo) for higher temperatures treatments. As the risks of distortion become higher when the overall pressing increases, we gain from a powder pressed at a high rate and homogenously (by CIC primarily). An alternative is to carry out a "pre-sintering" providing sufficient cohesion to the part to make its handling possible, and then to coat it powdered glass which, at sufficient temperature, will become viscous enough to coat the piece with an impermeable layer. This will make it possible for HIP to take place without the pressurized gas being able to penetrate the open porosity.

HIP as post-sintering operation

This involves sintering the part until the inter-connected open porosity is eliminated (which requires a densification of about 95%) and then subjecting this part to a secondary HIP treatment. The greatest advantage is avoiding the need for an envelope (cost, complexity, restrictions on the possible forms, necessity to clean the end product to eliminate the envelope). It is furthermore possible, for manufacturers who do not have an HIP equipment, to sub-contract this stage to a specialized partner. There are HIP chambers whose size is more than one meter, which makes it possible to treat large parts or a great number of small parts.

The densification of metallic powders ("powder metallurgy") involves HIP much more frequently than the densification of ceramic powders: a search on the Web shows that most sites dealing with HIP relate to metallic products (the term taken in its largest sense and including cermets).

3.8.4. Densification/conformity of shapes in HIP

Densification

The densification of the parts by HIP implies primarily three phenomena: i) fragmentation of the particles and rearrangement, ii) deformation of the inter-particle areas of contact and iii) elimination of the pores. The first process is transitory and hardly contributes to the overall densification, at least if the initial forming (for example, by CIC) has been correctly carried out. The second process brings into play effects of plastic deformation by movement of dislocations and diffusion phenomena that are similar to those indicated in the case of uniaxial pressure sintering. Lastly, by considering the final reduction of porosity, we can write phenomenologically:

$$(1/\rho)(d\rho/dt)_i = B_i f_i (\rho) \qquad\qquad [3.16]$$

where ρ is the relative density, B_i constant kinetics (implying the terms relating to the material and those relating to the characteristics of the HIP process) and $f_i(\rho)$ a geometrical function that depends only on the relative density. Each process i is described by specific expressions for B_i and f_i [LI 87]. For example:

$$K_i = 270\delta D_{jg}\Omega P/kTR^3 \quad \text{and} \quad f_i (\rho) = (1-\rho)^{1/2} \quad \text{if } \rho > 90\% \qquad [3.17]$$

for grain boundary diffusion (Coble), if δ is "the thickness" of the boundary, D_{jg} the corresponding diffusion coefficient, Ω the volume of the atom that diffuses and R the radius of the grain assumed to be spherical, k, T and P having their usual meaning.

Ashby *et al.* [LI 87] have developed the approach of "HIP maps", where, for a material under given conditions, the areas in two-dimensional space (relative density depending on the pressure), in which the predominant phenomenon that controls the densification has been identified, are traced (in particular the grain size and temperature). These maps make the pendant of the "creep maps" and "deformation maps" also credited to Ashby *et al.* (see Figure 7.2 in Chapter 7). The principle of these maps is certainly attractive, but their applicability requires three conditions: i) having a sufficient number of experimental data, ii) establishing, for each of these data, the nature of the predominant mechanism, and lastly iii) verifying the similarity of the treated cases (for example, the fact that the powders used contain the same impurities as the powders used for tracing the maps). The application of the maps is therefore qualitative more than quantitative. Let us use an example to illustrate this comment: when we compare the case of a metal with that of a ceramic, we observe that the mobility of dislocations in the former material is much higher than it is in the latter. This means that the relationship between the effect of an increase in temperature and that of an increase in pressure is higher for ceramics than it is for metal, which suggests different managements of the parameters T and p for the two categories of material.

Conformity of the shapes

The key point for HIP, which is an expensive treatment and therefore dedicated to high added value products, is to obtain parts whose final dimensions are as close as possible to the desired dimensions. However, this conformity of dimensions requires a perfect control of the shrinkage: it must occur particularly in a homothetical way, starting from the shape of the raw part until the consolidated and stripped part. However, this "homothetic shrinkage" is affected by various causes, including the envelope effect (in the case where there is not post-densification HIP) and the manner in which the consolidation front develops.

As regards the envelope effect: even if the "compact" is overheated perfectly homogenously throughout the HIP cycle, the various areas of the part do not offer the same resistance to the effects of isostatic pressure. Geometrical compatibilities require that the volume deformations should be accompanied by shearing strains, a requirement which introduces distortions. For the simple example of a cylindrical part (see Figure 3.14), the presence of the envelope causes a distortion of the "corners". The numerical calculation methods like finite elements are used widely for the study of such distortions in order to eliminate them by redrawing the envelope [NCE 00].

As regards densification: this progresses from outside the part towards the core, causing the formation of a consolidated crust whose thermal conduction is higher than that of the core that is not yet consolidated. The heat fluxes thus provoked lead to heterogenities in temperature, which lead to the accentuation of the shell effect of the crust with respect to the core. The effect is all the more marked the bulkier part is.

As an extension of pressureless sintering HIP confirms that a major concern for the production of ceramic parts – "traditional" ceramics as well as "technical" ceramic – is the maintenance of the shape and dimensions of the parts. As we said previously: the ceramist works on the product at the same time as he works on the material and therefore his efforts must be devoted to both sides of the problem.

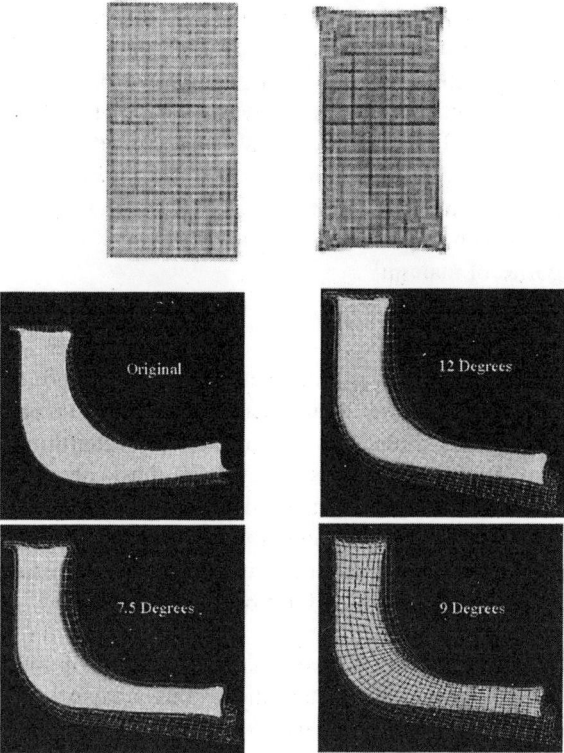

Figure 3.14. *HIP: at the top, distortion due to envelope effects; at the bottom, example of an iterative approach to determine the shape of the envelope, which allows the correction of the distortions [NCE 00]*

3.9. Bibliography

[ASH 75] ASHBY M.F., "A first report on sintering diagrams", *Acta Metall.*, 22, p. 275, 1975.

[BAE 94] BAE S.I. and BAIK S., "Critical concentration of MgO for the prevention of abnormal grain growth in alumina", *J. Am. Ceram. Soc.*, 77 (101), p. 2499, 1994.

[BER 93] BERNACHE-ASSOLLANT D. (ed.), *Chimie-physique du frittage*, Hermès, 1993.

[BOC 87] BOCH P. and GIRY J.P., "Preparation of zirconia-mullite ceramics by reaction sintering", *High Technology Ceramics*, Materials Science Monographs 38, Elsevier, 1987.

[BOC 90] BOCH P., CHARTIER T. and RODRIGO, "High purity mullite by reaction sintering", *Mullite and Mullite Matrix Composites*, Ceramic Transactions, Vol. 6, The Am. Ceramic Society, p. 353, 1990.

[DAV 91] DAVIS R.F., "Hot isostatic pressing", in Brook R.J. (ed.), *Concise Encyclopedia of Advanced Ceramic Materials*, Pergamon Press, p. 210, 1991.

[GER 96] GERMAN R.M., *Sintering Theory and Practice*, J. Wiley, 1996.

[HAR 84] HARMER M.P., "Use of solid-solution additives in ceramic processing", *Advances in Ceramics*, Am. Ceram. Soc., Vol. 10, p. 679, 1984.

[HAR 91] HARMER P.P., "Hot pressing: technology and theory", in Brook R.J. (ed.), *Concise Encyclopedia of Advanced Ceramic Materials*, Pergamon Press, p. 222, 1991.

[HER 50] HERRING C., "Diffusional viscosity of a polycrystalline solid", *J. Appl. Phys.*, 21(5), p. 437-445, 1950.

[KIN 76] KINGERY W.D., BOWEN H.K. and UHLMANN D.R., *Introduction to Ceramics*, 2nd edition, John Wiley and Sons, 1976.

[KUC 49] KUCZYNSKI G.C., "Self-distribution in sintering of metallic particles", *Trans. AIME*, 185, p. 169, 1949.

[LEE 94] LEE W.E. and RAINFORTH W.M., *Ceramic Microstructures*, Chapman & Hall, 1994.

[LI 87] LI W.B., ASHBY M.F., EASTERLING K.E., "On densifiaction and shape-change during hot isostatic pressing", *Acta Metallurgica*, 35, p. 2831-2842, 1987.

[NAD 97a] NADAUD N., KIM D.Y. and BOCH P., "Titania as Sintering Additive in Indium Oxide Ceramics", *J. Am. Ceram. Soc.*, 80(5), p. 1208-1212, 1997.

[NAD 97b] NADAUD N., "Relations entre frittage et propriétés de matériaux à base d'oxyde d'indium dopé à l'étain (ITO)", Thesis, Paris-6 University, 1997.

[NCE 00] National Center for Excellence in Metalworking Technology, CTC, 100 CTC Drive, Johnstown, Pa, USA.

[PHI 85] PHILIBERT J., *Diffusion et transport de matière dans les oxydes*, Editions de Physique, 1985.

[RIN 96] RING T.A., *Fundamentals of Ceramic Powder Processing and Synthesis*, Academic Press, 1996.

[TAI 88] TAI W.T. and WATANABE T., "Fabrication and mechanical properties of Al_2O_3 – WC–Co composites by vacuum hot pressing", *J. Am. Ceram. Soc.*, 81(6), p. 1673-1676, 1998.

Chapter 4

Silicate Ceramics

4.1. Introduction

Silicate ceramics are generally alumino-silicate based materials obtained from natural raw materials. They exhibit a set of fundamental properties, such as chemical inertia, thermal stability and mechanical strength, which explain why they are widely used in construction products (sanitary articles, floor and wall tiles, bricks, tiles) and domestic articles (crockery, decorative objects, pottery). They are often complex materials, whose usage properties depend at least as much on microstructure and aesthetics as on composition. Silicate products with an exclusively technical application (refractory materials, insulators or certain dental implants) will not be explicitly discussed in this chapter.

To distinguish silicate from technical ceramics, it is useful to qualify these products as traditional ceramics. This term refers to the centuries-old tradition that still strongly influences the classification of this type of materials and the vocabulary attached to them. However, it does not reflect the considerable evolution of a sector of activity where progress relates more to the manufacturing technologies (raw material mixtures, drying, sintering, etc.) than to the products themselves.

These products of terra cotta, earthenware, sandstone, porcelain or vitreous china are generally widely marketed materials. They represent a predominant share in the total sales turnover of the ceramic industry. In 1994, the fields of roof tiles and bricks, wall and floor tiles, crockery and ornamentation, and sanitary products

Chapter written by Jean-Pierre BONNET and Jean-Marie GAILLARD accounted for, respectively, 28, 14, 13 and 13% of the turnover of the French ceramic industry (technical + refractory + traditional) [LEC 96].

4.2. General information

Silicate ceramics can be formed in various ways: by casting in a mould aqueous suspension called slip, by extrusion or jiggering of a plastic paste, or by unidirectional or isostatic pressing of slightly wet aggregates. The quantity of water contained in the sample therefore depends on the method of forming. Generally, water is eliminated during a specific drying treatment. The raw part is transformed into ceramic by sintering, also called firing, carried out under suitable conditions of temperature, heating rate and atmosphere. Depending on the application considered, this ceramic, also called shard, can be dense or porous, white or colored.

Clay is the basic raw material for these products. Mixed with water, it can form a plastic paste similar to the one used by the potter on his wheel. Although easy to form, this paste often exhibits insufficient mechanical strength to enable handling without damaging the preform. Owing to the clay colloidal nature, a relatively pure paste is low in solid matter. It thus shrinks significantly during drying and sintering, which makes it difficult to control the shape and dimensions of the final piece. To limit all these effects, non-plastic products known as tempers can be added to the paste. They then form an inert and rigid skeleton that enhances the mechanical strength of the preform, favors the elimination of water during the drying stage and limits sintering shrinkage. Among the commonly used tempers, we can mention sand, feldspars, certain lime-rich compounds or grog (a paste sintered and ground beforehand).

Given the complexity of the composition of argillaceous raw materials, the appearance of a viscous liquid during firing can be observed. The addition of fluxes to the starting mixture amplifies this phenomenon. These compounds, which also behave as tempers, generally contain alkaline ions (Na, K, sometimes Li). The examination of the phase diagram of Al_2O_3-SiO_2-K_2O represented in Figure 4.1 highlights the role of flux of a potassium feldspar (orthoclase with the composition $K_2O,A_2O_3,6SiO_2$) with respect to the deshydroxylation product of kaolinite, whose composition in equivalent oxides $Al_2O_3,2SiO_2$ is symbolized by point MK. At equilibrium, the addition of a small quantity of feldspar leads to decrease the solidus temperature from 1,590 to 985°C. Under certain temperature and composition conditions, iron oxides and a few calcium-rich compounds, such as chalk, can also contribute to the formation of a liquid phase. In the presence of a sufficient quantity of molten matter, the heat treatment can be pursued until the almost complete disappearance of the porosity. The shards thus obtained are rich in vitreous phase and exhibit good mechanical strength. The quantity of liquid formed during partial vitrification must remain sufficiently low or its viscosity must be high enough so that the piece does not become deformed under its own weight.

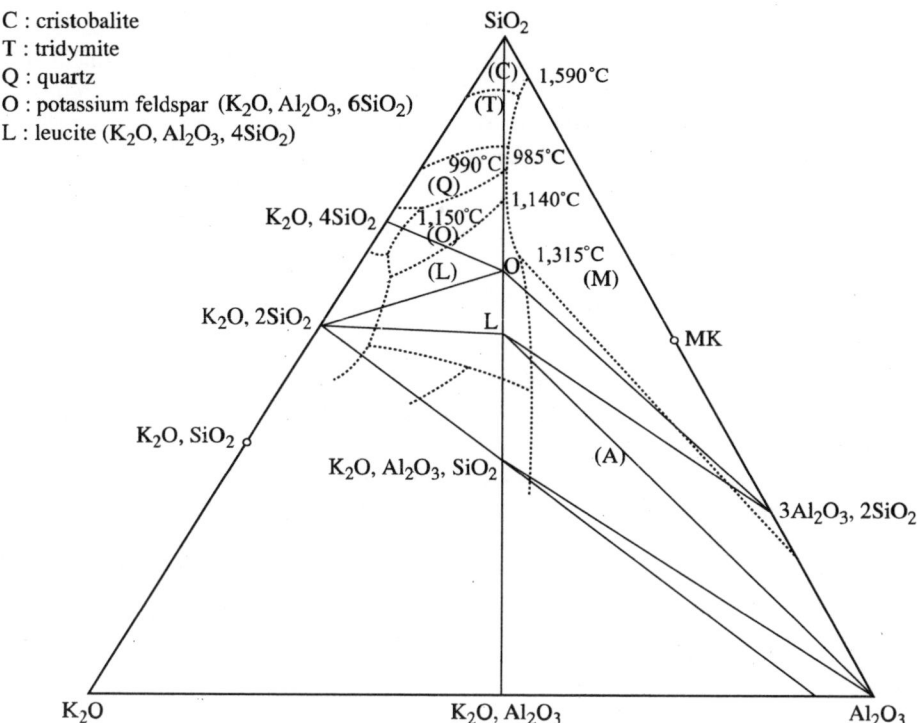

C : cristobalite
T : tridymite
Q : quartz
O : potassium feldspar $(K_2O, Al_2O_3, 6SiO_2)$
L : leucite $(K_2O, Al_2O_3, 4SiO_2)$

Figure 4.1. *Phase diagram Al_2O_3-SiO_2-K_2O [LEV 69]*

Some products are covered with a vitreous enamel film intended to modify the appearance of the ceramic and/or to waterproof it. This layer can be deposited on an engobe whose role is to mask the color of the shard and/or to facilitate the adhesion of the enamel. Depending on the case, the enameling operation is carried out on a green support, on a partially fired part during a so-called bisque firing (maximum temperature lower than that of enamel firing) or on a biscuit (completely fired shard at a temperature higher than that of enamel firing). The low temperature enamel intended for the protection of porous ceramics, such as earthenware and potteries, is also called glaze. Transparent glaze is the name used to denote enamel obtained by melting at the sintering temperature of the porcelain shard or the underlying stoneware. The enamel coloring is obtained using metallic oxides.

4.3. The main raw materials

4.3.1. *Introduction*

Each mineral raw material has a specific influence on the rheology of the paste, the development of the microstructure, the phases formation during the heat treatment and the properties of the finished product. The manufacture of all silicate ceramics requires such a large number of raw materials, which cannot be discussed here. Only those most commonly used, i.e. clays, feldspars and silica, will therefore be described.

4.3.2. *Clays*

4.3.2.1. *Common characteristics*

Clays are hydrated silico-aluminous minerals whose structure is made up of a stacking of two types of layers containing, respectively, aluminum in an octahedral environment and silicon in tetrahedral coordination. Their large specific surface (10 to 100 m^2g^{-1}), their plate-like structure and the physicochemical nature of their surface enable clays to form, with water, colloidal suspensions and plastic pastes. This characteristic is largely used during the manufacture of silicate ceramics insofar as it makes it possible to prepare homogenous and stable suspensions, suitable for casting, pastes easy to manipulate and green parts with good mechanical strength. By extension, the term clay is often used to denote all raw materials with proven plastic properties containing at least one argillaceous mineral. The impurities present in these natural products contribute to a large extent to the coloring of the shard.

4.3.2.2. *Classification*

All clays do not exhibit the same aptitude towards manipulation and behavior during firing. Ceramists distinguish vitrifying plastic clays, refractory plastic clays, refractory clays and red clays.

Vitrifying plastic clays, generally colored, are used for the remarkable plasticity of their paste. They are made up of very fine clay particles, organic matter, iron and titanium oxides, illite (formula $Si_{4x}Al_x)(Al,Fe)_2O_{10}(OH)_2K_x(H_2O)_n$) and micaceous and/or feldspathic impurities. These clays are also characterized by a high free silica content; sand can represent up to 35% of the dry matter weight. The product called "ball clay" is widely used for its plasticity and its particularly low mica content. Although it contains the same argillaceous mineral as kaolin, this clay has much higher plasticity because of the much smaller size of the kaolinite particles [CAR 98].

Refractory plastic clays are rich in montmorillonite (formula $(Si_{4-x}Al_x)(Al_{x-v}R_x)O_{10}(OH)_2M_{2v}(H_2O)_n$ with R = Mg, Fe^{2+} and M = K, Na), kaolinite or halloysite ($Si_2Al_2O_5(OH)_4(H_2O)_2$).

Refractory clays are used in high temperature processes. Their composition is rich in alumina. Kaolins are the most refractory among these clays. Always purified, they contain little quartz, generally less than 2% alkaline oxides in combined form and a small quantity of mica. Their plasticity is ensured by kaolinite and, if necessary, a little smectite or halloysite [CAR 98]. Very low in coloring element, they are particularly suited for the preparation of products in white shard.

Red clays used for the manufacture of terra cotta products are actually natural mixtures with a complex composition. They generally contain kaolinite, illite and/or other clays rich in alkaline, sand, mica (formula $Si_3Al_3O_{10}(OH)_2$), goethite (FeO(OH)) and/or hematite (Fe_2O_3), organic matter and, very often, calcium compounds. The latter, just like the micas and the other alkaline-rich compounds, help lower the firing temperature of the shard.

4.3.3. Kaolinite

4.3.3.1. Structure of kaolinite

Kaolinite, $Si_2Al_2O_5$ (OH)$_4$ or $Al_2O_3,2SiO_2,2H_2O$, is the most common among the argillaceous minerals used in ceramics. A projection of its crystalline structure is represented in Figure 4.2. It consists of an alternate stacking of $[Si_2O_5]^{2-}$ and $[Al_2(OH)_4]^{2+}$ layers, which confer to it a lamellate character favorable to the development of plates. The degree of crystallinity of the kaolinite present in clays is highly variable. It depends largely on the genesis conditions and the content of impurities introduced into the crystalline lattice.

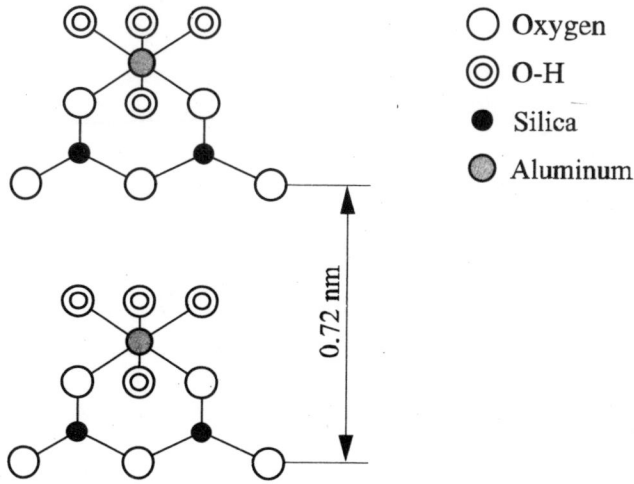

Oxygen
O-H
Silica
Aluminum

0.72 nm

Figure 4.2. *Projected representation of the structure of kaolinite*

4.3.3.2. *Evolution of the nature of phases during heat treatment*

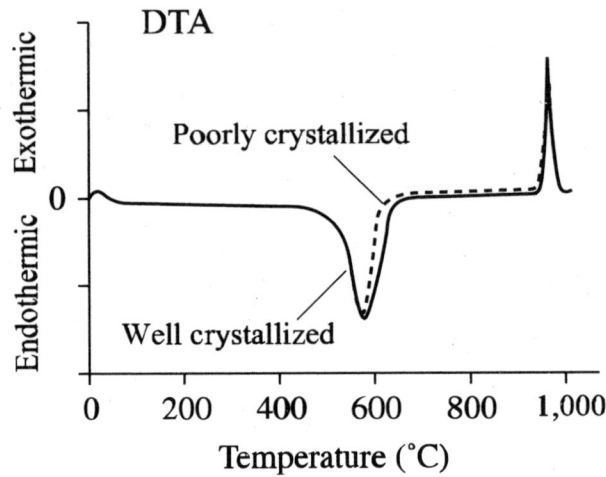

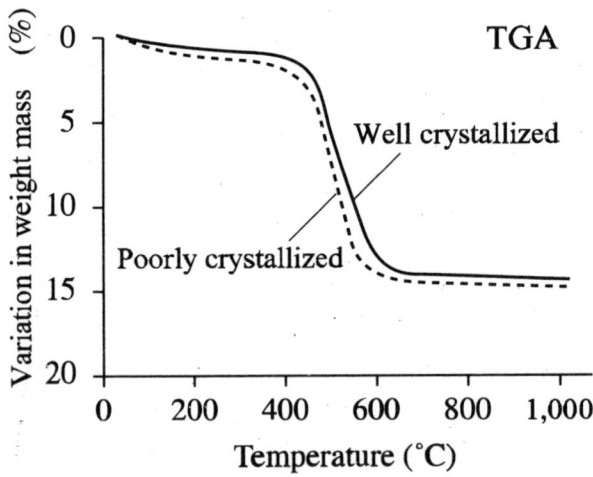

Figure 4.3. *Differential thermal analysis (DTA) and thermogravimetric analysis (TGA) of two kaolinites with different degrees of crystallinity*

During the heat treatment, kaolinite undergoes a whole series of transformations. The variations of exchanged heat and the corresponding mass changes are indicated in Figure 4.3. The departure of water, which occurs from 450°C onwards, is a very endothermic phenomenon. The amorphous metakaolin, Al_2O_3, $2SiO_2$ then formed, exhibits a structural organization directly derived from that of kaolinite. The exothermic transformation observed between 960 and 990°C is a structural reorganization of the amorphous metakaolin, sometimes associated with the formation of phases of spinel structure like $Al_8(Al_{13,33}\square_{2,67})O_{32}$ (γ variety of Al_2O_3) or Si_8 $(Al_{10,67}\square_{5,33})O_{32}$. In these formulae, $\square$ represents a cation vacancy. Between 1,000 and 1,100°C (often around 1,075°C), these phases are transformed into mullite stoichiometry ranging between $3Al_2O_3,2SiO_2$ and $2Al_2O_3,SiO_2$. During this reaction, amorphous silica is released. The surplus amorphous silica starts to crystallize in the form of cristobalite from 1,200°C onwards. It should be noted that the impurities present, the degree of crystallinity (see Figure 4.3) and the speed of heating influence each of these transformations.

4.3.4. Feldspars

Four feldspathic minerals are likely to enter the composition of silicate ceramic pastes. They are:
- orthoclase, a mineral rich in potassium with the composition $K_2O,Al_2O_3,6SiO_2$;
- albite, a mineral rich in sodium with the composition $Na_2O,Al_2O_3,6SiO_2$;
- anorthite, a mineral rich in calcium with the composition $CaO,Al_2O_3,2SiO_2$;
- petalite, a mineral rich in lithium with the composition $Li_2O,Al_2O_3,8SiO_2$.

Orthoclase and albite, which form eutectics with silica, respectively, at 990 (see Figure 4.1) and 1,050°C, are widely used as flux. Anorthite is rather regarded as a substitute to chalk. The use of petalite, especially owing to its negative expansion coefficient, is marginal [MAN 94].

Potassic feldspar is particularly appreciated by ceramists because its reaction with silica leads to the formation of a liquid whose relatively high viscosity decreases slightly when the temperature increases. This behavior is considered as a guarantee against the excessive deformation of the pieces during the heat treatment.

Natural feldspars used for the preparation of ceramics are mineral mixtures. Thus, the commercial potassium products can contain between 2.5 and 3.5% of albite mass, whereas anorthite and a small quantity of orthoclase, between 0.5 and 3.2%, are often present in the available sodium feldspars [MAN 94]. They can also be incorporated into the paste in the form of feldspathic sand. When these natural products are heated, mixed and homogenous feldspar is formed. This compound, called sanidine, occurs at a temperature which varies according to the sodium/potassium ratio and ranges between 700 and 1,000°C. Then, in the presence of silica, the formation of the liquid takes place. Above 1,200°C, mullite is formed in the still solid part of the feldspar grains.

A rather recent trend among stonewares and porcelain manufacturers consists of replacing feldspars by nepheline syenite with average composition $(Na,K)_2O,Al_2O_3,2SiO_2$. This rock, made up of nephelite (composition: $K_2O,3Na_2O,4Al_2O_3,9SiO_2$) and a mixture of potassium and sodium feldspars, is a powerful flux which makes it possible to decrease the sintering temperature of ceramics and increase the alkaline content of the vitreous phases [CAR 98].

4.3.5. *Silica*

Silica, SiO_2, is a polymorphic raw material found in nature in an amorphous (opal, pebbles) or crystallized form (quartz, cristobalite and tridymite). Sand contains between 95 and 100% of quartz mass. It is the most frequently used temper in the ceramic industry. To contribute significantly to the mechanical strength of the raw parts, it must consist of much coarser particles than those of clay. In the modern manufacturing processes of stonewares and porcelains, it is customary to use relatively fine sand grains (20 to 60 μm).

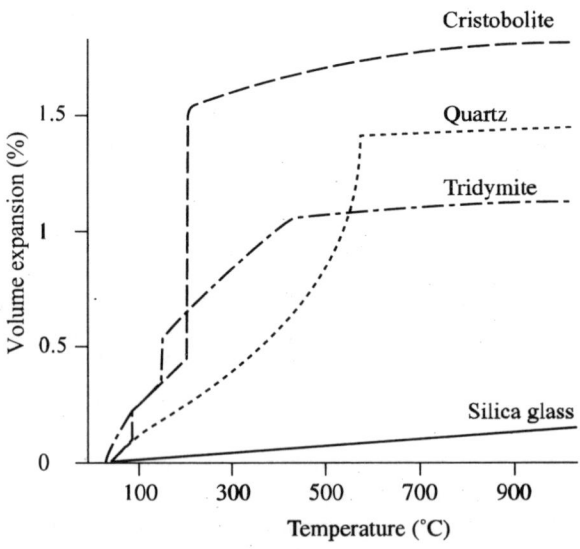

Figure 4.4. *Influence of temperature on the expansion of the various forms of silica*

When a ceramic is fired, the sand can react, particularly with the fluxes. This reaction is seldom complete. The transformation of residual quartz into cristobalite can then start from 1,200°C onwards. It is favored by the rise in temperature, the use of fine grained sand, the presence of certain impurities and a reducing atmosphere [JOU 90].

The form in which silica is found determines the thermal properties of silicate ceramics. Thus, quartz and cristobalite do not have the same influence on the expansion of the shard (see Figure 4.4). Quartz can also cause a deterioration of the mechanical properties of the finished product owing to the abrupt variation in dimensions ($\Delta L/L \cong -0.35\%$) associated, at 573°C, with the reversible transformation quartz $\beta \rightarrow$ quartz α. As the crystal of cristobalite formed from the flux are usually small, the transition cristobalite $\beta \rightarrow$ cristobalite α, which occurs at about 220°C (see Figure 4.4), often causes less damage to the finished product. It can even contribute to the shard/enamel fit by compressing it after cooling at room temperature.

4.4. Enamel and decorations

4.4.1. *Nature of enamel*

The enamel layer deposited on the shard generally has a thickness ranging between 0.15 and 0.5 mm. Its purpose is to mask the porosity and/or the color of the shard, to make the surface of the piece smooth and brilliant and to improve the chemical resistance of the ceramic. This layer, transparent or opaque, white or colored, is obtained from a silica-rich ceramic composition capable of developing glass during the heat treatment. The composition of the enamel also contains many other constituents, in particular alkaline and alkaline-earth oxides. They help to adjust the melting point, the thermal expansion coefficient, the surface tension and the viscosity to the enameling conditions, and ensure the wetting and adhesion of the enamel on the shard. The enamels used for sheet enameling have many common points with those described here [STE 81].

The properties of the enamel are often analyzed by considering that it is made up of a combination of acid oxides responsible for the vitreous structure (mainly SiO_2 and B_2O_3), amphoteric oxides (Al_2O_3) and basic oxides (K_2O, Na_2O, CaO, MgO, PbO). It should be noted that the role of flux, traditionally reserved for basic oxides, is now increasingly played by acid or amphoteric oxides, such as B_2O_3 or Bi_2O_3.

Enamel is obtained from a mixture of raw materials mineral and/or frits. The raw materials used are mainly feldspars, kaolin, quartz and chalk or dolomite. Frits are close mixtures of components prepared by melting several compounds at high temperature (T > 1,400°C). After quenching in air or water, the product, markedly vitreous in character, is ground. Combined in this manner, water soluble salts or volatile oxides can be used without harm in the composition of enamel. We can distinguish raw enamels formed only from natural raw materials and sintered enamels. The latter are particularly suitable for low temperature applications that require flux bases, richer in basic elements, which are non-existent in nature. The role of frits in the composition of the enamel is all the more important as the firing temperature reduces and the heat treatment is shortened. Frits are widely used for the enameling of the tiles in the fast sintering process [ENR 95].

It is customary to classify the various types of enamel, based on the nature of the flux used. Thus, we can distinguish lead enamels (PbO rich enamels), boron oxide enamels, alkaline enamels, alkaline-earth enamels, zinc enamels and bismuth oxide enamels [STE 85]. Lead enamels, historically the oldest, were the most commonly used for a long time. Because of the toxicity of lead, their future hinges on the evolution of legislation relating to the leaching of this element. They tend to be replaced by enamels containing a very small quantity of bismuth oxide (< 5% mass) and, especially by alkaline borosilicate products.

4.4.2. Enamel/shard combination

For the enamel to remain strongly attached to the shard, the interdiffusion must be effective and the thermal expansion coefficients of these two parts of the ceramic must be compatible. When the expansion coefficient of the shard is lower than that of the enamel, the latter is subjected to tension stresses when the piece cools. The stresses generated at the interface can cause the formation of cracks in the enamel. This flaw, also called crazing, is all the more important the more significant the difference between the expansion coefficients and the higher the modulus of elasticity of the enamel. To avoid this, it is customary to try stabilizing in the support, high expansion coefficient phases. On the other hand, when the shrinkage of the shard on cooling is the highest, the enamel is placed under compression and its mechanical strength is thereby reinforced. This positive effect occurs only if the difference between the expansion coefficients is sufficiently low to prevent the enamel from falling apart due to compression and from flaking off.

4.4.3. Optical properties of enamel

When the vitrification is complete, i.e. after the various components have melted completely, the enamel is generally brilliant, smooth and transparent. In most cases, opaque enamel is desired. This is achieved by favoring the formation of crystallized, vitreous or gas inclusions with an index of refraction different from that of the vitreous matrix. The differences between the indexes of refraction, the size and the form of the inclusions are then decisive parameters. The formation of crystallites can be favored by the presence in the enamel of mineralizers such as ZrO_2 and SnO_2 Vitreous inclusions occur when a decomposition takes place during the total fusion, as in the case of compositions like SiO_2-B_2O_3-MO (M = Pb, Ca, Zn, Mg).

A mat appearance and opacity owing to the diffusion of light on asperities can be observed when the surface of the enamel is slightly rough. This phenomenon occurs when the melting is incomplete or when the viscosity of the formed liquid is high. It can be favored by increasing the contents of SiO_2, Al_2O_3, CaO and ZnO.

4.4.4. Decorations

The pigments used for the production of decorations generally consist of colored frits or stain mixtures crystallized in a vitreous silico-aluminous phase. The main products used as coloring are oxides of antimony, chromium, copper, cobalt, iron, manganese, nickel, praseodymium, selenium, titanium, uranium, and vanadium [HAB 85].

In order to be applied on the parts, the ground pigments are mixed with liquid organic substances (for example, turpentine oil) which facilitate their adhesion. The nature of the process of decorating the enamel depends on the desired quality and the complexity of the decoration. The decalcomania technique is the most efficient, insofar as a very complex decoration, involving up to 20 colors, can be carried out by serigraphy. Processes making it possible to print directly on the enamel (direct transfer through a membrane) or decorate it without firing in the kiln (lazer sintering) can also be used.

An additional firing is generally necessary to fix the decorations. Depending on the application envisaged for the piece, it can be carried out below 800°C (low fire firing) or at round 1,200°C (high fire firing). Low fire firing makes it possible to obtain a very broad pallet of colors; high fire firing is especially used to fix decorations likely to change in a highly aggressive environment, a dishwasher for instance. In view of the interactions between phases existing at high temperature, the pallet of colors is therefore considerably reduced.

4.5. The products

4.5.1. Classification

Based on the criteria taking mainly into account open porosity and/or the coloring of the shard, it is customary to distinguish, among silicate ceramics, terra cotta products, earthenware, stoneware, vitreous china and porcelains. The materials treated at higher temperatures or in the presence of a large quantity of flux are generally the least porous. Whiteness is primarily the result of the use of raw materials free from iron and titanium or containing only small contents of transition metals.

The representation given in Figure 4.5 helps locate each of these families. Terra cotta products and earthenwares are characterized by a porous shard. The strong coloring in the mass of the terra cotta products has given them the name "red products". These porous ceramics can be used just as they are (bricks and tiles) or be covered with enamel (earthenwares). Among the dense products, stonewares shard is more colored than porcelain shard. Vitreous china forms an intermediate group between these two families. Many products are on the border between two of these groups; their name, which very often differs from one country to another, depends on the custom and the envisaged application.

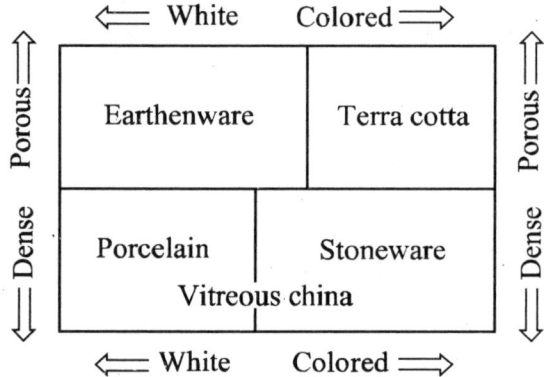

Figure 4.5. *Representation of the various traditional ceramic families*

The nature of the raw materials used for the manufacture and the chemical and mineralogical compositions of the shards can also be used as additional criteria for classification.

4.5.2. *Terra cotta products*

We are referring here to potteries or construction products such as roof tiles, bricks, flues, drainage pipes or some floor tiles. Terra cotta products were obtained a long time ago by modeling, drying and firing common clays. Nowadays, the compositions are more complex; they combine clays and additives, such as coloring, tempers or agents which make it possible to improve the manufacturing behavior or the final characteristics. The raw materials added to water form a plastic paste whose rheology must be adapted to the shaping process (extrusion possibly completed by pressing). The raw parts are dried in a ventilated cell or a tunnel dryer. The temperature at the end of firing usually ranges between 900 and 1,160°C.

Terra cotta products are porous and mechanically resistant. They are marketed raw, enameled or covered with a glaze realized at low temperature, between 600 and 900°C, called varnish. They are appreciated for their esthetic quality, their stability through time and their hygrothermic and acoustic properties. They represent a highly automated industrial sector which is the scene of continual technological developments.

The coloring of terra cotta shards can vary from yellowish white to dark brown. The variety in the tonality of the tiles present on roofs illustrates the extent of the pallet available. For roof tile manufacturers, the mastery over colors represents a commercial stake, insofar as they often constitute a regional specificity or a decorative element. The coloring of the shard depends on the bonding of iron ions with inhibitors, such as the calcium ions or with additional coloring (titanium and manganese oxides). The crystallized phases that are formed during the firing of a terra cotta product can be described using a ternary system defined by the major oxides Al_2O_3, SiO_2 and CaO. This includes primarily wollastonite (CaO, SiO_2), gehlenite ($2CaO$, Al_2O_3, SiO_2) and anorthite (CaO, Al_2O_3, $2SiO_2$). A high temperature heat treatment favors the formation of anorthite to the detriment of the other two phases. The Fe^{3+} ions dissolved in the anorthite confer a yellow coloring to it. Hematite, Fe_2O_3, is brown-red. The presence of compounds containing Fe^{2+} ions favors bluish or greenish tonalities. The final coloring of the shard is a combination of these three effects. In an oxidizing medium, a strong concentration of iron leads to the formation of a significant quantity of hematite and a brown red shard. The abundant presence of CaO in the starting mixture is favorable to the formation of anorthite and thus to the evolution of coloring towards yellow. A treatment at an excessively high temperature or the use of an atmosphere that is too low in oxygen can involve the formation of Fe^{2+} ions and a green or black coloring of the shard. Based on these considerations and experimental observations, the following rules may be laid down:

– when the Al_2O_3/Fe_2O_3 mass ratio is lower than 3, the shard is red;

– when the Al_2O_3/Fe_2O_3 mass ratio ranges between 3 and 5, the shard is pink;

– when the Fe_2O_3/CaO mass ratio is less than 0.5, a suitable heat treatment (high temperature and a sufficiently oxygen-rich atmosphere) yields a yellow shard;

– when the CaO/Al_2O_3 mass ratio is close to 1, the color of the shard is particularly dependent on all the parameters likely to affect the formation of anorthite. It is also significantly influenced by the other impurities present; MnO produces, for instance, black reflections.

As an atmosphere that is excessively oxidizing is detrimental to the formation of anorthite, the stacking density of the parts and the temperature of the final stage of firing can assume considerable importance, particularly in the last two cases. Thus, for a given composition, iron can be in the form of hematite at 1,000°C (pink coloring), dissolved in anorthite at 1,050°C (yellow coloring) and partially reduced at 1,100°C (coloring turning to green). The tile color therefore depends on the composition of the raw materials and the firing conditions (temperature, atmosphere and setting load of the kiln). Today it is often modified by using mineral coloring deposited, sometimes directly on the surface of the raw parts (colored engobe).

4.5.3. *Earthenwares*

4.5.3.1. *General characteristics of earthenwares*

We call earthenware the ceramic products made up of a porous shard covered with a glaze. This enamel makes it possible to mask the appearance of the shard and to remedy the high permeability due to the existence of an open porosity ranging between 5 and 20%. Although present in the form of objects of imagination and crockery, earthenwares are especially used as wall tiles.

These products are prepared from one or more clays to which quartz, chalk, feldspar or ground glass are added. Earthenwares are primarily shaped by slip casting, jiggering of plastic paste and atomized powder pressing. After drying, the raw product is subjected to a heat treatment called biscuiting, carried out at a temperature ranging between approximately 900 and 1,230°C. The deformation and the shrinkage of the shard during this stage are limited because of the refractory nature of the raw materials used. The porous biscuit obtained is then enameled during an enamel firing carried out at a temperature lower than or sometimes equal to that of biscuiting. The third firing, at a lower temperature, is necessary to fix some decorations deposited on the glaze, in particular those containing gold or platinum and those known as "low fire" decorations

4.5.3.2. *Common earthenwares*

Common earthenwares are found especially in old products. Their production is very limited nowadays. They are primarily glazed potteries and stanniferous earthenwares.

4.5.3.2.1. Potteries glazed with low melting temperature argillaceous paste

Potteries glazed with fusible argillaceous paste are very close to terra cotta products. Just like them, they are obtained from common argillaceous soils, relatively fusible, and naturally containing a certain quantity of sand. Although their sintering, carried out between 900 and 1,060°C, takes place in the presence of a significant quantity of liquid, their shard is still porous.

These products, used in construction (enameled bricks and tiles), for domestic uses and as crockery (jugs, pots, etc.), are generally covered with an engobe whose pores are finer than those of the shard. This engobe constitutes a smooth and regular surfacing intended to mask the coloring of the shard and to be used as decoration base.

4.5.3.2.2. Stanniferous earthenwares with low melting temperature argilo-calcareous paste

To produce certain decorative objects, it is customary to use an argilo-calcareous paste obtained by mixing argillaceous marls, limestone and often sand. Magnesium carbonate ($MgCO_3$) and dolomite (($Ca,Mg)CO_3$) can also be used. The biscuits are sintered at a temperature between 900 and 1,060°C. They are generally covered with a glaze opacified by tin dioxide, which explains the name stanniferous earthenwares.

4.5.3.3. Fine earthenwares

Fine earthenwares are characterized by a white or very lightly colored shard, a thin and regular texture, high mechanical strength and the brightness and durability of their glaze. They are widely used as decorative objects and crockery, fields where the quality of their enamel is highly appreciated.

The argillaceous component of the paste consists of a mixture of kaolin and clays. The kaolin increases the refractory character of the paste, whereas the plastic clays contribute to the mechanical strength of the raw parts. Although the selected clays are very poor in colorings, the presence of very small quantities of impurities in them, such as Fe_2O_3 and TiO_2, can be sufficient to slightly color the shard. The kaolin/clay ratio must therefore be adjusted in order to obtain an acceptable compromise between the whiteness of the biscuit and the resistance of the raw parts. Kaolin generally represents between 25 and 50% of the mass of all the argillaceous raw materials.

Grog, silica in its various forms and chalk (case of calcareous earthenware) can be used as tempering raw materials. The overall content of quartz, primarily introduced with the clays in the form of sand, is generally very high in fine earthenware pastes (30 to 40% of the mass). The biscuit is fired in an oxidizing atmosphere, at a temperature ranging between 950 and 1,150°C. The presence of a large quantity of residual quartz increases the shrinkage of the shard on cooling, thus reinforcing the mechanical strength of the enamel.

4.5.3.4. *Feldspathic earthenwares*

Feldspathic earthenwares are obtained by firing, at a temperature between 1,140 and 1,230°C, a mixture containing, for example, kaolin (40 to 70% of the mass), quartz (25 to 58%) and feldspar (3 to 14%). This latter component favors the formation of a viscous liquid during the high temperature treatment. After cooling, the biscuit has an increased solidity by virtue due to a significant quantity of vitreous phase formed during the solidification of the liquid. The porosity, 10 to 15%, is generally lower than the one observed in other earthenwares. The enamel is fired at a temperature between 1,000 and 1,140°C. At these high temperatures, it is possible to obtain glazes that cannot be easily scratched by steel. The increase in the silica content in the paste, the use of finer silica favorable to the formation of cristobalite or the reduction in the porosity of the shard contribute to the improvement of the shard/enamel combination.

Owing to their very high solidity, their particularly scratch-resistant enamel and their low open porosity, feldspathic earthenwares are particularly suitable for applications in the tiling field.

4.5.4. *Stonewares*

4.5.4.1. *General characteristics of stonewares*

Stonewares have a vitrified, opaque, colored and practically impermeable shard (0 to 3% open porosity). They are obtained from a mixture of vitrifying plastic clays and flux, sometimes supplemented by sand or grog. They are formed by extrusion (pipes, bricks, etc.) or by granulated powder pressing (tiles, slabs, etc.). The firing temperature generally ranges between 1,120 and 1,300°C and it forms a critical parameter. In fact, sintering at an insufficient temperature (non-firing) results in the persistence of a significant open porosity and a treatment at too high a temperature leads to the deformation of the pieces because of the excessively large quantity and the low viscosity of liquid formed. If usage requires it, stonewares can be enameled. A salt glaze during firing can also be carried out (traditional salt-glazed stonewares).

Stonewares are known for their unchangeability, excellent mechanical performances and resistance to erosion and chemical agents.

4.5.4.2. *Natural stonewares*

Natural stonewares are obtained from natural vitrifying clays, i.e. capable of forming a significant quantity of liquid at high temperature. They are used just as they are or are modified only by adding a kaolinitic refractory clay. Fe_2O_3 can represent up to 3% of the mass of the composition of these raw materials.

Irrespective of the firing atmosphere, mullite occurs in an acicular form between 1,000 and 1,100°C and continues to be formed up to 1,200°C. During this treatment, the viscous liquid dissolves the finest quartz grains. The solidification of this liquid on cooling leads to the formation of a significant quantity of vitreous phase. In an oxidizing atmosphere, the color of the shard can vary from ivory to dark brown. This coloring depends, in this case again, on the iron content and the nature of the other impurities present in the clays. Thus, titanium dioxide tends to color the shard of natural stonewares light yellow, whereas manganese oxide favors the development of darker colors. To avoid the appearance of blisters due to the presence of sulphates in the clays, the firing of natural stonewares must often be carried out in a reducing atmosphere. The Fe^{3+} ions are then reduced above 570°C. The ferrous oxide formed confers on the shard a grayish color and acts as a very active flux. As this action compounds that of the alkaline derivatives, a high iron content can lead to a marked softening and the deformation of the pieces during firing.

The production of natural stonewares is primarily traditional, insofar as the clays necessary for the manufacture of this type of product are seldom available in large quantities and their firing is often difficult.

4.5.4.3. *Compound or fine-grained stonewares*

Fine-grained stonewares are different from natural stonewares because the grains of flux are no longer contained in the clay, but added in the form of feldspars. They are obtained from clay very poor in coloring, kaolin, ball clay and a mixture of orthoclase and albite. Colorings are sometimes added to the paste to develop a particular color in the mass of the product. Fine-grained stonewares are used as crockery, walls or floor tiles, antacid tiles and sanitary pipes.

During firing, the deshydroxylation of clays occurs from 450°C onwards. Shortly before 1,000°C, orthoclase starts to react with silica and the liquid occurs. The mullite formation begins between 1,000 and 1,100°C. In this temperature range, certain micaceous phases contained in the clays can start to react with the products of the decomposition of metakaolin. The interaction of albite with silica begins from 1,140°C. The maximum firing temperature ranges between 1,250 and 1,280°C, so the amorphous silica derived from the kaolinite and undissolved in the liquid can be transformed into cristobalite. The degree of crystallization of SiO_2 in the end product depends on the thermal past of the stonewares, the nature of the flux and the mineralizing impurities.

4.5.4.4. *Porcelain stonewares*

Porcelain stonewares are characterized by an open porosity of less than 0.5%. This characteristic gives them remarkable mechanical properties and an excellent resistance to frost and corrosive agents. These products have experienced a rapid development as floor tiles. The world production rocketed from a few million m^2 in the 1980s to more than 150 million in 1997. This growth is linked to the use of new processes of grinding (wet process), forming (more powerful presses), sintering (fast mono-layer sintering roller-hearth kilns) and decoration (polishing, simultaneous pressing of several layers of enamel powder). These new technologies have reduced production costs and have considerably improved products' esthetics.

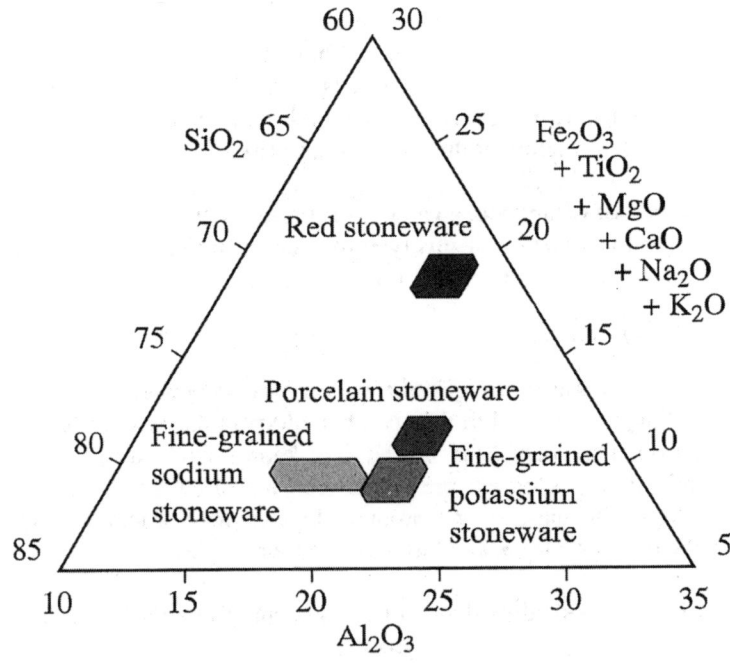

Figure 4.6. *Ranges of chemical composition of various types of stonewares tiles (mass %)*

The paste used to manufacture stonewares tiles generally consists of a mixture of plastic clays, kaolin, feldspathic sand, sodium or potassium feldspar and small quantities of talc, dolomite and/or chlorite. The overall chemical composition of the shard is generally less pure than that of fine-grained stonewares tiles, and it is also richer in Al_2O_3 (see Figure 4.6).

Expressed in mass % of oxide, it generally corresponds to 66 to 69% of SiO_2, 20 to 23% of Al_2O_3, 0.5 to 5% of MgO, 1.2 to 1.8% of CaO, 2.5 to 3.6% of Na_2O, 1.7 to 2.8% of K_2O, 0.7 to 1.3% of Fe_2O_3, 0.4 to 0.9% of TiO_2 and 0 to 2% ZrO_2 [DON 99].

The maximum temperature of the heat treatment generally ranges between 1,120 and 1,200°C. The evolution of the phases during this firing is very close to the one described in the case of fine-grained stonewares. The quantity of mullite formed represents only 50% of what is expected for this type of composition. The porcelain stonewares shards are mainly made up of mullite, amorphous phase and quartz. Their composition, after cooling, belongs to the field represented on Figure 4.7. These shards are free from open porosity and exhibit between 7 and 13% closed porosity [DON 99]. This microstructure confers on these materials a high Young's and rupture modulus (about 75 GPa and 85 MPa respectively) compatible with their use as floor tiles. These moduli increase with the Al_2O_3 content, the quantity of mullite formed and the compactness of the shard (reduction in closed porosity).

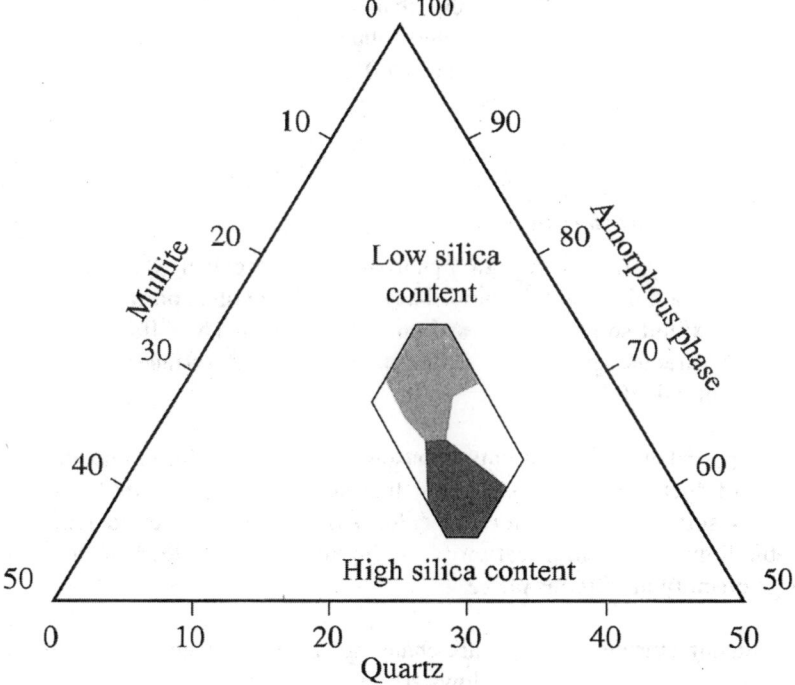

Figure 4.7. *Representation of the composition range of porcelain stonewares shards in the ternary mullite-quartz-vitreous phase diagram (mass %)*

4.5.4.5. *Grogged stonewares*

Grogged stonewares are obtained from a paste made up of vitrifying clay sometimes rich in kaolinite or quartz, a small proportion of flux and a large quantity of grog and/or ground shard (40 to 60% of the mass). It is generally shaped by casting a slip in porous plaster moulds. The use of large-sized grog grains, up to 0.8 mm, increases the permeability of the rigid skeleton and the speed of drying. It reduces the capillary forces responsible for the formation of cracks and then slits.

The dry products are engobed and then enameled before firing. The engobe and the enamel are deposited successively by dipping or pulverization. The operation must be carried out in several layers in order to obtain a sufficient thickness and avoid excessively rewetting the dry part. After a final drying, it is fired in single thermal cycle (process known as single firing). The rigid lattice made up of grog grains does not allow sufficient shrinkage to eliminate all porosity during sintering. After a treatment between 1,250 and 1,280°C, open porosity remains considerable (8 to 15%). The presence of an opaque engobe masks the coloring of the shard, levels out the surface imperfections and facilitates the fixing of the enamel.

Grogged stonewares, easier to dry than most other products, are particularly suited for the manufacture of bulky and robust products. They are widely used in sanitary plumbing (sinks, shower basins, etc.).

4.5.5. *Porcelains*

4.5.5.1. *General characteristics*

Thanks to the purity of the raw materials used, porcelain shards are white and translucent beneath the low thickness. They do not have open porosity (< 0.5%), but are likely to exhibit some large closed pores (air holes). Their fracture are brilliant and have a vitreous appearance. After enameling, the surface of the pieces is remarkably smooth and brilliant.

When porcelain is fired, a liquid phase surrounds the solid grains and dissolves the finest of them (< 15 μm). During this stage, known as "pasty fusion", the viscosity is sufficiently high for the deformation of the pieces to remain within acceptable limits. The solidification of the liquid on cooling leads to the formation of a large quantity of vitreous phase.

The manufacturing processes are changing constantly (see section 5.6.2). Thus, when the geometry of the parts allows it, pressure casting and shaping by isostatic pressing gradually replace jiggering and casting in plaster molds. Fast firing techniques are increasingly used for enamel and decorations. They improve the quality of the parts by limiting the risks of deformations [SLA 96]. A tendency has also been observed to decrease the sintering temperature [LEP 98].

4.5.5.2. *Hard-paste porcelains*

Hard-paste porcelains are obtained from mixtures made up almost exclusively of kaolin, quartz and feldspars. A little chalk ($\approx$ 2% of the mass) can be added to favor the formation of the viscous liquid. This mixture is very similar to the one used to prepare fine earthenware. It differs from it only because of the almost exclusive use of kaolin as clay and the proportions of the various components.

Sintering is carried out at a temperature ranging between 1,350 and 1,430°C. The use of a reducing atmosphere, by favoring the reduction of the Fe^{3+} ions to Fe^{2+}, guarantees being able to obtain a white shard (possibility of bluish reflections) [SLA 96]. In a twice-firing process, this treatment is preceded by a bisque firing, carried out in oxidizing atmosphere at a temperature ranging between 900 and 1,050°C. The product then has a sufficient rigidity and the high open porosity necessary for its enameling. The development of passage kilns with several heating zones, each with its own atmosphere, has made it possible to prepare certain types of hard-paste porcelains in a single firing. To prevent carbon monoxide, present in the reducing atmosphere, from depositing carbon (Boudouard equilibrium) in the still porous paste, the heating zone ensuring the rise in temperature up to about 1,050°C is traversed by an oxidizing atmosphere. The reducing atmosphere therefore circulates only in the sintering zone.

Because of the high compactness of the shard after firing, enameling is done at the same time as sintering. The enamel is then a glaze with a feldspathic and calcium composition. The high firing temperature and the reducing atmosphere diminish the pallet of possible colors during this treatment to green, blue and brown only. Most decorations are therefore painted or deposited on enamel by decalcomania and then fired between 800 and 900–950°C.

Hard-paste porcelains are particularly used in the field of crockery (Limoges porcelain) and as technical ceramics (insulators).

4.5.5.3. *Soft-paste porcelains*

Soft-paste porcelains differ from the above porcelains by their greater translucidity and lower sintering temperature. Chinese porcelains and porcelains for dental implants belong to this category. English porcelains, known as bone china, constitute a particular class.

Fine bone china

These are the most expensive porcelains on the market, particularly appreciated for their esthetics. They owe their name to the bone ash added to the mixture of raw materials. The composition of a paste is typically: 37 to 50 % of the mass of bone ash, 22 to 32% potassium feldspar, 22 to 41% kaolin and 0 to 4% quartz. Hydroxyapatite, $Ca_5(PO_4)_3OH$, present in the bone ash contributes to the formation of a low viscosity liquid, whose quantity increases very rapidly when the solidus temperature is reached. Above 1,200°C, this liquid dissolves the free quartz gradually. At the maximum firing temperature of the biscuit, between 1,250 and 1,280°C, the material is made up of a paste that contains only calcium phosphate, $Ca_3(PO_4)_2$ and liquid. Sintering shrinkage, highly dependent on the quantity and viscosity of liquid formed, is much more sensitive to the heat treatment conditions than in the case of hard-paste porcelains and vitreous china (see Figure 4.8). Mastery over the dimensions and deformation of the pieces requires the temperature conditions and sintering time to be strictly controlled. Deformation can be controlled by placing the raw products in a bed of alumina powder [SLA 93]. As anorthite is formed on cooling, the shard of a bone china porcelain is primarily made up of calcium phosphate (35 to 45% of the mass), vitreous phase (27 to 30%) and anorthite (25 to 30%).

The fixing of enamel presents difficulties inherent to the absence of porosity of the shard. Its firing is carried out at high temperature, ranging between 1,120 and 1,160°C. In order to avoid the appearance of efflorescence due to the decomposition of the enamel, this second firing must be done in a strictly oxidizing atmosphere. The brilliance of the enamel obtained is highly dependent on the lead oxide content.

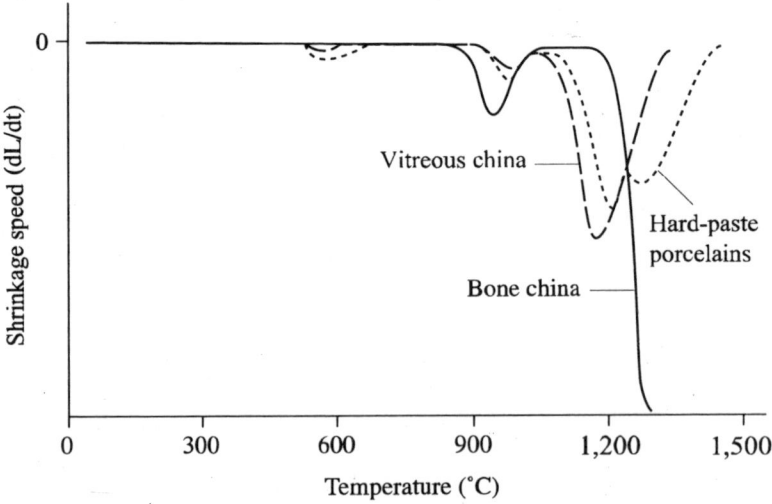

Figure 4.8. *Temperature influence on the shrinkage speed of a hard-paste porcelain, a bone china and a vitreous china*

4.5.5.4. *Aluminous porcelains*

The composition of the paste of aluminous porcelains can contain up to 50% of Al_2O_3 mass, about half of which can be introduced in the form of calcined alumina or corundum. The fluxes used are based on alkaline-earth oxides (CaO, MgO and/or BaO); lithium ions are sometimes added in the form of spodumene $(Li_2O,Al_2O_3,4SiO_2)$. Sintering and enameling are carried out in single firing at a temperature ranging between 1,280 and 1,320°C.

There are also products called extra-aluminous porcelains whose Al_2O_3 content can range between 50 and 95% of the mass. In order to facilitate their forming, 3 to 5% of very pure plastic clays and/or organic binders are added to the paste. When enameling is necessary, the raw parts are generally covered by dipping in the enamel suspension. After sintering at a temperature between 1,430 and 1,600°C the shard exhibits very little vitreous phase.

These porcelains are used for their electrical characteristics (insulators, disconnecting switches, spark plugs, dielectrics for high voltage) and their hardness (cutting tools, wire guides, grinding ball). These are technical ceramics whose dimensions can be subject to considerable constraints because of low tolerances (machining after firing) or the large size of the piece. Consequently, the conditions imposed for producing the 4 m height and 1 m diameter aluminous porcelain insulators used for very high voltage electricity transmission have had to be mastered.

4.5.6. *Vitreous china*

The term "vitreous china" denotes dense products obtained from pastes close to those used to manufacture feldspathic earthenwares. The feldspar content of these pastes is increased in order to produce, during the firing, a sufficient quantity of liquid to eliminate open porosity (< 0.5%). Used more particularly to manufacture sanitary articles and very robust crockery (wash basin, crockery for communities), vitreous materials are in the middle between white paste stonewares and porcelains.

These products are formed by jiggering, casting or isostatic pressing. A good mastery of the raw materials and shaping process makes it possible to obtain raw pieces with a mechanical strength sufficient to withstand the application of an enamel paste. Sanitary products are generally vitrified and enameled in a single treatment, carried out in oxidizing atmosphere at a temperature between 1,200 and 1,280°C. A twice-firing treatment is usually used for crockery. The first firing is thus carried out between 900 and 950°C. The elimination of open porosity and the formation of the enamel occur during the second firing at a temperature ranging between 1,200 and 1,250°C. The enamel thus obtained, generally opacified, can allow a large variety of decorations.

4.6. Evolution of processes: the example of crockery

4.6.1. *Introduction*

Although at the origin of most of the advanced processes of manufacturing massive technical ceramics (casting, extrusion, injection, pressing, etc.), the silicate ceramic industry has for a long time remained aloof from technological progress, basing its development on tradition and empiricism. A modernization was launched after 1960. The globalization of the market and the resulting increasingly fierce competition gave a momentum to the change of mindsets and methods. The development of new technologies (atomization/drying, isostatic pressing, microwave drying, etc.) and the improvement of refractory materials and kilns furnaces (roller-hearth kilns, fibrous cellular kilns with thermal inertia allowing short thermal cycle, high temperature kilns, etc.) increased the rates of output, improved quality, diversified the products and reduced the costs. Today, it is an industrial sector under constant change where progress continues as much in the mastery of processes as in the development of new products.

The changes affecting, on a global scale, the processes of crockery manufacture well illustrate the innovative state of mind that drives traditional ceramic producers.

4.6.2. *The case of crockery*

For a long time, crockery was shaped by modeling a plastic paste prepared from a suspension of raw material particles. In the variant of this process used today, one of the first stages consists of eliminating a part of the water from the slip by filtration in a filter/press device. The cakes obtained are extruded in the form of clots. The use of vacuum chambers eliminates the trapped air bubbles, thus improving the plasticity of the pastes. The deaerated clots are cut to a suitable length and then modeled into the desired form. The pieces obtained still contain a large quantity of water, typically between 12 and 18% of the mass. This water is eliminated during a drying treatment that generates mechanical stresses, because of the inevitable humidity gradients. This method of forming crockery pieces presents disadvantages due to, amongst other things, the inhomogenities existing in the paste and the duration of the drying stage. As a consequence, it is gradually being replaced by more sophisticated and especially better mastered processes.

The use of atomization/drying techniques represents a breakthrough. It is thus possible to have spherical granules with a well defined size, made up of a very homogenous mixture of raw materials. These granules contain about 3% water mass and sometimes the binders necessary for their cohesion. They have given rise to the development of the shaping process of certain parts by isostatic pressing. This process uses the fact that the granules of the raw material mixture can behave like a very viscous liquid under strong pressure and recover their stiffness when the stress is relieved. In a sufficiently flexible mould (polymer-based diaphragms) and surrounded by a pressurized liquid, the compression is uniform in all the directions. The isotropic distribution of the residual stresses and porosity in the raw parts, therefore, helps control sintering shrinkage. In the case of forming dishes or plates, the pressure is applied only on two sides (quasi-isostatic pressing). Compared to jiggering, this technique presents, among others, the advantage of significantly decreasing the risk of deformation of the plates. The raw parts are then ready for enameling and firing, without prior stage of drying. The conditions to be met in order to obtain a good reproducibility of the parts are largely described in [OBA 82]. This process is particularly suited for mass productions and relatively flat parts. For more complex forms, pressure casting in porous resin molds is increasingly used [GAI 99, SLA 96].

The processes using a wet method for forming parts are still widespread. The duration of the drying proves to be an obstacle for the development of production lines continuously functioning. The recent progress made in the field of infrared and microwave drying should allow a rapid evolution. Indeed, it has been shown that it is possible to dry a cup containing more than 22% mass of water in less than 30 minutes [CUB 93].

New grinding technologies have modified the methods of preparing coloring and the enamel. The use of fine powders considerably improves the covering effect and the brilliance of the enamel, as well as the quality and reproducibility of the decorations.

All these processes, the availability of new types of kilns and a better understanding of the phenomena have made it possible to reduce by up to 5 to 10 times the duration of the production cycle of certain types of crockery while guaranteeing better reproducibility of the parts.

4.7. Conclusion

Silicate ceramics are very old materials that are very widely used in society thanks to their undeniable assets, namely their high mechanical strength, excellent chemical inertia and esthetic qualities. A common feature is that they are obtained from a mixture of natural raw materials containing at least one argillaceous mineral, silica and, very often, alkaline oxides in a combined form. The finished products differ primarily in porosity, color, the quantity of vitreous phase, the nature of the crystallized phases (mullite, anorthite, wollastonite, gehlenite or tricalcium phosphate), the allotropic variety of silica (quartz or cristobalite) and the presence of enamel.

Apart from esthetics, which increasingly calls upon designers, the main characteristics of silicate ceramics available today are not different from those of the best performing products marketed 50 years ago. It is however an industrial sector undergoing rapid technological development, where efforts are made to reduce costs, improve the reproducibility of the pieces, adapt to an evolving demand and especially resist to the competition from products containing glass, concrete, metals and polymers. The progress made relies on the development of new equipment, the use of mineral or organic additives and a better knowledge of the parameters governing the rheological behavior and thermal evolution of ceramic pastes. In one decade, this approach has made it possible to reduce the scrap rate of pieces from 40% to a few percent.

Currently, the defects responsible for the downgrading of certain parts are primarily spots due to impurities contained in the raw materials. High magnetic field purification helps reduce by two the number of these scraps. Non-magnetic coloring impurities still cause uncertainties in quality. A few ppm of certain elements are enough to make 10% of the parts non-compliant with the standard. To achieve the zero defect objective, great effort must be made, in order to master the purification and the selection of raw materials and the constancy of their properties.

4.8. Bibliography

[CAR 98] CARTY W.M. and SENAPATI U., "Porcelain-raw materials, processing, phase evolution, and mechanical behavior", *Journal of American Ceramic Society*, vol. 81, p. 3-20, 1998.

[CUB 93] CUBBON R.C.P., "Review of new process technology in tableware manufacture. Part I-shaping processes", *Interceram*, vol. 42, p. 379-381, 1993.

[CUB 94] CUBBON R.C.P., "Review of new process technology in tableware manufacture. Part II-glazing and decoration", *Interceram*, vol. 43, p. 143-145, 1994.

[DON 99] DONDI M., ERCOLANI G., MELANDRI C., MINGAZZINI C. and MARSIGLI M., "The chemical composition of porcelain stonewares tiles and its influence on microstructural and mechanical properties", *Interceram*, vol. 48, p. 75-83, 1999.

[ENR 95] ENRIQUE J.E., AMOROS J.L. and MORENO A., "Evolution of ceramic tiles glazes", *Proceeding of Fourth Euro Ceramics*, Gruppo Editoriale Faenza, Editions S.p.A. (Italy), p. 121-134, 1995.

[GAI 99] GAILLARD J.M., "Le coulage sous pression. Présentation de l'état de l'art de cette technologie", *Actes des journées technologiques sur le coulage sous pression*, Limousins Technologie editions, Limoges (France), p. 1-9, 1999.

[HAB 85] HABER R.A., WILFINGER K.R. and MCCAULEY R.A., *Stains and coloring agents*, Ceramic Monographs, Handbook of Ceramics, Verlag Schmid GmbH, Freiburg (Germany), 1985.

[JOU 90] JOUENNE C.A., *Traité de céramiques et matériaux minéraux*, Septima editions, Paris, 1990.

[LEC 96] LECAT D., *Le secteur de l'industrie céramique en France*, ANVAR editions, Limoges (France), p. 1-9, 1996.

[LEP 98] LEPKOVA D. and PAVLOVA L., "Low-temperature porcelain synthesis and properties", *Interceram*, vol. 47, p. 369-372, 1998.

[LEV 69] LEVIN E.M., ROBBINS C.R. and MCMURDIE H.F., *Phase diagrams for ceramists*, American Ceramic Society, Colombus (USA), 1969.

[MAN 94] MANDT P., "The importance of feldspar as a flux in the ceramics industry", *Interceram*, vol. 43, p. 21-23, 1994.

[OBA 82] O'BARTO L.C., "Isostatic dry pressure of flatware", *Technological innovations in whitewares*, Editions Alfred University Press, Alfred (USA), p. 195-202, 1982.

[SLA 93] SLADEK R., "Bone china – a high quality soft porcelain as an enrichment of modern table art", *Interceram*, vol. 42, p. 100-102, 1993.

[SLA 96] SLADEK R., "Whitewares technology at the turn of the millennium", *Interceram*, vol. 45, p. 71-74 and 147-152, 1996.

[STE 81] STEGMAIER W., *Enamel and enamelling*, Ceramic Monographs, Handbook of Ceramics, Verlag Schmid GmbH, Freiburg (Germany), 1981.

[STE 85] STEFANOV S., *Ceramic glazes and frits*, Ceramic Monographs, Handbook of Ceramics, Verlag Schmid GmbH, Freiburg (Germany), 1985.

Chapter 5

Ceramic Forming Processes

5.1. Introduction

Although the term "ceramic" relates to a wide range of compounds as well as very different applications (building, culinary, sanitary, electric, magnetic, optical, mechanical, thermal, etc.), these materials have common properties because of their primarily ionic-covalent bonds (see Chapter 1):

– high melting point;

– high hardness;

– absence of ductility at low temperatures;

– brittleness, low toughness.

These properties reveal that conventional methods of manufacturing metallic parts are not adapted to ceramics. The absence of ductility precludes shaping by plastic deformation. High hardness and brittleness limit the possibilities of machining to cutting and grinding by means of expensive diamonds tools. With the exception of glasses, high melting points and possible decompositions make the process of fusion-solidification difficult. Ceramic objects are therefore generally obtained by high temperature consolidation (sintering) of a granular structure (green part) elaborated by applying a ceramic process (see Figure 5.1).

Chapter written by Thierry CHARTIER.

The physical (size and shape of the particles, grain size distribution) and chemical (surface of the particles) characteristics of the powders will have to be adapted beforehand to the process, in particular in terms of flow during the shaping and arrangement of the particles in the green part (see sections 5.2 and 5.3).

The choice of the shaping process depends on several parameters related to the piece, such as size and shape, surface quality, dimensional tolerances and microstructural characteristics, but also on economic considerations like productivity and cost of the equipment. The most widely used ceramic processes are casting (section 5.4), pressing (section 5.5), injection molding and extrusion (section 5.6). Although less traditional, methods of ceramic material deposition (vapor deposition, plasma spraying) on a support can be regarded as ceramic processes (section 5.8). Most ceramic manufacturing processes require liquid (water, organic solvents) and/or organic additives (dispersants, binders, plasticizers, lubricants, etc.), in order to confer on the ceramic powder the desired rheological and cohesion properties during shaping. These components must obviously be eliminated before sintering during a critical stage of drying or debinding (see section 5.7) while preserving the integrity and homogenity of the piece.

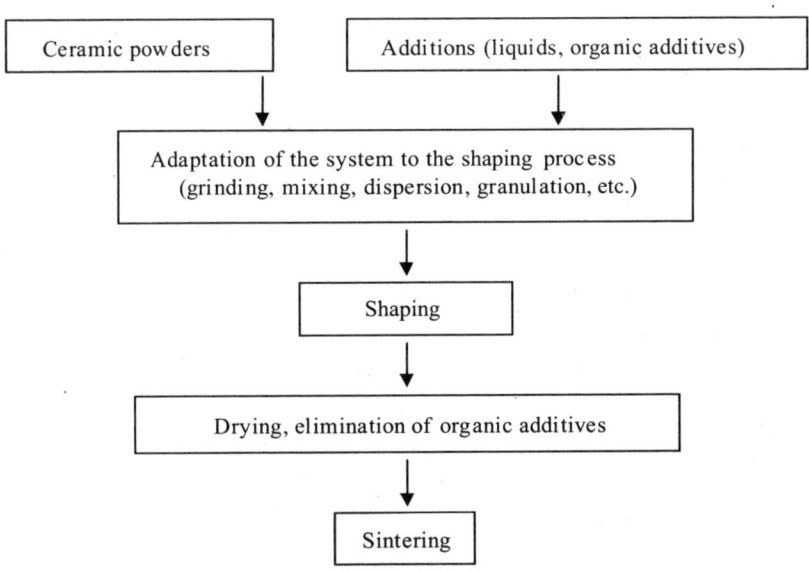

Figure 5.1. *General flow chart of the manufacturing of a ceramic piece*

Regardless of the process, the production of a ceramic piece implies a sequence of operations that modify the chemical and physical characteristics of the system.

Enhancing the properties of the final object and their reproducibility (reliability) therefore requires the identification of these characteristics at each stage of the process. In this context, the science of ceramic processes seeks to understand the mechanisms that govern shaping and naturally associates material science and process engineering in an integrated research, spanning from powder to usage properties.

5.2. Ceramic powders

A large variety of powders is available today in the ceramic industry. Powders for traditional ceramics (clay, feldspar, kaolin, silica, talc, etc.) and powders for technical ceramics (alumina, zirconia, nitrides, carbides, titanates, etc.) exhibit very diverse chemical and physical characteristics, which must be controlled to manufacture parts with the desired properties in a reproducible way. Powder preparation is therefore an important stage in the shaping processes. The objective is to obtain a powder which, on the one hand, yields the desired microstructure, generally dense and homogenous, during shaping (problems of particle stacking, dispersion and rheology of the mixtures) and, on the other hand, ensures a satisfactory densification during sintering (problem of powder reactivity). In this context, the size and size distribution of the particles, their shape and state of agglomeration, their specific surface area, their degree of purity and the chemical nature of their surface exert a determining influence.

After an overview of these characteristics, we will mainly present their influence on the stacking of particles and on the formation of the microstructure, and will develop the dispersion of particles in the mixtures and rheology in section 5.3. Finally, we will deal with the techniques that are used to adapt these characteristics to the shaping process.

5.2.1. *Chemical characteristics*

These relate mainly to the degree of purity of the powders and the nature of the particles' surface. Purity depends on the source of the raw materials and the transformation processes that give rise to impurities (iron, heavy metals, salts, carbon, etc.). It will determine to a large extent the sintering reactivity, with the possible formation of a second intergranular phase and the final properties of the piece (mechanical, chemical, electric, etc.). The surface properties of the particles determine the mechanisms of species adsorption and dissolution. They will control the dispersion properties, homogenity and the rheological behavior of the suspensions and ceramic pastes (section 5.3).

5.2.2. *Physical properties*

The physical characteristics of a powder are: i) the state of agglomeration, ii) shape and size of the particles, iii) density and iv) specific surface area.

5.2.2.1. *The state of agglomeration*

A powder particle is a solid and discrete unit of material. It can consist of one or more crystalline phases and internal pores. Depending on their size, the nature of their surface and their environment, these particles can be individualized, form slightly bound (agglomerates) or strongly bound (aggregates) groups. Ceramic particles, whose surface is generally hydrated, tend to agglomerate under the influence of Van der Waals forces. Agglomerates and aggregates must be avoided, because they give rise to heterogenities in particle packing during the shaping, which results in differential shrinkages during sintering and the formation of pores. Their elimination requires a preliminary stage of de-agglomeration and dispersion, generally carried out by milling (see section 5.3).

5.2.2.2. *Size and shape of the particles*

The grain size of ceramic particles can vary significantly according to the purpose of the products. In the case of refractory or construction materials, the particle size varies between a micrometer and a few millimeters. On the other hand, certain chemically synthesized powders have a size close to 10 nanometers. In general, the scale varies from a few microns to a few ten micrometers in the case of traditional ceramics, whereas it ranges between 0.1 and 10 μm for technical ceramics.

A powder is characterized by the extent of the particle size distribution and by the average size (denoted d_{50}), size for which 50% of the particle population has a size lower than d_{50} (see Figure 5.2). These characteristics are determined by sedimentation methods (detection by x-ray absorption) or by lazer diffraction methods. These techniques generally calculate an equivalent diameter of spherical particles. In case of spherical or equiaxed particles, the calculated value is therefore close to the actual value, while in case of anisotropic particles we have an approximate theoretical value.

In addition to their size, the particles also differ by their shape. Their shape is determined by the nature of the atomic lattice (crystalline cell) and by the processes used to obtain the powders. The particles can thus be spheroid, equiaxed, in the form of plates, fibers or needles. We can define an anisotropic coefficient which corresponds to the ratio of the greatest size of the particle to the smallest. Optical or scanning electronic microscopy helps us to observe the shape and to characterize the anisotropy of the particles. Particles with high anisotropic coefficient have a weak stacking aptitude (porous structure) and result in a shear-thickening behavior of suspensions for small powder concentrations. We can, however, take advantage of a preferential orientation of these particles to obtain a property (mechanical, magnetic, etc.) enhanced in a particular direction.

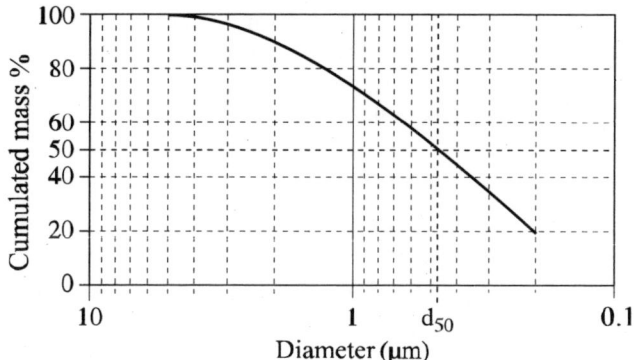

Figure 5.2. *Grain size distribution of a kaolin obtained by RX granulometer*

5.2.2.3. Density and specific surface area

The theoretical density of the powders refers to the mass per unit of volume of the dense material or particles without porosity. Density in particular makes it possible to determine the degree of densification after sintering (density of the sintered part compared to the theoretical density). We must then consider the densities of the phases formed during sintering. Furthermore, the knowledge of the densities of the various materials entering a mixture is necessary to assess the possible risks of segregation during shaping and their consequences on the homogenity of particles stacking.

The specific surface area of a powder refers to the developed surface per unit of mass. It reflects the shape of the particles and the roughness of their surface. The comparison between the specific surface area and the measured size of the particles informs us about the state of agglomeration of the powder. The grain size measurement technique takes into account the porous agglomerate, whereas the specific surface measurement takes into account the surface of the particles constituting these agglomerates. Specific surface area can vary from a few tens of cm^2g^{-1} to a few hundred m^2g^{-1}, the average value of commonly used ceramic powders ranging between 3 and 15 m^2g^{-1}. The higher the specific surface of a powder, the stronger the tendency to agglomeration, but the greater the sintering reactivity.

5.2.3. *Particle packing*

The stacking of particles formed during the shaping stage (green microstructure) will determine the microstructure of the sinter and consequently its properties. A non-homogenous stacking in the green part will lead to flaws in the sintered part (pores, cracks, residual stresses). If the stacking is not compact, the dimensional variations (shrinkage) will be considerable during sintering and the density of the sintered product will be low. The homogenity and density of the stacking of particles are therefore two parameters that must be taken into account simultaneously to adapt the characteristics of the powders to the process.

5.2.3.1. *Theoretical principles of packing*

The theoretical studies on stacking have been primarily undertaken on rigid spheres. A packing is characterized by the coordination number (number of close neighbors in contact) and by the packing density (volume of particles/(volume of particles + volume of pores)).

5.2.3.1.1. Stacking of monomodal spheres (same diameter)

In this case, five types of regular packing are possible (see Figure 5.3) [JOU 84]. Rhombohedric packing yields the highest packing density (0.74), in theory independent of the diameter of the spheres. In practice, sphere packing is random and the maximum packing density is close to 0.64. This value depends on the diameter of the spheres in question. The smaller the diameter, the less the spheres position themselves on the ideal sites of compact stacking, because the interaction forces and frictions between spheres restrict the displacements.

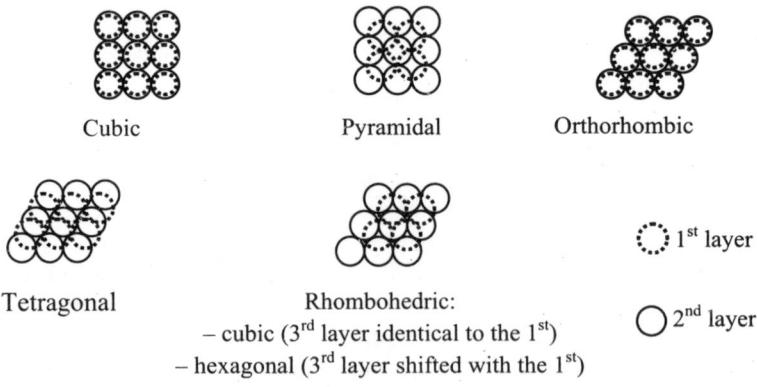

Figure 5.3. *Five types of packing of spheres with the same diameter*

5.2.3.1.2. Packing of mixtures of multimodal spheres (different diameters)

In mixtures with two different populations of monomodal spheres, the smallest fill the interstices between the largest, which increases the packing density. The latter therefore depends on the ratio of the diameters and the proportion between the two populations (see Figure 5.4).

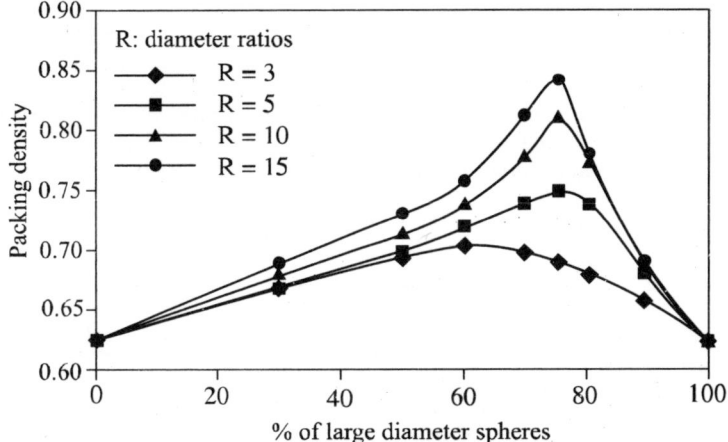

Figure 5.4. *Evolution of the packing density of two populations of monomodal spheres as a function of the diameter ratio and the fraction of large diameter spheres (glass beads) [RAH 95]*

The most compact stacking is achieved when the small spheres perfectly fill the interstices without disorganizing the network of the largest spheres. The maximum packing density is then equal to 0.87.

The more we increase the number of populations of spheres with different diameters, the more the maximum packing density increases (0.95 for a ternary mixture, 0.98 for a quaternary mixture) (see Table 5.1). In practice, these very high values cannot be reached for the same reasons as in the case of stacking of monomodal spheres.

5.2.3.1.3. Non-spherical particles

The particles deviation from the spherical shape exhibit a low aptitude for arrangement, which leads to a low green density as well as a non-homogenous microstructure (see Figure 5.5).

Type of mixture	Composition (mass %)	Diameters ratio
Binary (L + S)	85/15 (L/S)	1: 7 (S: L)
Ternary (L + I + S)	75/14/11 (L/I/S)	1: 7: 49 (S: I: L)
Quaternary	72/14/10/3	1: 7: 49: 343
(L + I1 + I2 + S)	G/I2/I1/F	(S: I1: I2: L)

Table 5.1. *Examples of diameter ratios and compositions of sphere mixtures resulting in high packing density (L: spheres with large diameters, I: spheres with intermediate diameters, S: spheres with small diameter) [MUT 95]*

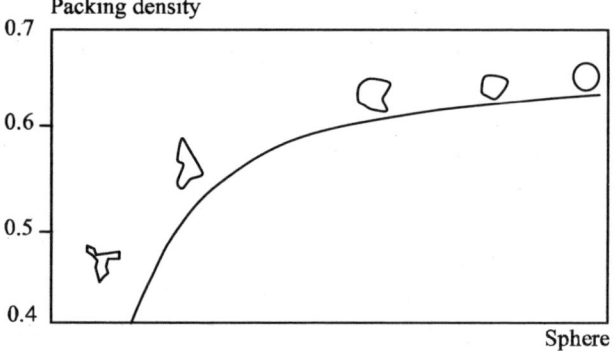

Figure 5.5. *Influence of the shape of the monosized particles on the packing density [GER 89]*

The case of the mixture of spherical particles and fibers has also been studied for producing reinforced ceramic materials (composites). The higher the length/diameter ratio (L/d) of the fibers, the weaker the packing density. A compact stacking, with a homogenous distribution of fibers, can be achieved for a fiber volume concentration close to 25%, with fibers having an L/d ratio close to 15 and a length at least ten times lower than the diameter of the spheres [GER 89].

5.2.3.1.4. Mixtures of spheres with a continuous size distribution

This situation is similar to the one seen with ceramic powders that generally exhibit a continuous size distribution. The packing density then depends mainly on the form of the size distribution and its width. Packing is all the more compact as the width of the distribution becomes higher.

5.2.3.2. *Stacking of ceramic particles*

Based on the principles described above, it is desirable to have a ceramic powder whose particles are as spherical as possible, or at least equiaxed and with low roughness, in order to facilitate their arrangement during shaping. Moreover, the particle size and the size distribution must ensure at the same time a compact, homogenous and reactive stacking on sintering.

In the case of mixtures requiring low sintering reactivity (building products, refractory products) it is possible to use binary or ternary powder mixtures, such as those defined in Table 5.1, with a substantial proportion of large diameter particles. We obtain, in this case, dense and sufficiently homogenous stackings that are then simply consolidated during sintering.

In the case of technical ceramics, the introduction of large diameter particles must be avoided because they present a low reactivity and preferentially grow during sintering to the detriment of fine particles. The powders are generally continuous size distributions of particles with a narrow distribution and an average diameter less than a micrometer, in order to guarantee a homogenous microstructure of the sintered object. These powders, however, result in a considerable shrinkage during sintering, since the packing densities seldom exceed 0.6. The study of powder dispersion and the choice of shaping additives assume great importance here, in order to maximize the volume fraction of the ceramic powder in the suspensions and mixtures.

5.2.4. *Influence of powders on the rheology of mixtures*

The rheological properties of concentrated suspensions and filled mixtures are very sensitive to the size distribution of the particles as well as their shape and specific surface (this aspect is discussed in section 5.3.8.4). If we consider a suspension of monomodal spherical particles, viscosity increases sharply for a powder volume concentration higher than 0.5 and tends towards the infinite for the maximum concentration of 0.74. A binary mixture of two populations of monomodal spheres decreases the viscosity of the suspension considerably (see Figure 5.13). The minimum viscosity is reached for a mixture containing 36% of particles of smaller diameter. This example again underlines the compromise to be found between packing aptitude, ease of shaping and sintering reactivity.

5.2.5. *Modifications in the characteristics of a powder*

The adaptation of the chemical characteristics of the particles surface is discussed in the dispersion section (see section 5.3). The adaptation of the physical characteristics is mainly achieved by milling, granulation and selection.

5.2.5.1. *Grinding techniques*

The main milling mechanisms are impact rupture and abrasive wear (attrition). The main milling techniques are listed in Table 5.2. Except the case of air jet milling, which utilizes the impact between the particles accelerated by an air blast, the milling techniques use milling media, particularly balls. The final average size of the particles depends on the technique used, the characteristics of the milling media (material, shape, size), the milling time, the milling environment (dry milling, in aqueous or non-aqueous medium, with or without dispersants) and load ratio (weight of milling media to powder weight). Beyond a certain milling time, the particle size does not change further because the energy provided is no longer sufficient to mill increasingly fine particles which contain fewer and fewer critical cracks. The contamination of the powders by milling media can be minimized by suitably choosing the material of the milling objects and the coating of the mills. Lastly, it must be noted that attrition milling generates more equiaxed particles that the other milling techniques.

Technique	Milling mechanism
Ball mill	Impact
Swing hammer mill	Impact
Planetary mill	Impact
Jet mill	Impact
Attrition	Friction, shearing

Table 5.2. *Commonly used milling techniques*

5.2.5.2. *Granulation*

Granulation produces spherical agglomerates of similar size, made up of elementary particles, which have good castability characteristics, while preserving a suitable sintering reactivity. Granulation is generally carried out by spray-dying, a technique described in section 5.5.

5.2.5.3. Selection

Sieving, generally assisted by pneumatic vibration, makes it possible to obtain grain size classes of down to 40 μm. For finer particles, we can use a classifier (air or water turbine) which is efficient down to a size of about 5 μm.

5.3. Ceramic particle suspensions

5.3.1. *Introduction*

The deagglomeration and dispersion of ceramic particles are fundamental stages in the shaping processes, be it by liquid, plastic or dry method. The aim is to achieve a homogenous and stable system of elementary particles, or partially agglomerated in the case of casting of traditional ceramics, in a liquid or a phase, generally organic, of higher viscosity. We must therefore understand and control the state of dispersion of the particles, which will determine the structure of the mixture, characterized by rheology. The suspensive liquid is usually called solvent, although the solubility of ceramic powders is generally not significant in these liquids.

The dispersion of a powder in a solvent involves three stages:

– wetting of the powder surface by the liquid, which results in the substitution of the solid/gas interfaces by solid/liquid interfaces;

– deagglomeration, i.e. rupture of the "soft" agglomerates into elementary particles or into "hard" aggregates of elementary particles. The simple capillary pressure of the solvent in the pores can destroy certain agglomerates (weak bonds), but the contribution of an additional mechanical energy, obtained by means of impacts with milling media or a shock wave, is generally necessary;

– stabilization of the state of dispersion with respect to sedimentation and reagglomeration.

The dispersant, or surface active agent, can play a role in these three stages. In the first, it can lower the solid/liquid interface energy by improving wetting. It also makes the milling stage more effective. Lastly, it helps obtain a stable suspension by avoiding reagglomeration and sedimentation.

5.3.2. *Surface charge of oxides in water*

Several mechanisms can cause surface charges of oxide particles [JOL 94] [HUN 87]. The main ones are i) the reaction of hydroxyl groups present at the surface of oxides, ii) the adsorption of specific ions or charged polyelectrolytes called dispersants or deflocculants and iii) the release of ions into water, for instance alkalines (K^+, Na^+) from the clay surface, resulting in a negative surface.

5.3.2.1. *Dissociation of hydroxyl surface groups*

The surface of oxide particles consists of hydroxyl groups (M-OH$_{surface}$) whose amphoteric character releases or captures protons, depending on the pH. The exchange reactions on the hydrated oxide surface, which result in the establishment of surface electric charge, are:

$$M\text{-OH}_{surface} + H^+ \leftrightarrows M\text{-OH}_2^+{}_{surface} \tag{5.1}$$

$$M\text{-OH}_{surface} \leftrightarrows M\text{-O}^-{}_{surface} + H^+ \tag{5.2}$$

These two equations show that, depending on the pH of the medium, the charge of the particles can be positive or negative. A characteristic point of an oxide is the point of zero charge (PZC). It reflects the acidic-basic character of the surface. The PZC can be defined in terms of pKa of reactions [5.1] and [5.2]:

$$PZC = 1/2(pKa_{[5.1]} + pKa_{[5.2]}) \tag{5.3}$$

The value of the PZC is linked to the nature of the oxide and will depend directly on the polarization of the OH group by the cation M of the oxide. A cation of high charge (Si^{4+}) will result in an acid surface and a weak PZC. The PZC of silica (SiO_2) is about 2. The PZC of alumina (Al_2O_3), with a trivalent cation (Al^{3+}) is about 9.

5.3.2.2. *Specific adsorption*

It is important to understand the phenomena that occur at the oxide-solution interface in the presence of an electrolyte, as they will govern the behavior of the suspensions. Certain ions present in the solution, like charged polyelectrolytes used as dispersants, exhibit a particular affinity for the surface and develop specific interactions with the hydroxyl groups. The same surface species $M\text{-OH}_2^+$ and $M\text{-O}^-$ occur both in acidic-basic balances and specific adsorption balances. Specific adsorption therefore affects the surface charge, which can be reversed, and displaces the ZPC (see Figure 5.6). Ammonium and sodium polyacrylates (Figure 5.7) are widely used to develop a high charge at the particle's surface.

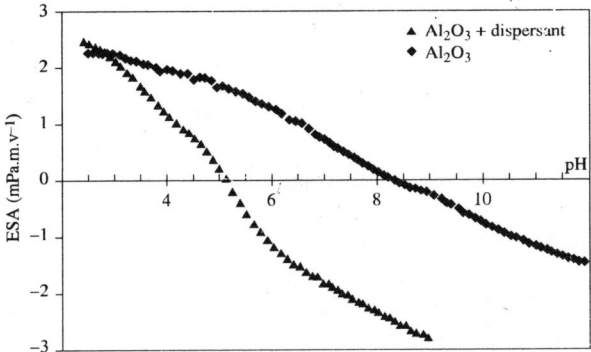

Figure 5.6. *Evolution of ESA signal, measured by acoustophorometry (see section 5.3.5.3), depending on the pH for an alumina powder and for an alumina powder with addition of ammonium polymethacrylate. The ESA signal is proportional to the zeta potential*

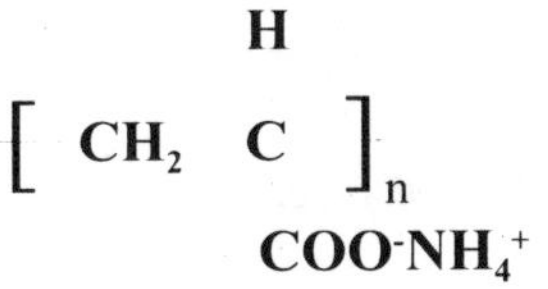

$$\text{H}$$

$$\left[\text{ CH}_2 \quad \text{C } \right]_n$$

$$\text{COO}^-\text{NH}_4^+$$

Figure 5.7. *Molecule of ammonium polyacrylate*

5.3.3. *Oxide/solution interface*

The ionization of the oxide particle surface in water modifies the distribution of the other ions in the environment. The double layer model is based on the electrostatic interactions between the surface and the oppositely charged ions (counter-ions), and on thermal agitation which tends to distribute these charges in space and destroy the order established by the electrostatic forces. The double electric layer governs the phenomena at the interface between the dispersed phase and the surrounding environment (see Figure 5.8). $M\text{-OH}_2^+$ and MO^- groups practically constitute a part of the solid. They are located in an average 0-plane assimilated to the particle surface, where they develop an average charge σ_0. The specifically adsorbed ions are located in a first compact layer which constitutes a very highly structured solvent zone (Stern layer). We assign them an average β plane, identified as the inner Helmholtz plane (IHP) carrying the charge density σ_β. The distance between the IHP plane and the 0-plane is in the order of an ionic radius (about one angström). Electric neutrality is ensured by a second layer (Gouy and Chapman diffuse layer), made up of the solvent ions distributed under the combined action of electrostatic interactions and thermal agitation. The average plane (outer Helmholtz plane (OHP)) constitutes the limit of the structured solvent zone (Stern layer). It is near this plane that the slipping of the solvent occurs and the electric potential on the outer Helmholtz plane (ψ_d) is assimilated to the electric potential (zeta electrokinetic potential) of the moving particle in the solution under the action of an electric field. The zeta potential, which reflects the mutual repulsion of the particles, is the primordial parameter to assess the properties of the system (agglomeration of the particles, stability and rheology of the suspensions).

The evolution of the potential, when we move away from the particle surface, is calculated by expressing the electrochemical potential difference between the particle surface and the solution. In the structured part of the double layer (Stern layer), the ions are supposed to be located on distinct planes. The decreases in the potential between the surface and the IHP, and between the IHP and the OHP are therefore as linear as inside a condenser. The potential decreases exponentially with the distance in the diffuse layer starting from the OHP. The electrically disturbed zone extends to about 100 angströms. Its thickness depends highly on the concentration and the charge of the electrolyte ions. The greater the charge and the concentration, the more the diffuse layer is compressed.

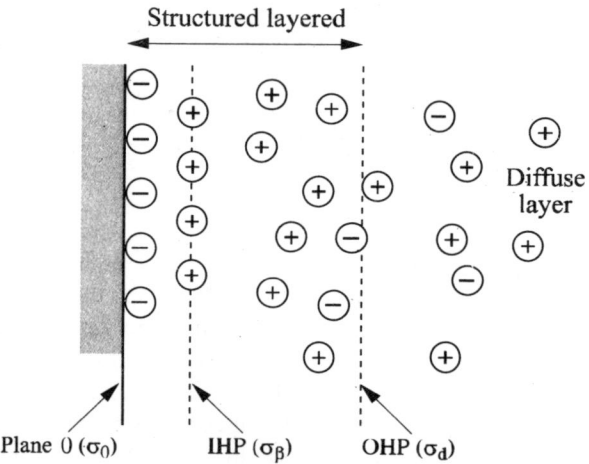

Figure 5.8. *Double layer model: distribution of charges and surface potentials*

5.3.4. *The zeta potential (ζ)*

A charged particle constituting an electrokinetic entity and subjected to an electric field (E), will move at a speed v (electrophoretic velocity). The electrophoretic mobility (μ in $m^2s^{-1}V^{-1}$) is defined, for spherical particles, by:

$$\mu = v/E \qquad\qquad [5.4]$$

The zeta potential is calculated using the following expression:

$$\xi = \frac{f_h \eta \mu}{\varepsilon_r \varepsilon_0} \qquad\qquad [5.5]$$

where η is the viscosity of the solution and ε_r the dielectric constant of the liquid. The Henry constant f_h is a function of the ratio of the particle radius to the thickness of the diffuse layer. The f_h constant is equal to 1.5 (Hückel equation) for a value of this ratio lower than 1 (broad diffuse layer). The particle is then regarded as an isolated charge. If the contrary happens (ratio higher than 100), which is the case of environments with low dielectric constant, the diffuse layer is regarded as a plane state and the f_h dielectric constant is equal to 1 (Helmholtz-Smoluchowski equation).

The value of the zeta potential is often correlated with the stability of the suspensions. At room temperature, an absolute value higher than 40 mV provides a satisfactory electrostatic stabilization.

5.3.5. *Measurement of the zeta potential (ξ)*

5.3.5.1. *Micro-electrophoresis*

Electrophoresis is widely used to characterize the stability of ceramic suspensions. The migration rate of an elementary particle, subjected to an electric field, is measured. This technique is applicable only to highly diluted systems, non-representative of ceramic suspensions.

5.3.5.2. *Electrophoretic mass transport*

An "average" mobility of suspended particles is measured with the mass transport technique. An electrode is placed in the suspension, whereas the other electrode is placed at the bottom of a small cell filled with the same suspension. The electrophoretic migration under electric field varies the solid concentration in the cell. The "average" mobility can then be calculated by:

$$\mu = \frac{\Delta W \lambda}{t I \phi_s (1 - \phi_s)(\rho_p - \rho_l)} \qquad\qquad [5.6]$$

where ΔW is the mass variation in the cell, λ the conductance of the suspension, t the time of application of the field, I the current applied, ϕ_s the volume fraction of powder in the suspension, ρ_p and ρ_l the respective densities of the solid and the liquid.

5.3.5.3. *Acoustophorometry*

Acoustophorometer is a measurement technique based on an electroacoustic effect which occurs when a high frequency alternate electric field (1 MHz) is applied to two electrodes immersed in a suspension of charged particles. The field applied periodically deforms the distribution of the mobile charges of the double electric layer of each particle and produces an acoustic pressure variation of the same frequency as the applied electric field. Its amplitude depends on the displaced charges and can be related to the zeta potential [O'BR 88].

5.3.6. *Electrostatic stabilization*

The stability of the dispersion characterizes, in the common meaning of the term, the quality of the dispersion of the particles in the system. Instability indicates the agglomeration, coagulation or flocculation of the particles. The particles are subjected to the repulsive forces resulting from the interaction of the double electric layers, the attraction of Van der Waals forces and the Brownian movement if they are small-sized. We must therefore know whether the particles, under the effect of the Brownian movement, can approach one another up to a distance sufficiently small to form permanent associations or not. The problem is addressed by using the DLVO theory (Dejarguin, Landau, Verwey, Overbeek) [DER 41] [HUN 87] [VER 48].

5.3.6.1. *Van der Waals forces*

These attractive forces result from dipolar interactions at the molecular level. At the scale of the particles and in the case of two spheres of radius a, whose centers are distant from D, the interaction potential energy is written as:

$$V_A = -\frac{A}{6}\left[\frac{2a^2}{D^2-4a^2}+\frac{2a^2}{D^2}+\text{Ln}\frac{D^2-4a^2}{D^2}\right] \qquad [5.7]$$

where A is the Hamaker constant which depends on the nature of the particles and the solvent. For particles (1) in a dispersive liquid (2), the Hamaker constant is written in an approximate way $A = (\sqrt{A_1}-\sqrt{A_2})^2$. The value of A is of $3.7.10^{-20}$ J for water and generally about 3.5 to 8.10^{-20} J for oxides.

For small distances $(D - 2a)/a \ll 1$, the interaction energy is simplified to:

$$V_A = - Aa/12(D - 2a) \tag{5.8}$$

The potential attraction energy therefore depends, in a first approximation, only on the nature of the material and the dispersion environment (Hamaker constant), on the size of the particles and their distance.

5.3.6.2. Electrostatic forces

They result from the interaction of the double electric layers. For identical particles, they are repulsive. When two particles approach one another, the diffuse parts of the double layers repel. If they are compressed too much, the Stern layers also enter into interaction. The calculation of the interaction potential energy V_R, based on the duration of collisions and the relaxation times of the layers, is complex. If the overlapping of the layers is weak, the approximate expression of the potential interaction energy is given by the relation:

$$V_R = 2\pi\varepsilon_r\varepsilon_0 a\psi_d^2 \exp\left[-\kappa(D - 2a)\right] \tag{5.9}$$

where κ is the inverse of the Debye length, which represents the extent of the electrostatic interaction and is usually assimilated to the thickness of the diffuse layer. The two parameters that have a major influence on V_R are the potential on the OHP slip plane (ψ_d assimilated to the ξ potential), the concentration and nature of the ions through the term κ, as well as the dielectric constant of the solution.

5.3.6.3. Total potential energy

The total potential energy V_T is the sum of the attraction and repulsion energies ($V_T = V_A + V_R$). The evolution of V_T according to the separation distance of the particles is represented in Figure 5.9. The total potential energy has, in general, one maximum and two minima. If the maximum is rather high, the collisions cannot provide a sufficient energy to overcome this potential barrier and the suspension is stable. Coagulation can occur at the second minimum, but it is weak and reversible.

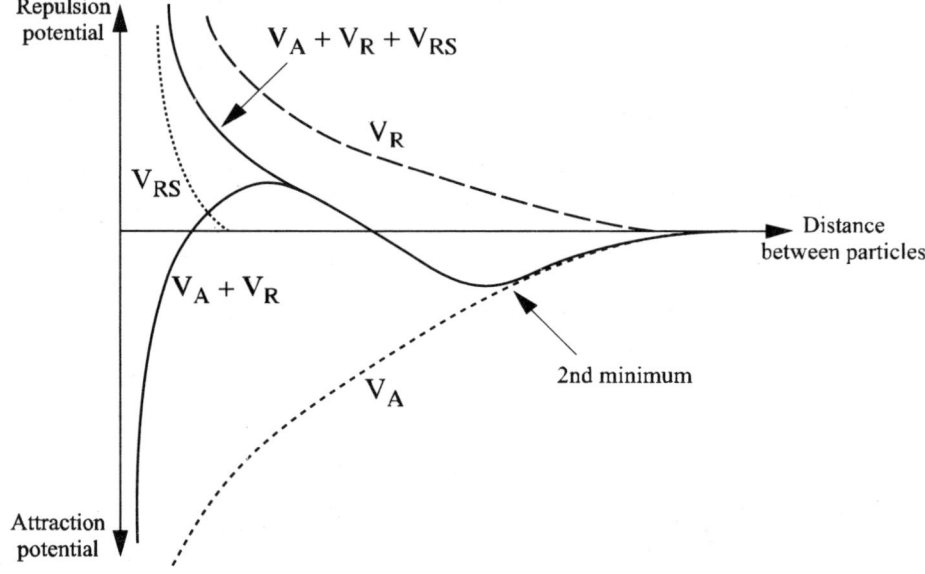

Figure 5.9. *Evolution of potential energies depending on the particle distance and the various contributions*

According to the DLVO theory, under isothermal conditions and in the absence of mechanical agitation, stabilization by electrostatic repulsion depends primarily on the following factors:

– the concentration of the electrolyte (dispersant) and its charge. The increase in the concentration (see Figure 5.10) or the charge causes the reduction in V_R by lowering the Debye length $1/\kappa$. An electrolyte will be all the more flocculating the higher its concentration and the charge of its ions. Likewise, the use of a liquid with lower dielectric constant will lower the stabilization;

– the value of the electric surface potential. A high value of ψ_d increases the value of V_R. The value of ψ_d depends directly on the phenomena at the origin of surface charges; that is why the pH, the concentration and the charge of adsorbed ions are among the most influential factors;

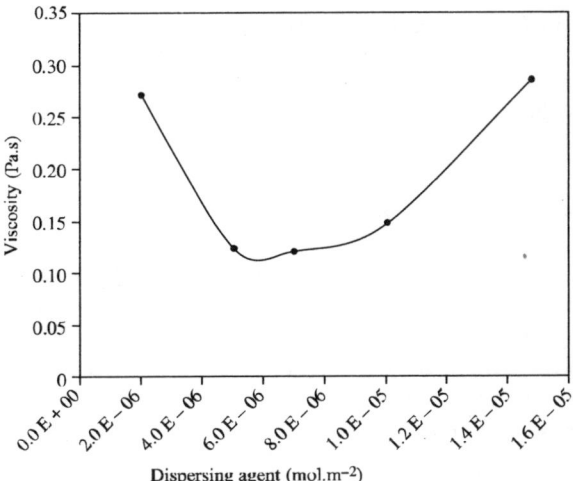

Figure 5.10. *Variation of the viscosity of an alumina suspension according to the dispersant concentration (polyelectrolyte)*

5.3.7. *Steric and electrosteric stabilizations*

An important factor that is not taken into account in the DLVO theory is adsorption, on the particle's surface, of long polymeric chains. The adsorption of a non-ionic polymer or a polyelectrolyte on the solid surface can cause, not only a modification of the zeta potential, but also a critical difference between the value of the zeta potential and the state of dispersion. Steric repulsion is associated with the obstruction effect of these polymers that are capable to form a sufficiently thick layer to prevent the particles from approaching one another in the distance of influence of the Van der Waals attractive forces. Steric stabilization will therefore depend on the adsorption of the polymeric dispersant and the thickness of the layer developed. Several interpretation models for stabilization by steric effect have been put forward. They rely either on a statistical approach, or on the thermodynamics of solutions. Steric stabilization is particularly useful in organic, fairly non-polar or non-polar environments, as in the case of tape casting (see section 5.4.3).

The adsorption of charged polymers (polyelectrolytes) leads to a mixed steric/electrostatic mechanism (electrosteric mechanism). The expression of total potential energy becomes:

$$V_T = V_A + V_R + V_{RS} \qquad [5.10]$$

where V_{RS} is the steric interaction potential.

Total potential energy exhibits only the second minimum (see Figure 5.9). Steric contribution avoids the contact between short distance particles (about a few nanometers), whereas the existence of a potential barrier, due to the double layer, is effective at higher distances.

5.3.8. *Rheology of ceramic systems*

Rheology is the science of the deformation and flow of the matter. The knowledge of the rheological behavior of ceramic systems is fundamental in the shaping stage. Ceramic suspensions and pastes are multi-constituent systems, often presenting complex rheological behaviors, which will depend on the shape of the particles, their concentration, their grain size distribution, but also on the interaction forces between these particles.

The rheological equation of a fluid state links the deformation γ with the shear stress τ. The rheogram generally used to represent this equation of state is:

$$\tau = f(\dot{\gamma}) \qquad\qquad [5.11]$$

where $\dot{\gamma} = d\gamma/dt$ is the deformation rate or shear rate.

Viscosity, dynamic or apparent, (η) is the ratio of the shear stress (τ) on the shear rate ($\dot{\gamma}$), its unit is Pa.s. Viscosity represents the flow resistance of a fluid.

The respective influences, on the rheological behavior, of the ceramic and suspensive phase (liquid, organic phase) are often determined using relative viscosity:

$$\eta_r = \eta/\eta_s \qquad\qquad [5.12]$$

where η and η_s are the respective apparent viscosities of the fluid and the suspensive phase.

The various rheological behaviors are described below.

5.3.8.1. *Rheological behaviors (see Figure 5.11)*

5.3.8.1.1. Newtonian fluids

Normal or Newtonian fluids exhibit a linear relation between the shear stress and the shear rate ($\eta = \tau/\dot{\gamma}$). Viscosity does not depend on the shear stress. The rheogram of a Newtonian fluid is a straight line passing through the origin and with a slope equal to the viscosity. Most ceramic systems are non-linear or abnormal.

5.3.8.1.2. Shear-thinning or pseudo-plastic fluids

The apparent viscosity of a shear-thinning fluid decreases with the shear. This frequent behavior in ceramics is due to the orientation of the particles, particularly anisotropic particles, and of the polymeric chains in the direction of the flow, which reduces flow resistance. This behavior is described by a power law:

$$\tau = K\dot{\gamma}^n \tag{5.13}$$

where K is an index of consistency and the exponent n $(n < 1)$ expresses the deviation from the Newtonian behavior $(n = 1)$.

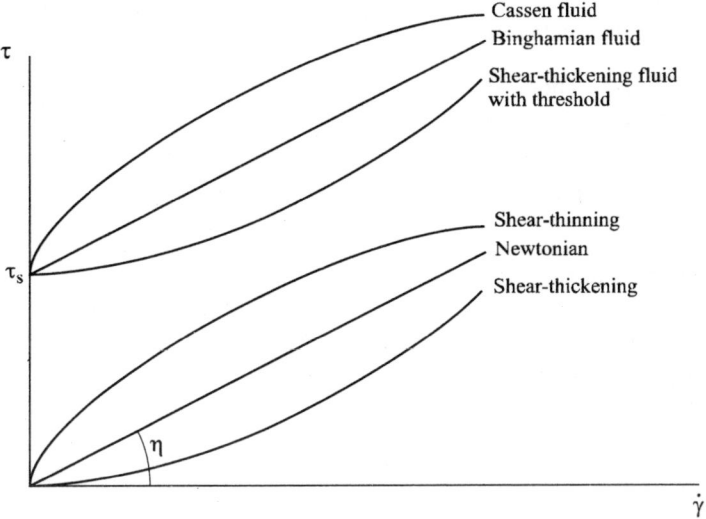

Figure 5.11. *The various rheological behaviors of ceramic compositions*

5.3.8.1.3. Shear-thickening fluids

The apparent viscosity of a shear-thickening fluid increases with the shear. This behavior can block the shaping equipment (extrusion machine, injection molding press) and must therefore be prohibited in ceramic processes. This behavior is attributed to a rupture of liquid film around the moving particles. Another explanation is the expansion in the volume of the fluid subjected to shear by the overlapping of the particles on one another (shear-thickening fluids are often called dilatants). It has been observed in the systems that are: i) highly concentrated, ii) containing agglomerates and iii) with angular particles. This behavior is described by the power law [5.13] with an exponent n greater than 1.

5.3.8.1.4. Plastic fluids

These are fluids which flow only from a threshold stress τ_s called yield stress or yield value. The suspensions having a three-dimensional structure with strong bonds between molecules and particles exhibit a plastic behavior. The flow threshold corresponds to the stress necessary to break this cohesion.

– Bingham fluids. These fluids behave like Newtonian fluids, but only above a threshold stress. Below this threshold stress, the behavior is that of an elastic solid. They are described by:

$$\tau < \tau_s: \dot{\gamma} = 0 \text{ and } \tau > \tau_s: \tau = \tau_s + \alpha\dot{\gamma} \qquad [5.14]$$

where α is a constant called plastic viscosity;

– Casson fluids. These fluids exhibit a fluidifying plastic behavior. They are described by the Casson equation:

$$\tau < \tau_s: \dot{\gamma} = 0 \text{ and } \tau > \tau_s: \tau^{0.5} = \tau_s^{0.5} + \left(\beta\dot{\gamma}\right)^{0.5} \qquad [5.15]$$

where β is a constant also called plastic viscosity.

A frequently used model to describe this behavior is the Herschel-Bulkley model:

$$\tau = \tau_s + K\dot{\gamma}^n \qquad [5.16]$$

The rheological behaviors described do not depend on the history of the fluid. Rheograms are unique and reversible. In other cases (thixotropic and rheopexic fluids), the destructuring of the fluid and its recovering depend not only on the stress, but also on the time required to reach equilibrium. The rheograms then exhibit hysteresis. The flow characteristics will therefore be influenced by the earlier treatments. Thixotropic and rheopexic fluids are similar, respectively, to the shear-thinning and the shear-thickening fluids, but with a sufficient rest period after shear relaxation to regenerate their original structure.

5.3.8.2. *Characterization of rheological behavior*

Steady state rheometers with rotational cylinders or cone/plate (see Figure 5.12) are the most widely used to characterize the rheological behavior of ceramic systems with the determination of rheograms $\tau = f(\dot{\gamma})$. They consist of shearing the fluid between a surface at rest and a mobile surface. The advantage of a cone/plate rheometer with a low-angle cone is that it leads to constant τ and $\dot{\gamma}$ parameters at any point of the fluid. These rheometers can typically measure viscosities ranging from 10^{-3} to 10^4 Pa.s. Extrusion and injection molding pastes, with viscosity going up to 10^7 Pa.s, are generally characterized using a capillary rheometer. Shearing is imposed here by a pressure difference between the two ends of a cylindrical tube with low cross-section, applied by a piston.

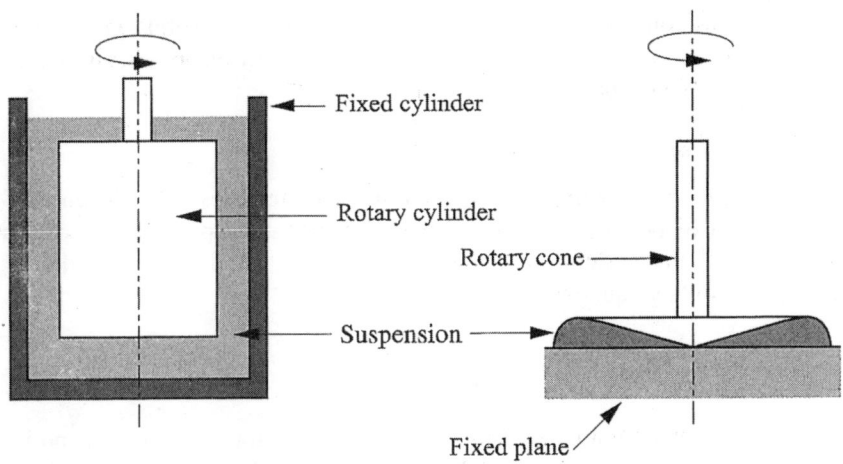

Figure 5.12. *Operating principles of steady state rheometers*

It is interesting, from a fundamental point of view, to characterize the viscoelastic behavior of a fluid, which reflects the forces between the particles and hence the structure of the fluid. This characterization is done dynamically by applying on the fluid a sinusoidal stress with frequency N and low amplitude, so as not to break the structure of the fluid (linear viscoelastic mode). The response is a deformation of the same frequency. Elastic systems have a response in phase with the stress. In a Newtonian system, the shear stress is proportional to the strain rate and the sinusoidal strain response is dephased by 90° compared to the sinusoidal stress applied. The phase angle δ ($0 < δ < 90°$) is therefore a characteristic of viscoelastic behavior. We use the formalism of complex numbers to write:

$$\overline{\tau}(t) = \overline{G}(\omega)\overline{\gamma}(t) \tag{5.17}$$

with: $\overline{\tau}(t) = \tau_0 e^{i(\omega t + \delta)}$ and $\overline{\gamma}(t) = \gamma_0 e^{i\omega t}$, where ω is the pulse ($\omega = 2\pi N$), t is time, τ_0 and γ_0 are the maximum amplitudes of stress and strain.

The complex modulus of rigidity $\overline{G}(\omega)$, determined by oscillation measurements, depends on the pulse but is independent of time:

$$\overline{G}(\omega) = G'(\omega) + iG''(\omega) \qquad [5.18]$$

where $G'(\omega)$ is the storage modulus (elastic component) and $G''(\omega)$ the loss modulus (viscous component). For a perfect elastic behavior (solid), $G'(\omega)$ is equal to the modulus of rigidity G and $G''(\omega) = 0$. In a Newtonian viscous behavior, $G'(\omega) = 0$ and $G''(\omega) = \eta\omega$.

5.3.8.3. Influence of particle concentration on viscosity

Several equations (empirical) have been proposed to assess the influence of the volume fraction of powder (ϕ_s) on viscosity. For spherical particle concentrations representative of the systems used ($\phi_s > 0.5$), Krieger and Dougherty [KRI 59] propose the following equation:

$$\eta_r = (1 - \phi_s / \phi_{max})^{-2,5} \qquad [5.19]$$

where ϕ_{max} is the maximum volume fraction of powder for which flow is no longer possible. We can also mention the Chong equation [CHO 71], which agrees well with experimental results when ϕ_s/ϕ_{max} is close to one:

$$\eta_r = \left(1 + \frac{0.75\,\phi_s/\phi_{max}}{1 - \phi_s/\phi_{max}}\right)^2 \qquad [5.20]$$

5.3.8.4. Influence of grain size distribution

Mixtures of various grain size classes of particles offer the advantage of reducing the quantity of the second phase necessary for the flow (liquid, organic phase) and leading to a more compact stacking, which reduces sintering shrinkage. On the other hand, a high variation in particle size is detrimental to obtaining a homogenous microstructure without grain growth during sintering. The theories on the viscosity of systems containing multimode powders were developed by Lee [LEE 69] and Farris [FAR 68]. For spherical particles volume fractions higher than 50%, a substantial increase in concentration can be achieved at constant viscosity (see Figure 5.13). However, a fine/coarse ratio equal to 36/64 leads to minimal viscosity.

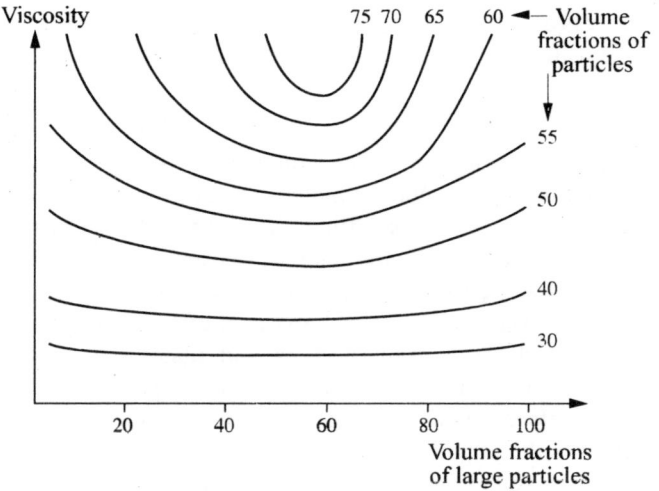

Figure 5.13. *Evolution of the viscosity of a bimodal particle suspension for various fine/large ratios and powder volume fractions [FAR 68]*

5.4. Casting

5.4.1. *Slip casting*

Slip casting is the most conventional method of producing very varied pieces that can have complex forms (culinary, sanitary, refractory materials, technical ceramics). This method consists of casting a suspension (slip) in a porous mold, generally made of plaster. The capillary migration of the liquid into the pores of the mold results in the formation of a consolidated layer of particles on the mold surface. The main advantages of slip casting are: i) the complexity of the forms that can be produced, ii) its low cost and iii) the use of perfectly dispersed suspensions in technical ceramics leading to dense and homogenous green microstructures. Its major disadvantage is its low production capacity. Aqueous suspensions are the most common but hydrolysable materials (MgO, CaO, La_2O_3, etc.) and non-oxides (SiC, Si_3N_4, AlN, etc.) are cast in an organic environment (alcohol, ketone, thrichlorethylene). The surfaces of these non-oxides can nevertheless be "hydrophobated" using with a surface active agent, which allows their suspension in water.

In the case of the open air casting (see Figure 5.14), the external form of the piece is defined by the mold and when the thickness of the consolidated wall is sufficient (setting time), the suspension is emptied. The green part, after drying, must have sufficient cohesion to be demolded. A low drying shrinkage facilitates this stage. The walls are thin and of constant thickness. Pieces produced by casting between two molds (or between two plasters) (see Figure 5.15) have the shape of the mold cavity. The walls can be thick and have variable thicknesses.

5.4.1.1. *Molds*

The mechanism of absorption of the liquid in the suspension is linked to the capillary suction effect of the porous mold. The porous network of the mold is therefore of vital importance for the setting rate and the characteristics of the green part. The most widely used molds are made of gypsum ($CaSO_4$, $2H_2O$) formed by reaction of plaster ($CaSO_4$, $0.5H_2O$) and water. We will use the common word plaster, wrongly employed, to indicate the material of the mold. Plaster presents an inter-connected network of gypsum needles and plates which confer on it its mechanical strength. Plaster allows the easy and low-cost manufacture of complex molds with good surface quality, high porous volume (40 to 50%) and pore size lower than 5 μm. On the other hand, plaster has a low abrasion resistance and high solubility in water, as well as a drop of its mechanical properties due to dehydration, above 40°C, which limits its necessary drying between two castings so as not to increase the setting time. The surface of the mold can be covered to facilitate the demolding and to reduce attacks by acid suspensions (silica) or by organic suspensions (alcohol), for instance by alginates, talc or by graphite.

Other materials are used for the production of casting molds with high mechanical strength and high hardness, for instance, epoxy resins filled or non-filled with ceramic powders.

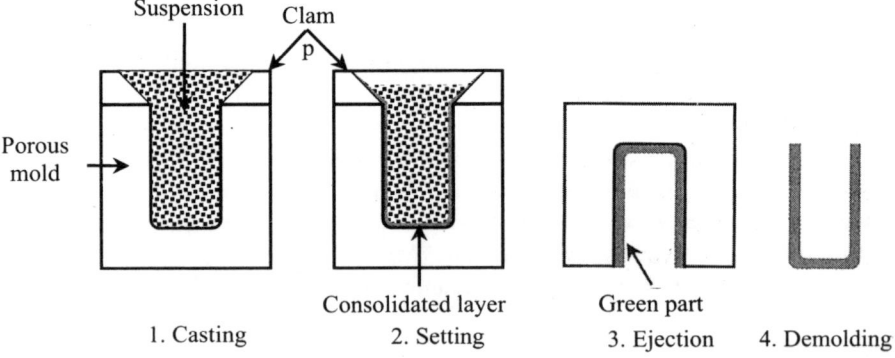

Figure 5.14. *Open-air casting*

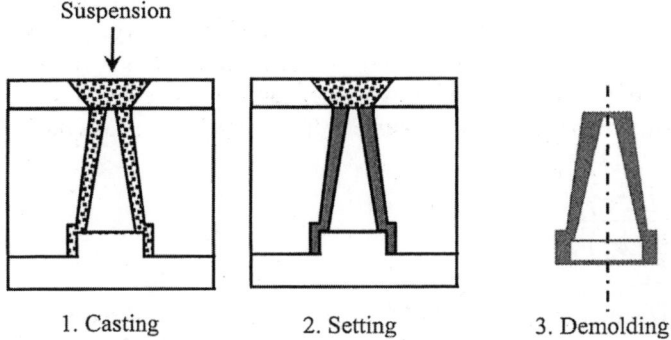

Figure 5.15. *Casting between two molds*

5.4.1.2. *Kinetics of setting*

As the casting process is an operation of filtration with formation of deposit, its kinetics is generally evaluated using Darcy's law [TIL 86]. By considering that rheological properties of the suspension remain constant during filtration, that there is no sedimentation and that the deposit is incompressible, its thickness *e* can then be expressed by:

$$e = \frac{1}{\phi_d \rho_p} \left[\left(\left(\frac{R_m}{R_d} \right)^2 + \frac{2\Delta P \phi_s \phi_d \rho_p}{\eta R_d (\phi_d - \phi_s)} t \right)^{0.5} - \frac{R_m}{R_d} \right] \qquad [5.21]$$

where ϕ_s and ϕ_d are respectively the solid volume fractions in the suspension and the deposit, ρ_p the density of the particles, ΔP the pressure difference of the liquid through the deposit (capillary pressure), η the viscosity of the transported liquid, R_m and R_d the respective specific transport resistances of the mold and the deposit.

The limiting stage in the setting kinetics of suspensions containing low-diameter particles is the transport of liquid through the porous structure of the deposit. Evaluation models therefore generally neglect the migration resistance of the liquid in the porous structure of the mold, but nevertheless assimilate the pressure drop through the solidified layer to the suction pressure of the mold. Plaster produces a suction pressure of 0.1 to 0.2 MPa [ADC 57]. Equation [5.21] becomes:

$$e^2 = \frac{2\Delta P}{\eta R_d \phi_d \rho_p} \frac{\phi_s}{(\phi_d - \phi_s)} t \qquad [5.22]$$

The setting rate will be all the higher as the permeability of the deposited layer increases (low R_d). The specific resistance of the deposit R_d is generally determined by the Kozeny-Carman model [CAR 56], which expresses this resistance according to the porosity of the deposit (porous volume, surface of the pores per unit volume S_v, and tortuosity represented by a parameter T, generally taken as equal to 5):

$$R_d = \frac{TS_v^2 \phi_d^2}{(1-\phi_d)^3}$$
[5.23]

The porous structure of the solidified layer, and therefore the grain size of the powder (size and distribution) and the state of dispersion of the suspension, will directly influence the setting kinetics. This model considers that all the pores take part in the same way in the liquid flow and that the porous structure is uniform, which is generally not the case. On the one hand, casting of traditional ceramic suspensions, partially flocculated, leads to a deposit with a bimodal porosity. The pores between agglomerates then constitute the principal ways of transport. In addition, the deposits are compressible with a rearrangement of the porous structure during the casting.

The setting time can vary from a few minutes to produce a thin wall with a partially coagulated porcelain suspension, to one hour in the case of a perfectly dispersed suspension of submicronic particles. It can take several days for the casting of very thick refractory parts (10 cm).

5.4.1.3. *Casting suspensions*

Obtaining ceramic parts, with satisfactory properties in a reproducible way by slip casting requires a judicious choice of the grain size and the control of the particle surface chemistry. The rheological behavior and the viscosity of the suspensions in fact depend directly on the grain size of the powders, the inter-particle interactions (state of dispersion) and the particle concentration.

The powder grain size distribution, which will influence the arrangement of the particles during the consolidation, varies considerably according to the type of ceramic and the final properties desired. A significant concentration of fine particles (in the order of a micron) is nevertheless necessary to control rheology and sedimentation. In the case of technical ceramics, the particle size is generally low with a narrow distribution to ensure satisfactory sintering reactivity and a homogenous microstructure. On the other hand, in the case of traditional ceramics, the mixture is typically made up of fine clay platelets, as well as silica and feldspar particles of about a few tens of micrometers.

Traditional clay ceramic suspensions and technical ceramic suspensions differ by the degree of particle dispersion (see Figure 5.16).

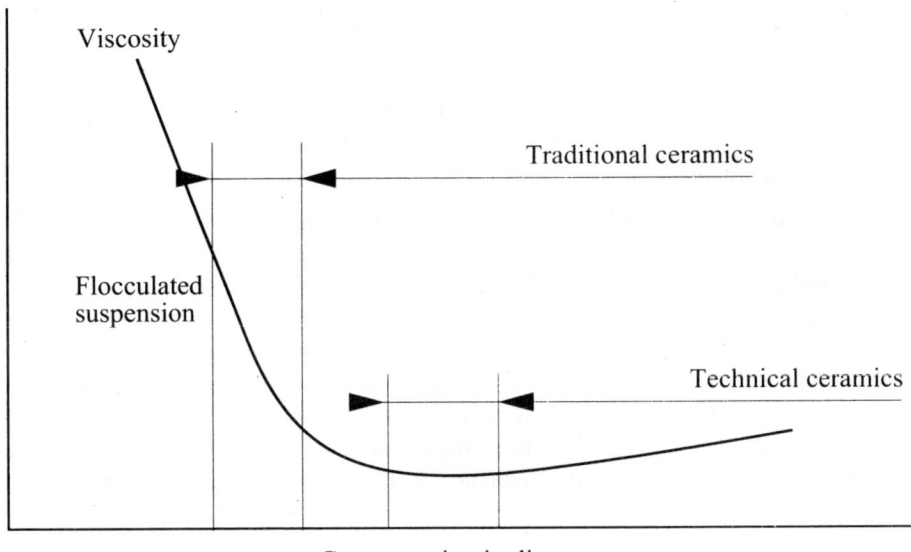

Figure 5.16. *Degree of particle dispersion for various suspensions*

A traditional ceramic suspension is partially flocculated with a viscosity higher than the minimal viscosity. This partially coagulated state makes a quick setting rate possible thanks to a high pore diameter and porous volume in the formed layer. Moreover, the high water retention of this layer will give to the green part a plasticity favorable to its handling after demolding. As traditional compositions develop a vitreous phase during sintering, the green microstructure obtained by this partially agglomerated state will not have a significant influence on the final properties. In the case of casting of technical ceramics, the objective is very different, namely, obtaining a high green density with a homogenous microstructure. The suspension will thus be completely dispersed with a minimal viscosity and a high particle concentration. The suspensions are formulated with a shear-thinning behavior and a viscosity lower than 1 Pa.s for a shear rate ranging between 1 and 10 s^{-1} during casting. They typically contain 40 to 60 vol.% of mineral matter and 0.2 to 1% by mass of dispersant on the dry matter basis. A high viscosity at low shear rate and/or a sufficiently high yield stress is favorable for avoiding the sedimentation of the particles at rest, before the casting, but especially in the mold during the setting. The minimal value of the threshold stress τ_s which balances the sedimentation forces of particles with diameter d and density ρ_p in a suspension of density ρ_s is expressed by:

$$\tau_s = \frac{2}{3} d \left(\rho_p - \rho_s \right) g \qquad\qquad [5.24]$$

5.4.1.4. *Emptying–demolding*

The consolidated layer must have sufficient mechanical strength so that it will not flow out with the suspension during the emptying of the mold and allow its demolding. In partially coagulated systems containing clay (traditional ceramics), the coagulation forces give a sufficient cohesion on the piece. The cohesion of pieces realized with dispersed suspensions (technical ceramic) can be increased by the addition of binders (carboxymethyl cellulose, ammonium or sodium alginate) which can also contribute to the dispersion. The addition of binders nevertheless presents the disadvantage of increasing the setting time through an increase in the viscosity of the suspension and a reduction in the permeability of the consolidated layer.

5.4.1.5. *Flaws*

The main flaws encountered in a green part obtained by slip casting are: i) the presence of large pores (pinholes) due to a poor degassing of the suspension, ii) a preferential orientation of anisotropic particles (clay platelets, mica) to a low thickness on the piece's surface which results in a differential shrinkage and therefore in stresses during drying and sintering, and iii) a non-homogenous microstructure due to a sedimentation of coarse particles.

5.4.2. *Pressure casting*

Equation [5.22] shows that the supply of an additional pressure, compared to the low capillary pressure of the mold (< 0.2 MPa for plaster), decreases the setting time. In this respect, the pressure casting consists of applying a pressure, generally lower than 5 MPa, to the suspension in the porous mold. The pressure gradient thus created (ΔP) will force the fluid through the porous network and the formed layer, considerably reducing the setting time compared to traditional casting (see Figure 5.17).

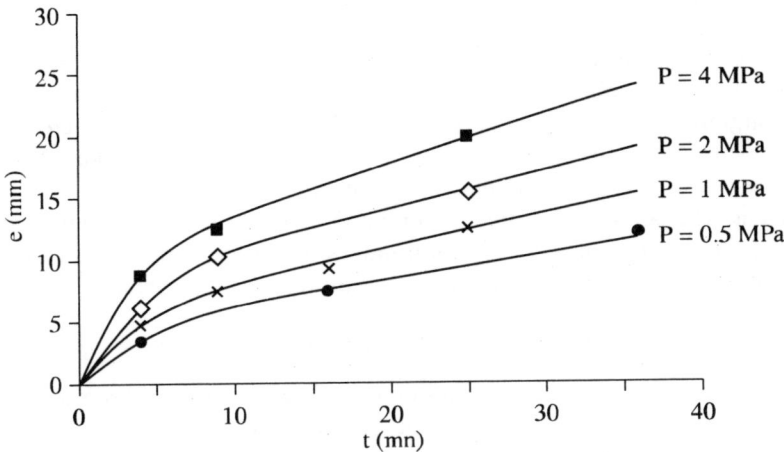

Figure 5.17. *Casting kinetics for various applied pressures [GRÜ 72]*

Pressure casting reduces the water content in the green part and increases its density and its cohesion. The mono- or multi-cavity molds are made of plaster or polymer with a mechanical strength and porosity greater than plaster, as well as an elasticity allowing a water tightness of the mold under low clamping force. They do not require drying between the various castings. This computerizable process provides high productivity. It is now quite widely used in the industry for production of ceramic sanitaryware and tableware. On the other hand, the application of this process to technical ceramics is still relatively new.

Let us finally mention two other casting methods that also significantly reduce the setting time compared to traditional casting: vacuum casting and centrifugation casting. The first method is similar to pressure casting, but here the suspension is sucked through the porous mold. In the second method, the particles are driven by centrifugation resulting in a high density deposit. Centrifugation casting is used in the field of technical ceramics.

5.4.3. *Tape casting*

Tape casting [CHA 94], [MIS 78] produces ceramic sheets of low thickness (25 to 1,000 micrometers) and large surface, whose homogenity, surface quality and mechanical strength in the green state are satisfactory for electronic applications such as substrates and multilayer capacitors.

5.4.3.1. *Casting benches*

Tape casting consists of depositing a suspension of powder and organic components on a support. The evaporation of the liquid, called solvent as it allows the dissolution of the organic phase (except in the case of emulsions where it ensures its suspension), provides a tape that will be separated from the support. The drying determines, to a large extent, the microstructure and properties of the sintered component. The deposit is created by a relative movement of a tank and a support. Two solutions are possible: the reservoir moves on a fixed support (non-continuous casting) or the support moves under the reservoir (continuous casting).

5.4.3.1.1. Non-continuous casting (see Figure 5.18)

A doctor blade spreads the suspension on a fixed support (glass, stainless steel, plastic film). For a given suspension, the height of the blade and the speed of the reservoir will determine the thickness of the tape. This casting technique is suitable for the realization of high thicknesses (substrates).

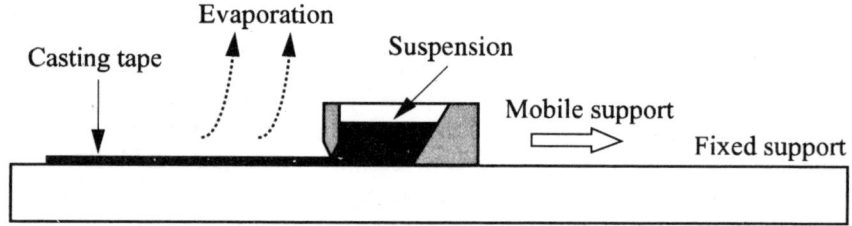

Figure 5.18. *Principle of non-continuous tape casting*

5.4.3.1.2. Continuous casting (see Figure 5.19)

The mobile support (stainless steel tape, plastic film) moves under a fixed reservoir in which the level of the suspension is maintained constant. The casting reservoir is fixed horizontally on the mobile support (coating similar to the manufacture of paper) or tangent at a drive cylinder (surface application). In this second configuration, the thickness is adjusted by the speed of the supporting tape and the angular position of the tank with respect to the cylinder. The speed of the support is generally controlled by a continuous measurement of the thickness and the suspension's viscosity is maintained constant during the casting. The longest part of the casting bench forms the area of evaporation of the solvent. A counter-current ventilation allows the elimination of the solvent homogenously with a controlled kinetics. At the end of the bench, the dry tape is separated and stored. The coating technology is largely used in the production of low thicknesses (dielectric films for multilayer capacitors).

The casting must be performed under constant conditions of temperature and hygroscopy to ensure the maintenance of the rheological and evaporation properties.

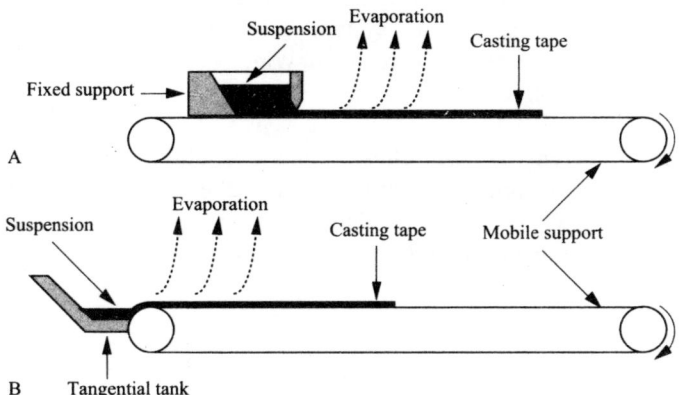

Figure 5.19. *Principles of continuous tape casting: A) coating, B) surface application*

5.4.3.2. Tape casting suspensions

Tape casting suspensions are complex mixtures of ceramic powders dispersed using a dispersant in a system containing a solvent, binders and plasticizers [MOR 92]. These components affect not only the rheological behavior of the suspension, but also the drying behavior (particle arrangement and sensitivity to cracking) and the capability for thermocompression [CHA 92]. The formulation of the casting suspension will therefore determine the properties of the green tapes and the end product.

5.4.3.2.1. Ceramic powder

As we saw earlier (section 5.2), the grain size distribution will directly influence the arrangement of the particles during casting and evaporation of the solvent, and consequently shrinkage during sintering. Shrinkage control is fundamental for multi-layer technology. Powders with high specific surface and low-diameter particles yield stable suspensions and exhibit a high sintering reactivity. On the other hand, they require a high quantity of organic phases, which reduces the green density and increases shrinkage during sintering. The particle size for tape casting is therefore a compromise between ease of shaping and sintering reactivity. It generally ranges between 1 and 4 µm for specific surface areas from 2 to 6 m^2g^{-1}.

5.4.3.2.2. Solvent

The solvent ensures the suspension of ceramic particles and the solubilization of the various organic components, or their dispersion in the case of emulsions.

Although flammable and toxic, organic solvents are largely used in tape casting because of: i) their low viscosity, ii) their weak surface energy which enhances the wetting of the particles and iii) their high vapor pressure resulting in a quick drying of the tapes. Nevertheless, aqueous systems have started to catch up on organic systems for obvious environmental reasons. Water has the major disadvantage of having a low evaporation rate.

A mixture of solvents is often chosen for the solubility of the various organic compounds. The azeotrope proportions thus make evaporation with constant composition possible. The most widely used solvents are 2-butanone/thanol azeotrope (60/40 vol.) and trichlorethylene/ethanol (72/28 vol.).

5.4.3.2.3. Dispersant

The role of the dispersant and dispersion mechanisms were discussed in section 5.3. Deagglomeration and the perfect dispersion of the particles are fundamental for the realization of homogenous tapes of low thickness. As the organic solvents used are generally not very polar (dielectric constant of about 20) and the suspensions are concentrated, the best dispersion is achieved by a combination of electrostatic and steric repulsions. Phosphoric esters are thus very effective. Fish oils (fatty acid) are also used, but less effectively. In an aqueous environment, traditional polyelectrolytes like ammonium polyacrylates are used.

5.4.3.2.4. Binder

After the evaporation of the solvents, the binder ensures the cohesion of the green tape. Most organic binders are long polymeric chains. These molecules are adsorbed on the particle surface and form bridges between one another, which produce mechanical cohesion. In the case of aqueous emulsions (latex), the evaporation of water results in the formation of a three-dimensional polymer network around the particles. The binder is the main constituent of the suspension which will determine the viscosity and rheological behavior of the system. In general, it increases viscosity and confers a shear-thinning behavior by orienting the molecules subjected to the shear rate imposed during casting. This behavior will favor the spreading on the support while limiting the sedimentation of the particles at rest during storage and evaporation immediately after the casting. The main binders in organic systems are vinyl butyrals, like polyvinyl butyral (PVB) and acrylic resins, like polymethyl methacrylate (PMMA). Acrylic latexes, cellulose derivatives and polyvinyl alcohols (PVA) are used in aqueous medium.

5.4.3.2.5. Plasticizer

Most binders require the addition of plasticizers to increase flexibility and to allow the easy handling of dry tapes during the subsequent stages of the process (cutting, stacking, lamination). These plasticizers are polymers with low molar mass that will reduce the glass transition temperature (Tg) of the binder. The addition of plasticizers increases the plasticity but decreases the rupture strength. The commonly used plasticizers are glycols, like polyethylene glycol (PEG) with low molar mass (typically 300) and phthalates, like dibutylphthalate (DBP). In aqueous medium, acrylic latexes with high Tg, ensuring cohesion, can be mixed with latexes with low Tg which provide flexibility.

Examples of alumina organic and aqueous tape casting formulations are given in Table 5.3. Let us conclude by saying that UV reagents photopolymerizable systems (monomer and photoinitiator), have been cast for the realization of substrates [CHA 99]. These systems have the advantage of not using any solvent and eliminate the critical stage of evaporation.

5.4.3.3. *The preparation of suspensions*

A tape casting suspension is prepared in two stages. The first (grinding) consists of deagglomerating the powder in the solvent with the help of the dispersant. The second allows the solubilization and homogenization of all the components. The preliminary deagglomeration of the powder in the solvent in the presence of the dispersant is necessary: i) to avoid adsorption competitions of the dispersant and the binder on the particle surface, which results in a state of poor dispersion and an evolution of the rheological properties of the suspension over time (aging), ii) not to deteriorate the long polymeric binder chains during grinding, and iii) to ensure an effective milling in a low viscosity system without organics. The suspension is then stabilized for one to two days and then deaerated before casting.

Constituent	Function	Organic system	Aqueous system
Al_2O_3	Ceramic	30.6	46.0
Butanone 2/Ethanol Water	Solvent	57.6	44.7
Phosphate ester Ammonium polyacrylate	Dispersant	0.8	0.5
Polyvinyl butyral Latex	Binder	4.6	8.8
Polyethylene glycol Dibutylphthalate	Plasticizer	2.9 3.5	

Table 5.3. *Examples of alumina organic [CHA 95]*
and aqueous tape casting formulations (vol.%) [DOR 98]

5.4.4. *Consolidation of concentrated suspensions*

To remedy the disadvantages of traditional shaping techniques of complex pieces such as slip casting or injection molding, the consolidation of concentrated suspensions, by destabilization or gelation in a non-porous mold, provides an innovating alternative to manufacture at a low cost in a reliable and reproducible way, near net shape pieces of complex geometry with final shapes and dimensions.

5.4.4.1. *Coagulation casting [GRA 94]*

This method consists of coagulating a concentrated and stable suspension, cast in a non-porous mold, by destabilizing it *in situ* to produce a green part with sufficient cohesion to allow its demolding. We saw in section 5.3 that the adsorption of electrolytes on the hydroxyl groups of the ceramic particle surface produces stable suspensions with pH very distant from the value of the ZPC, by creating a strong repulsive potential between the particles. The destabilization of the suspension in the mold consists of decreasing the repulsive forces so that the attractive Van der Waals forces become predominant, which causes the coagulation of the suspension. It is therefore necessary to trigger a reaction in the cast suspension which either makes the pH of the suspension tend towards the ZPC, or releases ions, thus increasing the ionic force of the environment (see Figure 5.20). For instance, ammonia can be used to release hydrolysis of urea by a rise in temperature (approximately 50°C) and to coagulate an alumina suspension (ZPC = 9) dispersed beforehand in acid environment.

5.4.4.2. *Gel casting [OMA 91]*

With regard to gelation, an organic monomer is added when the suspension is prepared. The rise in the temperature of the mold or the addition of an initiator in the environment, right before the casting, catalyzes the monomer's polymerization reaction and thus creates a rigid network around the ceramic particles. The monomer content in the suspension must be adjusted to maintain a viscosity suitable for the casting, while providing sufficient solidification. The initiator content helps control the time available to cast the suspension.

The success of these two methods lies firstly in the preparation of a very concentrated suspension (> 60 vol.%), because the granular network that results from the coagulation or gelation must be cohesive and must be formed with an isotropic shrinkage of low value (1 to 2%). The low concentrations of organic additives (1 to 3% in mass), necessary for the implementation of these processes, facilitate the debinding and sintering stages.

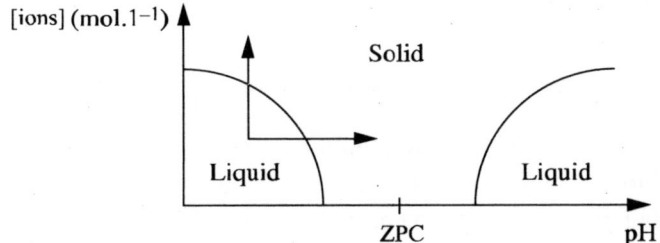

. **Figure 5.20.** *Destabilization of a suspension by modification of the pH to approach the zero point of charge (ZPC) and/or by increase of the ionic force*

5.5. Pressing

5.5.1. *Introduction*

Pressing consists of a compression of a powder or granules in a rigid matrix (uniaxial pressing) or in a flexible mold (isostatic pressing). Pressing is certainly the most widely used method for shaping ceramic pieces, because it allows the manufacture of relatively complex parts with tight dimensional tolerances and high productivity. Moreover, the later stages of drying and debinding, crucial in other shaping processes like injection and extrusion, are eliminated or at least simplified here. A broad range of ceramic pieces is produced thanks to the pressing technique: tiles, plates, refractory materials, abrasives, cutting tools, as well as various electroceramic, magnetic and dielectric pieces.

Uniaxial pressing in a metal matrix, with one (single effect) or two (double effect) pressing pistons, is used for the production of pieces whose thicknesses is higher than 0.5 mm, with a high surface/thickness ratio and exhibiting reliefs only in the pressing direction. Isostatic pressing in a flexible mold allows the manufacture of complex shapes with reliefs in three directions, as well as elongated shapes like tubes. Semi-isostatic pressing consists of a combination of a metal matrix and a flexible mold. Lastly, ceramic sheets with a thickness of about 1 mm can be obtained by rolling between rollers.

Pressing requires a good flow of the powder and a homogenous filling of the matrix or mold, in order to achieve uniform densities in a reproducible way. Spherical particles with diameter higher than 50 µm exhibit good flow capability. It is therefore necessary "to granulate" ceramic powders in the form of agglomerates, called granules, in order to confer on them the properties required for pressing.

5.5.2. *Granulation*

The granulation operation consists of increasing the diameter of the particles of a powder in order to confer on it the properties required for flow and mold filling. Several granulation processes are used in the ceramic industry [BAK 96]; they can be classified briefly in the following way:

– granulation by pressing/milling of the powder or extrusion of granules;

– granulation by agitation/evaporation of a suspension in a blade mixer;

– granulation by drying of a suspension: freeze-drying or spray-drying.

The granules must have two contradictory properties. On the one hand, they must be resistant enough to enable their handling and their transport, but on the other hand, they must be "soft" (ductile) enough to be deformed easily under the action of pressing. This is necessary in order not to introduce a macroporosity which cannot be resorbed during sintering and which will be detrimental to the properties of the parts. Such a compromise can be achieved only by using organic additives, like binders and plasticizers.

The method widely used for the production of granules with controlled density and grain size, and with good flow and compaction capability is spray-drying.

5.5.2.1. *Spray-drying*

Spray-drying consists of pulverizing a suspension, in the form of droplets, in a hot air (or inert gas) current to produce practically spherical granules (diameter of 50 to 500 μm), with a relatively smooth external surface (see Figure 5.21). Spray-drying is a continuous, economical and reproducible process.

Figure 5.21. *Granules obtained by spray-drying*

The formation of droplets is generally carried out by centrifugation using a turbine turning at high speed or by shearing the suspension injected under pressure trough a low diameter tube (see Figure 5.22). The configuration of the hot air flow will determine the drying of the droplets. If the hot air is injected co-current, the temperature of the granules remains low, thus avoiding the deterioration and/or evaporation of the organic components. In a counter-current configuration, the droplets gradually meet the hot and dry air. High temperatures can then be reached on the granule surface. Mixed flow configuration is a compromise between the two systems above. The advantage is to increase the residence time in the drying chamber whose size can be reduced.

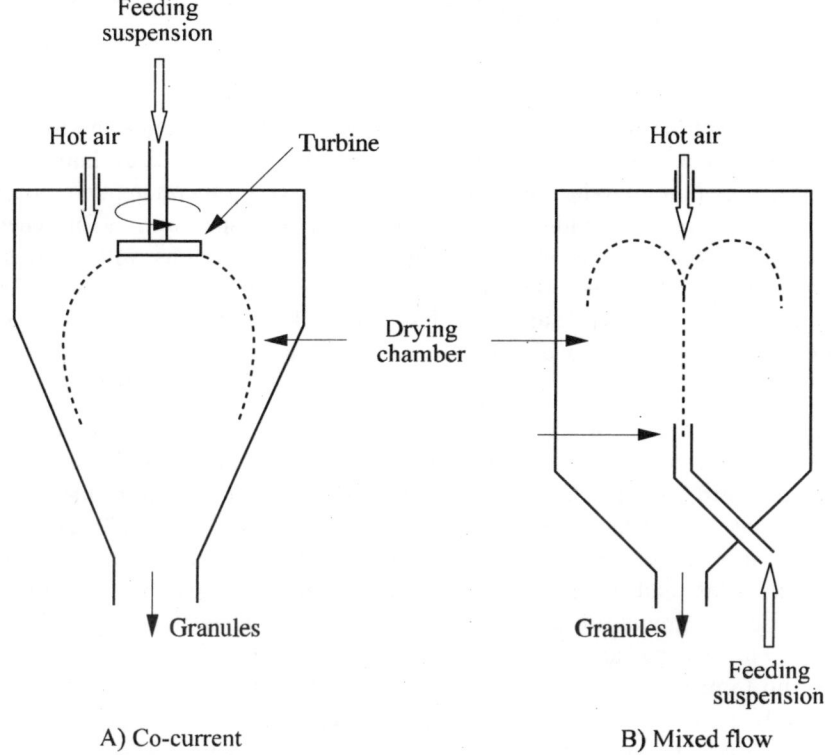

Figure 5.22. *Principles of suspension spray-drying processes:
A) co-current, B) mixed flow*

The migration of soluble species or fine particles towards the granule surface, during the evaporation of the solvent, is a well-known phenomenon [BAK 96]. The spray-drying binder (see section 5.5.2.2), which is generally water-soluble and to a large extent not adsorbed on the surface of the ceramic particles, is likely to migrate with the solvent as the droplets dry in the spray-dryer. This migration results in the formation of a shell on the granule surface, rich in polymer, whose thickness represents a few unit percent of the granule radius. The heterogenity of the binder distribution in the granules will obviously have a considerable impact on the mechanical properties of the pressed products.

5.5.2.2. *Spray-drying suspensions*

Most spray-drying suspensions are aqueous systems; however, organic solvents are sometimes used to avoid ceramic/water reactions (hydrolyzable, non-oxide materials). The granulation of powders by spray-drying implies the use of several organic additives, each fulfilling a specific function (dispersant, binder, plasticizer, lubricant, wetting agent, anti-foaming agent).

5.5.2.2.1. Dispersant

As in all ceramic shaping processes, the powder must be, before the formation stage itself, perfectly deagglomerated in order to produce homogenous microstructures in a reproducible way (see section 5.3). Furthermore, spray-drying, which consists of evaporating the solvent from a suspension, is an energy-consuming technique. It is therefore obvious that suspensions with high solid concentration must be used. Obtaining a suspension with a concentration of non-agglomerated and low viscosity particles, compatible with the spraying system requires the use of a perfectly dispersed system using a dispersant.

5.5.2.2.2. Binder

Binders must be added to clay-less ceramic compositions in order to give the pressed part sufficient mechanical strength to allow its ejection from the pressing mold, its subsequent handling and possibly machining (see Figure 5.23). The binder must also provide sufficient mechanical strength to the granules to allow their storage in silos and their transport. The binder must satisfy the following compromise: the granules must be sufficiently "soft" to be easily deformed during pressing, but also sufficiently "hard" to preserve their integrity between spray-drying and forming by pressing. The most commonly used spray-drying binders in the ceramic industry are: i) water-soluble polymers like vinyl polymers (polyvinyl alcohol (PVA), polyvinylmethylether, polyvinylpyrrolidone), acrylic polymers (polyacrylamide, polydimethylacrylamide, polyisopropylacrylamide), polyimines and polyoxides (polyethylene imine, polyethylene glycol (PEG)) [BAK 96], ii) emulsion polymers (latex), and iii) polymers of natural origin like substituted celluloses. The nature, content and molar mass of the binder used for spray-drying have a great influence on the compression behavior of the atomized powders, on the problems of sticking on the matrix and the mechanical resistance of the green part. An increase in the molar mass of the binder, in its glass transition temperature or in its concentration in the powder results in a reduction in the relative density of the pellets. The binders are generally introduced with contents ranging between 0.5 and 5% in mass on the ceramic powder mass basis.

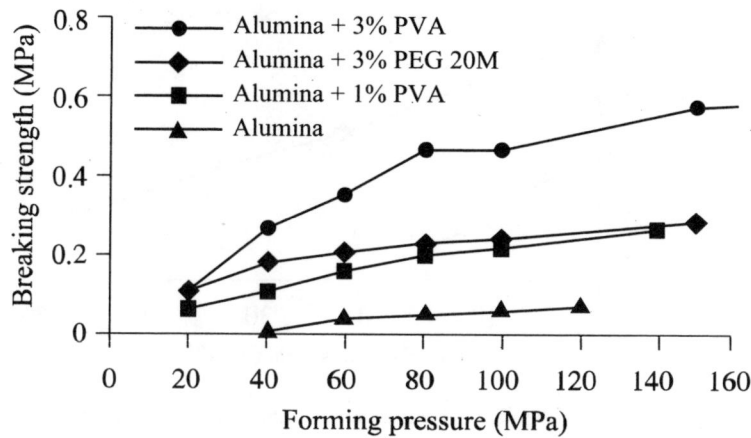

Figure 5.23. *Influence of the binder on the strength at rupture of pressed pieces*

5.5.2.2.3. Plasticizer

It is believed that the organic phase used to ensure the cohesion of the granules will exhibit an aptitude for deformation during pressing only if it is ductile. However, water-soluble polymers generally have a glass transition temperature higher than room temperature and consequently exhibit, in the solid state, a fragile behavior. Plasticizers are used to lower the glass transition temperature of the binder at least to approximately pressing temperature [REE 95]. Several types of plasticizers can be used:

– polyols like glycol, glycerol: these small-sized molecules are supposed to insert themselves into the organic matrix and thereby facilitate the relative movement of the polymeric chains. Their disadvantage is their tendency to evaporate during spray-drying;

– polyethylenes glycols (PEG): they are polymers of average molar mass ($\approx$ 500-1,500) with a lower vapor tension than simple polyols. Their presence decreases the glass transition temperature of most water-soluble binders, such as polyvinyl alcohol (see Figure 5.24);

– water is also a plasticizer of PVA (see Figure 5.24). The compression of a powder spray-dried with PVA will therefore be influenced by the relative humidity rate of the storage and working atmosphere. Water also plays the role of plasticizer in compositions containing clay (pressing of tiles, refractory materials, porcelain, etc.).

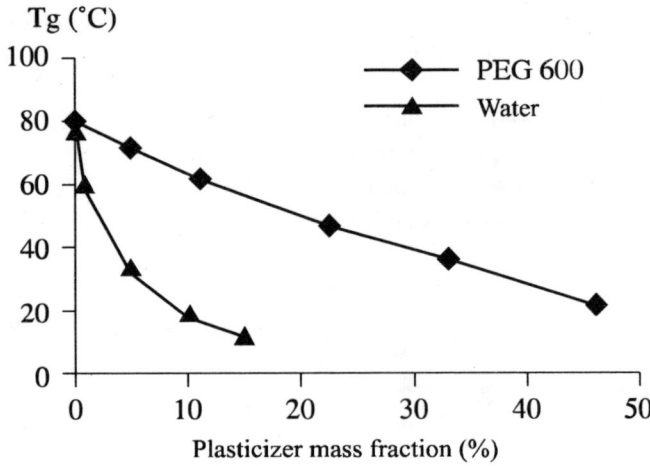

Figure 5.24. *Evolution of the Tg of PVA depending on the PEG 600 and water concentration*

5.5.2.2.4. Lubricant

The role of lubricant is to reduce piece/matrix frictions during pressing and ejection. The lubricant is introduced either directly into the spray-drying suspension, or subsequently by coating of the granules. Stearates (Zn, Mg), waxes, clays or talc are used as lubricants.

5.5.3. *Uniaxial pressing*

5.5.3.1. *Compression behavior*

It is generally believed that the sequence of stages occurring during the pressing of spray-dried powder is as follows [REE 88]:

– stage I: rearrangement of the granules;

– stage II: deformation or fragmentation of the granules, elimination of porosity between the granules (intergranular macroporosity);

– stage III: elimination of the microporosity present initially inside the granules (intragranular microporosity), by rearrangement or fragmentation of the particles.

These stages are illustrated in Figures 5.25 and 5.26.

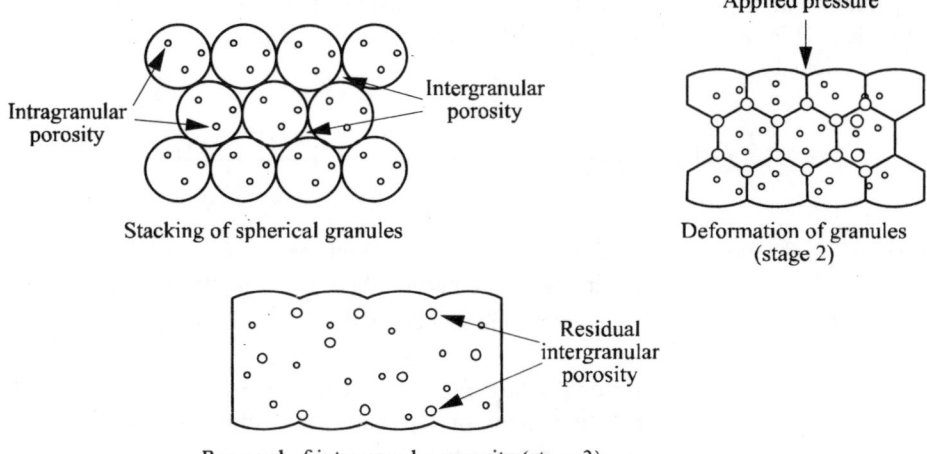

Figure 5.25. *Evolution of intergranular and intragranular porosity during the pressing of a spray-dried powder*

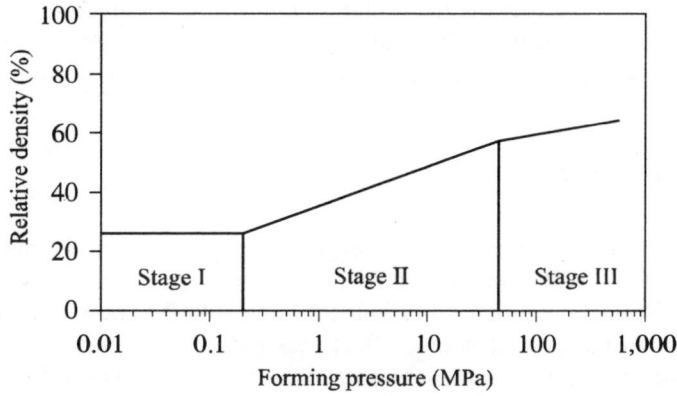

Figure 5.26. *The three stages in pressing according to Reed [REE 95]*

The variation in the relative density of the compact, under the action of the applied pressure, therefore results from two generally successive phenomena: the progressive removal of intergranular porosity, followed by a reduction in microporosity, initially present in the granules between the particles that constitute them. Many expressions describing the evolution of the stacking compactness formed by the granules depending on the applied pressure have been proposed in other works. The model put forward by Baklouti [BAK 94] takes into account the progressive hardening of the granules during the pressing generated by the appearance of triple points that block the extension of the contact zones between granules. This model correctly predicts the evolution of relative density Γ up to a relative density of about 87% [BAK 94]:

$$P = H[\frac{2}{3} f_d(\Gamma) \mathrm{Ln} \frac{(1-\Gamma_0)}{(1-\Gamma_0)-(\Gamma-\Gamma_0)\Gamma} + (1-f_d)\frac{\Gamma^2(\Gamma-\Gamma_0)}{1-\Gamma_0}] \qquad [5.25]$$

where P is the pressure applied, H a parameter which expresses the "hardness" of the granule, Γ_0 the relative density of the particle bed in the matrix before pressing and f_d the volume fraction of the zones hardened by the presence of triple points.

On the other hand, no model takes into account the microstructural hardening which results from the rearrangement of the particles within the granules and contributes to the reduction of microporosity.

5.5.3.2. Wall effects

During uniaxial pressing, a fraction of the applied load is transmitted to the walls of the matrix. The powder/matrix frictional forces lead to pressure gradients and therefore density gradients within the green part. The average axial pressure P_h transmitted at a depth h of the piece can be expressed according to the pressure applied P by:

$$P_h = P \exp-(f\sigma_{r/a} S_f / S_p) \qquad [5.26]$$

where f is the powder/matrix friction coefficient, $\sigma_{r/a}$ the ratio of the radial stress to axial stress and S_f/S_p the ratio of the friction surface to the pressing surface. In the case of single effect pressing of a cylindrical piece with thickness e and diameter d, $S_f/S_p = 4e/d$. In the case of double effect pressing, the pressing surface is double ($S_f/S_p = 2e/d$) and the minimum pressure is located at mid-thickness. Equation [5.26] highlights that, the longer and narrower a part is along the pressing direction (high e/d), the less dense it will be. Uniaxial pressing is consequently restricted to the realization of pieces with a high surface/thickness ratio. The use of a lubricant, either on the matrix walls, in the spray-drying formulation or by coating of the granules, will provide a better transmission of the pressure by reducing the terms f and $\sigma_{r/a}$. The choice of the material and the roughness of the surfaces in contact with the powder are also parameters that must be taken into account.

Figure 5.27 schematizes the pressure profiles within a compressed piece starting from a non-granulated powder, for which particle mobility is low. The density gradients that result will therefore lead to differential shrinkages during sintering ("diabolo" effect in single effect pressing).

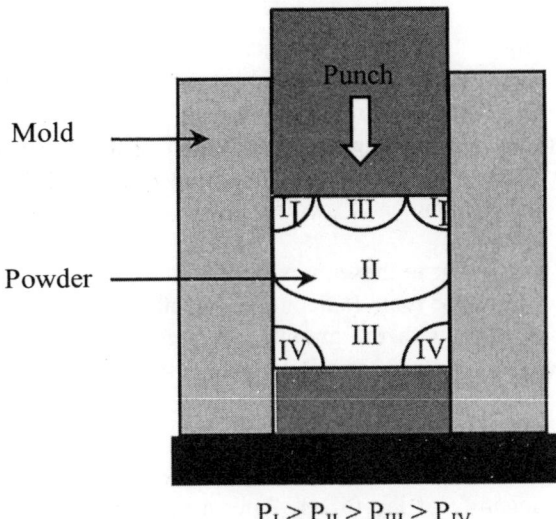

$$P_I > P_{II} > P_{III} > P_{IV}$$

Figure 5.27. *Pressure profiles within a compressed piece starting from a non-granulated powder [REE 88]*

5.5.3.3. Springback

The elastic energy stored in the green part during compression leads to an expansion of the piece, called springback, at the time of ejection. The springback increases with the pressure and is higher for a pressing temperature below the Tg of the organic phase. The addition of plasticizers will therefore reduce the rebound.

5.5.3.4. Flaws

The main flaws encountered during pressing are the lamination perpendicular to the pressing direction and the rising of the upper part of the piece in contact with the piston (end capping). They are mainly due to the pressure gradients within the piece, the differential springback between the part outside the matrix and the part still stressed by the matrix during ejection, and the piece/matrix friction. The solutions consist of: i) increasing the ductile behavior of the granules (plasticizer), ii) reducing the piece/matrix friction (lubricant) and iii) increasing the mechanical strength of the green part (binder).

5.5.4. *Isostatic pressing*

Isostatic pressing is used for the production of pieces that are difficult to obtain by uniaxial pressing: elongated pieces (tubes), with complex shapes and/or large volume (spark plugs, cast metal refractory tubes). This pressing method offers the advantage of leading to a homogenous distribution of the pressure inside the piece and thus is also used to produce pieces requiring a high and very uniform green density (bearing and grinding balls, medical prostheses).

The granules are similar to those prepared for uniaxial pressing, but are generally more ductile. A deformable mold (silicone, polyurethane), with the shape of the part to be produced, is filled with the granules. A pressure of about 150-200 MPa is applied to this flexible envelope via a fluid, generally oil.

5.5.4.1. *Wet mold pressing (see Figure 5.28a)*

The filling of the mold is carried out initially. The use of a vibratory table improves the homogenity of the filling. After a vacuum deaeration, the molds are immersed in the fluid contained in the compression chamber and are pressed. This process is reserved for bulky pieces and those with complex shapes.

5.5.4.2. *Dry mold pressing (see Figure 5.28b)*

The mold is filled here directly in the press. The pressure is applied only radially by a fluid between the deformable mold and a rigid carcass. This process allows a higher output rate than wet mold pressing and is used for the manufacture of small pieces.

5.5.5. *Semi-isostatic pressing*

Semi-isostatic pressing is a combination of isostatic and uniaxial pressing (see Figure 5.29). It is widely used for the production of flat pieces like plates and dishes which have a high diameter/thickness ratio with thicknesses lower than 5 mm. A metal punch applies a uniaxial pressure to the simplest surface of the part (hollow part of the plate) and a flexible membrane applies, using a fluid, an isostatic pressure to the other surface of more complex shape. The filling, pressing and extraction carried out simultaneously on revolving tables allow high rates of output. The main advantages compared to shaping by plastic deformation (see section 5.6) are the elimination of drying and great dimensional accuracy.

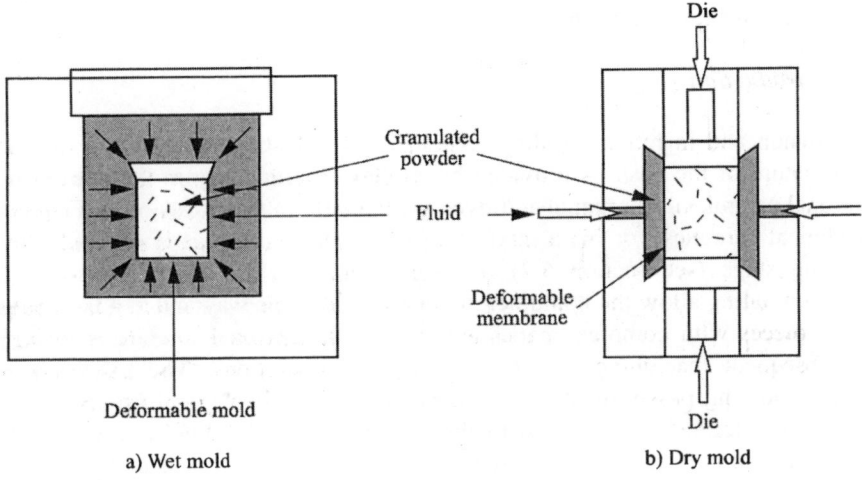

Figure 5.28. *Principles of isostatic pressing: a) wet mold, b) dry mold*

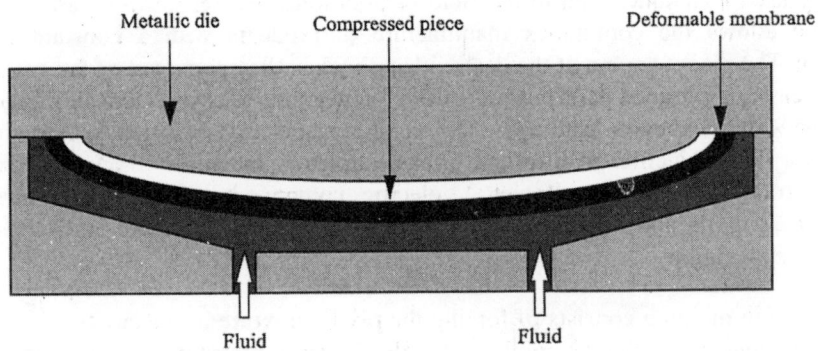

Figure 5.29. *Principle of semi-isostatic pressing*

5.5.6. *Roller compressing*

The granulated powder can be continuously compressed between rollers for the production of sheets with thickness of about 1 mm.

5.6. Extrusion-injection molding

5.6.1. *Introduction*

Extrusion and injection molding techniques are plastic shaping processes. The plastic nature of the paste is provided by: i) clays (particularly in the extrusion of traditional ceramics), ii) organic additives (particularly in the extrusion and injection of technical ceramics), or iii) a mixture of both. The green part is sintered after a debinding stage (see section 5.7) if organic additives are used. Extrusion and injection molding allow the manufacture, in a reproducible way and at a high output rate, of pieces with complex shapes and narrow dimensional tolerances requiring few subsequent machining and surface-finishing operations. The extrusion and injection molding pastes must exhibit flow properties (rheological properties) that are perfectly defined and adapted to the process used. This behavior will in fact govern the quality of the piece (microstructural homogenity, presence of defects), as well as the dimensional variations (shrinkage) from shaping to sintering.

Extrusion consists of forcing the plastic mixture through a die. Extrusion is a widely used technique, both in the field of traditional and technical ceramics. This method allows the continuous manufacture of products with a constant cross-section. The pieces are cut at the desired length when they come out of the extruder. The weight of extruded parts can vary from a few grams to several tens of kilograms and the wall thicknesses from a few to several centimeters. For instance, extrusion is used for the production of tiles and bricks, refractory materials (bricks, protective tubes, rotating furnace walls, etc.), electric components (insulators, resistors), catalyst supports and heat exchangers with a honeycomb structure and tubes for various applications.

Injection molding consists of forcing the plastic mixture in the cavity of a mold. The technique of injecting ceramic materials is a more recent development than the extrusion process. Based on the technological knowledge in the field of plastics, the mass production of ceramic pieces by injection molding became widespread first in the field of textiles (thread guides) and electronics (packages) before being used for the design of thermomechanical ceramics (valves, rocker arm shoes) and refractory materials (nozzles). In the field of traditional ceramics, however, the attempts towards the industrialization of this manufacturing technique have been hampered until now by the additional costs for equipment and organic additives, compared to the traditional methods of casting and pressing. The pieces produced are generally limited in size and have a thickness lower than 1 cm, with the notable exception of refractory parts that can have a thickness of 2 to 3 cm.

These two shaping techniques using plastic pastes bring into play the following main stages: the choice of components (ceramic phase, organic phase) leading to the desired properties (flow, cohesion of the green part, debinding, sintering), the preparation of a homogenous mixture, the shaping by extrusion or injection molding, the drying and/or the debinding.

5.6.2. *The choice of components*

Extrusion and injection molding mixtures consist of a mineral phase and an aqueous and/or organic phase.

The ceramic phase is made up of one or more ceramic powders, characterized by the shape and the size distribution of their particles, as well as by the chemical nature of the surface of these particles. The choice of these characteristics must lead, during shaping, to the most compact stacking possible of individualized particles and to a sufficient cohesion of the green part in order to minimize the quantity of water or binder to be extracted and the associated risks of deformation. We must keep in mind here the desired reactivity of the powder to subsequent sintering. A ceramic phase adapted to extrusion or injection molding therefore ideally consists of a population of fine particles (with a size close to one micron) and other larger grain size populations, in well defined proportions [MUT 95].

The aqueous or organic phase must confer on the mixture a suitable rheological behavior and allow the formation of a dense and homogenous arrangement of the ceramic load during shaping. Four main types of additives are used.

5.6.2.1. *Dispersant*

Its role and dispersion mechanisms were described in detail in section 5.3. The dispersant decreases the viscosity of the mixture and increases the volume concentration of ceramic particles.

5.6.2.2. *Binder*

It ensures the cohesion of the green part. It is the binder that generally imposes the rheological behavior to the mixture. In the case of traditional ceramics, clay plays the role of binder, whereas polymers are introduced into the compositions of technical ceramics. The organic binder is generally a mixture of a major and a minor binder. The major binder is a polymer with high molecular weight. It ensures the cohesion of the green part during the shaping and debinding, but gives a high viscosity to the mixture. It is therefore necessary to lower the viscosity while maintaining acceptable mechanical properties. This is the role of the minor, polymeric binder with lower molecular weight, which, on the one hand, supports the flow of the major binder by inserting itself between the long chains and, on the other hand, facilitates its elimination during debinding by opening the porosity before the departure of the major binder.

5.6.2.3. *Plasticizer*

Plasticizer is used to modify and adapt the intrinsic rheological behavior of the binder. Water is the plasticizer of clays. Organic molecules with low molar weight decrease the glass transition temperature of the polymers and make them more ductile at the shaping temperature.

5.6.2.4. *Lubricant*

Lubricant minimizes frictions between the mixture and the shaping ools. Generally oils, paraffins or stearates are used as lubricants.

Examples of binders, plasticizers and dispersants for extrusion and injection are indicated in Tables 5.4 and 5.5, respectively.

The final choice of this organic phase will be completed after having characterized the rheological behavior of the mixture, for example by using a capillary rheometer (see section 5.3). The rheological behavior of the paste must allow the plastic flow through the extrusion die to obtain the desired form. In the case of injection molding, it must on the one hand ensure the plastic flow through the injection orifice of low cross-section and on the other hand, ensure the filling of the injection mold. The rheological behavior must also confer sufficient mechanical strength on the extruded or injected green part to avoid its deformation and make its handling possible. A shear-thinning behavior with a flow yield stress meets these conditions. Such a behavior allows the paste:

i) to be viscous during the mixing stage (moderate speed rate), which ensures a good dispersion under the effect of high shear forces;

ii) to be more fluid during the extrusion and injection stage (high shear rate) to ensure a satisfactory flow and filling of the mold;

iii) to regain a high viscosity after shaping (zero shear rate).

Examples of the rheological behaviors of extrusion and injection molding mixtures are given in Figures 5.30 and 5.31, respectively.

5.6.3. *Mixing*

This stage must produce a homogenous mixture, without agglomerates and with a high volume concentration of ceramic particles. Mixing has a decisive influence on the microstructure of the green part and the possible presence of flaws (cracks, residual stresses) that could not be resorbed during sintering [SPU 97]. In the case when generally thermofusible organic additives are used, the organic phase is melted prior to the mixture. This liquid phase then coats the ceramic particles and gradually forms a viscous mixture which becomes homogenous under the influence of the high shear stresses imposed by the mixer. The water physically absorbed on the surface of the ceramic particles must be previously eliminated in order to facilitate their homogenization in the molten hydrophobic organics.

Aqueous systems	Binder	Polyethylene glycol with high molar weight, cellulose derivatives (methyl cellulose, ethyl cellulose, hydroxyethyl cellulose), polyvinyl alcohol
	Plasticizer	Glycols with low molar weight, polyethylene oxide, traditional pastes: water
	Dispersant	Polyelectrolytes, sodium carbonate, sodium silicate, stearates
Non-aqueous systems	Binder	Polyethylene polypropylene, polystyrene, waxes
	Plasticizer	Vegetable waxes, beeswax, dibutyl phthalate
	Dispersant	Fatty acid, octadecanoic acid, stearic acid, oleic acid, fatty amines

Table 5.4. *Examples of additives for extrusion shaping (aqueous and non-aqueous systems)*

Binder	High pressure	Polyethylene, polypropylene, polystyrene, polyvinyl alcohol, polyethylene glycol, ethyl vinyl acetate, polyacetal, paraffins, microcrystalline waxes
	Low pressure	Paraffins, microcrystalline waxes, ethyl vinyl acetate, emulsified waxes.
Plasticizer		Vegetable waxes, beeswax, dibutyl phthalate
Dispersant		Fatty acids, octadecanoic acid, stearic acid, oleic acid, fatty amines

Table 5.5. *Examples of additives for injection molding shaping*

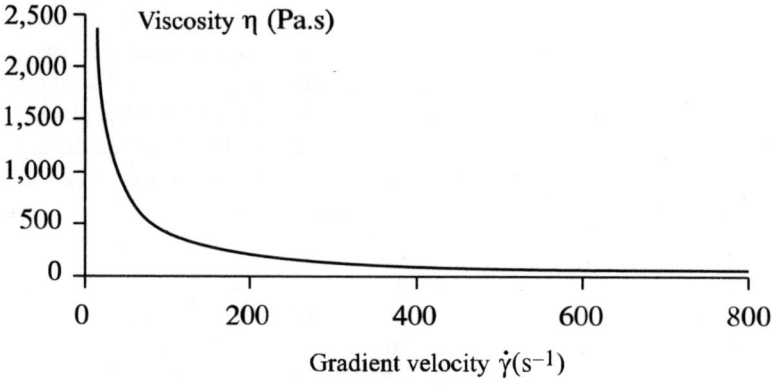

Figure 5.30. *Rheogram of an extrusion paste (clay + water)*

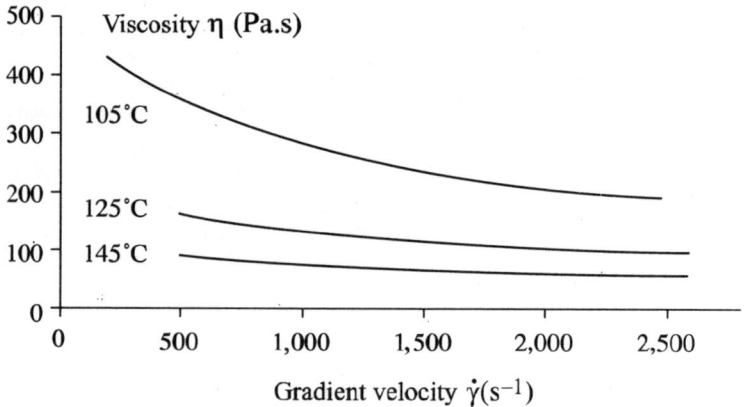

Figure 5.31. *Rheograms of an injection paste at various temperatures*
(Si_3N_4 system + functional polyolefins)

Three types of mixers are particularly used (see Figure 5.32): Z-blade mixers, roller mixers (in which the mixture is forced between two heating rollers, with variable spacing and turning at different speeds) and double screw mixers, mainly used in extrusion techniques.

Mixing can be done under vacuum in order to remove air bubbles. The main experimental parameters are the sequence in which the constituents are added, the energy supplied by the mixer and the residence time as well as the possible precoating of the ceramic powder by the dispersant.

A)

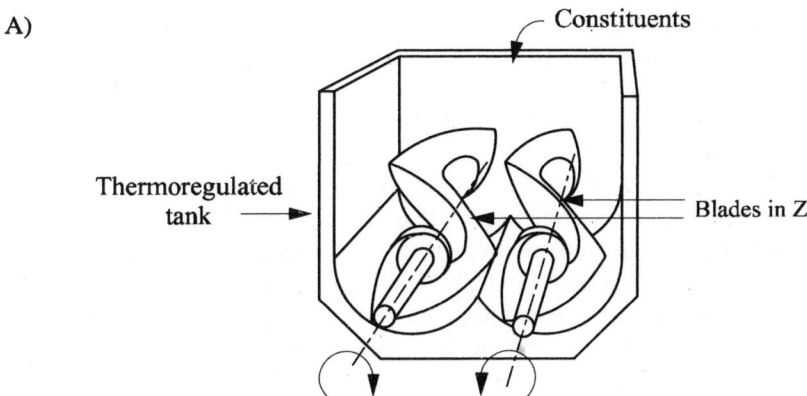

B)

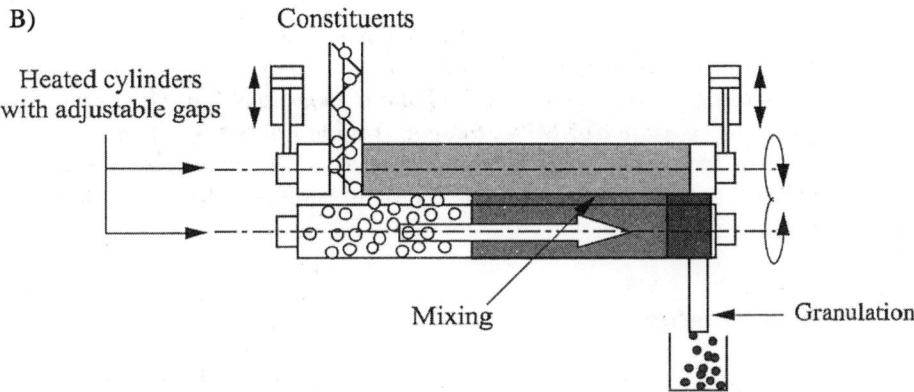

C)

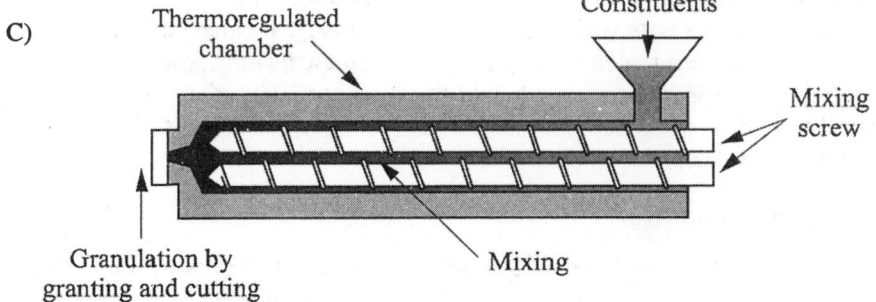

Figure 5.32. *The main types of mixers: A) Z-blade mixer, B) roller mixer,
C) double screw mixer*

Once the formulation and mixing stages are mastered, it is possible to effectively increase the volume of powder contained in the injection molding paste [WRI 90]. In practice, the volume concentrations of particles are about 40 to 50%, even 60% in the case, for example, of the injection molding of technical ceramics.

The mixtures obtained at the end of the mixing are generally granulated by extrusion through a grid or by crushing. The granules thus obtained are easy to handle and will be used to feed extrusion and injection molding machines using hoppers. In the case of very fluid mixtures (low pressure injection molding process for instance), it is sometimes preferable to maintain the mixture in temperature in a fluid state and to pour it into the tank of the injection molding machine. Finally, we may note that the mixing stage favors the introduction of impurities caused by the abrasion of the metallic parts. The use of lubricants (petroleum jelly oils, olein, stearates) in the formulation reduces this pollution.

A particular technique of mixture preparation is also used in the industry of traditional ceramics: filter-pressing. The various components (raw materials, dispersants, water) are mixed to form a fluid and homogenous slip. This slip is then filtered, under low pressure (0.5 MPa), through flexible filters aligned on a frame to eliminate a large fraction of water and soluble species. The plastic paste obtained, with a small percentage of residual humidity, then exhibits a rheological behavior suitable for extrusion.

5.6.4. *Shaping the mixture*

Extrusion and injection molding require a homogenous mixture obtained by plasticization right before the forming. If an organic phase is used, plasticization is the passage from the solid state (granules resulting from the mixing) to a homogenous viscous state with a viscosity and rheological behavior adapted to the processes. The temperature range is located between the melting point of the organic phase and the temperature at the start of the decomposition of the polymers. This temperature range is sometimes narrow and a strict control of the mixture's temperature is necessary.

5.6.4.1. *Extrusion molding*

The extrusion stage consists of pushing the paste that has been plasticized and previously deaerated through a given geometric die, using a piston or a screw.

Piston extruders (see Figure 5.33) work in discontinuous mode and are primarily used on a laboratory scale. The paste is introduced into the chamber, which is then closed again. This chamber can be connected to a vacuum pump to ensure the deaeration of the mixture and heated if necessary. In the case of traditional compositions containing clay, or if a binder in aqueous phase is used (cellulose derivatives), the extrusion is carried out at room temperature and the cohesion of the pieces is ensured by the high viscosity of the paste. When a thermoplastic binder is used (case of technical ceramics), the paste is heated in the chamber during the deaeration/compression phase then extruded. The cohesion of the piece is then ensured by the solidification of the organic phase during cooling. The piston extruder enables extrusion at high pressures and has the advantage of limiting the contacts between the paste and the tools, thus decreasing the risks of pollution.

Screw extruders (see Figure 5.33) working in continuous mode are used in production. They can treat substantial loads going up to 100 tons per hour. These extruders generally comprise a first feeder chamber into which the paste is introduced in the form of granules, then mixed and deaerated. This mixture is then transported and compressed by one or two feeder screws before being finally extruded through the die. The feeder chamber and the compression screws are maintained at room temperature for traditional and cellulose pastes or heated for thermofusible pastes. The extrusion pressures generally range between 4 and 15 MPa. They result from a balance of the pressures and flow-rates between the flows in the screw/cylinder air-gap on the one hand, and the extrusion die on the other. The frictions generated between the paste and the metal walls of the screw extruders are considerable and can results in the fast wear of the screws as well as a pollution of the paste.

The extrusion of hollow profiles is made possible by the use of dies comprising of a central core attached to the die by two or more bridges, called spiders (see Figure 5.34) [REE 95]. The paste is separated as it passes through these bridges and is divided into several flows through the ports. It must then join downstream from the spider under the effect of the compression imposed from the die. After extrusion, the pieces leaving the die are cut at the desired length and then placed on suitable supports during the drying and/or debinding stages.

5.6.4.2. *Injection molding*

Injection molding consists of filling a mold whose shape is that of the piece to be manufactured by introducing into it under pressure the mixture, called feedstock, previously plasticized. This mixture, generally thermofusible, is heated in a chamber and then forced through a low diameter tube into the mold, whose temperature is lower than the melting point of the mixture. After solidification, the piece is ejected from the mold.

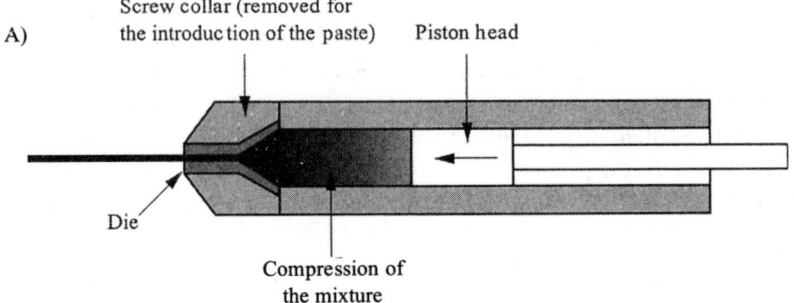

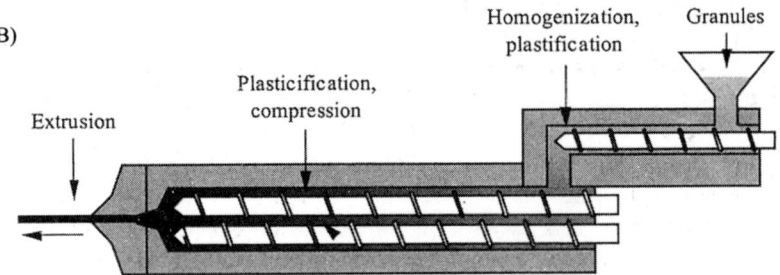

Figure 5.33. *The two main types of extruders A) piston, B) screw*

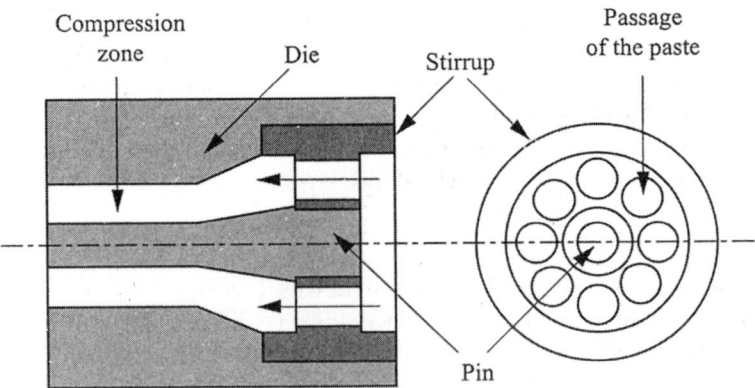

Figure 5.34. *Example of a die for the extrusion of hollow tubes*

The value of the pressure determines two injection techniques: high pressure and low pressure injection molding. The high pressure injection molding process [GER 90], which is used most often, is directly inspired from the technology of injection molding of plastics. Thermoplastic polymers with high melting point confer on the paste a high viscosity, and high pressures are necessary (between 50 and 300 MPa). The pressure is applied mostly by a screw/piston system (see Figure 5.35). The screw, with a suitable profile (pitch, thread height, compression ratio), ensures the plasticization of the mixture (mainly by friction), its transport from the feed hopper to the screw head and its compression at the screw head. This compression makes the screw move back. When the mixture volume necessary for the injection is plasticized and compressed at the screw head, the rotation is stopped and the screw then acts as a piston to inject the mixture into the mold through the feed nozzle. The injection pressure and the imposed injection flow-rate result in a very high shear rate of the paste in the feed nozzle ($1,000 \ s^{-1}$). Immediately after the injection phase, the screw maintains a residual pressure often amounting to 50% of the value of the injection pressure (Figure 5.36). In fact, the pressure decreases quickly in the mold because of the shrinkage of the mixture on cooling. It is therefore necessary to apply a residual pressure during cooling in order to provide the additional matter for filling the mold. The temperature of the mixture, higher than the melting point of the binder at the time of injection, is about 100 to 220°C depending on the polymers used. The mold is generally heated to 60-80°C in order to avoid a too rapid cooling of the external envelope of the injection molded parts in contact with the metal walls.

Because of the temperature-sensitivity of polymers and in order to avoid their premature degradation, the injection molding technique requires a strict control of the pressure and temperature cycle. In addition, the abrasive nature of ceramics calls for the use of abrasion-resistant materials (surface treatment of the injection screws, cylinders and molds) in order to limit the pollution of the ceramic composition. The cost of the high pressure injection equipment is therefore relatively high and remains a major disadvantage of this technique.

Low pressure injection molding (medium or low) calls for the use of binder families different from those used for high pressure injection molding. These binders allow the manufacture of ceramic pieces under less restrictive pressure and temperature conditions. The organic phase mainly consists of paraffins, waxes and other binders with low melting point and low viscosity, such as ethyl-vinyl-acetate (EVA) [KOS 97]. In the case of medium pressure, the parts are obtained by injection molding under a pressure from 1 to 5 MPa applied by a piston (see Figure 5.37). In the case of low pressure (0.6 to 0.8 MPa), the mixture is forced into the mold by compressed air (see Figure 5.37b). In both cases, the injection temperature does not exceed 130°C and the mold is, or is not, controlled in temperature.

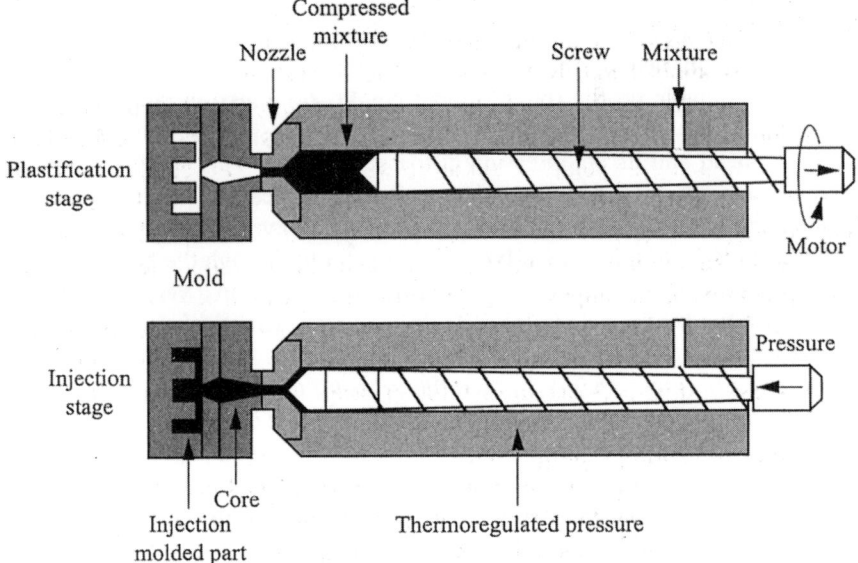

Figure 5.35. *Principle of high pressure injection molding*

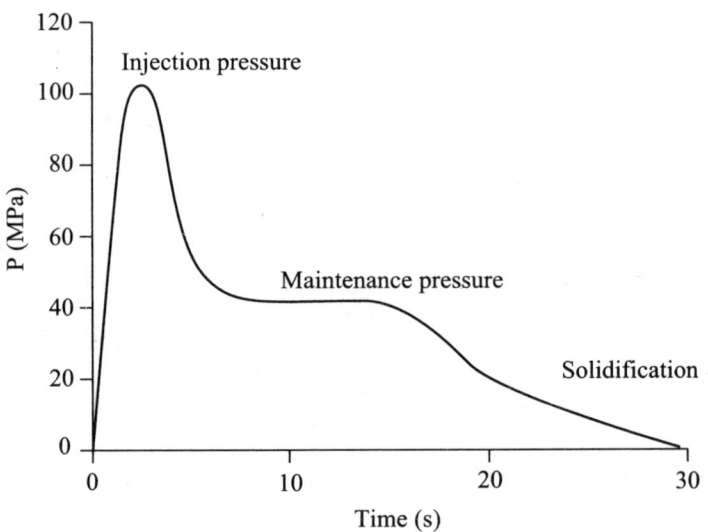

Figure 5.36. *Typical evolution of the pressure in the injection mold during a manufacturing cycle of a piece*

A)

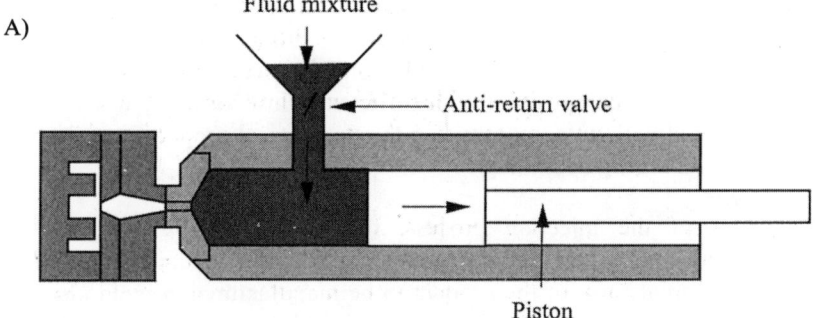

1) Suction of the mixture by pulling back the piston

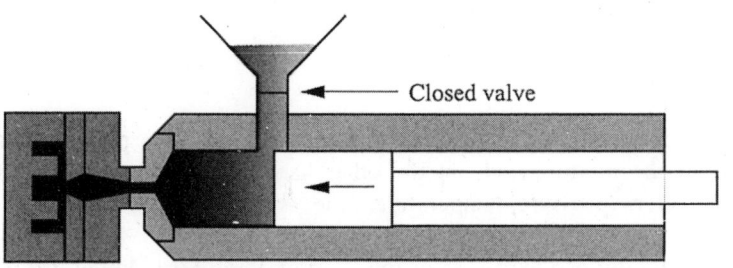

2) Compression and injection by pushing forward the piston

B)

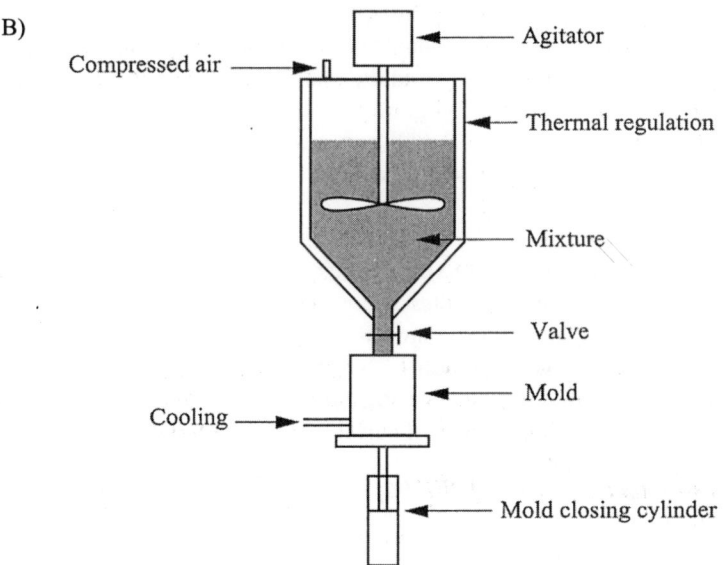

Figure 5.37. *Operating principle of low pressure injection molding machines:*
A) medium pressure, B) low pressure

This process uses a much less expensive material than high pressure injection molding, because the temperature and pressure conditions are lower and the stresses imposed on the material are less considerable. Its main disadvantage lies in its capacity to develop a composition with sufficiently low viscosity to ensure a good filling of the mold without compromising the mechanical properties of the injection molded piece and during debinding.

Regardless of the injection process, the molds can be complex systems comprising many parts and moving pieces. In addition to the parts of the mold defining the external form of the product to be manufactured, a mold also includes ejectors to separate and evacuate the injection molded product as well as the cores used to obtain relatively complex hollow forms.

5.6.5. *Common flaws*

The most common flaws are due, on the one hand, to a poor definition and preparation of the mixture and, on the other hand, to an improper choice of the shaping parameters (pressure, temperature, time, shape of the tools, etc.).

An incorrectly defined extrusion or injection molding formulation, for instance a poor wetting of the powder by the organic phase, can result in a powder-binder separation during the plasticization stage, in a migration of the binder inside the part during extrusion in the die or during the filling of the injection mold, or in an insufficient cohesion of the formed product. Powder-binder separation gives rise to excessive frictions between the mixture and the metal walls, even the blocking of the extruders and injection machines. Migration inside the part will result in the formation of areas rich in binders that will generate stresses during cooling. These stresses could relax during drying and/or debinding, causing a cracking and/or a deformation of the piece.

An unsatisfactory sequence of mixing results in the presence of agglomerates which will lead to differential shrinkages and will cause the appearance of residual stresses. An insufficient deaeration of the mixture during its plasticization and its compression gives rise to the formation of large pores within the piece. Lastly, a poor surface quality of the extrusion or injection screws, the extrusion dies and injection molds result in the appearance of surface of flaws on the shaped products.

5.6.5.1. *Flaws specific to extrusion [REE 95]*

These flaws are associated with the extrusion parameters (temperature, pressure) and the geometry of the extrusion die. Temperature of course has an influence on the consistency of the paste, in particular in the case of thermofusible systems. An excessively low temperature will not make the extrusion of excessively viscous pastes possible, whereas an excessively high temperature will result in the bonding of the paste to the shaping tools or the immediate warpage of the extruded profile. Pressure influences the cohesion of the extruded part and the possible presence of voids due to insufficient compression. It also influences the elastic rebound at the die outlet. The die, perfectly centered on the extruder head, must have a well defined geometry and must allow a matter flow at a constant speed in any point. If not, the extruded profile will exhibit deformations and surface appearances defects. The risk of deformation and depression can become significant for hollow profiles with thin walls. Lastly, a poor bonding of the paste after passing through the stirrups leads to the presence of a crack all along the piece.

5.6.5.2. *Specific defects to injection molding [ZHA 89]*

These result from a poor definition of the injection molding parameters (injection pressure, temperature of the mixture, mold and injection screw) [TSE 98] and from a poor design of the molds. The mixture can then, for instance, solidify upstream from the mold before completely filling it, or else the injection front can remain apparent on the piece if the filling is too slow compared to the cooling of the mixture in the mold. The application of too high a pressure, associated with a high temperature, can lead to a migration of the organic phase within the piece or an intrusion of the mixture in the mold weld surfaces. A poor geometry of the mold or a poor positioning of the injection entrances will also compromise the uniform filling of the mold. This can therefore result in a paste flow in the form of a wire whose pattern will remain visible on the surface of the pieces if the compression is insufficient. An insufficient de-airing of the mixture or the lack of vents on the mold for expelling the air during filling can lead to the presence of air bubbles in the part.

5.6.6. *Other plastic shaping techniques of traditional ceramics*

The rheology of plastic pastes for traditional ceramics (containing clay) allows their shaping by pressing between two fixed molds (plastic pressing) or between a fixed and a rotary mold (roller jiggering). These techniques advantageously replace shaping by casting the suspensions in porous mold (see section 5.4). The plastic paste pressing is used for the manufacture of large, non-circular and deep shapes, whereas roller jiggering is more suitable for the manufacture of circular parts with thin walls (less than 5 mm in thickness), like plates.

In these two processes, the paste previously homogenized and deaerated by extrusion is placed on a mold (metallic or porous). In the case of pressing, the paste flows between the two parts of the mold during applying pressure until the desired thickness is obtained. The part is then separated, generally pneumatically from the mold. In the case of roller jiggering, it is the rotary tool that plastically deforms the paste, which is rolled between the mold (plaster) and the shaping tool (metallic). After drying, the piece is separated from the mold.

The defects observed with these shaping techniques are similar to those associated with extrusion (section 5.6.5.1). We can add to these other flaws such as the tearing off of the paste and of surface appearance in the roller technique, if the speed of the rotating part of the mold or its surface quality are not suitable, or its lubrication has not been done correctly. Lastly, the part can be subjected to a differential drying between the core and its surface during the shaping stage, which can result in the appearance of cracks.

5.7. Extraction of organic shaping additives

5.7.1. *Introduction*

The elimination of organic shaping additives, called debinding, is one of the most critical stages in ceramic processes. In fact, if it is not controlled perfectly, this phase in the process generates flaws in the green part that will be detrimental to the final properties. It can lead to internal stresses, the appearance of cracks, grain displacements, as well as the formation of carbonaceous residues. Debinding is all the more delicate to carry out the thicker the walls of ceramic parts become, or as the parts exhibit important variations in section. In this last case, the thin walls are debinded, and consequently weakened, more quickly than the more massive parts.

Today, the most commonly used debinding technique is thermal debinding, which is based on the pyrolytic degradation of organic additives.

5.7.2. *Thermal debinding*

Any organic species exhibits a kinetics of transformation and specific degradation depending on the temperature, the atmosphere and the nature of the ceramic powder. Thermal debinding is therefore performed by placing the parts in a furnace and by subjecting them to a given thermal cycle. The debinding can be carried out in air or a particular atmosphere (neutral or oxidizing).

The rise in temperature results in the softening and, as the case may be, melting of the binders, followed by the degradation of the molecules into volatile species which are eliminated by transport in gas phase. Thermogravimetric measurements (see Figure 5.38) of the mass loss of a sample subjected to a thermal cycle bring to light the critical phases in the degradation process. The parameters of the thermal cycle (heating rates temperatures and duration of the levels) must therefore be determined depending on these transformations related to physical and chemical mechanisms.

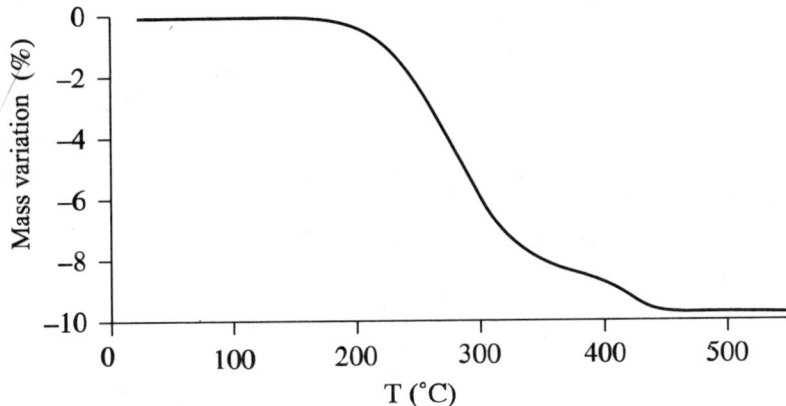

Figure 5.38. *Example of a thermogravimetric curve of a green piece containing 9.8wt.% in mass of paraffin (melting point = 52°C)*

5.7.2.1. *Physical mechanisms*

During the rise in temperature of the thermal cycle, as the organic compounds in the pores of the green part become liquid, they are subjected to the action of capillary forces. This capillary migration of the organic compounds, from the large diameter pores towards the low diameter pores, will lead to a redistribution of the organic phase within the porosity.

When the debinding temperature reaches higher values, generally beyond 180°C, the polymeric chains of the binders are subjected to chemical degradation reactions. To the displacement of the binder in the liquid state is added the diffusion of the gas species within the porosity of the sample. The volatile species formed with low molecular mass (H_2O, CO, CO_2, CH_4, light unsaturated hydrocarbons, etc.) are eliminated by diffusion and surface evaporation.

The flaws that are likely to be created by all these physical mechanisms are varied. At the beginning of the debinding cycle, when the binders are still in a solid state, the residual stresses resulting from shaping (for example, during the non-homogenous cooling of injection molded parts) can be relaxed and result in the deformation of the piece or the formation of cracks. The organic binder molecules can also be reorientated leading to a stress relaxation. The softening of the organic additives and the redistribution of the liquid organic phase within the porosity bring about the deformation of the part. This capillary migration can also displace fine particles.

5.7.2.2. Chemical mechanisms

The degradation of organic additives occurs at temperatures ranging between 180 and 600°C, depending on the nature of the organic compounds and the ceramic powder, as well as the atmosphere. Three main reactions are observed: random rupture of polymeric chains, depolymerization and elimination of radical fragments. The random scission of the chains produces gas and liquid fragments. The radicals formed in this way then tend towards a thermodynamically more stable state, by elimination of a monomer (depolymerization). This reaction takes place for example in the case of degradation of polystyrene:

$$(-CH_2-CH(C_6H_5)-)_n \longrightarrow (-CH_2-CH(C_6H_5)-)_{n-1} + (CH_2=CH(C_6H_5))   [5.27]$$

Lastly, when the lateral bonds are weaker than the internal bonds of the carbonaceous skeleton of the molecule, the scission occurs between this skeleton and the lateral groups. Vinyl polybutyral (PVB), used for tape casting, is degraded according to this mechanism (see Figure 5.39) [BAK 83].

The reaction products resulting from these three types of mechanisms have a structure generally close to that of hydrocarbons. These hydrocarbons, except for methane and ethane, are thermally unstable beyond 400°C compared to their basic elements (C, H_2) and so the thermal activation generates the rupture of C-C and C-H bonds. As the energy of the C-C bonds (345 kJ mol[-1]) is weaker than that of the C-H bonds (413 kJ mol[-1]), the rupture of C-C bonds is more frequent.

At this stage of thermal debinding, the main cause for the formation of flaws is the poor evacuation of the volatile products of degradation, which is likely to generate overpressures within the porosity and a microfissuring or even the bursting of the piece. In order to balance the quantity of volatile species produced by thermal degradation with the quantity evacuated by diffusion and evaporation, the thermal cycle must be adapted, particularly with very slow heating rates. Pinwill [PIN 92] has plotted graphs indicating the various flaws and the zones in which they occur according to the debinding kinetics (see Figure 5.40). These graphs were obtained with a mixture of alumina, polypropylene and wax. The admissible maximum heating rates between 150 and 200°C are very slow (lower than 6°C h[-1]), even for low thicknesses of 3 or 6 mm.

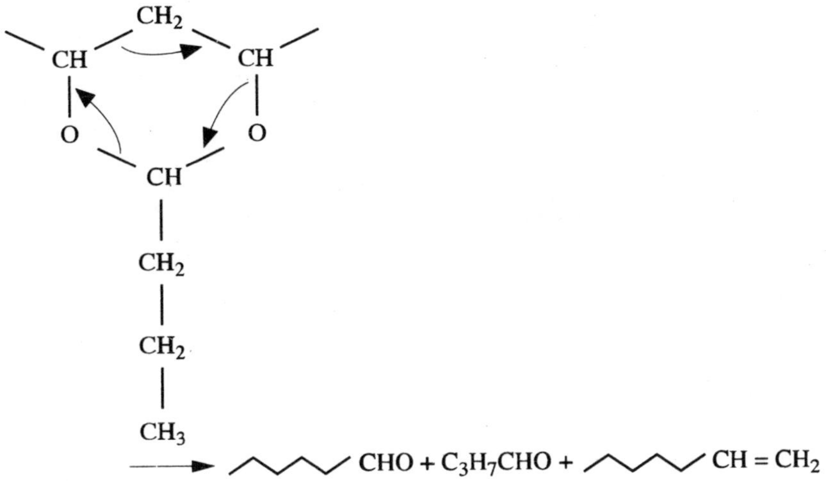

Figure 5.39. *Example of the rupture of vinyl polyvinyl butyral chains (PVB) during debinding [BAK 83]*

5.7.2.3. External factors affecting thermal debinding

The thermal degradation of organic additives is influenced by the debinding atmosphere, but also by the nature of the ceramic powder. A debinding in air, for example, causes the oxidation of polymers and increases the kinetics of degradation, but can also result in the recombination of the pyrolysis products and the formation of molecules that are more difficult to extract. The nature of the surface of ceramic particles (ceramic oxides in particular) modifies the degradation reactions and influences the presence of undesirable carbonaceous residues. In fact, in the case of an incomplete burning of the organic species, the degradation products derived from the starting polymers form, between 500 and 700°C, carbonaceous residues which decompose to form carbon at temperatures higher than 1,000°C.

5.7.3. Other debinding techniques

In order to eliminate flaws inherent to thermal debinding and to shorten the duration of this stage, other techniques for the extraction of organic shaping additives have been developed. These techniques are based on an under- or over-pressure of the treatment atmosphere, on microwave heating, capillary migration of molten binder, sublimation of a binder in aqueous phase, or solubilization by catalytic reaction or solvents.

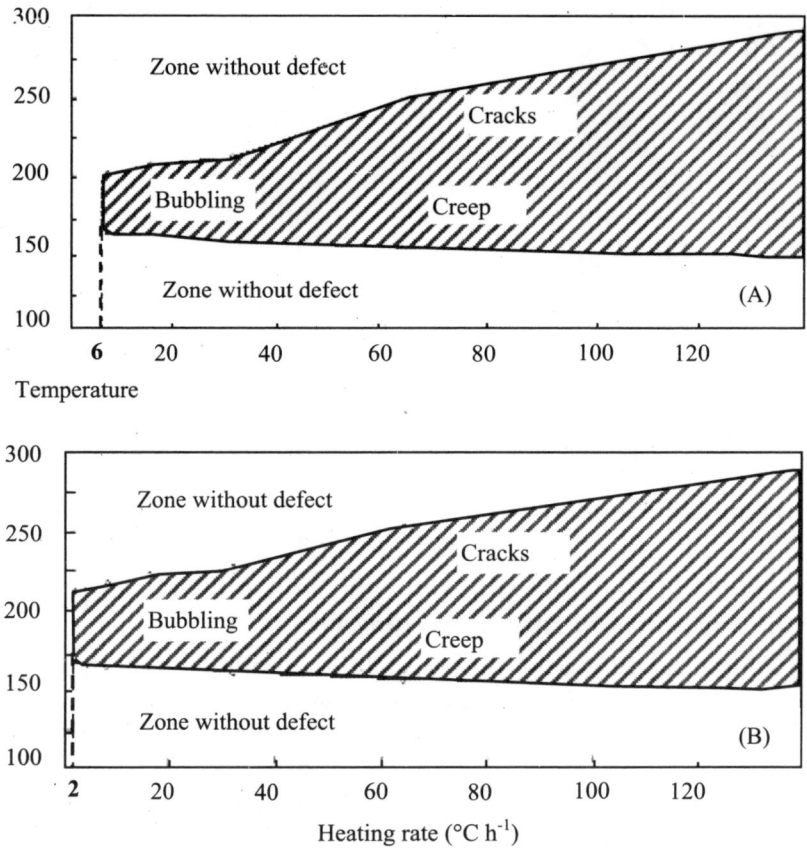

Figure 5.40. *Diagrams giving the debinding flaws depending on the heating kinetics (alumina + polypropylene + wax system): A) pieces with 3 mm thickness, B) pieces with 6 mm thickness [PIN 92]*

5.7.3.1. *Assisted thermal debinding*

a) By under-pressure or over-pressure

The under-pressure method eliminates the binder by evaporation without passing through the liquid state, which avoids problems of capillary redistribution. The overpressure method makes it possible to use heating rates higher than in the case of traditional thermal debinding in atmospheric pressure, because the degradation of the binder molecules is shifted towards high temperatures and the degradation products then diffuse in a binder with very low viscosity.

b) By microwaves

Microwaves have the benefit of directly heating the core of the pieces to be debinded, thus avoiding the waste of time due to problems of heat transfer towards the core of the parts and furthermore reducing gradients of temperature. The advantage of using microwaves is therefore primarily that we have, at any moment, a binder in an identical state from the surface to the core of the piece, which limits excessively high composition gradients between the debinded zone and the zone containing binder, thus minimizing delamination flaws.

c) By capillary migration [BAO 91]

The pieces are placed in a powder bed and are subjected to a slow heating rate in order to sufficiently evacuate the binder by capillary migration in the powder bed before the start of thermal degradation. The quantity of binder evacuated depends on the permeability of the powder bed, which itself depends on the size of the particles forming the bed, their arrangement and the degree of compaction. This last parameter is important because it must ensure a high evacuation by high capillarity without, however, obstructing the departure of the gas species formed. In addition to its "wicking" function, the support bed avoids the deformations of the pieces. The disadvantages of this technique arise from the rearrangement of the fine particles in the piece to be debinded if the liquid binder flow becomes too high, and from problems of the reprocessing and storage of the powders saturated with organic materials.

5.7.3.2. Debinding by sublimation [NOV 92]

This process uses a mixture whose organic phase is made up of a plasticizer and a binder with low molecular weights, dispersed in a solvent representing more than 80% in volume of this phase. The parts are obtained by pressure injection molding (50 MPa) in a mold cooled to a temperature less than 0°C, which ensures the cohesion of the piece by the freezing of the intergranular phase. Debinding is then carried out in two stages: a first phase of elimination of the liquid by vacuum sublimation at low temperature, then a phase of pyrolysis of the remaining additives.

5.7.3.3. Debinding with liquid solvent [KIM 96]

This technique relies on the solubility of organic additives in solvents such as trichloroethane, pentane, hexane, toluene, acetone or petroleum ether. The ceramic pieces are immersed in a solvent bath which can be preheated to temperatures of about 40 to 50°C. The solvent penetrates into the pores, solubilizes the binders and evacuates them by migration towards the surface. After the treatment, the solvent present in the porous structure of the piece is evaporated and the non-solubilized binder fraction is eliminated during the sintering cycle. As the soluble binder is eliminated from the surface towards the inside according to an abrupt reaction front (shrinking core model), we must introduce into the formulations a non-soluble binder which ensures the mechanical strength of the piece and limits the risk of shedding of the grains on the surface.

5.7.3.4. *Catalytic debinding [TER 93]*

This technique applies to particular binders that exhibit specific degradation (catalytic) mechanisms in certain atmospheres. BASF (Germany) has developed an injection formulation containing polyacetal binders, which are eliminated at 110°C in acid atmosphere (98% nitrogen + 2% nitric acid). This atmosphere leads to the sublimation of a part of the polyacetal binder and the crystallization of the remaining part. However, sublimation vapors require a reliable reprocessing before their evacuation into the air and the debinding furnaces must have an anticorrosion treatment in order to resist nitric acid fumes.

5.7.3.5. *Debinding by supercritical fluid [DEL 99]*

This debinding process is based on the use of supercritical CO_2, which is a good paraffin solvent under certain conditions of temperature and pressure. In the supercritical state (T > 31.2°C, P > 7.38 MPa), CO_2 combines solvent capacity to satisfying high transport properties in finely divided porous environments. Furthermore, the possibility of significantly varying the solvent capacity of the supercritical fluid by means of its density, which depends on temperature and pressure conditions, allows a selective elimination of the organic species and the recycling of dissolved binders after treatment (see Figure 5.31). The pieces to be debinded are placed in a thermoregulated chamber and then subjected to a flow of supercritical CO_2 at a temperature ranging between 40 and 120°C, under a pressure of about 28 MPa, for one to three hours depending on the thickness of the treated pieces. The fluid loads itself with dissolved organic compounds, and is then expanded in several separators that allow the recovery of the non-degraded binders as well as the recycling of CO_2. Such a process allows the debinding of the pieces without flaws in a period six times shorter less than thermal debinding.

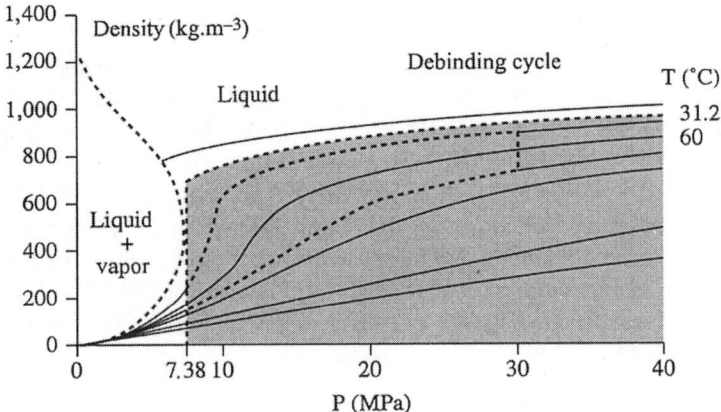

Figure 5.41. *Supercritical domain of CO_2 and typical debinding cycle*

5.8. Deposition techniques

The techniques of depositing ceramic materials on a support (substrate) are generally excluded from books dealing with ceramic shaping processes. They are nevertheless processes involving ceramics. For this reason, the main deposition techniques are briefly described in this section.

The objective is to deposit a ceramic layer on a support (metal, ceramic, organic) to obtain surface properties different from the massive material. The applications of these coated materials are numerous (tribology, heat insulation, corrosion resistance, electronics, optics, optoelectronic, biomedical, etc.). The thickness of the layers can vary from a few atomic layers for techniques of vacuum deposition (they are called thin layers) to a few millimeters in the case of plasma spraying.

5.8.1. *Vacuum deposition [RIC 94]*

A large variety of ceramics can be deposited in vacuum (oxides, nitrides, carbides, carbon-diamond). Vacuum deposition can be classified under two techniques: chemical vapor deposition (CVD) and physical vapor deposition (PVD).

The formation of the thin layer can be broken down into three main stages which can be separated or superimposed depending on the process:

– production of the species to be deposited;

– transport of the species on the substrate;

– deposition on the substrate and growth of the layer.

5.8.1.1. *Chemical vapor deposition (CVD)*

The techniques of chemical vapor deposition consist of producing a layer, on a heated substrate, by chemical reaction starting from reactive species in gas phase. The reactive species can be directly gases or generated from the decomposition of gas precursors. The depositions are homogenous and dense and have a good adhesion on the substrate. The microstructure of the layer and the crystalline orientation can be controlled during the deposition, which justifies the use of CVD for the deposition of silicon epitaxed layers on single-crystal silicon substrates (wafer) in the manufacture of semiconductors. The depositions are generally carried out with dichlorosilane (SiH_2Cl_2), as source of silicon, in hydrogen at a temperature of around 1,000°C:

$$SiH_2Cl_2 \leftrightarrows Si + 2HCl \qquad [5.28]$$

The limitation of this technique lies in the fact that the chemical reaction must be capable of taking place at a temperature compatible with the material of the substrate. In order to make chemical reactions on substrates at lower temperatures possible, they can be activated by electric discharges in plasma: this is called plasma assisted chemical physical vapor deposition (PACVD). We can thus obtain silicon nitride depositions (Si_3N_4) at 300°C, from silane (SiH_4) and ammonia in argon or nitrogen. On the other hand, reaction gases cannot be desorbed easily from the layer formed at low temperatures and the deposition of pure materials is difficult.

5.8.1.2. *Physical vapor deposition (PVD)*

The two basic techniques of physical vapor deposition are evaporation and spraying (sputtering). There are also variants derived from these two techniques.

5.8.1.2.1. Evaporation

The material to be deposited is evaporated by heating, electronic bombardment, electric arc or lazer beam. The evaporated species then condense on the surface of the substrate.

Reactive evaporation consists of making evaporated species react with a reactive gas to form a compound, either in gas phase or on the substrate. The reaction can be activated by ionization of the gas mixture in vapor phase. Let us take the example of the coating of a metal piece by a hard titanium nitride deposition (TiN). Titanium (Ti) is evaporated by means of an electron gun. The titanium vapors react, in the presence of a plasma surrounding the piece, with the reactive gas, i.e. nitrogen (N_2) introduced into the chamber with an inert gas in order to form the TiN deposition [PEY 97].

5.8.1.2.2. Spraying

Spraying consists of ejecting atoms from a target of the material to be deposited by impact of the ionized atoms of a gas, generally argon and then transferring these atoms to the substrate by an electric field. It is the method traditionally used to metallize ceramic samples, non-electric conductors, for observation under the scanning electronic microscope. Here again, the evaporated species can react with a reactive gas to form a compound. In order to increase the speed of deposition, the discharge between the target and the substrate creating the plasma can be confined to the vicinity of the cathode using a magnetic field (magnetron system).

A variant of spraying is "ion plating" which consists of bombarding, by inert gas ions (argon), the surface of the substrate before deposit and then the deposited layer during the treatment or at least during the formation of the interfacial layer. This technique yields a better adhesion of the deposit. It is largely used in the field of optics.

5.8.2. *Thermal spray deposition [FAU 00]*

The principle of this deposition technique is the spraying on a substrate, generally metallic, of droplets of the material to be deposited, molten or in a plastic state using a high temperature source. The thickness of the depositions obtained can vary from a hundred micrometers to a few millimeters. The applications of thick depositions produced by thermal spraying are mainly thermomechanical (thermal barriers: partially stabilized ZrO_2, anti-wear: Al_2O_3, ZrO_2, Cr_2O_3, Al_2O_3-TiO_2, WC-Co), but also electric insulation (Al_2O_3). Ceramic oxides are sprayed in free air, whereas carbides and nitrides require a controlled atmosphere.

The high temperature source used for the deposition of ceramics, materials with high melting points, is primarily thermal plasma produced by electric arc. The plasma arc sprays all the materials whose melting and vaporization points differ by at least 300°C. The gas is injected into a permanent electric arc established between two concentric electrodes to produce the plasma that corresponds to a gas where the molecules, excited and/or ionized atoms and electrons coexist (see Figure 5.42). The cathode is typically made of thoriated tungsten and the anode of electrolytic copper. These parts are cooled by water circulation. The gas mixture used for the spraying of ceramic materials is generally an argon/hydrogen mixture with a percentage of hydrogen varying from 5 to 20%. Other gases, such as nitrogen, nitrogen-hydrogen and argon-helium can also be used for specific applications. The plasma jet is ejected at high temperatures (about 12,500 K) and high speeds (about 2,000 $m.s^{-1}$, subsonic speed at these temperatures).

The sprayed powders, generally obtained by fusion-milling, have sizes ranging between 22 and 45 μm. They can also be prepared by agglomeration-sintering, coating, atomization or mechanofusion.

The substrates are sanded before spraying in order to ensure a satisfactory mechanical anchorage.

The deposition obtained by plasma spraying consists of a microcracked stacking of splats with a porosity varying from 1 to 20%. The quality of the deposition is controlled by a large number of experimental parameters, which makes the process difficult to master. However, it is clear that the trajectory and thermal history of the particles in the plasma jet, as well as the surface temperature during deposition, are the critical parameters that must be controlled.

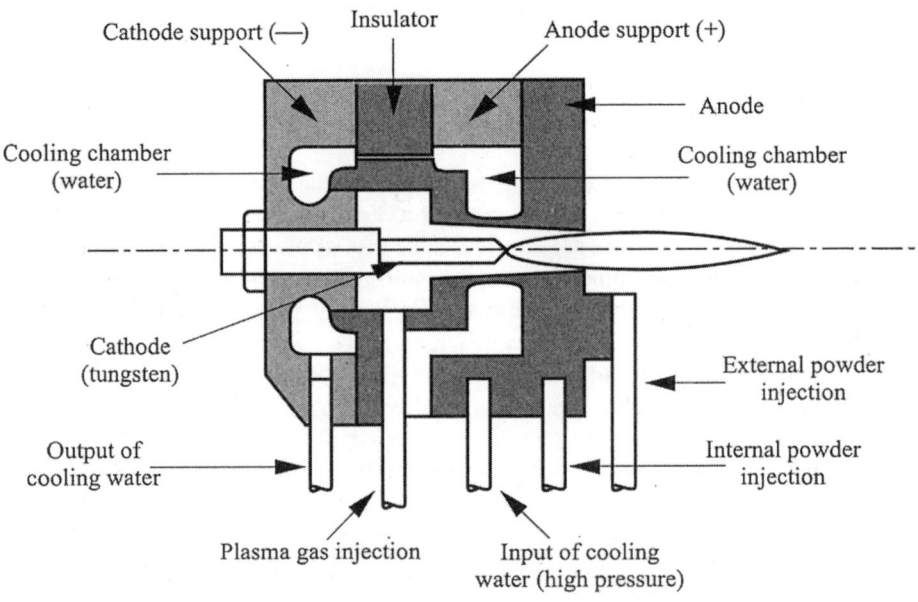

Figure 5.42. *Diagram of the plasma spraying torch*

5.9. Bibliography

[ADC 57] ADCOCK D.S. and MC DOWALL I.C., "Mechanism of filter pressing and slip casting", *J. Am. Ceram. Soc.,* 40(10), p. 355-62, 1957.

[BAK 83] BAKHT M.F., "Thermal degradation of copolymers of vinyl alcohol and vinyl butyral", *Pakistan J. of Science and Industrial Research*, 26, 1983.

[BAK 96] BAKLOUTI S., "Pressage de poudres d'alumine atomisée: Influence des liants organiques sur l'aptitude au presage", Thesis, Limoges University, 1996.

[BAO 91] BAO Y. and EVANS J.R.G., "Kinetics of capillary extraction of organic vehicle from ceramic bodies, piece I: flow in porous media, piece II: partitioning between porous media", *J. Eur. Ceram. Soc.,* 8, p. 95-105, 1991.

[CAR 56] CARMAN P.C., *Flow of Gases through Porous Media*, London, Butterworths, 1956.

[CHA 92] CHARTIER T., STREICHER E. and BOCH P., "Preparation and characterization of tape cast aluminum nitride substrates", *J. Eur. Ceram. Soc.*, 9, p. 231-42, 1992.

[CHA 94] CHARTIER T., "Tape casting", *Encyclopedia of Advanced Materials*, Pergamon Press, p. 2763-2768, 1994.

[CHA 95] CHARTIER T., MERLE D. and BESSON J.L., "Laminar ceramic composites", *J. Eur. Ceram. Soc.*, 15, p. 101-107, 1995.

[CHA 99] CHARTIER T., HINCZEWSKI C. and CORBEL S., "UV curable systems for tape casting", *J. Eur. Ceram. Soc.*, 19, p. 67-74, 1999.

[CHO 71] CHONG J.S., CHRISTIANSEN E.B. and BAER A.D., "Flow of viscous fluid through a circular pipe", *J. of Applied Polymer Science*, 15, p. 369-79, 1971.

[DEL 99] DELHOMME E., "Déliantage, par CO_2 supercritique, de matériaux céramiques réfractaires mis en forme par injection basse pression", Thesis, Limoges University, 1999.

[DER 41] DERJAGUIN B.V. and LANDAU L.D., *Acta Physiochim. USSR*, 14, p. 633, 1941.

[DOR 98] DOREAU F., TARI G., PAGNOUX C., CHARTIER T. and FERREIRA J.M., "Processing of aqueous tape casting of alumina with acrylic emulsion binders", *J. Eur. Ceram. Soc.*, 18, p. 311-321, 1998.

[FAR 68] FARRIS R.J., "Prediction of the viscosity of multimodal suspensions from unimodal viscosity data", *Trans. Soc. Rheol.*, 12, p. 281-301, 1968.

[FAU 00] FAUCHAIS P., VARDELLE A. and DUSSOUBS B., "Quo vadis thermal spraying", *J. of Thermal Spray Technology*, 2000.

[GER 89] GERMAN R.M. (ed.), *Particle Packing Characteristics*, Metal Powder Industries Federation, Princeton NJ, 1989.

[GER 90] GERMAN R.M. (ed.), *Powder Injection Molding*, Metal Powder Industries Federation, Princeton NJ, 1990.

[GRA 94] GRAULE T.J., BAADER F.H. and GAUCKLER L.J., "Shaping of ceramic green compacts directly from suspensions by enzyme catalyzed reactions", *Deutsche Keramik Gesellschaft*, 71(6), p. 317-323, 1994.

[GRÜ 72] GRÜNEWALD H. and KARSCH K.H., "La formation du tesson de barbotine de céramique fine en coulage sous pression jusqu'à 40 atm", *Bulletin de la Société Française de la Céramique*, 97, p. 75-80, 1972.

[HUN 87] HUNTER R.J., *Foundations of Colloid Science*, Vol. 1, Oxford, Oxford University Press, 1987.

[JOL 94] JOLIVET J.P., *De la solution à l'oxyde*, CNRS Editions, Paris, 1994.

[JOU 84] JOUENNE C.A., *Traité de céramiques et matériaux minéraux*, Editions Septima, Paris, 1984.

[KIM 96] KIM S.W., LEE H., SONG H. and KIM B.H., "Pore structure evolution during solvent extraction and wicking", *Ceram. Int.*, 22, p. 7-15, 1996.

[KOS 97] KOSMAC T. and JANSSEN R., "Low pressure injection molding of SiC platelet reinforced reaction bonded silicon nitride", *J. Mat. Sc.*, 32, p. 469-474, 1997.

[KRI 59] KRIEGER I.M. and DOUGHERTY T.J., "A mechanism for non-Newtonian flow in suspensions of rigid spheres", *Trans. Soc. Rheol.*, 3, p. 137-152, 1959.

[LEE 69] LEE D.I., "The viscosity of concentrated suspensions", *Trans. Soc. Rheol.*, 13, p. 273-288, 1969.

[MIS 78] MISTLER R.E., SHANEFIELD D.J. and RUNK R.B., "Tape casting of ceramics", in *Ceramic Processing before Firing*, Wiley Science, New York, p. 411-488, 1978.

[MOR 92] MORENO R., "The role of slip additives in tape casting technology", *Am. Ceram. Soc. Bull.,* Part I, 71(10), p. 1521-1531, Part II, 71(11), p. 1647-1657, 1992.

[MUT 95] MUTSUDDY B.C. and FORD R.G., *Ceramic Injection Molding*, Chapman and Hall, London, 1995.

[NOV 92] NOVITCH B.E., SUNDBACK C.A. and ADAMS R.W., "Quickset injection molding of high performance ceramics", *Ceram. Trans.*, 26, p. 157-164, 1992.

[O'BR 88] O'BRIEN R.W., "Electroacoustic effects in a dilute suspension of spherical particles", *J. of Fluid Mechanics*, 190, p. 71-86, 1988.

[OMA 91] OMATETE O.O., JANNEY M.A. and MENCHHOFER P.A., "Gel casting of alumina", *J. Am. Ceram. Soc.*, 74(3), p. 612-618, 1991.

[PEY 77] PEYRE J.P. and TOURNIER C., "Revêtements PVD, CVD, bains de sels", *Manuel des traitements de surface à l'usage des bureaux d'études*, CETIM, Senlis, 1977.

[PIN 92] PINWILL I.E., EDIRISINGHE M.J. and BEVIS M.J., "Development of temperature-heating rate diagrams for the pyrolytic removal of binder used for powder injection molding", *J. Mat. Sc.*, 27, p. 4381-4388, 1992.

[RAH 95] RAHAMAN M.N., *Ceramic Processing and Sintering*, Marcel Dekker Inc., New York, 1995.

[REE 95] REED J.S., *Principles of Ceramic Processing*, 2nd ed., John Wiley and Sons, New York, 1995.

[RIC 94] RICHARDT A. and DURAND A.M., *Le vide: les couches minces, les couches dures*, In Fine, Paris, 1994.

[SPU 97] SPUR G. and MERZ P., "Compounding of feedstocks for injection molding", *Ceram. For. Int./Ber.*, 74(7-8), p. 371-375, 1997.

[TER 93] TER MAAT J.H.H. and EBENHÖCH J., "Feedstock for ceramic injection molding using the catalytic debinding process", *3rd Euro Ceramics Proceedings*, Faenza Editrice Iberica S.L., vol. 1, p. 437-442, 1993.

[TIL 86] TILLER F.M. and TSAI C.D., "Theory of filtration of ceramics: I, Slip casting", *J. Am. Ceram. Soc.*, 69, p. 882-887, 1986.

[TSE 98] TSENG W.J. and LIU D.M., "Effect of processing variables on warping behaviours of injection-molded ceramics", *Ceram. Int.*, 24, p. 125-133, 1998.

[VER 48] VERWEY E.J.W. and OVERBEEK J.Th.G., *The Theory of Stability of Lycophobic Colloids*, Elsevier, Amsterdam, 1948.

[WRI 90] WRIGHT J.K., EDIRISINGHE M.J., ZHANG J.G. and EVANS J.R.G., "Particle packing in ceramic injection molding", *J. Am. Ceram. Soc.*, 73(9), p. 2653-2658, 1990.

[ZHA 89] ZHANG J.C., EDIRISINGHE M.J. and EVANS J.R.G., "A Catalogue of Ceramic Injection Molding Flaws and their Causes", *Industrial Ceramics*, 9(2), p. 72-82, 1989.

Chapter 6

Alumina, Mullite and Spinel, Zirconia

6.1. Alumina, silica and mullite, magnesia and spinel, zirconia

In Chapter 1, we recalled the abundance of silicates on the Earth's crust and indicated that they were the basic components of non-metallic inorganic materials: glasses, cements and ceramics. We restricted ourselves to ceramics and said that almost all traditional ceramics are silicate materials. This is not necessarily the case with technical ceramics. For which it is alumina that constitutes the most "useful" compound and the most widespread. We can say that alumina is for a ceramist what iron is for a metallurgist and, to pursue the analogy further, that the alumina-silica diagram is as important for ceramics as the iron-carbon diagram is for metals.

Underlining the importance of the Al_2O_3-SiO_2 diagram does not imply that technical ceramics are assimilated to silicate ceramics. This is first of all because it is high purity alumina, free from silica, which is the first of technical ceramics [JAC 94] and then because high purity silica products do not have very many uses as such apart from refractory applications. In the crystallized state, and if we exclude the considerable field of the piezoelectricity of quartz, silica is first and foremost a raw material. In the vitreous state, all the wonderful products that are derived from silica fall under glass industries, and not under ceramic ones.

The examination of the Al_2O_3-SiO_2 diagram highlights that the two components combine to form a single crystallized compound: mullite, whose stoichiometry corresponds to $3Al_2O_3$ $2SiO_2$. In fact, mullite crystallizations develop in silicate

---

Chapter written by Philippe BOCH and Thierry CHARTIER.

ceramics. The usage spectrum of mullite is limited, even if this material is appreciated for refractory applications and, while we cannot consider alumina and mullite on equal footing, we cannot speak of the former while ignoring the latter. This chapter, primarily devoted to alumina therefore also treats mullite, though more briefly.

The Al_2O_3-MgO diagram is also a useful diagram for ceramists: magnesia MgO is an essential compound in refractory industries and spinel Al_2O_3-MgO plays in the Al_2O_3-MgO binary a role similar to the one played by mullite in the Al_2O_3-SiO_2 binary. We will therefore say a few words about magnesia and spinel.

Finally, zirconia exhibits very original electric and mechanical properties. In this chapter we will discuss only mechanical properties, as electric properties are covered in Chapter 11.

6.2. Alumina

Alumina is produced primarily from bauxite rocks treated through the Bayer process (see section 6.2.2). Aluminum production consumes about 85% of the bauxite used and the non-metallurgical usages use up the remaining 15%: 10% in the form of alumina and 5% in the form of calcined bauxite, which is not transformed into alumina. The non-metallurgical applications of alumina in 1998 accounted for about five million tons [BAC 99]. The refractory industry is the largest consumer of alumina, followed by abrasive industries, technical porcelains (including spark plug bodies), ceramics for mechanical and electronic use, and chemistry (catalyses in particular). The prices vary in a very broad range: currently, the difference is of two orders of magnitude between metallurgical alumina (about €200 per ton) and high purity alumina, which is used for the preparation of sapphire monocrystals (about €20,000 per ton).

6.2.1. *Alpha alumina*

Aluminum oxide, or alumina, crystallizes in the *corundum* structure (mineralogical term) to form sapphire monocrystals. We are referring here to white sapphire, like the one used for the manufacture of scratchproof "glasses" for watches, whereas in gemmology sapphire is the blue sapphire, which is alumina with a little Ti^{4+}; ruby being alumina colored in red by Cr^{3+}. This is the *alpha* variety (Al_2O_3-α), which is the stable variety, but we will see that the preparation of alumina involves various metastable varieties: hydrated aluminas and transition aluminas.

The corundum structure belongs to the rhombohedric system, space group $R\bar{3}c$, the lattice parameters at room temperature being, in hexagonal axes, a = 0.4759 nm and c = 1.299 nm, with Z = 6. The crystal can be described as a compact hexagonal stacking of O^{2-} anions, in which the two-thirds of the octahedral interstices are occupied by Al^{3+} cations (see Figure 6.1) [KIN 76 and 83, KRO 57]. We note that this reference to anions and cations insists rather misleadingly on the ionicity of the bond: it is considered that the alumina bond is ionic for two-thirds and covalent for the remaining third part. The considerable amount of formation enthalpy ($\approx$ 1,600 kJ mole^{-1}), which explains the absence of natural deposits of metal aluminum, makes alumina one of the most tightly bonded compounds, resulting in very high hardness (9 in the Mohs hardness scale, which goes from 1 (talc) to 10 (diamond), and where quartz is classified at 7) and high melting (2,050°C) and boiling temperatures (3,500°C). The density of sapphire is very close to 4 g.cm^{-3}, i.e. a little more than half the density of iron.

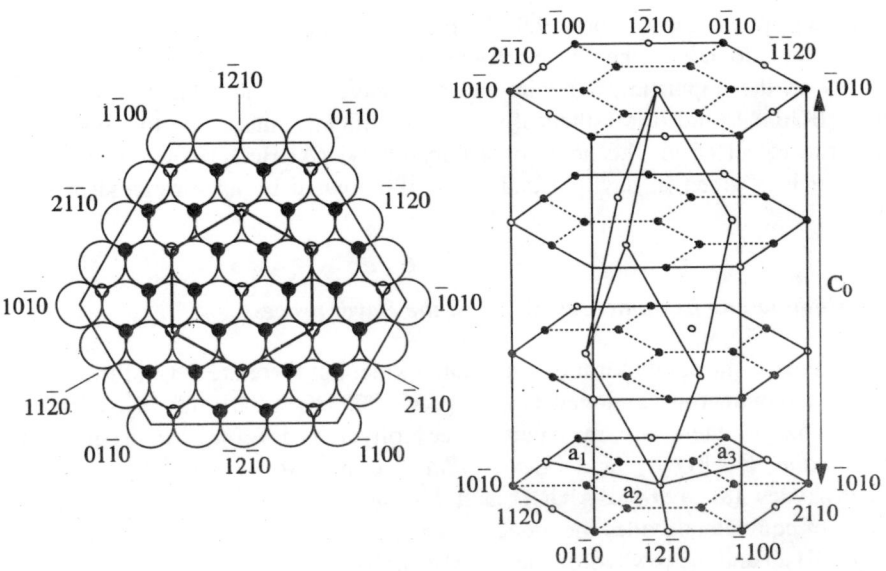

Figure 6.1. *Structure of α alumina. On the left, seen in the base plan: the large circles represent the anions, the small full circles the cations, the small empty circles the vacant octahedral interstices. On the right, the view of the cation sublattice: the full circles represent the full cations and the circles the vacant sites [KRO 57]*

The strong iono-covalent bonds and the rhomboedric structure combine to restrict the dislocation movement [HEU 84]: the sapphire monocrystal remains brittle until very high temperatures and the basal slip (0001) < 11$\bar{2}$0 > is only possible beyond 900°C. The plasticity of the polycrystal comes primarily from the effects of intergranular diffusion or flow of vitreous phases. These are mechanisms that depend on the purity of the material and whose occurrence temperature thresholds are lower than 800°C for impure aluminas and higher than 1,100°C for aluminas with purity exceeding 99%.

Alumina is one of the best electrical insulators, hence its dielectric applications. Electric conduction depends primarily on impurities which act as acceptors (for example, Mg, Fe, Co, V or Ni) or as donors (for example, H, Ti, Si, Zr or Y), the balance between electronic and ionic contributions depending on the temperature and the partial pressure of oxygen [KRO 84]. Among the impurities, sodium deserves a particular mention because the synthesis of alumina by the Bayer process brings into play sodic mediums; sodium aluminate ($Na_2O11Al_2O_3$), also called beta alumina, is an ionic conductor with very high conductivity, which is why this compound is envisaged for solid electrolyte applications [WES 90].

With an optical gap of about 9.9 eV, pure alumina is transparent in the visible and colorless domain. Like the electrical properties, the colored effects depend primarily on the impurities. Sapphire and ruby have been the subject of numerous studies [BUR 84] and it is esthetically and scientifically interesting to mention the alexandrite effect (from the name of a chrysoberyl (Al_2BeO_4) containing a little Cr^{3+}) which is green in daylight and red if illuminated by an incandescent lamp [NAS 83].

6.2.2. Alumina and its numerous varieties: the Bayer process

α alumina is the stable form of aluminum oxide, but there are several metastable varieties and numerous hydrated phases [CAS 90, DOR 84, GIT 70]. For these hydrated forms, there are in nature three trihydroxides (*bayerite* α-Al(OH)$_3$, *nordstrandite* β-Al(OH)$_3$ and *hydrargillite* or *gibbsite* γ-Al(OH)$_3$) and two oxyhydroxydes (*diaspore* α-AlOOH and *boehmite* γ-AlOOH). By adopting the notation of equivalent oxides mentioned in Chapter 1, the trihydroxydes correspond to $Al_2O_33H_2O$ and the oxyhydroxydes to $Al_2O_3.H_2O$.

Bauxite rocks contain 40 to 60% of equivalent alumina in the form of hydrargillite, boehmite, diaspore and silico-aluminous minerals, 10 to 20% iron oxide and approximately 5% quartz sand and various impurities. The Bayer process consists of attacking the rock crushed by a hot caustic detergent (NaOH solution at a temperature of 150–160°C, under a pressure of 0.5 MPa) in order to dissolve alumina in the form of aluminate and to precipitate the impurities (Fe_2O_3, SiO_2, TiO_2). These red deposit are eliminated by filtration and, after cooling, the hydrated sodium aluminate solution $NaAl(OH)_4$ is seeded with hydrargillite germs which leads to a massive precipitation of this latter phase.

Heat treatments of Bayer hydrargillite initially cause dehydration, then a recrystallization to many types of *transition aluminas* (khi, kappa, rho, eta, theta, delta, gamma, etc.) that can be classified under the collective name of gamma aluminas. The distinctions within transition aluminas are based on macroscopic criteria (particularly the specific surface of the products) and crystallographic criteria (particularly the presence of numerous defects: point defects as well as extended flaws).

The lion's share of the Bayer production is used for the production of metal aluminum by electrolysis of the molten mixture of alumina plus cryolite. Apart from metallurgical alumina, hydrated aluminas are used in paper mills, water treatment, chemical industry, and act for instance as soft abrasives in toothpastes. Alumina gels among other applications are used as precursors for the formation of abrasive grains and in the pharmaceutical industry where they are used as gastric treatment. Thanks to their considerable specific surface area (up to 1,000 $m^2.g^{-1}$) and their surface reactivity, transition aluminas are among the basic catalysis products, but for ceramic applications, which are our concern here, alumina is always in its alpha form. However, the morphology of the crystals is not the same. The three principal categories of products are:

i) calcined aluminas obtained by heat treatment of transition aluminas at temperatures of about 1,100–1,250°C and then milling;

ii) tabular aluminas prepared by sintering at very high temperature (1,950°C) to obtain pellets which aggregate grains in the form of plates with large lateral size (50 to 300 μm) but low thickness, from which the name "tabular" stands for;

iii) molten aluminas or corundum which result from a fusion in electric furnace (temperature of around 2,100°C).

Purities and grain sizes vary considerably from one product to another, as is shown in the catalogues of alumina producers [ALC 99 or PEC 99]. Typically, sintered ceramics require fine (grain size of about one micrometer) and "reactive" powders, like calcined alumina at rather low temperatures. An increase in the calcination temperature augments the size of the crystals, which disadvantages sintering. Refractory materials require coarse grains (several hundred micrometers) to limit creep and to reduce reactions with the environment, and therefore prefer tabular alumina. Finally, abrasives use electromelt grains, although the search for enhanced performances is fuelling the development of sol-gel abrasive grains [SCH 91].

The reactivity of alumina powders that must be used for the sintering of ceramics has not been fully understood, especially since it is covered by the shroud of the industrial secrecy in powder preparation treatments. The challenge is to obtain a densified ceramic very close to the theoretical density and with a well-controlled microstructure, but sintered at a temperature as low as possible: high temperatures are expensive, in energy consumption as well as in investment costs.

The first way to decrease the sintering temperature is to use liquid phase sintering by additives that produce eutectics at low melting temperatures: we thus use aluminas with purity of 80 to 98% approximately, in particular 96% pure aluminas, the remaining 4% corresponding to additions of SiO_2, CaO or MgO. The Al_2O_3-MgO-SiO_2 system contains a eutectic that melts at 1,355°C and the Al_2O_3-CaO-SiO_2 system a eutectic that melts at 1,170°C. Low sintering temperatures made possible by the vitreous flow have the disadvantage of a microstructure of the polycrystal where the alumina grains are embedded in an amorphous phase, which considerably degrades performances at high temperatures (creep). However, this is not necessarily a drawback for certain electric applications; there is even the advantage that the "anchoring" of a metallized layer on a ceramic substrate can be favored by the emergence of these intergranular segregations on the surface.

However, the most demanding applications – refractory materials (creep strength), mechanics (resistance to corrosion under stress and other effects of delayed failure), optics (perfect transparency and absence of coloring) or electronics (high frequency uses) – prohibit the presence of parasitic phases, therefore the purity of the alumina must be high (in general higher than 99.7%), but with the need for to add a few hundreds ppm of additives in order to limit grain growth that hampers sintering (see section 6.4). We must, however, stress that knowing only the purity, the nature of the impurities and the grain size distribution expressed in equivalent diameter of particles assumed to be appreciably spherical is not enough to explain the "reactivity" of a powder that allows sintering at around 1,500°C. This is a low temperature for a compound where the atomic diffusion is slow and whose fusion occurs beyond 2,000°C, whereas another seemingly equivalent powder can require a treatment at 1,600°C. The necessary condition is to use fine (micronic) and well deagglomerated particles, but we cannot be sure that this is a sufficient condition: grain size distribution, particularly the proportion of the fine ones, as well as surface flaws also have an impact. However, these parameters depend on the calcination and milling stages, milling being a process whose scale effects are not clearly established. This explains the difficulty of a laboratory study, from a few grams of powders, of the attrition mechanisms involved in large industrial grinders.

For specific uses, particularly for the preparation of ultrapure powders used for the synthesis of sapphire monocrystals, the Bayer process makes room for other methods, including *ammonium alum method* (dissolution of Bayer hydrargillite in excess sulphuric acid, neutralization in ammonia in $NH_4Al(SO_4)_2.12H_2O$ form, calcination at about 1,000°C and then milling) or the *alcoholate method* (based on an aluminum isopropylate).

In addition to the use of the term "reactivity" according to the meaning given by ceramists (it then means "possible sintering at low temperature"), we must define the chemical reactivity of alumina. Transition aluminas (Al_2O_3-γ), with metastable structure and very high specific surface area, are soluble in aqueous, acid or basic solutions, alumina being thus regarded as an amphoteric oxide. Alpha alumina is much less reactive than gamma alumina, but it dissolves heated in alkaline media or eutectics with low melting points like those of the CaO-SiO_2 system. Highly reactive metals (calcium or magnesium) can reduce alumina beyond $\approx 900°C$.

Table 6.1 illustrates the main ceramic applications of alumina powders, shows the desired properties and indicates the recommended powders.

6.3. Alumina ceramics

This book does not cover non-ceramic applications of alumina (filler in paper or in polymers, gel, catalyses, etc.). The applications where alumina grains are used primarily will be only touched upon (abrasives) or commented on elsewhere in the book (refractory materials). This chapter therefore primarily considers alumina ceramics for structural and functional uses.

6.3.1. *Structural applications of alumina*

Alumina ceramics owe to the stability of this oxide and its strong atomic bonds their mechanical and thermal performances which imply – but do not necessarily combined in a single type of microstructure – high hardness, high modulii of elasticity, satisfactory mechanical strength, wear resistance and good tribological properties and refractarity. Giving accurate values would be useless because mechanical properties are sensitive properties which vary with the microstructure, and the hot properties depend highly on the temperature and chemical reactions when the environment is aggressive. To illustrate this point, a dense fine-grained alumina ceramic has a Young's modulus of ≈ 400 GPa (twice the modulus of steel), a Poisson's ratio of ≈ 0.25, a Vickers hardness of 20 GPa and a mechanical bending strength σ_F of ≈ 300 to 500 MPa – the mechanical strength being a value that expresses a particular sensitivity to measurement conditions (see Chapter 8). For high temperature applications, in the absence of corrosion, sintered alumina pieces allow long duration uses, at temperatures that can exceed $1,600°C$, but at stress levels that should not exceed a few MPa.

Applications	Main desired characteristics	Sodium content/reactivity
High voltage insulators	Mechanical and electric strength, resistance to arc effects, purity, resistance to thermal shocks	Normal/
Spark plugs	Mechanical and electric strength, controlled shrinkage, durability, green workability, resistance to thermal shocks	Normal to low/
Ceramics for electronics	Resistance to high temperatures, high electric resistance, thermal conductivity, ease of metallization	Low/
Tiles, enamels	Mechanical and chemical properties, controlled fusion and viscosity, adjustment of surface effects	Intermediate/
Metallized substrates	Good thermal conductibility, ease of metallization	Low/reactive
Catalysis support	Preparation of cordierite, thermal conductivity, surface control	Normal/
Ceramic filters for molten metal	Surface control, good thermal conductivity, resistance to thermal shocks	Normal/
Crockery, sanitary	Mechanical and chemical resistance, whiteness	Normal/
Ceramics for mechanics	Mechanical resistance, thermal conductivity, resistance to corrosion, abrasion and thermal shocks	Normal to weak/reactive
Tribology	Uniform microstructure, compromise between hardness and toughness	Normal/

Table 6.1. *Recommendations for choosing alumina powders [ALC 99]*

Toughness at 20°C is equal to ≈ 3.5 MPa m$^{1/2}$, which is a respectable amount for a ceramic, but does not make alumina a tough material, especially when compared to metals. Ductility is modest. The coefficient of thermal expansion between 20°C and 1,000°C ($\alpha_{20\text{-}1,000}$) is 8.5 10^{-6} K^{-1}. This relatively high thermal expansion for a ceramic combines with a high Young's modulus in order to lower the first parameter of resistance to thermal shocks (R $\approx \sigma_F/E\alpha$): alumina ceramics are not very resistant to thermal shocks. The expansion anisotropy is low (9.10^{-6} K^{-1}

parallel to the ternary axis of the crystal and 8.3. 10^{-6} K^{-1} perpendicular to this axis) and therefore the residual stresses at the grain boundaries are reduced. In order to complete the comment on a good resistance at high temperatures under weak mechanical load, it must be noted that we observe a poor creep resistance under substantial load. Alumina ceramics are best suited for mechanical uses (excluding shocks) or refractory uses, but they are not thermomechanical materials, if this word means combination of mechanical and thermal performances.

Mechanical applications

Used in the grain state – as opposed to sintered ceramics – alumina is the most widely used abrasive besides silicon carbide. The performances of an abrasive depend on a complex set of qualities, where hardness is not the only parameter in question. In addition to tribochemical effects (which, for example, prohibit the use of diamond for the grinding of metals capable of forming carbides, such as ferrous alloys), the toughness and the morphology of the degradation facets of the abrasive grain also come into play: wear should not blunt it but maintain cutting sides. Friction terms or thermal conductivity must also be considered. The industrial products are very varied [NOR 99] and range from free abrasives (powders or pastes) to agglomerated abrasives (grinding stones), passing through abrasives on flexible supports ("emery paper"). White and brown corundum are the usual products, but increased performances are obtained with globular alumina, alumina-zirconia with high toughness, or alumina obtained by precursor methods [SCH 91].

Tabular alumina grains constitute the filler of many refractories, whether they are fashioned as bricks or not fashioned: castable refractory concretes whose binder is cement, itself with high alumina content [BOC 99].

The tablets of alumina cutting tools, if necessary filled with titanium carbide to increase thermal conductivity or with zirconia to increase toughness, certainly do not have a market as broad as WC-Co cemented carbides. However, they can make impressive cutting speeds possible and they are adapted to the machining of hard metals or metals coated with oxidized layers. As for tribological applications, they are at the heart of the structural uses of alumina. We are referring to thread-guides for textile industry, spray nozzles, rollers and other supports for the paper industry, extrusion dies and seals between parts which turn or slide on one another. We can mention automobile water pump seals, but especially mixing tap seals because this application is the most developed – at least in Europe and in Japan because, curiously, the North Americans continue to prefer the old technique of flexible seals, fibers or elastomers. In our "modern" taps, the lever displaces two parts of polished alumina bored with ports, which help regulate hot and cold water flows (see Figure 6.2). The surface quality of the discs is decisive for the sealing of these taps, whose lifespan can exceed 10 years.

Figure 6.2. *In a mixing tap, hot and cold water flows are regulated by rotating two alumina discs bored with ports [SAI 99]*

The uses of alumina as a *biomaterial* (hip prostheses or dental implants; see Chapter 12) also highlight the tribological performances of alumina ceramics.

Finally, we must mention shieldings (bulletproof jackets, protective coatings of helicopters or tanks), which must stop high speed projectiles ($\approx 1,000$ m s^{-1}) or the dart of melting metal at very high speed ($\approx 10,000$ m s^{-1}) from hollow charges. The mechanisms of shock resistance have not been clearly established, but they are primarily due to the fact that the shock wave is propagated at a speed higher ($\approx 10,000$ m s^{-1}) than the speed of growth of cracks ($\approx 3,000$ m s^{-1}). Therefore, the brittle fracture cannot follow the displacement of the shock wave and the "resistance" of ceramic then tends towards its theoretical value (about one-tenth of Young's modulus). We need hard products, with high modulus and high mechanical strength under compression. Even if it is outclassed by ceramics like boron carbide and if light shieldings require polymers like *Kevlar*, alumina is still a good shielding material. The data on these uses are classified and are therefore difficult to collect.

We may lastly underline that thanks to the availability of alumina powder on the market with varied characteristics and the relative ease of sintering, alumina constitutes the first choice for ceramics with mechanical usages. It is only subsequently that insufficient performances for a given application compel us to consider another oxide, even a non-oxide, whose preparation and use have the annoying tendency of reminding us that we live in an oxidizing atmosphere!

Refractory applications

The performances of cutting tools, thread guides and bearings, tape cartridges, prostheses or shieldings require dense and fine-grained microstructures. On the other hand, the high temperature applications of alumina demand a porous and coarse-grained material, except for the case where a need for sealing justifies the choice of dense materials. Also resistance to thermal shocks is better for a porous material and creep strength is improved by coarse grains, but it is affected by porosity, and therefore we have to find a compromise.

In addition to the iron and steel industry, the industries that use large quantities of refractories are glass, cement and incineration: a chapter is devoted to these uses. We must also mention lining and conveying equipment for furnaces (for the firing of ceramics, in particular), crucibles, protections of thermocouples, etc. The insulating tubes of laboratory electric furnaces illustrate such uses. Placed in horizontal position and thus subjected to bending creep, the alumina tubes should not be used beyond 1,400°C; placed in a vertical position and therefore subjected to traction or compression, their use can exceed 1,700°C. Fibrous refractories often require alumina fibers and thermo-structural composites can also use Al_2O_3 fibers (but "thermal" fibers and "mechanical" fibers are quite different, as are glass fibers for insulation and glass fibers for reinforcement). Metal-ceramic composites continue to develop: the dissemination in a metal matrix of short fibers or ceramic particles, alumina in particular, widens the field of application of the metal matrices to higher temperatures – but at the price of a loss of toughness. *Radomes*, for example, prepared by fusion of alumina powder and spraying using a plasma torch on a preform, are an exotic application that relies on the thermal performances and abrasion resistance of alumina.

In short, alumina is one of the basic compounds for the refractorist, in addition to other oxides (for example, MgO in the iron and steel industry and "AZS": Al_2O_3-ZrO_2-SiO_2 in glassmaking) and a few non-oxides (primarily carbons and graphite and silicon carbides).

6.3.2. Functional applications of alumina

6.3.2.1. Electricity and electronics

Alumina ceramics are widely used by the electric and electronic industries.

Spark plugs for automobiles illustrate the oldest application using materials on the borderline between aluminous ceramics and alumina, the compositions containing here $\approx$ 94% Al_2O_3. We also find a number of insulation products, including those requiring tight ceramic-metal sealings.

For electronics, the main product is the insulating substrate on which conducting or resistive, even capacitive or inductive, circuits are deposited (see Chapter 11). The advantage of alumina is its very high resistivity (it may be recalled that for this it must be very pure, or at least free from impurities like sodium or transition metals), but also its mechanical properties (hardness, mechanical strength), because it can never be stressed enough that functional materials must, in addition to the specificities demanded by their function, have a sufficient level of mechanical performance: spectacles with broken glasses cannot correct vision! Another advantage of alumina is that it lends itself to metallization by various techniques which require not only chemical compatibility, but also excellent surface quality. The top-of-the-range consists of very pure aluminas ($\approx$ 99.7%) sintered close to the theoretical density, whereas the less demanding applications call for aluminas containing doping agents, therefore sintered in liquid phase. The primary mode of production of substrates is tape casting [BOC 88] (see Chapter 3) where a slip of alumina powder and polymeric binders are cast on a plane surface, which is reminiscent of paper techniques. The three limitations of alumina for the manufacture of dielectric substrates are:

i) a modest thermal conductivity ($\approx$ 30 W m^{-1} K^{-1}), which does not facilitate the dissipation of the heat produced by the Joule effect;

ii) a coefficient of expansion double that of silicon, which generates stresses at the substrate ("chip" interfaces);

iii) a marked permittivity ($\varepsilon \approx 10$), which induces capacitive couplings and decreases the transit times.

Aluminum nitride AlN is one of the competitors for the manufacture of high technology substrates, where it brings improvements with respect to the three drawbacks mentioned above, but its manufacture is expensive and delicate.

6.3.2.2. Optics

The remarkable combination of optical and mechanical performances explains the uses of monocrystals (sapphire, possibly ruby) [CAS 90, GIT 70]. The main manufacturing method is the *Verneuil method*, invented at the end of the 19th century. It consists of melting the alumina powder in the flame of an oxyhydrogen blowpipe, then solidifying the liquid by making it drip on a monocrystal which turns around its axis; the stalagmite thus formed will grow several centimeters per hour. The monocrystals (called *balls*, although their form is elongated) can be large (diameter expressed in centimeters and length in tens of centimeters). The other crystallogenesis processes (*Czochralski method* or more rarely *Bridgman method*) are more expensive, but yield crystals containing less growth defects. We can now produce monocrystals in the desired form [CEA 99] and it is also possible to manufacture fibers of approximately 100 μm diameter [LAB 80]. Jewelry (jewels, scratch-proof glass for watches), lazer matrices, substrates for certain electronic circuits and waveguides are the primary applications of sapphire monocrystals, as well as abrasion-resistant optical windows: for military use, but also barcode readers in supermarkets.

It is, however, at the polycrystal state that alumina has found its most widespread use: the internal, transparent tube of sodium vapor lamps used for street lighting. The difficulties which had to be overcome and the methods adopted to succeed have made this application the academic case for the sintering of ceramics. Whatever the compound considered, the vantage point for approaching its sintering is initially to try to transpose the case of alumina, or at least take advantage of the information that it has provided. We will therefore detail this historical example [COB 61].

6.4. Sintering of dense alumina

The discharge lamps used for street lighting use various atmospheres (mercury, xenon, argon, sodium, or various halides including tin and sodium). High pressure sodium vapor lamps are the most widespread. Their principle [RHO 91] is to start the discharge in a rare gas (xenon or neon) in order to overheat and vaporize a mercury-sodium amalgam, and then to allow the establishment of the discharge in the mercury vapor, then followed by sodium vapor. Plasma, at a temperature of 3,700°C, emits a radiation centered on the wavelength of 589 nm characteristic of the "d" transition of sodium. The sodium pressure can reach half the atmospheric pressure; the partial pressure of oxygen must be low to avoid the oxidation of the components; the temperature of the envelope that contains the plasma is about 1,200°C.

The material constituting the envelope must be colorless and transparent at the wavelengths considered, it must resist the chemical aggressions due to plasma and, lastly, it must be able to tolerate this temperature of 1,200°C for a very long lifespan (more than 20,000 hours). The performance of the most refractory glasses – like silica glass of halogen lamps – does not satisfy these specifications, so the only solution is to use a sintered ceramic: alumina has therefore become indispensable.

The optical applications of ceramics demand a material that is transparent and not just translucent. Any inclusion, even a transparent one, but with an optical index different from that of the matrix, causes a scattering of the light – an effect illustrated by depolished glass. Owing to the variation between the optical index of gases inside a pore ($n \approx 1$) and that of alumina ($n \approx 1.76$), a porous alumina ceramic is not transparent. The effect of porosity is extremely marked and a porosity of 0.3% is enough to reduce transparency by 90% compared to a dense body [KIN 76]: the manufacture of alumina envelopes for sodium vapor lamps consequently requires that the densification practically reaches 100%.

To obtain a perfectly dense material is the major difficulty in the sintering of ceramics. As indicated in Chapter 1, it is generally easy to eliminate opened porosity, whose canal morphology allows various matter transport mechanisms (surface diffusion, grain boundary diffusion, etc.). When porosity becomes closed (towards $\rho/\rho_0 \approx 92-94\%$), *intergranular* pores are relatively easy to resorb thanks to the grain boundary diffusion, but *intragranular* pores are practically impossible to eliminate, because on the one hand they can now benefit only from volume diffusion, and on the other hand the gas that they contain can have trouble migrating through solid.

To be transparent, alumina must be very pure or, at least, free from impurities that are sources of coloring, as is the case with transition metal oxides (iron, titanium, etc.). However, experience shows that the sintering of very pure alumina (99.7%) does not prevent granular growth: large-sized grains (see Figure 6.3, the micrography on the left where the measuring bar is equal to 30 μm) trap pores in intragranular position and the ceramic is not transparent. It is to Coble [COB 61] that we must give the credit of showing that the addition to alumina of about 500 ppm MgO limits granular growth considerably, which prevents the trapping of intragranular pores and yields a transparent polycrystal (see Figure 6.3, the middle view where the measuring bar is equal to 10 μm, and the right-hand side view almost on the same scale).

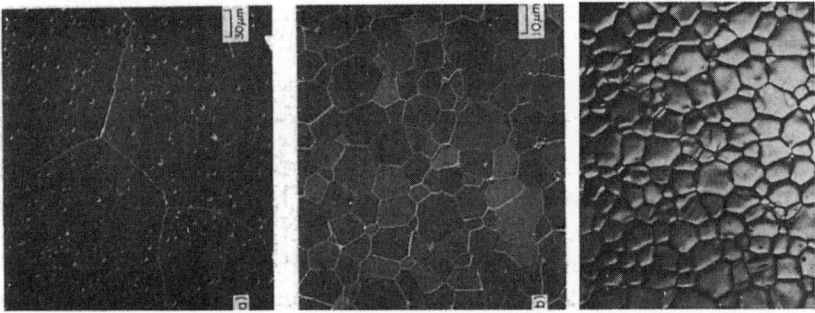

Figure 6.3. *Microstructures of sintered aluminas: on the left "pure" Al_2O_3; in the middle and right, Al_2O_3 doped with ≈ 500 ppm MgO [RHO 91]*

The exact role played by MgO in Al_2O_3 has given rise to numerous and sometimes contradictory explanations, but it has been conceded since 1985 that MgO acts at two levels: by increasing surface diffusion – which contributes to pore mobility – and by decreasing grain boundary mobility [BEN 85]. More recent studies stress more the reduction of boundary mobility than the increase in mobility, and they show that MgO could also "purify" the grain boundaries where silicate segregations exist, because the behavior of "very pure" aluminas (> 99.97%) is quite different from that of "ultrapure" aluminas (> 99.995%) [BAE 94]. We refer the reader to Chapter 3, which gives further information on this question.

The success of the sintering of dense alumina has been and continues to be a model for "optimizing" the sintering of other ceramics. The use of sintering additives has become a common practice, even if all ceramics do not benefit from this to the same degree and a dose of empiricism remains as regards the choice of the best additive: in fact, there is still no irrefutable explanation justifying why it is the magnesia – and not another oxide – which is most effective sintering additive for alumina.

6.5. Point defects and diffusion in alumina

The sintering of ceramics, like their creep and their reactivity, depends primarily on the diffusion processes, these processes themselves depending on the structure of the compounds, the defects that they contain and the microstructure.

The many studies devoted to the defects found in alumina and diffusion mechanisms indisputably highlight the fact that the "cleanest" works have been carried out on high purity sapphire monocrystals, but they are not easily usable for sintered ceramics – polycrystalline and seldom free from involuntary impurities or voluntary doping agents. To understand the behavior of ceramics, experiments on diffusion creep are often more useful [LAN 91]. The Nabarro-Herring creep, controlled by volume diffusion [HER 50], can determine the performances of coarse-grained materials, for example refractory products, but it is especially the Coble creep, controlled by grain boundary diffusion [COB 63], that elucidates the properties of sintered ceramics (Chapter 3 shows the creep maps).

For experiments conducted at temperatures ranging between 1,150 and 1,550°C, the typical temperatures of creep/sintering problems, the coefficients of volume expansion deduced from the creep of polycrystalline alumina show that the process is controlled by the diffusion of aluminum in the lattice [GOR 84]. Grain boundary diffusion also seems to be controlled by the cation. The diffusion of oxygen in the lattice implies oxygen vacancies $V_O^{\bullet\bullet}$ and boundary diffusion neutral interstitials O_i^x [KRO 84]. The grain boundary diffusion of oxygen seems very fast. Most doping agents increase cation mobility (in volume and at the boundaries), with particular mention of Mn, Ti and Fe. On the contrary, MgO does not seem to modify the diffusion kinetics significantly.

6.6. Al$_2$O$_3$-SiO$_2$ system mullite

Argillaceous minerals of the sillimanite group were used as early as in the first era of ceramics, but it is mullite $3Al_2O_3.SiO_2$ that is the only stable crystallized compound in the binary phase diagram of the Al_2O_3-SiO_2. However, mullite is very rare in the state of natural ore, except in some places such as the Scottish island of Mull, which explains its name.

The phase diagram Al_2O_3-SiO_2 has fuelled numerous debates. For Aramaki and Roy [ARA 62], mullite exhibits a narrow field of solubility close to $3Al_2O_3.SiO_2$ (Figure 6.4): this mullite is called 3:2, because of the alumina/silica ratio. It exhibits congruent melting, i.e. it melts to directly yield a liquid whose composition remains unchanged when the temperature continues to rise. This diagram accounted for most of the experiments, but did not explain certain anomalies and Aksay and Pask [AKS 75] modified it (Figure 6.5) by showing that mullite exhibits incongruent melting at $\approx 1,830°C$, with peritectic reaction Al_2O_3-α plus liquid, which forms this mullite with 52.3% in mass of Al_2O_3. However, the stable diagram often gives room to *metastable* extensions. For instance, a monocrystal that has grown in the absence of alumina germs does not have an alumina/silica ratio of 3:2 but 2:1, i.e. 77.2% in mass of Al_2O_3, and the solid solution can extend up to the composition 3:1, with 83.6% of Al_2O_3. A later fine-tuning of the diagram [KLU 90] confirmed the incongruent melting and the metastable extensions, but by modifying some of the limits. It is now admitted that when alumina germs are present, the peritectic reaction occurs and that there is incongruent melting, but for a homogeneous liquid free from these germs, the melting is congruent.

Mullite crystallizes in the orthorhombic system, space group Pbam. Its complex structure accommodates stoichiometry variations (from $3Al_2O_3.SiO_2$ to $3Al_2O_3.iO_2$) thanks to the presence of oxygen vacancies [EPI 91].

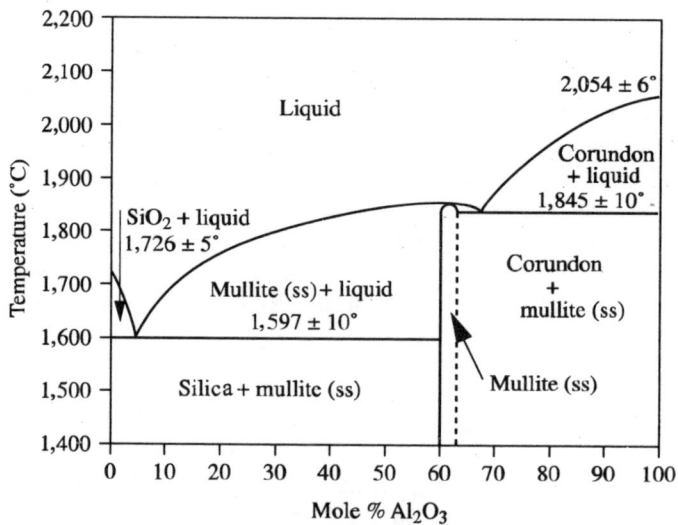

Figure 6.4. *Al_2O_3-SiO_2 diagram [ARA 62]*

In comparison with alumina, mullite is slightly lighter ($\rho \approx 3.2$ g cm^{-3}) and has lower values of hardness ($H_V \approx 14$ GPa), Young's modulus ($E \approx 250$ GPa), mechanical bending strength ($\sigma_F \approx 250$ MPa) and toughness ($K_c \approx 2.5$ MPa m$^{1/2}$). However, its thermal expansion is also less ($\alpha_{20-1,000°C} \approx 6 \cdot 10^{-6}$ K^{-1}), which improves resistance to thermal shocks, and its mechanical strength drops much less quickly when the temperature increases than in the case of alumina: at 1,300°C, most mullite ceramics have a mechanical strength close to the one at room temperature, and some mullites with vitreous segregations even have a peak mechanical strength at about 1,300°C. These characteristics make mullite a material of choice for refractory applications, particularly when it is necessary to consider attacks by silica or silicates, because then an alumina refractory would react to form mullite.

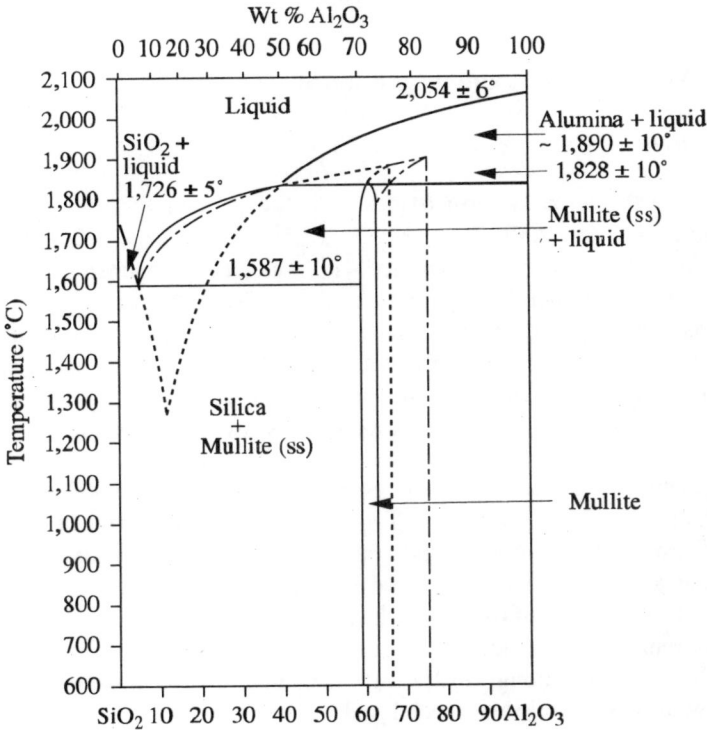

Figure 6.5. *Al$_2$O$_3$-SiO$_2$ diagram: the dotted lines show metastable extensions [AKS 75]*

The manufacture of sintered mullite is not easy. On the one hand, mullite powders are not easily available. Furthermore, the diffusion phenomena are slow in this compound, a characteristic which justifies the good mechanical hot properties – because less diffusion means less creep – but also more difficult sintering. The reactive sintering [BOC 87, 90, GIR 87, ROD 85] of a mixture of alumina and silica powders can be adopted with the use of a mixture of zircon ($ZrSiO_4$) and alumina powders making the preparation of a mullite toughened by zirconia dispersions (ZrO_2) possible. The difficulty is to avoid the presence of silica segregations along the grain boundaries, which would degrade creep resistance. The peak of mechanical strength (a variable measured by applying a load for a short period of time, while creep implies longer periods) at 1,300°C is due to such segregations, here beneficial, because they release the stress concentrations and thus attenuate the manifestations of brittleness.

Mullites with high mechanical performances have hardly made any progress over the last ten years [SOM 90]. The efforts that their development would require have perhaps been overshadowed by the domination of alumina.

6.7. Al₂O₃-MgO system: magnesia and spinel

6.7.1. *Magnesia*

Alkaline-earth elements form oxides with the general formula MO with high melting temperature ($\approx$ 2,800°C for MgO, $\approx$ 2,580°C for CaO, $\approx$ 2,430°C for SrO) [KIN 84, BRE 91]. Owing to reactivity with water and propensity to carbonation, the most common minerals that contain these elements are calcite $CaCO_3$ for calcium, magnesite $MgCO_3$ and its hydrated varieties for magnesium, and dolomite $(CaMg)CO_3$ which combines both cations. Sea water also contains an appreciable proportion of magnesium ($\approx$ 1.3 kg m^{-3}, in the form of chlorides and sulphates) and hydrated magnesia $Mg(OH)_2$ can be extracted by seawater treatment.

The archetype of oxides with ionic bond, magnesia MgO crystallizes in a NaCl-like structure (space group Fm $\overline{3}$ m). The simplicity of this structure, the very low difference from stoichiometry and the possibility of having of very pure monocrystals and perfectly densified polycrystals make MgO a "model material" to understand the properties of oxides, such as the dislocation movement. The plasticity of MgO is remarkable, at least for a ceramic compound: if, below 350°C, slips occurs only according along $(110) < 110 >$, at high temperature ($\approx$ 1,700°C) five independent slip systems are activated and rupture is no longer brittle, because the von Mises criterion is met.

The majority of magnesia produced is used for refractories in the iron and steel industry (see Chapter 10), where the basic oxide properties of the material are necessary (steel-works use the BOF method = basic oxygen furnace). Magnesia withstands the very high temperatures of converters (1,700°C), where it can dissolve several times its weight of iron oxide without melting and it effectively resists sealing and the attack of slag. The high reactivity of magnesia with water is a disadvantage for its use, but this reactivity decreases if we decrease the specific surface area of Chamottes by a high temperature treatment, which also makes impurities, including silicate impurities, form a less reactive skin on the grain surface. This is called dead-burnt magnesia. Calcined dolomite, which is an almost 50:50 mixture of $CaO + MgO$ (in moles), is less sensitive to hydration, but it is carbon products that ensure, in most cases, the protection of magnesia and also less wettability with respect to the attacking agents, such as slag. Magnesia-carbon refractories (about 30% carbon, in various forms) are the reference refractories in steel-works.

6.7.2. *Spinel*

Spinel $MgAl_2O_4$ has given its name to a crystalline structure adopted by several mineral phases. Based on the notation $Mg^{2+}Al_2^{3+}O_4$, Mg^{2+}, Mg^{2+} can be replaced by other divalent cations like Fe^{2+}, Mn^{2+} or Zn^{2+}, and Al^{3+} can be replaced by other trivalent cations like Fe^{3+} or Cr^{3+}. The spinel structure AB_2X_4 (space group Fd3m) is made up of a face-centered cubic stacking of X^{2-} ions (O^{2-} in the case of $MgAl_2O_4$) containing eight structural units per array, i.e. a total of 32 X^{2-} (eight units multiplied by four nodes per CFC array). The cations A and B are placed in the interstitial sites. There are 64 tetrahedral interstices of which eight (Wyckoff 8a positions) are occupied and 32 octahedral interstices of which 16 (Wyckoff 16d positions) are occupied. Based on the occupancy of the interstitial sites, we can distinguish two cases:

 – in a normal spinel, like $MgAl_2O_4$, the trivalent Al^{3+} ions are in octahedral sites and the bivalent Mg^{2+} in tetrahedral sites;

 – in an inverse spinel, like Fe_3O_4 (written as $Fe^{2+}Fe_2^{3+}O_4$), the trivalent Fe^{3+} ions are divided half/half between octahedral sites and tetrahedral sites, and the bivalent Fe^{2+} are in octahedral sites.

These two cases are borderline cases, but there can be various deviations from this distribution. Statistical spinel corresponds to complete disorder where trivalent and bivalent cations are distributed randomly in the A and B sites. Besides II-III spinels that we have just considered, the spinel structure can adapt to cations with other nominal oxidation states: II-IV spinels like Mg_2SiO_4, I-VI spinels like Na_2WO_4 and I-III spinels like $K_2Zn(CN)_4$. Gamma alumina (see section 6.2.2) has a lacunar spinel structure: the aluminum ions are all trivalent, but electric balance is respected by considering that a bivalent entity M^{2+} corresponds to $2/3Al^{3+}$ plus $1/3V_{Al}^x$ (by using the Kröger notation [KRÖ 74, WES 90] of point defects, where V_{AL}^x indicates a constitutional vacancy, whose effective charge compared to the normal occupation of the site is zero).

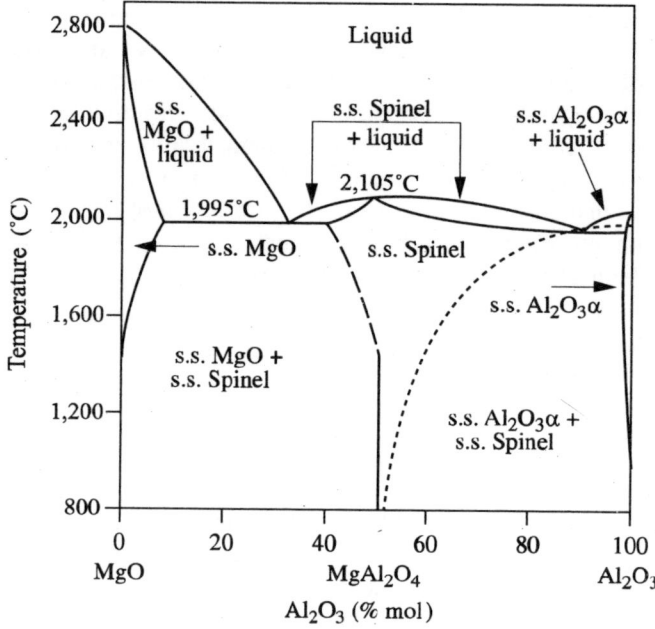

Figure 6.6. *MgO-Al₂O₃ diagram*

The magnetic properties of magnetite Fe_3O_4 (also written as $FeOFe_2O_3$) have been known since antiquity. This iron ferrite (different from "ferrite", the centered cubic form of iron) crystallizes into inverse spinel. There is an antiferromagnetic order between octahedral and tetrahedral sites, therefore the magnetic moments of Fe^{3+} ions cancel each other out and overall magnetization is due only to the sole contribution of Fe^{2+} ions: this is called ferrimagnetism. The magnetic properties of ceramics are the foundation for the tape recording industry (see Chapter 11).

The balance diagram of Al_2O_3-MgO (see Figure 6.6) shows that, at high temperatures, spinel is not stoichiometric, but allows a broad solid solution domain, in particular on the alumina-rich side. The fusion of $MgAl_2O_4$ occurs at ≈ 2,105°C, i.e. more than 50°C above the fusion of alumina. In addition to its refractarity, the interest of spinel as refractory chamotte is due to its possible formation by *in situ* reaction, for example, in concretes made up of magnesia grains bound by an aluminous cement [CHA 98].

Dense sintered spinel is transparent in a broad domain of wavelengths, which explains its use in the manufacture of optical windows (military applications in particular) [BOC 91].

6.8. Zirconia

The most striking characteristic of zirconia is that it exhibits phase transitions that were regarded for a long time as drawbacks for applications, but which, since 1975, have proved to be a rich source of possibilities [SOM 88].

6.8.1. *Polymorphism of zirconia*

Zirconium dioxide (or zirconia: ZrO_2) is found in natural state in the form of baddeleyite (primarily in South Africa), but is more frequently prepared from zirconium silicate sands (zircon: $ZrSiO_4$) by high temperature heat treatments, accompanied by chemical treatments, which eliminate the siliceous fraction from the zircon.

Zirconia is an oxide with very high melting temperature ($T \approx 2,880°C$), which solidifies in cubic phase (ZrO_2-c, group space $Fm\overline{3}m$), then transforms ($T < \approx 2,370°C$) to tetragonal phase (ZrO_2-t, $P4_2/nmc$) and finally, below $\approx 1,170°C$, becomes monoclinical (ZrO_2-m, $P2_1/c$). This last transition t→m is accompanied by considerable dimensional variations (shear strains of ≈ 0.16 and increase in volume of $\approx 4\%$), which largely exceed the maximum stress limit, resulting in a fragmentation of the material. A zirconia part sintered at temperatures that the refractarity of this oxide requires – let us say sintered at about 1,500°C – breaks up and is destroyed during cooling, during the t→m transition. This means that "pure" zirconia can be used only in powder form (for example, as starting product for the manufacture of ceramic enamels), and therefore for uses that do not require consolidation into a massive part. To produce zirconia sintered pieces, ZrO_2 must be combined with other oxides known as "stabilizers" (M_xO_y= primarily CaO, MgO or Y_2O_3): the ZrO_2-M_xO_y phase diagram is then modified favorably, which helps preserve (at the stable state or metastable state) a "stabilized zirconia", free from transitions in the entire useful temperature range – in practice from the sintering temperature to room temperature. In the ZrO_2-CaO diagram, for example, it is observed that for 20 mol% CaO, the material remains in cubic phase from room temperature to practically the melting temperature.

Among the uses of stabilized zirconia, denoted "SZ", we can mention four main fields:

– the production of monocrystals for jewelry, because the optical properties of zirconia are not very different from those of a diamond, at an incomparably lower cost;

– the manufacture of crucibles and other refractory parts, because of the high melting temperature and good resistance to corrosive mediums, including molten glass (refractories of A-Z-S system: Al_2O_3-ZrO_2-SiO_2);

– the manufacture of thermal barriers, for example deposited by plasma spraying using plasma torches for the internal protection of the combustion chamber of jet engines, because the thermal conductivity of zirconia is one of lowest ever known among non-metallic inorganic solids ($k \approx 1 \ W \cdot m^{-1} \cdot K^{-1}$, i.e. 30 times lower than alumina);

– the manufacture of ionic conductors. Although this last aspect is detailed in Chapter 11, we must mention it here because it is related to the same mechanism of "stabilization". ZrO_2-c has a fluorite structure (like, for example, UO_2), with the zirconium atoms at the nodes of a face-centered cubic lattice and the oxygen atoms occupying all the tetrahedral interstices of this lattice (four nodes for a CFC array and eight tetrahedral sites per array give the stoichiometry ZrO_2). The unit is very compact and we can say that the transitions that occur when we cool ZrO_2-c come from the need to decompress the structure. However, the introduction of bivalent (CaO and MgO) or trivalent (Y_2O_3) metal oxides that are used as stabilizers requires, for the charges to be balanced, that the substitution of Zr^{4+} by M^{2+} or M^{3+} is compensated by the presence either of interstitial cations or anionic vacancies. It is the second case that occurs here: a usual SZ, 20% CaO-80% ZrO_2 (in moles), therefore contains a 20% deficit of oxygen atoms: $Zr_{1-x}Ca_xO_{2-x}V_{Ox}$, in Kröger-Vink notation, where V_O represents oxygen vacancies and x is equal to 0.2. The anionic sublattice is thus decompressed and this considerable concentration in vacancies – more than ten orders of magnitude higher than the normal concentration of defects in thermodynamic equilibrium – allows a considerable mobility of residual oxygen ions, since we know that the diffusion proceeds primarily by a vacancy mechanism, like a puzzle. By virtue of these constitutional oxygen vacancies, the stabilized zirconia offers properties of ionic conduction that allow its application as solid electrolyte, particularly in oxygen sensors and in solid oxide fuel cells [SIN 00].

6.8.2. Ceramic steel?

SZ has rather modest mechanical properties, significantly less remarkable compared to alumina which, associated with higher density, higher thermal expansion (consequently greater sensitivity to thermal shocks) and markedly increased costs explain why these stabilized zirconias *a priori* do not have a mechanical application.

It is partially stabilized zirconias (PSZ) that have justified the resounding article ("Ceramic steel?") published in 1975 by Garvie *et al.* [GAR 75]. The title suggests that a ceramic can exhibit the high mechanical performances associated with steel, but also that toughening mechanisms recall those used by steel manufacturers. The t→m transformation of zirconia is a martensitic transformation, in analogy with the transformation used to obtain martensite in tempered steels, and the role of microstructural parameters in ZrO_2 is similar to what is observed in metals.

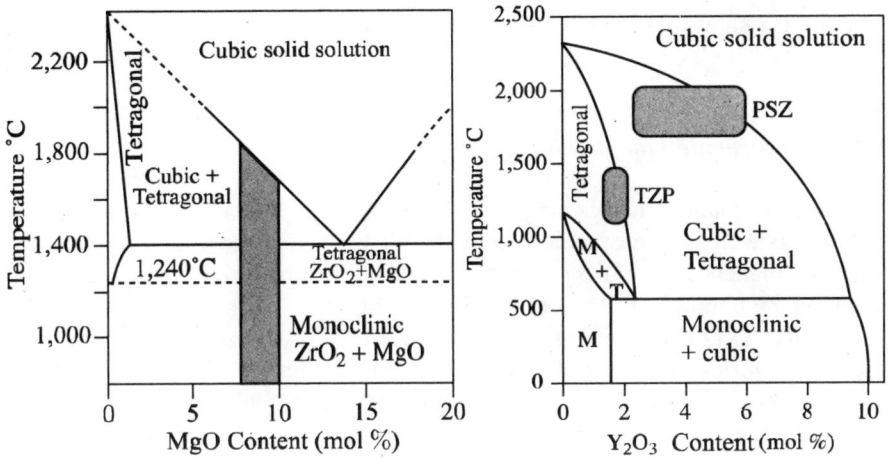

Figure 6.7. *ZrO₂-MgO and ZrO₂-Y₂O₃ diagrams; the hatched zones show the compositions normally chosen for ceramics with TT effect [HAN 00]*

The mechanical properties of zirconias with high mechanical performances – of which there are multiple varieties – constitute one of the themes that have inspired the greatest number of publications in the field of ceramics. [HAN 00] constitutes an excellent study on the subject.

Let us start by analyzing what it is about, by reasoning at room temperature. A non-stabilized zirconia (Z) is subjected, during its cooling from its processing temperature, to destructive phase transition t→m: its mechanical performances are therefore almost zero. A stabilized zirconia (SZ) is free from these transitions; it remains in general in cubic phase: its mechanical performances are modest. Lastly a partially stabilized zirconia (PSZ) can be made up of a matrix rich in stabilizers, therefore in the form of SZ, within which there is a dispersion of small precipitates of zirconia poor in stabilizers, which should be in monoclinic form but can subsist in a metastable state in tetragonal form. The propagation of a microscopic crack relaxes the stresses applied by the matrix on these precipitates, which enables them to switch to the monoclinic, stable state. Hence, local swelling (increase in volume of ≈ 4% associated with the t→m transformation) which "clamps" the crack and stops its propagation: we can thus considerably increase mechanical strength and toughness. However simplistic this explanation might be, it highlights the two essential points:

i) the t→m transformation, previously considered as a drawback, can become an advantage;

ii) the control of microstructural characteristics (for instance, grain size) is essential.

Figure 6.7 shows the ZrO_2-CaO and ZrO_2-Y_2O_3 phase diagrams, where the main PSZ materials are located.

6.8.3. Transformation toughening

In transformation toughening (TT), toughening indicates the increase in toughness as well as in mechanical strength. We have the following acronyms: ZTC (zirconia toughened ceramics), Ca-PSZ or Mg-PSZ (partially stabilized zirconia containing ≈ 8 mol% CaO or ≈ 9 mol% MgO), TZP (tetragonal zirconia polycrystals, Y-TZP containing typically 2-3 mol% Y_2O_3), etc. There are indeed various categories of materials depending on the nature of the stabilizer, its concentration, microstructure and various associated phases: ZTA (zirconia toughened alumina) is for example alumina toughened by a zirconia dispersion, prepared in such a manner that TT mechanisms are operational.

The application of TT mechanisms to manufacture TTCs (transformation toughened ceramics) still requires that we retain a fraction of metastable ZrO_2-t, capable of being transformed to ZrO_2-m, at the usage temperature (close to room temperature), under the effect of the applied stresses; recent works have shown that in addition to elongation and variation of volume, these stresses imply significant shearing.

As in steels, martensitic transformation t→m is an instantaneous transformation, displacive in nature, which develops when temperature decreases. In pure ZrO_2, cooling transformation starts at about 950°C (point known as M_S) and reversible heating transformation occurs beyond 1,150°C (A_s). We can summarize the crystallographic aspects by saying that the structures "t" and "m" derive from the fluorine structure "c" by various distortions, the most important of which is the one associated with the t→m transition, with a shearing of $\approx 9°$ parallel to the base plan of the array "t" to lead to an angle of the monoclinic cell $\beta \approx 81°$.

We have said that there are multiple categories of TTCs [CLA 85]. To limit ourselves to the three primary categories: i) in Mg-PSZ, the toughening is due to a dispersion of lenticular ZrO_2-t precipitates, ii) in TZP (Y-TZP or Ce-TZP), the material has very fine, monophase ZrO2-t grains and iii) in DZC (dispersed zirconia ceramics, like ZTA), the ZrO_2-t particles are dispersed within an alumina matrix (see Figure 6.8).

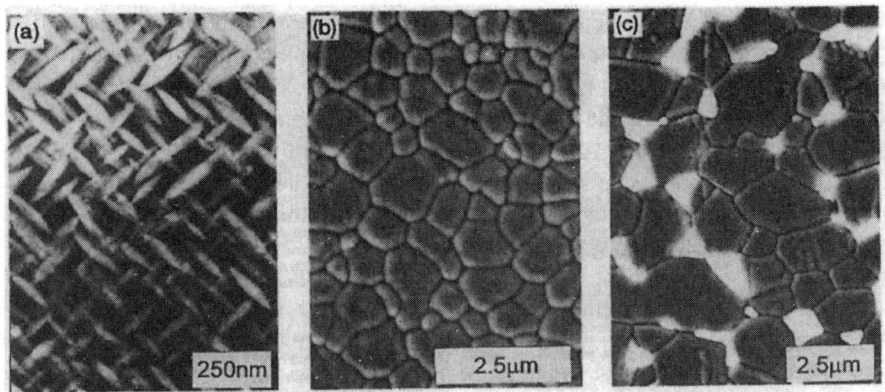

Figure 6.8. *Typical microstructures of three materials toughened by zirconia: (a) Mg-PSZ, (b) Y-TZP, (c) ZTA. The transformable precipitates in Mg-PSZ are lenticular in form; TZP is monophased, with very fine grains [HAN 00]*

Phenomenologically, we can write the toughness of a multiphase brittle material as:

$$K_{1C} = K_0 + \Delta K_C \qquad [6.1]$$

where K_0 is the toughness of the matrix and ΔK_C the contribution of specific toughening mechanisms.

For TT, the transformation of a metastable particle "t" into a stable particle "m" in the stress field associated with the propagation of a potentially dangerous crack originates increased toughness, because on the one hand the increase in volume due to the transformation places the crack front under compression and on the other hand the strain energy associated with the shearing components contributes to an increase in the rupture energy. Lastly, the expansion of the transformed particle can lead to local microcracking, hence the beneficial effects of crack bifurcation. The interaction of these effects results in three main mechanisms: i) ΔK_{CT} (transformation toughening), ii) ΔK_{CM} (transformation-induced microcrack toughening) and iii) ΔK_{CD} (crack deflection toughening).

The essential term is ΔK_{CT}, which can be written [HAN 00]:

$$(1\text{-}v)\Delta K_{CT} = \eta E^* e_T V_f h^{1/2} \qquad [6.2]$$

where η is a characteristic geometrical factor of the crack front and the stress field, E* the effective Young's modulus of the material used as matrix e_T the expansion strain, V_f the volume fraction of the transformed particles, h the half height of the transformed zone, and v the Poisson's ratio. The elasticity characteristics of the matrix play an important role. For example, an alumina matrix (E* $\approx$ 380 GPa, $v \approx 0.2$) places more stress on transformable particles than a zirconia matrix (E* $\approx$ 210 GPa, $v \approx 0.3$), causing a reduction in the thickness of the transformed zone h and the transformed volume V_f.

These effects result in microplasticity phenomena, which explain a non-linear stress-strain behavior in the zones close to the transformation (for example, close to the edges of a microhardness indentation).

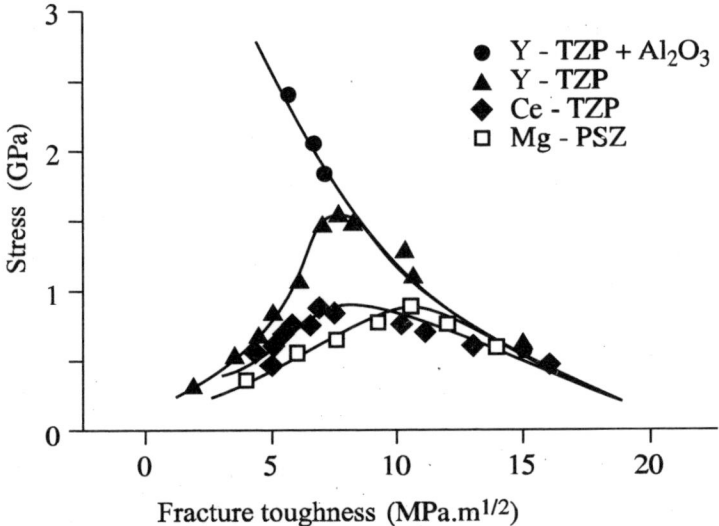

Figure 6.9. *Mechanical strength/toughness relation for various materials toughened by ZrO$_2$. The maximum of the curve corresponds to the transition from a rupture controlled by the size of the cracks to a rupture controlled by the triggering threshold of the t→m transformation [SWA 85]*

We cannot cumulate maximum increase in toughness and maximum increase in mechanical strength and we must choose between increasing one or the other. Figure 6.9 shows this compromise. For materials with $K_{1C} < 8$ MPa·m$^{1/2}$, it is the size of the critical crack that limits the mechanical strength σ_f (according to the usual law $\sigma_f \approx K_{1C} a_C^{-1/2}$, where a_C is the equivalent size of the critical crack). For $K_{1C} > 8$ MPa·m$^{1/2}$, it is the level of stress inducing the transformation that limits σ_f.

the transformation consumes, at low stress levels, the entire metastable fraction of zirconia, after which the rupture is controlled again by the size of the microcracks.

An interesting effect made possible by the toughening with the help of transformable zirconia is the effect known as "R curve", thanks to which non-linear phenomena close to the crack front lead to a growth in toughness K_{1C} when the crack propagates itself (Figure 6.10). In the absence of such an effect, the starting of a crack is generally catastrophic when the stresses that act on it reach the critical point. In the presence of an R curve effect, on the contrary, the development of damaged zones (process zone and shielding zone) can increase the toughness sufficiently for the crack to become "subcritical"; it is only a later increase in stress that would allow the continuation of its propagation.

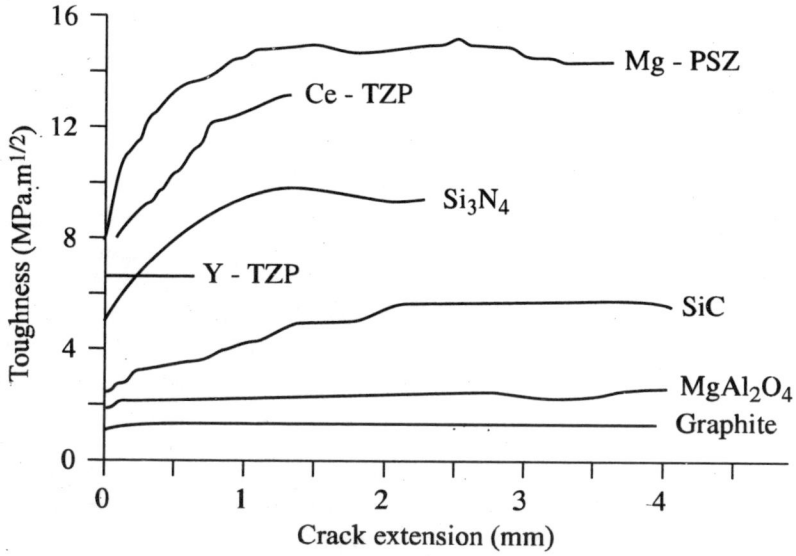

Figure 6.10. *Toughening depending on the propagation of the crack ("R curve" effect). This effect is pronounced in Mg-PSZ, Ce-TZP and Si₃N₄; less pronounced in SiC; non-existent in Y-TZP, spinel MgAl₂O₄ and graphite [HAN 00]*

6.8.4. *ZrO₂ based materials: treatments and microstructures*

The management of the toughening is delicate: excessively stabilized, a zirconia exhibits mechanical properties that are hardly improved; insufficiently stabilized, it can be subjected to destructive phases transformation. Furthermore, we must negotiate between the reinforcement of toughness and the reinforcement of mechanical strength. Other factors also intervene, like resistance to temperature effects or destabilization and economic givens (for instance, Y_2O_3 is a definitely more expensive stabilizer than MgO).

The stabilization of ZrO_2-t can be achieved by: i) additions of bivalent or trivalent metals, which introduce oxygen vacancies: Ca^{2+}, Mg^{2+}, Gd^{3+}, Fe^{3+}, Ga^{3+} and Y^{3+} and ii) trivalent cations but of size higher or lower than that of Zr^{4+} Ti^{4+}, Ge^{4+} and Ce^{4+}

PSZ materials are bi-phased, generally in CaO-ZrO_2 or Mg-ZrO_2 diagrams. Based on this last case (see Figure 6.7), we choose compositions with $\approx$ 9 mol% MgO, cooled quickly, which leads to a dispersion of lenticular precipitates of very small-sized ZrO_2-t (30–60 nm), therefore too stable (their temperature M_S is below room temperature). A subsequent treatment for the enlargement of precipitates (Figure 6.8) must then be carried out, at about 1,400°C, i.e. above the eutectoid transformation temperature. We can also proceed by adjusting the post-sintering cooling speed to a sufficiently low value to allow an enlargement of the precipitates, which avoids the subsequent heat treatment.

TZP materials generally use Y_2O_3 or CeO_2 as stabilizing oxides. For Y-TZP (see Figure 6.7), the compositions vary from 1.75 to 3.5 mol% (3; 5 to 8.7% in mass). A sintering performed at about 1,500°C can then yield a monophase polycrystal (100% tetragonal) made of very fine equiaxed grains (0.5–2 μm) (see Figure 6.8). Y-TZP materials exhibit the highest mechanical performances, but they are degraded when maintained in wet heat; from this point of view, Ce-TZPs exhibit much better properties. Figure 6.9 shows that Y-TZPs are unequalled as regards mechanical strength, but that Ce-TZPs have a much higher toughness (K_{CT}Maxi $\approx$ 10 MPa·m$^{1/2}$ for Y-TZP and 17 MPa.m$^{1/2}$ for Ce-TZP).

As regards non-zirconia matrices toughened by zirconia, the most widespread materials are based on Al_2O_3 (ZTA). When the zirconia fraction is not stabilized, its expansion causes a microcracking of the matrix (microcrack toughening); when this is a TZP fraction, it allows a TT mechanism. Cutting tools use these ZTA materials. Non-oxide materials (including SiC or Si_3N_4) have been the subject of studies, but have not yet experienced any industrial development. Lastly, reasoning on a large scale – i.e. beyond particulate composites that we have been discussing in order to speak about composite structures – we can use TT effects in, for example, laminates where y% (M_aO_b)(100-y)% ZrO_2 alternate layers and z% (M_aO_b) (100-z)% ZrO_2, M_aO_b being for example Al_2O_3 [BES 87].

To conclude on zirconias with mechanical uses, we must underline that TT mechanisms cease to operate at temperatures too close to the M_S point: TTCs therefore do not compete with non-oxide ceramics like SiC or Si_3N_4, which have been developed to offer high mechanical strength or creep resistance at high temperatures. Even at modest temperatures (T $\approx$ 200°C) the effectiveness of transformation toughening is reduced on the one hand, and TTCs can suffer from harmful microstructure deteriorations on the other hand: it is only around room temperature that these "ceramic steels" are at their best. Their uses are primarily intended for things that cut, rub or wear – because the tribological performances are very satisfactory – for instance, knife blades or scissors, scrapers, linings, valve seats or guides (automobile), wear parts (paper industry), extrusion nozzles or die (metallurgy), etc.

6.9. Other oxides

Alumina, materials of the binary systems Al_2O_3-SiO_2 and Al_2O_3-MgO, and zirconia do not exhaust the list of oxides with ceramic interest [RYS 85]; far from it.

We may repeat that SiO_2 and the compositions in the ternary system SiO_2-CaO-Na_2O are the basic ingredients in glass industries and that the quaternary system CaO-SiO_2-Al_2O_3-Fe_2O_3 contains the compositions of cement *clinkers*. Chapter 11 devoted to ceramics for electronics, considers oxides with magnetic performances (for example, ferrites with spinel structure) and also barium titanate $BaTiO_3$ and zinc oxide ZnO, not to speak of beryllium oxide BeO or superconductive cuprates of which $YBa_2Cu_3O_7$ is the archetype.

Chapter 12 underlines, once again, the merits of alumina, as a material with low biological reactivity and shows the possibilities of calcium phosphates, with high biological reactivity. Chapter 11 focuses on uranium and plutonium oxides.

Among the oxides with industrial interest that this book, on account of its constraints, has been forced to omit, or to which it has devoted only very little place, we can mention TiO_2, chromium oxides and oxides of other transition metals, tin oxide and indium oxide, yttrium oxide Y_2O_3 and rare earth oxides, highly refractory oxides like ThO_2 or HfO_2, and the various oxides developed in the form of monocrystals, for jewelry or professional uses (for example, radiation detectors). Either these oxides have very specialized uses (for example, ultra-refractories like HfO_2), or their main use falls outside the scope of ceramic industries (for example, TiO_2, which is used primarily as a pigment).

It may be recalled that the *Journal of the European Ceramic Society* and *Journal of the American Ceramic Society* are the most valuable sources for information on ceramics not dealt with in this book.

6.10. Bibliography

[AKS 75] AKSAY I.A. and PASK J.A., "Stable and metastable equilibria in the system SiO_2-Al_2O_3", *J. Am. Ceram. Soc.,* 58(11-12), p. 507-512, 1975.

[ALC 99] website: http://www.alumina.alcoa.com/applications.

[ARA 62] ARAMAKI S. and ROY R., "Revised phase diagram for the system SiO_2-Al_2O_3", *J. Am. Ceram. Soc.,* 45, p. 229-242, 1962.

[BAC 98] "Annual Mineral Review", *Am. Ceram. Bull.,* 78, 8, p. 118-121, 1999.

[BAE 94] BAE S.I. and BAIK S., "Critical concentration of MgO for the prevention of abnormal grain growth in alumina", *J. Am. Ceram. Soc.,* 77(101), p. 2499, 1994.

[BEN 85] BENNISON S.J. and HARMER M.P., "Grain growth kinetics for alumina in the absence of a liquid phase", *J. Am. Ceram. Soc.,* 68, C22-4, 1985.

[BES 87] BESSON J.L., BOCH P. and CHARTIER T., "Al_2O_3-ZrO_2 layer composites", *High Technology Ceramics*, Materials Science Monographs 38, Elsevier, 1987.

[BOC 87] BOCH P. and GIRY J.P., "Preparation of zirconia-mullite ceramics by reaction sintering", *High Technology Ceramics*, Materials Science Monographs 38, Elsevier Edit. 1987.

[BOC 88] BOCH P. and CHARTIER T., "Understanding and improvement of ceramic processes: the example of tape casting", *Ceramic Developments, Materials Science Forum*, Trans. Tech. Publ., p. 813, 1988.

[BOC 90] BOCH P., CHARTIER T. and RODRIGO, "High purity mullite by reaction sintering", *Mullite and Mullite Matrix Composites, Ceramic Transactions,* Vol. 6, The Am. Ceramic Society, p. 353, 1990.

[BOC 91] BOCH P. and RAYNAL P., "Fabrication of transparent magnesium aluminium oxide (MgAl2O4) ceramics", *Mat. Sc. Monographs 66D*, P. Vincenzini Edit., Elsevier Publ., p. 2343, 1991.

[BOC 99] BOCH P., LEQUEUX, MASSE S. and NONNET E., "High-performance concretes are exotic materials", *Z. Mettalk.,* 90, p. 12, 1-5, 1999.

[BRE 91] BRETT N.H., "Magnesium and alkaline-earth oxides", p. 283-286, in [BRO 91].

[BRO 91] BROOK R.J. (ed.), *Concise Encyclopedia of Advanced Ceramic Materials*, Pergamon Press, 1991.

[BUR 84] BURNS R.G. and BURNS V.M., "Optical and Mössbauer spectra of transition-metal-doped corundum and periclase", p. 46-61, in [KIN 84].

[CAS 90] CASTEL A., *Les alumines et leurs applications*, Nathan, 1990.

[CEA 99] THEODORE F., "Des saphirs haute technologie", *CEA Technologies*, 45, p. 2, 1999.

[CER 99] *Ceramics Monographs, Handbook of Ceramics*, updated by *Interceram*, Verlag Schmidt (since 1982).

[CHA 98] CHAN C.F. and KO Y.C, "Effect of CaO content on the hot strength of alumina–spinel castables in the temperature range of 1,000° to 1,500°C", *J. Am. Ceram. Soc.*, 81(11), p. 2957-2960, 1998.

[CLA 85] CLAUSSEN N., RUHLE M. and HEUER A.H., *Advances in Ceramics*, vol. 12, *Science and Technology of Zirconia II*, Am. Ceram. Soc., 1985.

[COB 61] COBLE R.L., "Sintering crystalline solids: I Intermediate and final stage diffusion models; II. Experimental test of diffusion models in powder compacts", *J. Appl. Phys.*, 32, p. 787-799, 1961.

[COB 63] COBLE R.L., "Model for boundary diffusion controlled creep in polycrystalline materials", *J. Appl. Phys.*, 34(6), p. 1679-1682, 1963.

[DÖR 84] DORRE E. and HUBNER H., *Alumina*, Springer-Verlag, 1984.

[EPI 91] EPICIER T., "Benefits of high-resolution electron microscopy for the structural characterization of mullites", *J. Am. Ceram. Soc.*, 74(10), p. 2359-2366, 1991.

[GAR 75] GARVIE R.C., HANNINCK R.H.J and PASCOE R.T., "Ceramic Steel?", *Nature*, 258, p. 703-704, 1975.

[GER 96] GERMAN R.M., *Sintering Theory and Practice*, John Wiley, 1996.

[GIR 87] GIRY J.P., Etude du frittage-réaction alumine-zircon: préparation et propriétés des céramiques zircone-mullite, Thesis, University of Limoges, February 1987.

[GIT 70] GITZEN W.H., "Alumina as a ceramic material", *The Am. Ceram. Soc.*, 1970.

[GOR 84] GORDON R.S., "Understanding defect structure and mass transport in polycrystalline Al_2O_3 and MgO via the study of diffusional creep", p. 418-437, in [KIN 84].

[HAN 00] HANNINCK R.H.J., KELLY P.M. and MUDDLE B.C., "Transformation toughening in zirconia-containing ceramics", *J. Am. Ceram. Soc.*, 83(3), p. 461-486, 2000.

[HAR 84] HARMER M.P., "Use of solid-solution additives in ceramic processing", *Advances in Ceramics*, Vol. 10, p. 679, Am. Ceram. Soc., 1984.

[HER 50] HERRING C., "Diffusional viscosity of a polycrystalline solid", *J. Appl. Phys.*, 21(5), p. 437-445, 1950.

[HEU 84] HEUER A.H. and CASTAING J., "Dislocations in α-Al_2O_3", p. 238-257, in [KIN 84].

[JAC 94] Special series on alumina, *J. Am. Ceram. Soc.*, 77(2), 1994.

[KIN 76] KINGERY W.D., BOWEN H.K. and UHLMANN D.R., *Introduction to Ceramics*, 2nd edition, John Wiley and Sons, 1976.

[KIN 84] KINGERY W.D. (ed.), "Structure and properties of MgO and Al_2O_3 Ceramics", *The Am. Ceram. Soc.*, 1984.

[KLU 90] KLUG F.L., PROCHAZKA S. and DOREMUS R.H., "Alumina-silica phase diagram in the mullite region", p. 750-759, in [SOM 90].

[KRO 74] KROGER F.A., *The Chemistry of Imperfect Crystals*, North Holland, 1974.

[KRO 84] KROGER F.A., "Electrical properties of α-Al$_2$O$_3$", p. 1-15, in [KIN 84].

[KRO 84] KROGER F.A., "Experimental and calculated values of defect parameters and the defect structure of α-Al$_2$O$_3$", p. 100-118, in [KIN 84].

[KRO 57] KRONBERG M.L., "Plastic deformation of single crystals of sapphire: basal slip and twinning", *Acta Metallurgica*, 5, p. 508-527, 1957.

[LAB 80] LABELLE H.E., "EFG, the invention and application to sapphire growth", *J. Cryst. Growth*, 50, p. 8-17, 1980.

[LAN 91] LANGDON T.J., "Creep", p. 92-96, in [BRO 91].

[LEE 94] LEE W.E. and RAINFORTH W.M., *Ceramic Microstructures*, Chapman & Hall, 1984.

[MOR 85] MORRELL R., *Handbook of properties of technical & engineering ceramics, Part 1, An introduction for the engineer and designer*, Her Majesty's Stationery Office, 1985.

[MOR 85] MORRELL R., *Handbook of properties of technical & engineering ceramics, Part 2, Data Reviews*, Her Majesty's Stationery Office, 1987.

[NAS 83] NASSAU K., *The Physics and Chemistry of Color*, J. Wiley, 1983.

[NOR 99] NORTON–SAINT-GOBAIN, website: http://www.nortadvceramics.thomasregister.com/olc/nortadvceramics/home.htm.

[PEC 99] Catalogue des alumines, Péchiney, 1999.

[RHO 91] RHODES W.H. and WEI G.C., "Lamp envelopes", p. 273-76, in [BRO 91].

[RIN 96] RING T., *Fundamentals of Ceramic Powder Processing and Synthesis*, Academic Press, 1996.

[ROD 85] RODRIGO P.D.D. and BOCH P., "High purity mullite ceramics by reaction sintering", *Int. J. High Technology Ceramics*, 1, p. 3-30, 1985.

[RYS 85] RYSHKEWITCH E. and RICHERSON D.W., *Oxide Ceramics*, Academic Press, 1985.

[SCH 91] SCHWABEL M.G., "Alumina abrasive grains produced by sol-gel technology", *Am. Ceram. Bull.*, 70, 10, p. 1596-1598, 1991.

[SIN 00] SINGHAL S.C., "Science and technology of solid-oxide fuel cells", *MRS Bull.*, 25, 3, p. 16-21, 2000.

[SOM 88] SOMIYA S., YAMAMOTO N. and YANAGIDA H., "Advances in ceramics", vol. 24A and B, Science and Technology of Zirconia III, *The Am. Ceram. Soc.*, 1988.

[SOM 90] SOMIYA S. (ed.), "Mullite and mullite matrix composites", Ceram. Trans. 6, *The Am. Ceram. Soc.*, 1990.

[SWA 85] SWAIN M.V., "Inelastic deformation of pg-psz and its significance for strength-toughness relationship of zirconia-toughened ceramics", *Acta Metall.*, 33, 11, p. 2083-2091, 1985.

[WES 90] WEST A.R., *Solid State Chemistry and its Applications*, J. Wiley, 1990.

Chapter 7

Non-oxide Ceramics

7.1. Introduction

Non-oxide ceramics essentially comprise carbides, nitrides, silicides and borides. These chemical compounds, considered until 1970 as materials that had high melting points, that were very stable thermally and that exhibited high chemical inertia [ALI 79, PAS 65], were primarily used in the iron and steel industry or in the chemical industry. During the last three decades, the rapid development of new processes of powder synthesis or shaping of parts, combined with a better knowledge of their mechanical, thermal and electric properties, have paved the way for their use as sintered materials, fibers, monocrystals or coatings. The needs for new high performance ceramics for various economic sectors (mechanical engineering, aeronautics, electronics, nuclear) have been the driving force for their very diversified development [FRE 90, SAI 88]. Thus, for example, silicon carbide powder, used for a very long time for its hardness as an abrasive, can be used when it is sintered as a structural ceramic or semiconductor for components operating at high temperatures. Boron carbide, which has a remarkable hardness, is used for high temperature thermo-electric conversion or in the nuclear industry as a protective barrier and neutron speed reducer.

Chapter written by Paul GOURSAT and Sylvie FOUCAUD.

7.2. Synthesis of non-oxides

7.2.1. *Powders*

7.2.1.1. *Carbides*

There are three main families:

– alkaline metal or alkaline-earth carbides;

– interstitial carbides with transition metals Ti, Zr, Hf, V, Nb, Ta, etc.;

– covalent macromolecular carbides with B, Al, Si.

Only compounds of the two last families are used for preparing ceramics. These products are synthesized industrially according to various methods, the choice of the process depending on certain criteria: purity, crystalline state, morphology of the powder.

Solid-solid reaction

Carbides can be obtained by direct synthesis at high temperature by mixing powders compacted beforehand:

$$Si + C \xrightarrow{T = 1,400°C} SiC$$
$$W + C \xrightarrow{T = 1,500°C} WC$$
$$Ta + C \xrightarrow{T = 2,200-2,400°C} TaC$$

After the carburization stage, the products are ground. This process, which is the oldest, does not yield fine powders.

The carboreduction of oxides is the most developed production method:

$$TiO_2 + 3C \xrightarrow{T = 2,00-2,200°C} TiC + 2CO$$
$$SiO_2 + 3C \xrightarrow{T = 1,000-1,300°C} \alpha SiC + 2CO$$

Grinding is also necessary, followed by a chemical treatment to eliminate oxide traces that very often play a harmful role during sintering. Thus, for the second example, after washing with hydrofluoric acid, the powder contains 98-99% of SiC, with iron as the main impurity.

Sol-gel processes developed recently, in particular in the synthesis of carbosiloxanes, yield ultrafine amorphous powders. The distribution on a nanometric scale of the Si, C, O elements increases their reactivity. Amorphous silicon carbide can be synthesized between 1,000 and 1,300°C. It then crystallizes into β-SiC at about 1,450°C.

Gas-gas reaction

A large number of gas phase reactions can be developed to manufacture carbides:

$$4BCl_3 + CH_4 + 4H_2 \xrightarrow{\text{1,300-1,500°C}} B_4C + 12HCl$$
$$SiH_4 + CH_4 \xrightarrow{T=800-1,500°C} SiC + 4H_2$$
$$CH_3 - SiH_3 \xrightarrow{\text{1,600°C}} SiC + 3H_2$$

During these last few years, synthesis methods "assisted" by the use of plasma or a lazer beam have been developed. They offer several advantages compared to conventional methods; in particular, they lead to ultrafine multi-element powders (mixed carbides) of uniform size (a few dozen nanometers).

7.2.1.2. Nitrides

Nitrides can, like carbides, be classified under three main families. Actually, only interstitial (TiN, ZrN, HfN, TaN) and macromolecular covalent (Si₃N₄, AlN, BN) nitrides have been developed industrially.

7.2.1.2.1. Solid-gas reaction

The oldest production method consists of nitriding an element by nitrogen, ammonia or N₂/H₂ mixtures:

$$3Si + 2N_2 \xrightarrow{T=1,300-1,400°C} Si_3N_4$$
$$2Al + N_2 \xrightarrow{T=1,300-1,500°C} 2AlN$$
$$2Ti + N_2 \xrightarrow{T=800-900°C} 2TiN$$

Extensive research has shown that the nitriding of ultrapure silicon is very slow because of the formation of the protective nitride coating on the grains. "Catalysts" are necessary, generally containing iron, to achieve a complete reaction [RIL 83].

As starting powders have high grain size in order to avoid excessive sintering and a blockage of the reaction, the products obtained are ground and then purified chemically.

The thermal carboreduction of oxides in the presence of nitrogen is also used. The overall reactions are expressed below:

$$3SiO_2 + 6C + 2N_2 \xrightarrow{T=1,400-1,500°C} Si_3N_4 + 6CO$$
$$2AlO_3 + 3C + 2N_2 \xrightarrow{T=1,400-1,500°C} 2AlN + 3CO$$

The yield and rate of the reaction can be enhanced by the use of ammonia or nitrogen/ammonia mixtures.

The powders manufactured are finer than with the previous process, but with a purity that varies with the conversion rate. In particular, the excess carbon necessary for the reaction to be complete must be eliminated by combustion in air at temperatures lower than 600-700°C to avoid a parasitic oxidation.

7.2.1.2.2. Gas-gas reaction

This refers to high temperature reactions between gas precursors, followed by the formation of solid particles by germination-growth:

$$BCl_3 + NH_3 \xrightarrow{T=1,200-1,500°C} BN + 3HCl$$

$$3SiCl_4 + 4NH_3 \xrightarrow{T=1,000-1,200°C} Si_3N_4 + 12HCl$$

$$AlCl_3 + NH_3 \xrightarrow{T=900-1,000°C} AlN + 3HCl$$

As for carbides, this process which leads to ultrafine powders has not yet reached a stage of industrial development.

7.2.1.2.3. Liquid-liquid reaction

Liquid medium reactions can synthesize, from mineral or organometallic compounds, precursors which by pyrolysis lead to amorphous nitrides:

$$Al-(C_2H_5)_3 + NH_3 \xrightarrow{solvent:\ benzene} (C_2H_5)_3\ AlNH_3$$

$$(C_2H_5)_3\ AlNH_3 \xrightarrow{T=20-400°C} AlN + 3C_2\bar{H}_6$$

$$NaBH_4 + (NH_4)_2CO_3 \xrightarrow{solvent:\ THF} 2NH_3BH_3 + Na_2CO_3 + 2H_2$$

$$NH_3BH_3 \xrightarrow{T=20-800°C} BN + 3H_2$$

If this remains a laboratory process for aluminum or boron nitride, for silicon nitride it has reached the stage of industrial production. Silicon tetrachloride dissolved beforehand in a cyclohexane-benzene mixture reacts at –40°C with liquid ammonia to form solid polymeric silicidimide:

$$x\ SiCl_4 + 6x\ NH_3 \longrightarrow [Si(NH)_2]_x + 4x\ NH_4Cl$$

Ammonium chloride is extracted by liquid ammonia washing and $[Si(NH_2)]_n$ recovered by filtration. Pyrolysis under nitrogen up to 1,200°C leads to nitride Si_3N_4, whose grain diameter is submicronic:

$$3Si(NH)_2 \longrightarrow Si_3N_4 + 2NH_3$$

Lastly, we must mention the precursor-organometallic method, which is similar to the sol-gel method for oxides. According to the degree of polymerization of the precursors, we can manufacture by pyrolysis the following: powders, coatings or long fibers. As regards powders, the advantage of this procedure lies in the possibility of obtaining multi-element powders: SiC, SiCO, SiCNO, SiBCN, etc. These alloys, which cannot be prepared by traditional methods, exhibit a variable state of crystallization according to the experimental conditions.

7.2.2. Fibers

The oldest method used to manufacture non-oxide fibers is a decomposition of precursors (Cl_3SiCH_3/H_2, BCl_3/CH_4) on a filament substrate (C, W) heated by the Joule effect. This method presents the disadvantage of yielding fibers with large diameters (80–150 μm) at a high cost.

In the last two decades, a new process has been developed, very similar to one used for carbon fibers [YAJ 76]. It consists of a thermal decomposition or pyrolysis of organometallic polymer fibers containing silicon or boron. This method is used to manufacture long weavable fibers with low diameter (8–15 μm), controlled composition and state of crystallization. This process generally includes five stages summarized below for silicon carbide fiber, which is currently the most developed.

Synthesis

Polydimethylsilane (PDMS) is synthesized by reaction in an organic solvent between dimethylchlorosilane and sodium:

$$x\ Cl-\underset{\underset{CH_3}{|}}{\overset{\overset{CH_3}{|}}{Si}}-Cl + 2xNa \longrightarrow \left[\underset{\underset{CH_3}{|}}{\overset{\overset{CH_3}{|}}{Si}}\right]_x + 2\ x\ NaCl$$

Polymerization

The polymerization of liquid PDMS can be done in an autoclave (T = 450°C, P = 1–5 MPa) and yields polycarbosilane (PCS) which is a fusible solid:

$$\left[\underset{\underset{CH_3}{|}}{\overset{\overset{CH_3}{|}}{Si}}\right]_x \xrightarrow[\text{autoclave}]{\text{T thermal}} \left[\underset{\underset{H}{|}}{\overset{\overset{H}{|}}{Si}}-CH_2\right]_x$$

PCS has a very complex connected structure with a very broad molar mass distribution. It is then treated by dissolution-filtration to eliminate the infusible macromolecules which could cause structural defects in the ceramic fiber.

Extrusion

The purified precursor is then molten in a die in inert atmosphere at 300°C to manufacture organometallic fibers by mechanical extrusion (diameter $\cong$ 20 μm).

Cross-linking-infusibility

As PCS fibers are fusible, they must be cross-linked to make them infusible before the ceramization stage. During this treatment, chemical reactions are developed to considerably increase the size of the macromolecules. The most widely used processes consist of a controlled oxidation in air of fibers between 175 and 200°C with formation of Si-O-Si and Si-O-C bridges:

$$2(- \overset{|}{\underset{|}{Si}} - H) + O_2 \longrightarrow 2(- \overset{|}{\underset{|}{Si}} - OH) \longrightarrow - \overset{|}{\underset{|}{Si}} - O - \overset{|}{\underset{|}{Si}} - \\ + H_2O$$

or by electronic bombardment, during which hydrogen is eliminated and Si-Si or Si-C bonds are created:

$$\equiv Si - H + H - Si \equiv \xrightarrow{\text{irradiation}} \equiv Si - Si \equiv + H_2$$

$$\equiv Si - H + H_3CSi \equiv \xrightarrow{\text{irradiation}} \equiv Si - CH_2 - Si \equiv + H_2$$

Pyrolysis

The ceramization heat treatment is carried out under mechanical tension in neutral atmosphere (vacuum/N_2) up to temperatures of 1,400°C, according to the desired state of crystallization. During this stage, the polycarbosilane precursor undergoes many structural transformations with ruptures of chemical bonds that are accompanied by variations in composition, mass and gaseous releases (see Figure 7.1). First of all, up to 500–550°C we observe a release of matter corresponding to the elimination of volatile oligomers, then, thermal decomposition occurs with the rupture of Si-H, C-H, Si-CH$_2$, Si-CH$_3$ bonds. This transformation of polycarbosilane starts with deshydrogeno-carbonation and dehydrogenation reactions. With the departure of hydrogen and alkanes, a three-dimensional lattice of Si-C bonds develops, the C/Si ratio tends towards 1 and C = C bonds appear in the structure. At 1,200°C, mineralization is practically completed (Figure 7.2).

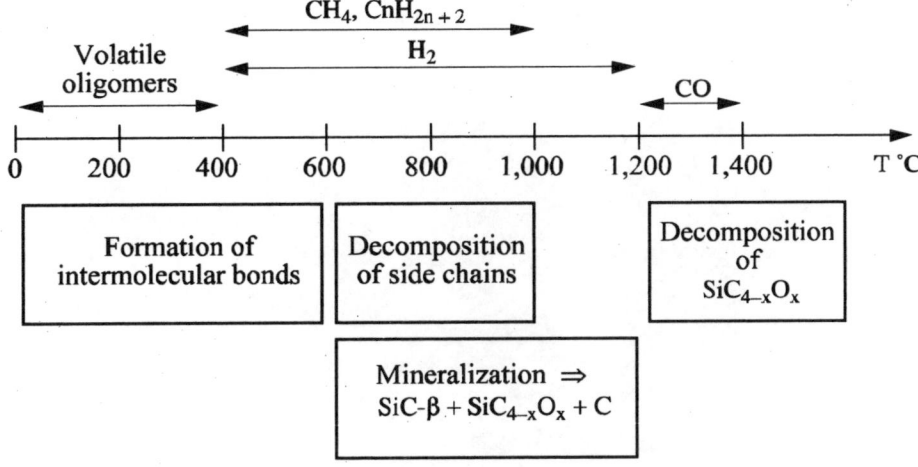

$$CH_4, CnH_{2n+2}$$

$$H_2$$

Volatile oligomers

CO

| 0 | 200 | 400 | 600 | 800 | 1,000 | 1,200 | 1,400 | T °C |

| Formation of intermolecular bonds | Decomposition of side chains | | Decomposition of $SiC_{4-x}O_x$ |

Mineralization $\Rightarrow$
$SiC\text{-}\beta + SiC_{4-x}O_x + C$

Figure 7.1. *Structural evolutions and gaseous releases*

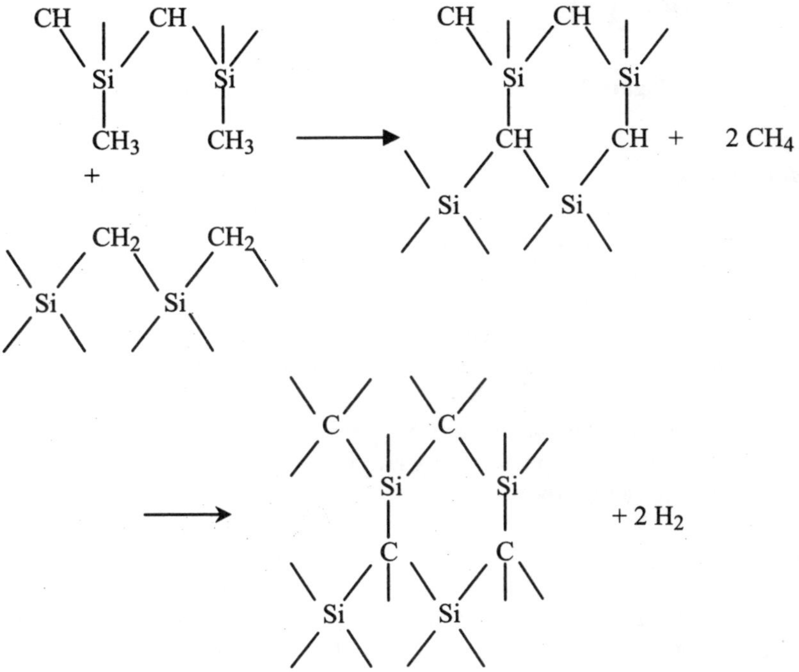

$$+ 2\ CH_4$$

$$+ 2\ H_2$$

Figure 7.2. *Schematic representation of the mineralization of PCS*

The fibers are then formed from β–SiC nanocrystals, carbon nanoparticles and an amorphous phase of SiO_xC_{4-x} type if the cross-linking has been carried out by controlled oxidation (Figure 7.3).

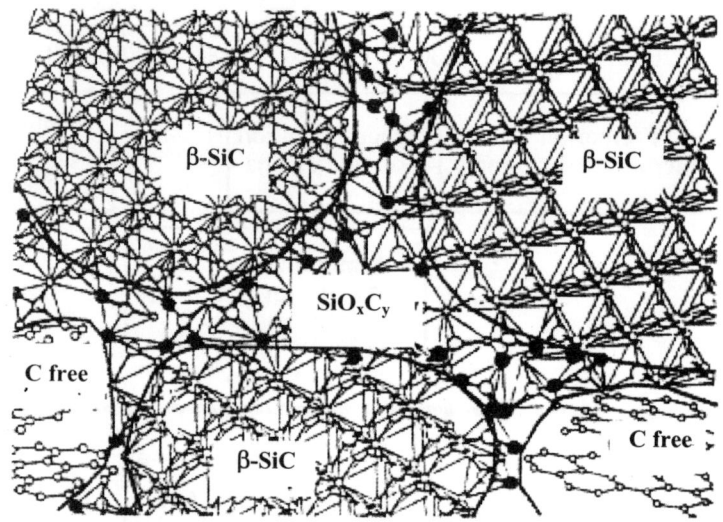

Figure 7.3. *Structural model of the fiber [LAF 89]*

7.2.3. *Monocrystals*

Monocrystals (β-SiC, β-Si$_3$N$_4$, B$_4$C) in the form of filaments (∅ < 10 μm, L < qcq mm) or plates (L < 200 μm) can be prepared by crystalline growth along a simple crystallographic axis. Several methods have been developed, but the purity of the raw materials strongly influences the characteristics and morphology of the monocrystals obtained. The manufacture of β-SiC whiskers is the most developed.

SiC monocrystals can be synthesized by coking at 700–800°C and pyrolysis at 1,500–1,800°C in inert atmosphere of organic matters, like rice, which contains a large quantity of silica and carbon:

$$SiO_2 + 3C \longrightarrow SiC + 2CO$$

The final product consists of a mixture of silicon carbide particles and monocrystals of low dimensions (∅: 0.2–0.5 μm, L < 100 μm).

The growth of monocrystals that results from a gas phase matter transport is controlled much better in the VLS method (vapor-liquid-solid) [SHY 71]. The synthesis is done in a heated reactor at 1,400°C. At this temperature and in the presence of gas mixtures CH_4/H_2, CO, silicon monoxide is generated by the reduction of silica by carbon:

$$SiO_2 + C \longrightarrow SiO + CO$$

On the internal walls of the reactor small particles containing iron are deposited, which at the temperature of 1,400°C are liquid. Silicon monoxide and carbon coming from the decomposition of methane dissolve in the liquid droplets until saturation and precipitation of SiC:

$$SiO + 2C \longrightarrow SiC + CO$$

As the droplets are constantly fed by the gas phase transfer, precipitation continues with the growth of SiC monocrystals along the crystallographic direction (111). The morphology varies with the composition of the gas phase. For low pressures of SiO, whiskers with circular cross-section ($\varnothing < 5$ μm – L ≈ mm) with a perfectly smooth surface can be manufactured.

7.2.4. Depositions-coatings

For many applications, surface properties, such as hardness, coupled with high chemical inertia are required. Non-oxide ceramics have the required characteristics, but their sintering necessitates temperatures that are often incompatible with the thermal stability of the support to be protected. Methods for synthesizing these materials in the form of depositions or coatings have been developed considerably during the last few decades [WEI 92]. They can be divided into two groups:

– chemical processes: CVD (chemical vapor deposition); PECVD (plasma enhanced chemical vapor deposition); OMCVD (organometallic chemical vapor deposition), etc.;

– physical processes: PVD (physical vapor deposition); ionic deposition; plasma depositions; spraying, etc.

From a gas or ionized phase, we synthesize a compound (carbide, nitride, boride) *in situ* on the part to be protected or to modify the surface properties. The energies brought into play (nucleation, growth) to form the solid are weaker than in the traditional method, which makes it possible to manufacture dense materials at much lower temperatures. Table 7.1 shows the various compounds with the temperature range for the production of coatings by CVD.

Depositions	Precursors	Deposition temperature range (°C)
SiC	CH_3SiCl_3/H_2	900–1,200
B_4C	BCl_3-CH_4/H_2	1,200–1,400
TaC	$TaCl_5$-CH_4/H_2	1,000–1,200
TiN	$TiCl_4$-N_2/H_2	900–1,050
Si_3N_4	$SiCl_4$-NH_3/H_2	1,000–1,400
	SiH_4-N_2/H_2	350–500
BN	BCl_3-NH_3/H_2	1,000–1,400
	$B_3N_3H_6$-Ar	450–750
	B_2H_6-NH_3/H_2	1,000–2,300
TaN	$TaCl_5$-N_2/H_2	800–1,500
AlN	$AlCl_3$-NH_3/H_2	800–1,300
	$AlBr_3$-NH_3/H_2	800–1,300
	$Al(CH_3)_3$-NH_3/H_2	900–1,150
TiB_2	$TiCl_4$-BCl_3/H_2	850–1,050

Table 7.1. *Non-oxide depositions*

7.2.4.1. *CVD method*

In view of the very large number of parameters that influence the kinetics of the reactions and the characteristics of the depositions, equipments are generally complex in order to control the uniformity of temperature in the reactor, the partial pressures of the precursors and the gas flows. There are two types of reactors. The first are cold wall reactors for which only the substrates are heated directly: they limit the volume where the reactions take place and avoid the formation of depositions on the walls. On the other hand, in the hot-wall reactors, the reactions between the gas species take place in the entire environment of the parts to be treated.

Vapor phase chemical reactions can be classified under various categories. The simplest is the thermal decomposition of a precursor which is often a chloride (thermal CVD):

$$CH_3SiCl_3 \xrightarrow{\;H_2\;} SiC + 3HCl$$

Coupled or displacement reactions are more difficult to control, in particular the composition of the deposition:

$$SiCl_4 + CH_4 \rightarrow SiC + 4HCl$$
$$TiCl_4 + NH_3 + \tfrac{1}{2}H_2 \rightarrow TiN + 4HCl$$

We can also carry out co-reactions to manufacture composite deposition (SiC, $TiSi_2$):

$$TiCl_4 + 2SiCl_4 + 6H_2 \rightarrow TiSi_2 + 12HCl$$
$$SiCl_4 + CH_4 \rightarrow SiC + 4HCl$$

When the depositions must be obtained at low temperature to avoid the degradation of the substrate, organometallic precursors are used. They already contain metal-non-metal bonds and therefore require lower energies (OMCVD). They are, for instance, silazanes (Si_3N_4 precursors) or carbosilanes (SiC precursors) which have the advantage of being liquids at room temperature.

The various phenomena brought into play in a chemical deposition are matter transfers, energy exchanges and the kinetics of the chemical reactions. They are generally broken down into seven main stages which are schematized in Figure 7.4:

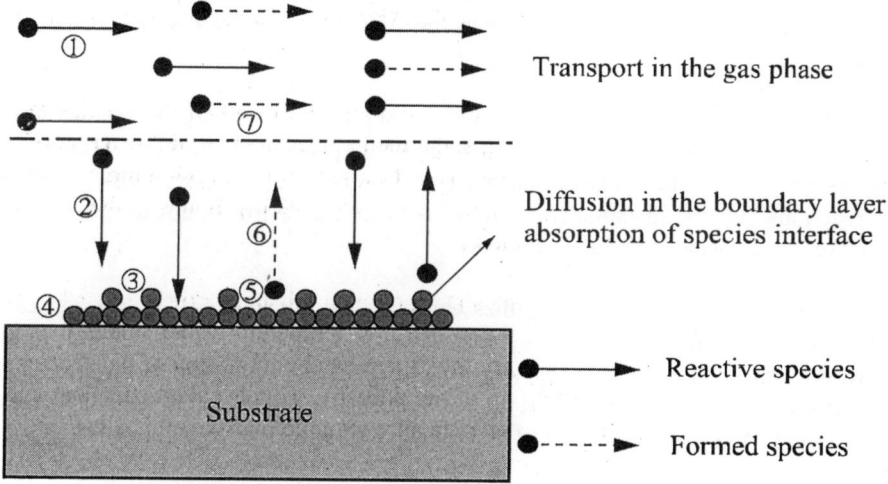

Figure 7.4. *Schematic representation of the various processes in CVD*

1) transport of reactive species (precursor) to the gas phase of the reactor;

2) migration or diffusion of reactive species to the boundary layer (or diffusion space) towards the interface;

3) adsorption of species on the substrate;

4) reaction between the species or surface diffusion and insertion of atoms into the deposition;

5) desorption of the formed species or reaction gases;

6) migration or diffusion of the species formed in the boundary layer towards the gas phase of the reactor;

7) transport of the formed species to the gas phase.

Given the complexity of the phenomena related to the dynamic system with gas flows, the short residence times of the species and a rarely achieved thermodynamic equilibrium, we must call upon thermochemical modeling and chemical kinetics to have indications on the feasibility and appropriate experimental conditions for each type of deposition.

The formation of the coating can be limited, either by transport of matter to the boundary layer or by the kinetics of chemical reactions at the interface. In most cases, the limiting stage in heterogenous gas/solid chemical reactions is of the first order in the reactants. The rate of formation of the deposition then depends on the temperature and partial pressures of the species. For simple systems, the constant rate of formation of the deposition obeys the Arrhénius law for distinct temperature ranges.

At low temperatures, the speed is very slow, the transport of the species to the boundary layer being fast; the limiting stage then corresponds to the reaction of the species adsorbed at the substrate surface (reaction regime). At high temperatures, as the reactions become instantaneous, the slow stage then corresponds to the diffusion of the reactive species (diffusion regime).

The existence of these two regimes can be taken advantage of in order to carry out a deposition inside a porous solid. At low temperature and under low gas pressures, as the reaction speed is very low, there can be migration of the species on long distances, allowing the filling up of the porosity. This is called chemical vapor infiltration (CVI), a method used to manufacture ceramic matrix composites.

7.2.4.2. Coatings and applications

Silicon nitride (Si_3N_4)

Anti-corrosive or anti-wear coatings are generally obtained by thermal CVD because of the possible control of the state of crystallization. A number of precursors are used $SiCl_4$, SiH_2, SiH_2Cl_2 with NH_3. As the temperature range for the formation of depositions in CVD is high, T > 600–700°C and can lead to interactions with the substrate, PECVD has been developed for the manufacture of films for electronic applications. Depositions whose composition can be modulated (N/Si ratio) are obtained between 300 and 500°C.

Silicon carbide (SiC)

SiC coatings prepared by CVD ($SiCl_4$, CH_4) are used for their oxidation resistance (protection of carbon), their hardness or their optical properties.

PECVD is developed to prepare films at lower temperatures and to introduce hydrogen into the structure, which modifies the optical and electric properties. These depositions are used to manufacture solar cells and photoreceivers.

Titanium nitride (TiN)

Titanium nitride depositions are industrially the most developed to modify the surface properties of materials, in particular hardness and wear resistance. As the temperature at which TiN is formed from $TiCl_4/N_2/H_2$ is about 1,000°C, CVD is reserved for the surface treatment of cemented carbides. For the protection or decoration of steel parts, PECVD which allows depositions at 400°C has been developed recently.

Aluminum nitride (AlN)

The main precursors used to manufacture AlN depositions by PECVD are $AlCl_3/NH_3/H_2$, $AlBr_3/NH_3/H_2$ and $Al(CH_3)_3/NH_3/-H_2$. The characteristics of films, in particular the state of crystallization and the crystalline orientation, can be modulated. The main applications in question are protective films or, in electronics, the production of substrates for integrated circuits.

7.3. Sintering and microstructure

Nitrides (Si_3N_4, BN, AlN) and carbides (SiC, B_4C) have strongly covalent chemical bonds and a crystalline structure with a low coordination number of elements which make the diffusion of atoms or vacancies extremely low. Contrary to oxides, natural sintering by heat treatment in inert atmosphere of these compounds is extremely difficult, even when ultrafine powders are used. Moreover, very high temperature treatments to increase atom mobility can, in some cases, cause a decomposition that inhibits sintering:

$$Si_3N_4 \longrightarrow 3Si + 2N_2$$

For all these products, it is therefore necessary to use sintering additives or apply a mechanical or gas pressure at high temperatures to enhance the redistribution of matter [CHE 80].

7.3.1. Silicon nitride

Silicon nitride can be sintered in five different methods:
– reaction sintering (reaction bonded silicon nitride: RBSN);
– natural sintering (reaction sintered silicon nitride: RSSN);
– pressure sintering (hot pressed silicon nitride: HPSN);
– hot isostatic pressing (hot isostatically pressed silicon nitride: HIPSN);
– nitrogen gas pressure sintering (gas pressure sintered silicon nitride: GPSN).

Apart from reaction sintering, all the other processes require sintering additives. This consists of nitriding powdery silicon compact at about 1,300–1,350°C and partially filling the porosity (final residual porosity $\cong$ 15%) at the time of the reaction:

$$3Si + 2N_2 \longrightarrow Si_3N_4$$

The most widely used sintering additives are oxides (MgO, Al_2O_3, Y_2O_3, rare earth) which form in the temperature range 1,450-1,750°C with silica or silicon oxynitride present as impurity on the surface of Si3N4 crystals, an oxynitride eutectic with the general formula MSiM'ON, with M and M' = Al, Y, Mg, etc. [HAM 85, LEW 89]. This liquid phase plays a major role in the sintering and allows a distribution of the matter thanks to the dissolution of the nitride crystals and the migration of the dissolved species. These phenomena can be accelerated by the use of a mechanical and/or gas pressure.

Silicon nitride exists under two crystalline structures of hexagonal symmetry: the low temperature α variety, whose crystals are equiaxed and the high temperature β variety, generally in the form of acicular crystals. The transformation $\alpha \rightarrow \beta$ is not reversible.

Sintering and the various mechanisms which take place during densification are well-known [BRO 77]. It is possible to modulate the microstructure (size and shape of the crystals) in a broad domain according to the desired mechanical properties or applications in question.

During the heat treatment of the granular compact (α-Si$_3$N$_4$ + additives: Al$_2$O$_3$, Y$_2$O$_3$, etc.), the eutectic flows into the porosity to generate capillary forces that cause shrinkage according to three different stages that overlap in the course of time.

In the first stage, the liquid phase located at the points of contact solubilizes the crystals of small diameter, which results in a rearrangement of the particles. Pressure resulting from the capillary forces causes a preferential dissolution at the points of contact of the α–Si$_3$N$_4$ crystals, thus generating concentration gradients. Then the second stage takes place with diffusion of the species in the eutectic and reprecipitation in the form of β–Si$_3$N$_4$ crystals, or solid solution β'–SiAlON when alumina has been used as densification additive. The flow via the intergranular liquid allows a redistribution of the matter and a macroscopic shrinkage. The comparison of the rate of transformation (Vr) α–Si$_3$N$_4$ $\rightarrow$ β–Si$_3$N$_4$ and the rate of densification (Vd) shows that there is a proportionality relation $\left(\dfrac{Vr}{Vd} = 1 \right)$ which confirms the existence of a diffusional regime.

Solid particles in a liquid exhibit a solubility that varies with their size. This difference in solubility forms the driving force of the crystalline growth and corresponds to the phenomena of Ostwald ripening. For two soluble grains with radius r_1 and r_2 ($r_1 < r_2$), the rate of growth of the coarsest grain is expressed in the relation:

$$\frac{dr_2}{dt} = \frac{2\gamma \; \Omega DC_\infty}{kTd} \left(\frac{1}{r_1} - \frac{1}{r_2} \right)$$

where C_∞ is the concentration at equilibrium in the liquid for a plane solid surface, γ the surface energy, Ω the molecular volume of the mobile species, T the absolute temperature, k the Boltzmann constant, D the coefficient of diffusion of the species in the liquid, and d the average distance between the two particles. For a liquid phase sintering, the grain growth kinetics is of the type $\bar{r}^n - \bar{r_0}^n = At$, where $\bar{r}$ is the average grain radius, n = 3 the exponent of the growth ratio, A a constant and t time.

For silicon nitride, the phenomena are more complex because of the irreversible transformation α – β. Grain enlargement is superimposed on the dissolution-diffusion-reprecipitation phenomenon to become dominating and forms the third stage when the rate of densification reaches 90–92%. The acicular growth of β-Si$_3$N$_4$ grains can be developed and controlled by playing on the duration of the heat treatment.

7.3.2. *Aluminum nitride*

Aluminum nitride can be densified in the presence of additives (Y_2O_3, YF_3, CaO, etc.) and in nitrogen atmosphere by natural or pressure sintering or hot isostatic pressing. In view of the thermal stability of the nitride, the first process is the most widely used, because it yields totally dense pieces at temperatures of about 1,750°C. As for silicon nitride, the additives react with alumina or aluminum oxynitrides present at the surface to form a liquid phase of an oxynitride aluminate. The three stages of sintering of a solid in a liquid phase, namely, grain rearrangement, dissolution diffusion-reprecipitation and enlargement, are observed.

After heat treatment, the microstructure is made up of aluminum nitride crystals and secondary phases of aluminates at the grain boundaries and at the triple points. The degree of crystallization and the distribution vary with the nature of the additives used and the thermal cycle. As pure aluminum nitride is a very good conductor of heat and aluminates of heat insulators, the development of the microstructure, in particular the location of the phases at the grain boundaries, have been the subject of extensive research. It transpires that yttrium-based additives (Y_2O_3, YF_3) yield sintered aluminum nitride substrates with the highest thermal conductivity (150–200 $W/m^{-1}h^{-1}$).

7.3.3. *Silicon carbide*

Silicon carbide exists in the form of a very large number of crystallographic varieties. The most widespread polytypes are the β-SiC (cubic, low temperature) and α–SiC (hexagonal, high temperature) varieties. SiC powder is densified by reaction sintering reaction, natural or pressure sintering [CHE 80].

-Reaction sintering

The process consists of compacting a mixture of SiC and carbon, which is then infiltrated with liquid silicon. During the infiltration heat treatment, the following reaction takes place:

$$Si_l + C_s \rightarrow SiC_s$$

However, some free silicon remains, as it is very difficult to achieve a total reaction.

Natural sintering

Industrially, natural sintering is the most widely used. It requires introducing a small quantity of additives, carbon and boron or carbon and aluminum. The mixture is formed, then the compact treated in inert atmosphere between 2,000 and 2,100°C. The chemical reactions and the densification mechanisms are complex and vary with the temperature and the nature of the sintering additives.

Carbon is generally incorporated using soluble or liquids precursors (polyvinyl alcohol, phenolic resin), in order to deposit a carbon film on the SiC grains. During the thermal cycle, carbon reduces (T > 1,250°C) the silica or silicon oxycarbide (SiC_xO_y) present at the grain surface, which increases the specific surface energy and facilitates sintering. As regards boron, very slightly soluble at high temperature in the carbide lattice, its presence decreases the specific energy of the grain boundaries. The granular compact is densified thanks to mechanisms of solid state diffusion at the grain boundaries in a phase rich in boron and carbon. This phase disappears gradually and partially by dissolution in the SiC lattice. The excess carbon at the grain boundaries can inhibit the growth of the crystals.

Other methods

The use of pressure sintering or hot isostatic pressing is limited to the production of completely dense pieces with very low additive contents.

Silicon carbide can also be densified between 2,000 and 2,100°C in the presence of a liquid phase formed from additives, such as Y_2O_3 and Al_2O_3. During the heat treatment, the following transformation occurs:

$$\beta\text{-}SiC \rightarrow \alpha\text{-}SiC$$

with an acicular growth of α-SiC grains. It is then possible, as for β-Si_3N_4 nitride, to modulate the development of the microstructure according to the desired properties [KLE 92].

7.3.4. *Boron carbide*

Boron carbide (B_4C) is particularly difficult to densify and only hot pressing completely eliminates the porosity [THE 79]. Like SiC, an additive must be used, for example, carbon. It is introduced in the form of soluble organic precursors to facilitate its distribution on the grain surface. Natural sintering at 2,200–2,300°C in inert atmosphere leads to slightly porous materials (1 to 3%). Only hot pressing under identical experimental conditions densifies the parts completely and avoids an exaggerated grain growth.

7.4. Chemical stability and behavior at high temperature

Non-oxide ceramics exhibit a very high intrinsic stability (see Tables 7.2 and 7.3) and very high melting points.

However, these materials react at high temperature with oxidizing atmospheres to create oxides, oxynitrides (or oxycarbides) according to the operating conditions. The understanding of the phenomena and the knowledge of the rate of degradation are essential for choosing the material best adapted to a given use. Non-oxides that have very high mechanical characteristics are differentiated by their resistance in an oxidizing atmosphere (air, O_2, H_2O, CO/CO_2, etc.). As their rate of oxidation depends on the covering of the formed oxide, we observe two types of behaviors:

– silicon-based ceramics (SiC, Si_3N_4, $MoSi_2$) which are covered with a protective oxide coating exhibit very good resistance compared to metals and alloys;

– other ceramics which contain boron or transition metals whose oxide has a low melting point ($T_f B_2O_3 = 430°C$) or is non-protective, become degraded from 600–700°C (Table 7.4) [GOG 92].

Compound	$\Delta H_f °_{298}$ (Kj/mol)	$S °_{298}$ (J/mol.K)	$\Delta G_f °_{298}$ (Kj/mol)
SiC	−71.6	16.5	−69.1
TaC	−144.1	42.4	−142.7
TiC	−184.1	24.2	−180.4
ZrC	−196.1	33.3	−193.3
AlN	−318	20.1	−287.0
BN	−250.9	14.8	−225.0
Si_3N_4	−744.8	113	−647.3
TaN	−252.3	41.8	−223.9
TiN	−337.7	30.2	−308.9
ZrN	−365.3	38.9	−337
HfB_2	−336	42.7	−332.2
TiB_2	−279.5	28.5	−275.3

Table 7.2. *Thermodynamic characteristics*

Compound	SiC*	TaC	TiC	ZrC	AlN*	BN*	Si3N4*	TaN	TiN
T_f(K)	3,100	4,270	3,290	3,805	2,570	2,600	2,150	3,360	3,220
Compound	ZrN	HfBr	TaB2	TiB2	MoSi2				
T_f(K)	3,220	3,650	3,370	3,190					

Table 7.3. *Melting or dissociation temperature**

Material	Oxidation starting temperature °C	Maximum usage temperature °C	
		Oxidizing atmosphere	Inert or reducing atmosphere
Si3N4	1,000	1,400	1,600
SiC	1,000	1,500	2,100
AlN	800	1,400	1,800
MoSi2	1,400	1,700	1,900
BN	700	700	2,200
TiN	500	500	2,000
B4C	600	600	2,000
TaC	500	500	2 000

Table 7.4. *Oxidation starting temperature or maximum usage temperature of non-oxide ceramics*

7.4.1. *Oxygen oxidation of Si₃N₄ and SiC*

Thermodynamic computations and experimental results show that there are two oxidation regimes depending on the oxygen pressure and the temperature.

For high pressures and moderate temperatures, the silica formed gradually covers and protects the substrate (SiC, Si₃N₄): this is "passive oxidation":

$$SiC_{(s)} + 2O_{2(g)} \rightarrow SiO_2 + CO_{2(g)}$$

$$Si_3N_{4(s)} + 3O_{2(g)} \rightarrow 3SiO_{2(s)} + 2N_{2(g)}$$

On the other hand, for low oxygen pressures and high temperatures, no covering product is formed, as silicon monoxide is volatile: this is "active oxidation":

$$SiC_{(s)} + O_{2(g)} \rightarrow SiO_{(g)} + CO_{(g)}$$

$$SiC_{(s)} + 2SiO_{2(s)} \rightarrow 3SiO_{(g)} + CO_{(g)}$$

$$Si_3N_{4(s)} + 3SiO_{2(s)} \rightarrow 6SiO_{(g)} + 2N_{2(g)}$$

Figure 7.5 represents the transition between the two domains corresponding to the formation of silica or silicon monoxide [SIN 77].

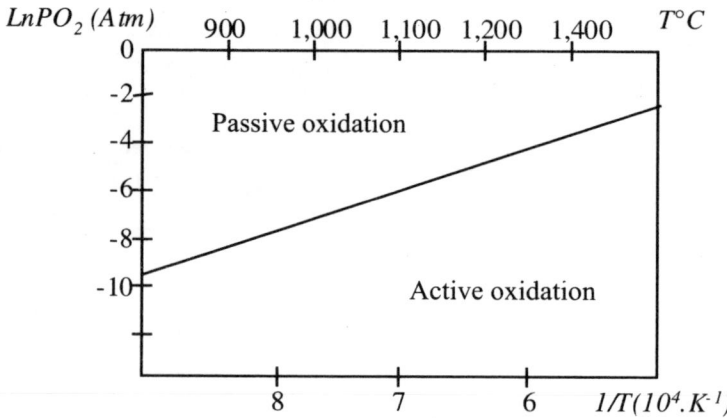

Figure 7.5. *Influence of oxygen pressure on the kinetic regime*

Products of the reaction

The study of the products of the oxidation reaction of powders or monocrystals (SiC) up to temperatures of 1,600°C shows that the silica layer is amorphous and then crystallizes gradually into cristobalite from 1,100°C onwards. For sintered materials that contain impurities or sintering additives, the nature of the formed products is much more complex (silicates, aluminosilicates, etc.). However, for all these materials, the oxidation reaction results in a volume expansion. The coefficient of expansion, which represents the ratio of the molar volume of silica or silicates on the molar volume of SiC (or Si_3N_4), is higher than one. The reaction product is covering and protective.

Oxidation kinetics of SiC or Si_3N_4 sintered under pressure

During a solid/gas reaction (SiC/O_2), the formation of the new solid (SiO_2) brings into play two processes, nucleation and growth, which correspond to the same chemical reaction but occur on different sites and according to different mechanisms. Nucleation is the creation of SiO_2 nuclei from SiC or Si_3N_4. Growth refers to the development of these seeds by progression of the SiO_2/SiC or SiO_2/Si_3N_4 interface.

From the point of view of kinetics, each process can be limited by interfacial reaction or a diffusion phenomenon. Moreover, one of the processes can be very rapid compared to the other. For example, the nucleation of SiO_2 at high temperature on SiC surface is almost instantaneous; the overall rate is then controlled by the growth of the covering layer (Figure 7.6).

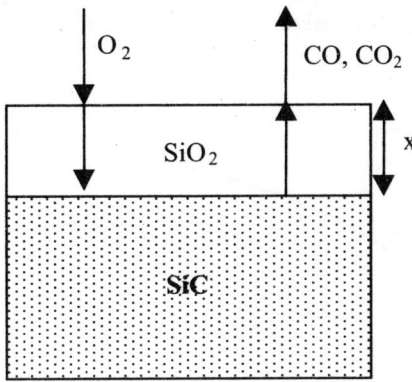

Figure 7.6. *Schematic representation of the passive oxidation of SiC*

The development of silica at the internal interface requires the migration of oxygen from the environment towards the substrate. There are then six elementary stages, the slowest of which constitutes the regulating process:

– diffusion of oxygen in the gas phase towards the surface of the sample;

– adsorption of oxygen;

– diffusion of oxygen in SiO_2 towards the substrate because of the concentration gradient;

– reaction at the internal interface;

– diffusion of the formed gas species (CO, CO_2) towards the outside;

– gas desorption.

As the diffusion of oxygen towards the substrate is the slowest stage, the quantity of oxygen that reacts decreases when the thickness of the silica layer increases. If the oxidation reaction occurs at constant temperature, we observe a monotonous decrease in the oxidation rate $\left(\dfrac{dx}{dt}\right)$ on the curve, giving the thickness of the silica layer according to time (see Figure 7.7).

Silica protects the SiC substrate from the oxygen, since the rate becomes very slow for long durations. The equation of the kinetics according to time is written:

$$x^2 = k_p t$$

with $k_p = V_{m(SiO2)}D(C_e - C_i)$ rate constant of the parabolic regime, where V_m is molar volume of silica, D the coefficient of diffusion (intergranular migration of molecular oxygen for T < 1,350°C, transport of ionic oxygen for T > 1,350°C) and C_e and C_i respectively the external or internal concentration of the chemical species.

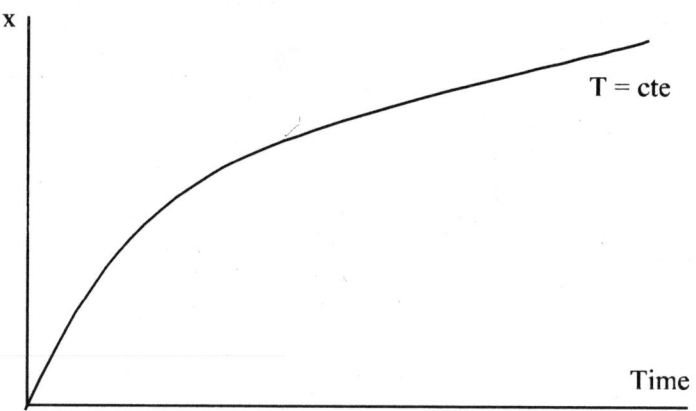

Figure 7.7. *Thickness of the SiO_2 layer according to time*

7.4.2. *Water vapor oxidation*

Non-oxide ceramics are seldom used in the presence of an atmosphere of dry oxygen. In this case, a passive oxidation takes place with the water vapor according to the equation:

$$SiC + 4H_2O \rightarrow SiO_2 + CO_2 + 4H_2$$

At high temperatures, the water vapor in the air plays a very important role in the kinetics of degradation. The parabolic regime is preserved, but for identical conditions of temperature and oxygen pressure, we observe that the water vapor considerably increases the rate of oxidation. There is molecular diffusion of H_2O, which constitutes an independent oxidant and acceleration of the diffusion of oxygen because of the structural modifications of SiO_2 caused by H_2O.

7.4.3. *Corrosion by molten salts*

Non-oxides are used to manufacture parts for uses in aggressive environments (electrolytic baths, heat exchanger, etc.) which contain molten salts. In most cases there is first an oxidation of the ceramic and then an interaction between the oxide formed and the molten salt. The dissolution of the oxide favors the oxidation of the substrate and triggers in fact the start of the corrosion.

For example, for SiC in the presence of molten Na_2SO_4, the overall reaction is written:

$$Na_2SO_4 + SiC + O_2 \rightarrow Na_2SiO_3 + CO + SO_2$$

Table 7.5 shows corrosion start temperatures in air by a few molten salts.

Salts	Corrosion start temperature (°C)		
	Si_3N_4	SiC	BN
Na_2CO_3	850	850	850
Na_2SO_4	1,000	950	900
NaOH	450	450	500
$PbCrO_4$	450	450	
$Na_2B_4O_7$	1,000	1,000	1,000
V_2O_5	750	750	

Table 7.5. *Corrosion resistance of various ceramics*

7.5. Properties and applications

Non-oxide ceramics are used in thermo-structural or electronic applications because of their specific properties: stiffness, mechanical resistance, toughness, resistance to corrosion by hot gases or molten metals (see Table 7.6).

These materials have experienced a constant development during the last few decades thanks to the progress made in the processes used to produce monoliths, composites or coatings and the great diversity of the properties associated with a possible modulation of their microstructure.

The difficulties encountered in obtaining dense parts and their high cost make their marketing very dependent on the application in question. Their use is justified when the required characteristics are much higher than those of other materials.

Compound	E Young's modulus (GPa)	σ Cantilever strength (MPa)	K_{IC} Toughness (MPa √m)	H_v Micro-hardness (GPa)	α Coef. of expansion $(10^{-6}K^{-1})$	K Thermal conductivity $(Wm^{-1}k^{-1})$	ρ Electric resistivity (Ωm)
AlN	400	400	3 – 3.5	13	4.8	250	$> 10^{11}$
α-BN	100	50		0.1	2.7	20	$> 10^{11}$
α-Si$_3$N$_4$	310						
β-Si$_3$N$_4$	290	800	6 –7	25	2.8	25	$> 10^{12}$
TiN	390	250		20	8.1	28	40.10^{-8}
ZrN	320	200		15	7	27	20.10^{-8}
B$_4$C	420	400	3 – 3.5	35	5	28	1.10^{-5}
HfC	480	300		25	6.8	16	40.10^{-8}
α-SiC	400	500	4 – 5	30	5.1	40	4.10^{-3}
β-SiC	430				4.3	40	
TaC	500	300		17	7.1	20	20.10^{-8}
TiC	490	400	2	20	7.9	17	20.10^{-7}
ZrC	490	500	2 – 3	20	7	27	80.10^{-5}
MoSi$_2$	440	400		12	8.2	50	20.10^{-5}

Table 7.6. *Main characteristics of non-oxide ceramics*

Silicon carbide

Among all non-oxide ceramics, silicon carbide, used for several decades now, ranks first for its industrial development (see Figures 7.8, 7.9 and 7.10). Many grades are marketed according to the usage temperature:

– 1,350°C: silicon carbide infiltrated with silicon;

– 1,600°C: sintered and recrystallized silicon carbide;

– 1,750°C: sintered silicon carbide.

The applications include a very large number of industrial sectors, because of its thermostructural properties or its semiconductor nature with negative temperature coefficient. For the last few years, it has been used in mechanics for machining or friction parts (Table 7.7).

In the field of thermostructural composites, silicon carbide occupies a privileged place, whether it is for the production of ceramic matrices or fibers. Silicon carbide fibers prepared from organometallic precursors are the most stable at high temperature in oxidizing atmosphere.

Industrial sector	Application
Ceramic and electrothermics industry	Furnaces: muffles, supports for parts to be sintered Electric furnaces: heating resistors
Mining and iron and steel industries	Equipments for ore preparation, pre-heating furnaces: coatings Pre-heating furnaces of steel slab: muffles, supports
Non-ferrous metallurgy	Mixers: internal coating Blast furnaces: internal refractories Wire drawing: wire guide Electrolyzers for aluminum: coating plates, protective sheaths Pumps for molten metals: pump housing elements
Chemical industry	Furnaces and devices for the drying of corrosive substances (fertilizer): tubes, wear parts, crucibles, nacelles Catalyst supports
Mechanical and automobile industries	Machines for machining or correction: grinding stones, abrasive or machining tools Water pumps of car engines: sealing linings, friction rings
Nuclear industry	Reactions: sheaths for heating elements
Aeronautical industry	Fiber composite parts turbines and thermomechanical protections

Table 7.7. *Main applications of silicon carbide*

Figure 7.8. *Parts for thermostructural applications
(courtesy of Refractories and Ceramic Concepts)*

Figure 7.9. *Pump sealing lining
(courtesy of Céramiques et Composites S.A.)*

Figure 7.10. *Supports for diffusion ovens of the semi-conductor industry (courtesy of St Gobain-Norton)*

Boron carbide

Boron carbide B_4C, whose hardness is close to that of diamond, is used as abrasive for machining, for wearing parts or for friction and sealing rings. As the isotope ^{10}B has a high section for capturing thermal neutrons, B_4C is used for the production of protective materials which absorb the neutrons resulting from nuclear reactors.

Boron nitride

Boron nitride α-BN (hexagonal variety) has a crystalline structure similar to that of graphite. It is an excellent heat insulator, which exhibits a low coefficient of thermal expansion and high chemical inertia. It is used to manufacture thermocouple cladding, tubes, pumps elements and crucibles, because it is hardly wetted by liquid metals. It is used as solid lubricant in joints or bearings functioning at high temperatures. Small machine parts subjected to very high thermal shocks are manufactured by the machining of blocks obtained by pressure sintering. Cubic boron nitride is the second hardest material after diamond.

Boron nitride acts as diffusion barrier when the reducing properties of carbon are detrimental.

Silicon nitride

Because of the good resistance to metals and molten salts, silicon nitride has been used for a very long time in the manufacture of refractories for the iron and steel industry. The considerable research undertaken during the last 30 years has made it possible to develop materials whose microstructure is flexible. Thus, the very high toughness of the grades obtained by gas pressure sintering (GPSN) compared to other ceramics, and their excellent resistance to thermal shocks explain their development in the automobile and mechanical engineering industry (see Figure 7.11). Likewise, the possibility of directly manufacturing complex parts by reaction sintering (RBSN) has opened up new applications in non-ferrous metallurgy or in industrial thermics. Table 7.8 presents some of these applications.

Figure 7.11. *Balls for silicon nitride bearings*
(courtesy of St Gobain Cerbec)

Industrial sector	Applications
Automobile	Turbocompressor rotors, valves, springs, guides, friction rings
Mechanical industry	Bearing balls, inserts of cutting tools, sealing rings, matrices for the molding of metals
Ceramic industry and industrial thermics	Furnaces: refractories, supports, crucibles, sheaths, parts for burners
Non-ferrous metallurgy	Sheaths, crucibles, supports

Table 7.8. *Examples of applications of silicon nitride*

Aluminum nitride

Among non-oxide ceramics, aluminum nitride developed recently. Its mechanical properties (hardness, breaking strength) are modest, but its thermal conductivity is very high. This conductivity associated with high electrical resistance, good dielectric property and a coefficient of expansion close to that of silicon, which varies linearly with the temperature, make it an excellent material for substrates for power micro-electronics (Figure 7.12).

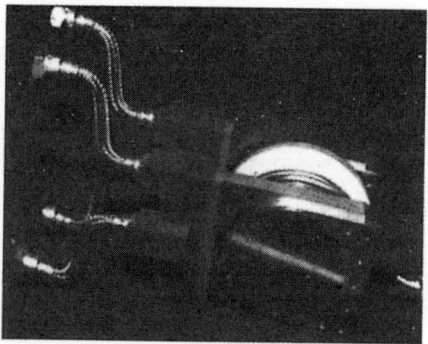

Figure 7.12. *Aluminum nitride metallized products for electronics (courtesy of Boostec Industries)*

7.6. Bibliography

[ALI 79] ALIPRANDI G., *Matériaux réfractaires et céramiques techniques*, Editions Septima, Paris, 1979.

[BRO 77] BROOK R.J., CARRUTHERS T.G., BOWEN L.J., WESTON R.J., "Mass transport in the hot pressing of α silicon nitride", *Nitrogen Ceramics*, Ed. Riley F.L., Noordhoff Inter. Publishing, p. 383-390, 1977.

[CHE 80] CHERMANT J.L., OSTERSTOCK F., "Préparation et façonnage de matériaux SiC, Si₃N₄, SiAlON et Silcomps", *Rev. Int. Hautes Temp. Réfract.*, 17, p. 295-315, 1980.

[FRE 90] FREER R., *The physics and chemistry of carbides, nitrides and borides*, Kluwer Academic Publishers, Dordrecht, 1990.

[GOG 92] GOGOTSI Y., LAVRENKO V.A., *Corrosion of high-performance ceramics*, Springer-Verlag, Berlin, 1992.

[HAM 85] HAMPSHIRE S., DREW R.A.L., JACK K.H., "Oxynitride glasses", *Phys. Chem. Glass*, 26, p. 182-186, 1985.

[KLE 92] KLEEBE H.J., "SiC and Si₃N₄ materials with improved fracture resistance", *J. Eur. Ceram. Soc.*, 10, p. 151, 1992.

[LAF 89] LAFFON C. *et al.*, "Study of nicalon-based ceramic fibres and powders by EXAFS spectrometry, X-ray diffractometry and some additional methods", *J. Mater. Sci.* 24, p. 1503-1512, 1989.

[LEW 89] LEWIS M.H., *Glass and glass ceramics*, Chapman and Hall, 1989.

[MUL 95] MULLOT J., LECOMPTE J.P., JARRIGE J., "High thermal conductivity aluminium nitride substrates with Y₂O₃ or YF₃ additives", *Fourth Euro Ceramics*, Part II, Ed. C. Galassi, p. 235-241, 1995.

[PAS 65] PASCAL P., *Nouveau traité de chimie minérale*, volumes 6 and 8, Masson, Paris, 1965.

[RIL 83] RILEY F.L., *Progress in nitrogen ceramics*, Martinus Nijhoff Publishers, Dordrecht, 1983.

[SAI 88] SAITO Sh., *Fine Ceramics*, Elsevier, New York, 1988.

[SHY 71] SHYNE J.J., MILEWSKI J.V., Method of growing silicon carbide whiskers, U.S. Patent, 3622272, 1971.

[SIN 77] SINGHAL S.C., "Oxidation of silicon nitride and related materials", *Nitrogen Ceramics*, Ed. Riley F.L., Noordhoff, Leyden, p. 607-626, 1977.

[THE 79] THEVENOT F., BOUCHACOURT M., "Le carbure de bore: matériau industriel performant", *Ind. Céram.*, 10, p. 655, 1979.

[WEI 92] WEIMER A.W., *Carbide, nitride and boride materials synthesis and processing*, Chapman and Hall, London, 1992.

[YAJ 76] YAJIMA S., OKAMURA S., HAYASHI K., OWARI J., "Synthesis of continuous SiC fibers with high tensile strength", *J. Amer. Ceram. Soc.*, 59, p. 324, 1976.

PART II

Properties and Applications of Ceramics

Chapter 8

Mechanical Properties of Ceramics

The peculiarities of the mechanical behavior of ceramics did not go unnoticed, even as far back as 800,000 years ago, by flint makers or, more than 14,000 years ago, by makers of knives in obsidian – a natural glass found on volcanic slopes. These primeval artisans knew that if on the one hand the materials that they were working with were brittle, they resulted on the other hand in efficient and durable tools: the edges of obsidian knives compete with the best present day steel edges. Later, they made use of the brittle nature of the stone for creating blocks with definite geometry and dimensions. In the 1st century BC, King Herod constructed a fortress on the rocks of Massada; the rocks were wedged out to create large tanks by introducing wooden pegs in the surface cracks of rocks and then swelling these by soaking. The stress intensity factor was not invented yet: its effects were already known!

Yet, the mechanics of brittle materials is still a young and living science. Even now, the making of ceramic pieces for use in mechanical design, that are subjected to dimensional tolerances similar to those used for steel, or of complex geometry, gives rise to difficult problems, which hinder the large scale development of these materials, despite their extraordinary intrinsic properties. Hardness and brittleness: two fundamental characteristics sufficient to describe a ceramic, contribute to its originality and its specificity, but are also the root cause of all their shortcomings. Ceramics are hard (HV > 1,000 Vickers), and it is therefore very difficult and costly

Chapter written by Tanguy ROUXEL.

to manufacture them, and their low toughness ($K_{Ic} < 10$ MPam$^{0.5}$) and means they have to be handled delicately.

Consequently, the excitement of the 1980s and the ambitious programs such as ceramization of motors and substitution of metallic alloys in aeronautics have given way to more realism with the emergence of more long-term projects and the development of complex polyphase materials. Recently, tougher and/or more ductile ceramics at reasonable temperatures (i.e. compatible with plastic deformation industrial tooling) have been developed. The ability for superplasticity has even been demonstrated by dense materials having a fine grained microstructure. Besides, the present day processes provide materials with a homogenous microstructure, controlled and therefore reproducible (Weibull modulus > 30) [BRA 92] [CHE 89] [KIN 76] [SCH 91].

8.1. Brittleness and ductility

Brittleness and ductility are not intrinsic properties of materials. The evaluation of the degree of brittleness or ductility of a material is associated to the relationship existing between the characteristic speed of the processes (kinetic) and the duration of observation, and depends on the temperature.

8.1.1. *Brittle behavior*

For ceramic materials, the term "brittleness" has two significances: absence of plasticity and poor resistance to impact. A brittle fracture takes place without prior plastic deformation during elastic loading, as soon as the stress locally reaches the critical threshold of interatomic decoherence. The fracture occurs by cleavage along defined crystallographic planes for crystalline materials (most of the time along compact planes with weak indices) and along a conchoidal fracture pattern for glasses (conchoidal pattern resembling a shell, is typical in flint). For an agglomerate, the fracture is generally transgranular or intergranular in fine-grained (grain size < 1 μm) materials. By neglecting kinetic effects, the energy supplied to a structure during the process of destruction corresponds to the energy for creation of the rupture surfaces, that is $2\gamma S$, where S is the area of one of the edges of the fracture surface and γ is the energy of the surface (Jm^{-2}). Brittle fracture therefore involves two stages:

– production of a crack (or of a flaw: porosity, inclusion, etc.);

– propagation of a crack (emanating from the most serious flaw existing).

An expression for the theoretical breaking strain σ_{th} of a solid with Young's modulus E was proposed by Orowan (1949) [ORO 49] and Gilman (1959) [GIL 59], by modeling the interatomic separating force with a simple sinusoidal function. The following result was obtained:

$$\sigma_{th} = \sqrt{\frac{E\gamma}{r_o}} \qquad\qquad [8.1]$$

where r_o is the interatomic distance in equilibrium.

Although this result is corroborated by measurements obtained on fibers or whiskers, experimental values obtained are very clearly below σ_{th}. In fact, in most cases, the fracture is initiated from a flaw located near an area where the stress attained is of the critical decoherence value.

8.1.1.1. *Energy approach [GRI 20]*

Griffith is credited with the proposition that glass structures contain a multitude of microscopic cracks: typically the surface of a glass window is covered with more than 10,000 flaws or scratches of less than 1/100th of a mm per cm^2! For ceramics, these flaws take the form of pores, of impurities or of microcracks arising from the processing method, or of surface flaws (scratches, scales, etc.) arising from the fabrication process or handling. Griffith showed, based on the works of Inglis [ING 13], that the breaking stress σ_r, for a solid containing a crack of half-length a_c (most critical flaw) is expressed as:

$$\sigma_r = \sqrt{\frac{2E'\gamma}{\pi a_c}} \qquad\qquad [8.2]$$

where E' = E in plane stress and E' = E/(1-v^2) in plane strain (v: Poisson's ratio).

NOTE.– when a_c tends towards r_o, we get an expression close to that obtained by Orowan and Gilman.

During sub-critical loading, the material stores up the potential energy W_e in the form of elastic energy:

$$W_e = \frac{1}{2}\int_{Volume} \left(\tilde{\sigma}:\tilde{\varepsilon}\right) dV \qquad\qquad [8.3]$$

where $\tilde{\sigma}$ and $\tilde{\varepsilon}$ are local stress and deformation tensors respectively.

Let us consider the closed system constituted by the cracked part (S: surface area of the crack) and the forces which are applied, exchanging neither heat nor work with the outside. The conservation of total energy of this system is written:

$$dW_f + dW_e + d(A + 2S)\,\gamma + dW_c = 0 \qquad [8.4]$$

where W_f is the work done by applied forces on the boundary of the part, W_e the elastic energy stored, A the external surface and W_c the kinetic energy.

Considering that $dA = 0$ and with $dP = dW_f + dW_e$, where P represents the total potential energy $W_f + W_e$, the preceding equation becomes:

$$dP + 2\gamma dS + dW_c = 0 \qquad [8.5]$$

from which:

$$G = 2\gamma + \frac{dW_c}{dS} \qquad [8.6]$$

where G is the expansion force of the crack, also known as the elastic energy release rate, defined by $G = -\partial P/\partial S$ (Griffith theory).

In equilibrium conditions and at rest, $dW_c = 0$ and $G = 2\gamma$; if G crosses this critical threshold, the crack propagates slowly as long as $G < G_c$ (see section 11.1) and then in a catastrophic manner for $G = G_c$. G_c is the critical energy of tearing per unit of area. If $G < 2\gamma$, the crack should close ($dS < 0$, $dW_c > 0$). In reality, different irreversible processes, and particularly surface hydrolysis, oppose the healing of the crack.

8.1.1.2. Concept of toughness (Irwin's stress analysis)

The stress intensity factor depicts the distribution of stresses and deformations near a crack. It was introduced by Irwin [IRW 57] for describing the stress field near the crack tip, depending on the type of stress (nature of loading of the crack). In the present case, which is a more dangerous one, of a type I opening, the elasticity analysis carried out by Irwin led to the following expression for a description in cylindrical coordinates (origin at the face of the crack with $\theta = 0$ on the propagation plane):

$$\begin{pmatrix} \sigma_{xx} \\ \sigma_{yy} \\ \tau_{xy} \end{pmatrix} = \frac{K_I}{\sqrt{2\pi r}}\cos\frac{\theta}{2}\left\{\begin{aligned} &1-\sin\frac{\theta}{2}\sin\frac{3\theta}{2} \\ &1+\sin\frac{\theta}{2}\sin\frac{3\theta}{2} \\ &\sin\frac{\theta}{2}\cos\frac{3\theta}{2} \end{aligned}\right\}$$

[8.7]

where $K_I = Y\sigma_\infty\sqrt{a}$ is the stress intensity factor, σ_∞ is the stress at "infinity" (far field), Y is the geometric factor depending on the geometry of the test specimen (function of a) and a the length of the crack.

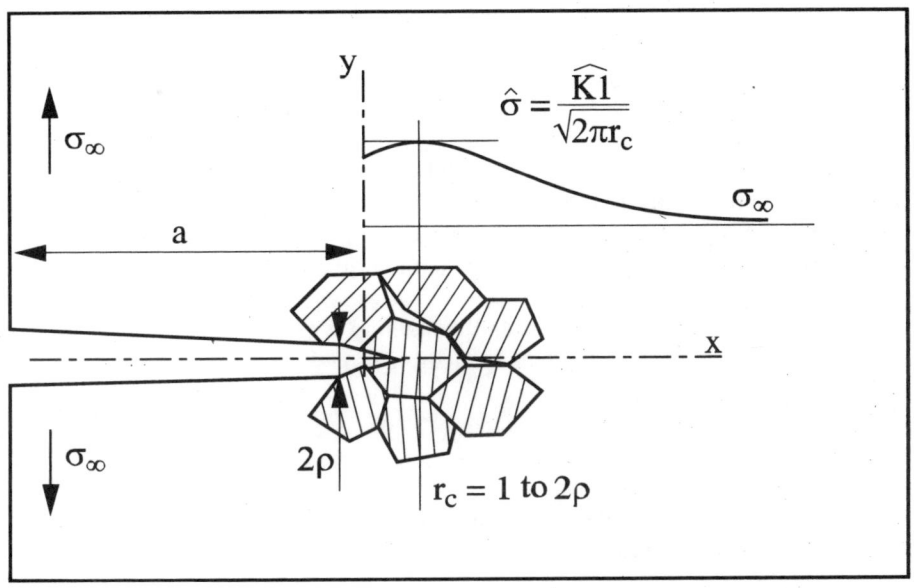

Figure 8.1. *Profile of normal stress σ_{yy} near the crack face*

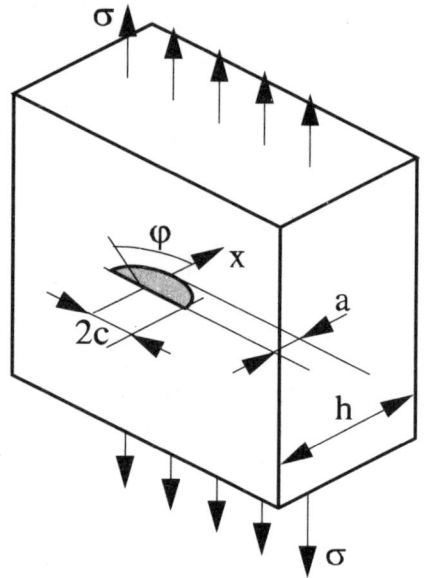

Figure 8.2. *Semi-elliptic surface crack*

Actually, it is at a distance r_c in front of the crack face that the fracture finds a favorable ground; the stress σ_{yy} attaining its maximum value: $\hat{\sigma}$ (see Figure 8.1). $\hat{\sigma}$ is localized at a distance r_c (t) = 1 to 2 ρ(t), where ρ is half the displacement of the opening at the crack tip (a few nanometers in the brittle field) [MCM 77].

For a plate subjected to tension and containing a crack with rectilinear face of depth "a" from the surface, the intensity factor of stresses will be:

$$K_I = 1.12\sigma_\infty \sqrt{\pi a} \qquad [8.8]$$

A better description of reality is obtained by modeling the surface flaw as a semi-ellipse in a flat plate (Figure 8.2):

$$K_{I\max} = 1.2\frac{\sigma_\infty \sqrt{\pi a}}{\phi(a/c)} \approx (1.2 - 0.43\frac{a}{c})\sigma_\infty \sqrt{\pi a} \qquad [8.9]$$

Materials (HP: Hot–Pressed)	σ_{th}(GPa) (equation [8.1])	$\sigma_{r\,max}$(GPa) (fibers or whiskers)	$\sigma_{rexp,\,tens}$ (solid test specimens) (GPa)	K_{Ic}(MPa$\sqrt{m}$)	G_{Ic}(Jm^{-2})
Silica	16	14		0.74–0.81	7.3–8.8
Sheet glass	14	5	0.15	0.72–0.82	6.7–8.6
Borosilicate	/	/	/	0.75–0.82	8.5–10.2
Oxynitride glass	/	/	0.15	1.18–1.22	4.8–5.0
Ice	/	/	/	0.2	3
Porcelain	/	/	/	1.0–1.3	6–22
Al_2O_3 (H.P)	47	29	0.4–0.8	2.5–3.5	15–70
SiC (H.P)	50–100	21	0.5–0.6	2.6–2.8	13–34
Si_3N_4 (H.P)	80	14	0.8–0.9	4.9–10	24–76
ZrO_2	/	/	/	6–12	200–600
Diamond	180–200	/	/	/	/
Fe	30–46	13	0.18	150	100,000
Steels	/	/	0.5–1.9	30–140	4,000–85,000

Table 8.1. *Common values for fracture characteristics of ceramics*

The toughness of ceramic lies typically in the range of 0.5–15 MPa $\sqrt{m}$ (Table 8.1) and that of glass between 0.1 (chalcohalide glass) and 1 MPa $\sqrt{m}$, that is about 50 to 100 times less that of metals. The measurement of breaking resistance of a structure and the knowledge of the toughness of the constituent material makes it possible in principle to estimate, from the above equations, the size of the critical flaw at the origin of the rupture. This size is situated generally between 0.1 and 10 microns and is often comparable to the grain size for polycrystalline ceramics.

8.1.1.3. *Principle of similarity*

By comparing equations [8.2] and [8.7], it appears that the rate of restitution of elastic energy G is of a value linked to the singularity of the stresses. We can demonstrate that in the case of a perfectly elastic linear behavior:

$$G = \frac{\left(K_I^2 + K_{II}^2\right)}{E'} + \frac{1+v}{E} K_{III}^2 \qquad [8.10]$$

where K_I, K_{II} and K_{III} respectively represent the stress intensity factors in cracking modes of types I, II and III (E and E' are defined in section 8.1.1.1).

Hence, in the brittle area and in linear elasticity, the fracture in type I openings can simply be described by the values K_{Ic} (critical stress intensity factor) or G_{Ic} (critical energy of tearing per surface unit). G and K ($K_{I,II \text{ and } III}$) are experimental parameters [SAK 87]. The development of microplasticity at the crack tip for most metals and polymers is not ambiguous, as G_c is in fact several times higher than 2γ. For ceramics at ambient temperature, the measured G_c values approach the predicted values starting from binding energies, and observations by electronic microscope in transmission seem to reinforce the hypothesis of a sharp crack tip on an atomic structural scale [LAW 83], [TAN 88], with a profile matching the linear mechanics of the fracture [IRW 57] [BAR 62].

NOTE.– for obtaining a fracture criteria in the absence of precise information on the behavior near the face of the crack, we have to resort to concepts of energy making it possible to escape from the uniqueness factor. The presence of plasticity (confined) or of creep has been studied by Rice [RIC 68] and Riedel [RIE 81] respectively.

Figure 8.3. *Transmission of an electron microscope view of a crack font in a sialon [TAN 88]*

8.1.2. The R-curve effect

On the microscopic scale, the resistance given by a material to the crack growth, or toughness (K_{Ic}), is a characteristic of the material constituting the solid under study. It is therefore a parameter independent of the size (a) of the flaw or of the growing crack (see section 8.1.1). However, when we try to evaluate the toughness, which poses a number of experimental problems, we often observe a variation (mostly an increase) of crack growth resistance with the extension of the crack: this is the R-curve behavior (resistance curve at cracking). In this case we denote K_R, the stress intensity factor, at the crack extension threshold ($K_R = K_R$ (a) and K_R (a = 0) = K_{Ic}) (Figure 8.4).

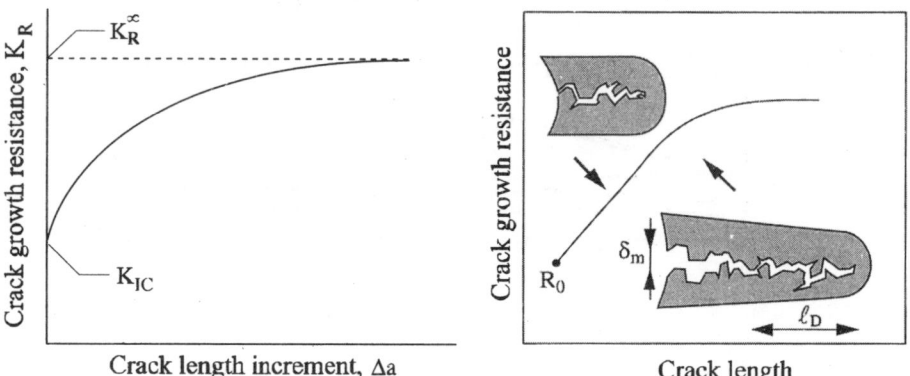

Figure 8.4. *Diagram illustration of the R-curve [SAK 87] – case of an alumina of coarse microstructure [STE 92]*

This phenomenon, discovered almost 30 years back in polycrystalline ceramics [PAB 72] [BUR 73], manifests itself in several refractory materials (see Figure 8.5) [STE 92] reflecting the existence of reinforcement mechanisms occurring on a small scale, near the crack face. We therefore distinguish three types of mechanisms:

i) mechanisms arising ahead of the crack face (case of the appearance of a unique zone where the behavior is micro plastic, or even of an energy dissipation zone by phase transformation);

ii) mechanisms of bridging by acicular grains or fibers, behind the crack face;

iii) wake mechanisms linked to friction between the lips of the crack behind the crack face or resulting from the modification of the material (or the microstructure) in the course of the crack (see Figure 8.6).

It is clear that if a material shows the R-curve effect, its behavior at fracture can no longer be defined by a sole parameter (K_{Ic} or G_{Ic} for example), but requires investigation of the resistance curve for crack extension. Such a curve is obtained from controlled (or stable) cracking tests. This type of test is difficult to conduct on ceramics which are materials that are both rigid and brittle; furthermore, the curve obtained is sensitive to the nature of tests conducted and dimensions of the test specimen.

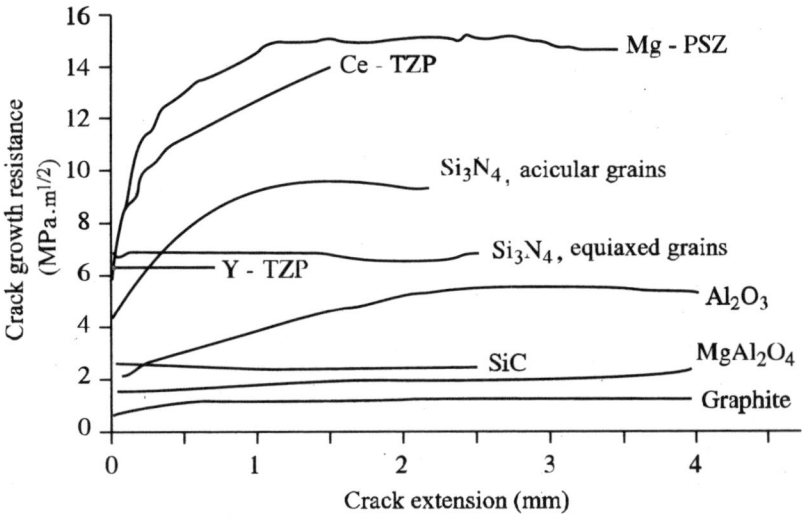

Figure 8.5. *R-curves obtained with different ceramics [STE 92]*

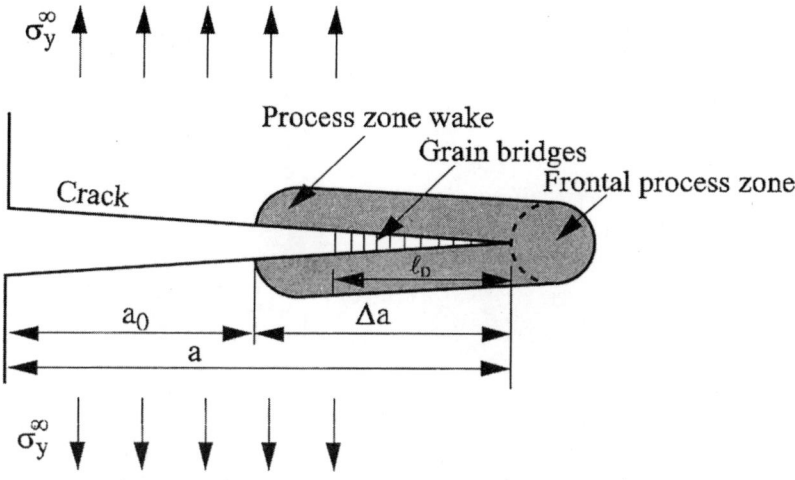

Figure 8.6. *Zones where different mechanisms at the origin of the R-curve effect occur [SAK 87]*

The R-curve effect is largely responsible for the dispersion of K_{Ic} values and the influence of the size of test specimens on the results. Nevertheless, acicular grain ceramics or fiber composites partly owe their attraction to this same effect. This property is in fact put to good use now in the inspection logic of microstructures for making the architecture of materials less brittle and therefore more attractive to the constructor [MIY 95]. Figure 8.7 is a collection of some examples of microstructures favorable to the appearance of the R-curve effect. Furthermore, it is noted that the materials for which this effect is pronounced have a Weibull modulus higher than when the resistance curve is flat.

The analysis of a resistance curve to the extension of a crack may basically be made in three different manners.

Phenomenological approach

The observations are defined in terms of shielding factor or resistance (known as SSC for stress shielding coefficient) and we then elaborate:

$$K_R = SSC_{fpz} \times SSC_{bridging} \times SSC_{wake} \times K_{Ic} \text{ (fpz: frontal process-zone)} \quad [8.11]$$

where $SSC \geq 1$ and K_{Ic} represents the threshold of initiation of an unstable cracking.

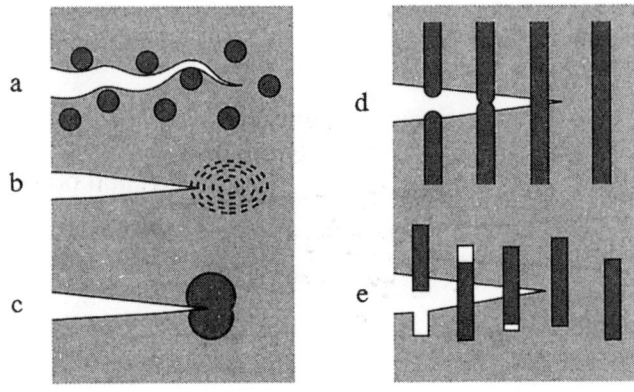

Figure 8.7. *Types of microstructures producing R-curve effect: a) dispersion of hard particles; b) microstructure causing multicracking; c) phase transformation inducing compressive stresses at crack tip (case of partially stabilized zirconia); d) reinforcement of the matrix by ductile fibers and e) reinforcement of the matrix by high resistance fibers [MEN 92]*

Approach through distribution of stresses near the crack face

We use the formalism introduced by Dugdale [DUG 60] and Barenblatt [BAR 32] for interpreting the phenomenon in terms of the closing forces:

$$K_R = K_{Ic} - \sqrt{\frac{2}{\pi}} \int_0^{\ell_D} \frac{\sigma(x)}{\sqrt{x}} \, dx \qquad [8.12]$$

where ℓ_D is the length of the Dugdale zone where closing forces are exerted (Figure 1.6), x the distance from the crack face and $\sigma(x)$ the closing stress ($\sigma(x) < 0$).

In this equation ℓ_D depends generally on the length of the crack [FET 93]. By assuming a uniform closing stress and supposing that this results from the phenomenon of bridging, the stress in the bridges will then be:

$$\sigma_{bridging} = -\sigma(x),$$

and equation [8.12] becomes:

$$K_R = K_{Ic} + 2 \sqrt{\frac{2\ell_D}{\pi}} \, \sigma_{bridging} \qquad [8.13]$$

At the beginning, while the opening of the crack is small, the length of the bridging zone behind the crack face increases as the crack progresses, and K_R

increases with a. Subsequently, the displacement of the crack opening attains a critical value dependent on the microstructure (length of bridging grains for example), above which the value of ℓ_D is constant. Finally, if the crack is near the free surface of the studied structure, the effect of phenomena taking place in front of the crack face dissipates little by little and K_R decreases.

Energy approach

The R-curve phenomenon can be evaluated from the energy point of view, by evaluating the contribution of nonlinear elastic processes resulting, for example, in the bridging of grains. This contribution is given by the difference $\gamma(K_R) - \gamma(K_c)$, where γ represents the fracture work done per surface unit and can be measured experimentally by graphical methods [SAK 86] [MAI 89].

8.1.3. *Statistical approach to fracture*

The disparities in the nature, size, geometry, location and orientation of flaws between nominally identical test specimens explain the great variations in the stress frequently observed.

8.1.3.1. *Weibull's statistical approach [WEI 51]*

Hypothesis: the rupture of a chain with N undifferentiated links is set off by the rupture of the weakest link. Let P_s be the survival probability of a link; the survival probability of the chain can be written as: $P_{sN} = P_s{}^N$, that is, by reasoning on the rupture probabilities: $1 - P_{rN} = (1 - P_r)^N$ or also:

$$P_{rN} = 1 - (1 - P_r)^N \tag{8.14}$$

which can be put in the form:

$$P_{rN} = 1 - \exp(-N\phi(x)) \tag{8.15}$$

where ϕ is an increasing and positive function of parameter x, comparable here to the breaking stress.

Weibull proposed the following expression (Weibull distribution) for $\phi(x)$:

$$\phi(x) = [(x - x_u)/x_o]^m \tag{8.16}$$

where x_u, x_o and m are three characteristics of the population studied. m is the Weibull modulus. The application for the study of a material is made by modeling the material by a chain whose number of links is identified with the volume (or the surface) in question:

$$P_{rV} = 1 - \exp[-\int_V [(\sigma - \sigma_u)/\sigma_{oV}]^m dV] \text{ if } \sigma \geq \sigma_u \qquad [8.17]$$

$$P_{rV} = 0 \text{ if } \sigma < \sigma_u$$

with σ: stress; σ_u: threshold stress; σ_{oV}: normalization constant.

We generally consider that the rupture starts in a region subjected to tension through stress σ, which reduces the previous expression to:

$$P_{rV} = 1 - \exp[-[(\sigma - \sigma_u)/\sigma_{oV}]^m V] \text{ if } \sigma \geq \sigma_u \qquad [8.18]$$

8.1.3.2. Methods for estimating m

a) Choice of a probability function

For each rupture stress σ_i, measured experimentally over a batch of N similar test specimens, a rupture probability P_i is assigned, the stresses being previously arranged as per increasing value.

We will find for example:

– $P_i = i/(N + 1)$: equation for the average value, the most used but also the most biased (in the probability sense of the term);

– $P_i = (i - 0.3)/(N + 0.4)$: equation for the median value;

– $P_i = (i - 0.5)/N$: the least biased of the four expressions if $N > 20$;

– $P_i = (i - 3/8)/(N + 1/4)$: expression adopted for a sample of small size ($N < 20$).

Each of these expressions allows us to draw up a graph collecting the coordinate points ($\ln \ln(1/(1 - P_i))$, $\ln \sigma_i$), from which we derive the value of m. We have in fact:

$$\ln \ln(1/(1 - P_r)) = \ln(V/\sigma_{oV}^m) + m \ln(\sigma - \sigma_u) \qquad [8.19]$$

where the stress σ_u is generally taken as 0.

The Weibull modulus m is therefore obtained by linear regression or by looking for the straight line passing through the barycenters of point groups, if the need arises.

b) Method of maximum of the probability function [TRU 79]

Let L be the probability density function; we can then write:

$$\left.\frac{\partial L}{\partial \sigma_o}\right|_m = 0 \text{ and } \left.\frac{\partial L}{\partial m}\right|_{\sigma_o} = 0 \tag{8.20}$$

which brings us back to the equation:

$$N/m - N(\Sigma_i\sigma_i^m \ln\sigma_i)/\Sigma_i \sigma_i^m + \Sigma_i \ln\sigma_i = 0 \tag{8.21}$$

This equation is easily solved by Newton's method. N > 20 (> 30) is desirable. It is the best method from a statistical point of view: standard deviation and minimum variation.

Typically, for metals, m > 50. For industrially produced ceramics, m lies actually in the interval [5–30].

8.1.3.3. *Transition of the specimen of volume V to volume V functional structure*

If the fracture probability P is associated with a breaking stress σ of a specimen of volume V, then a breaking stress Σ corresponds to the same probability value, for a volume structure V made up of the same material, processed and treated in strictly identical conditions, with:

$$P = 1 - \exp[-(\sigma/\sigma_{oV})^mV] = 1 - \exp[-(\Sigma/\sigma_{oV})^mV] \tag{8.22}$$

and hence:

$$\Sigma = \sigma(V/V)^{1/m} \tag{8.23}$$

8.1.4. **Brittle/ductile transition**

The temperature of the brittle-ductile transition can be viewed as the maximum service temperature or as the onset temperature for rapid creep. From a macroscopic point of view, ductility manifests itself by a deviation from elastic behavior on the load curves P-u (load-displacement), reflecting the appearance of a permanent deformation.

All materials possess a brittle-ductile transition zone in the plane (du/dt,T), with a transition temperature (BDT (brittle-ductile transition)), which is sharper when the behavior change is sudden and higher with greater speed [LOU 92].

Material	Grain size (μm)	Temperature of BDT (°C)	Interval of speed considered (s⁻¹)	Reference
Ice	7,500	−10	10^{-5}–6.10^{-5}	[BAT 93]
Window glass	/	490–580	10^{-5}–10^{-3}	[ROU 99]
β–spodumene (vitroceram)	0.9–2	1,000–1,100	10^{-4}–2.10^{-4}	[WAN 84]
Si (monocrystal)	/	500–800	$\dot{K}_I$ =10–300 MPa.m$^{0.5}$.s^{-1}	[HIR 91]
MgO	0.1–1	900–1,100	10^{-7}–10^{-5}	[CRA 80]
Al$_2$O$_3$	0.9	1,400–1,500	10^{-5}–10^{-4}	[YOS 92]
Y$_2$O$_3$	0.6–0.7	1,400–1,550	10^{-5}–10^{-4}	[ROU 95]
ZrO$_2$ (Y–TZP)	0.5	1,300–1,500	10^{-5}–10^{-3}	[WAK 89]
PbTiO$_3$	2.5	950–1,150	5.10^{-5}–10^{-4}	[WAK 91]
YBa$_2$Cu$_3$O$_{7-x}$	17	180–200	/	[CHA 93]
Ca$_{10}$(PO$_4$)$_6$(OH)$_2$	0.64	900–1,000	10^{-5}–10^{-4}	[WAK 91]
Si$_3$N$_4$	0.5	1,500–1,600	10^{-5}–10^{-3}	[CHE 90]
Si$_3$N$_4$/SiC(30%) (nanocomp)	0.5	1,580–1,630	10^{-5}–10^{-4}	[ROU 92]
SiC	1–2	1,650–1,800	10^{-7}–10^{-5}	[CAM 93]
Fe$_3$C/Fe (20%)	3–4	725–1,000	10^{-4}–10^{-2}	[KIM 91]

Table 8.2. *Brittle-ductile transition zone*

In the case of ceramics with intergranular vitreous phase (materials sintered with additives), the transition is smooth and allows the vitreous transition temperature of the secondary phase at a lower temperature limit (Tg ≈1,000°C for the sialons), [ROU 93]. Furthermore, the finer the grains, the earlier the appearance of BDT (low temperature). There is finally the effect of size, which is evident from the dependence of the BDT with respect to the structure studied and reflects the time it takes for a confined soft zone to spread on both sides of the structure [RIE 81]. On the normal timescale for plastic deformation forming processes (from a few seconds to a few hours), ductility appears as soon as the characteristic relaxation time τ gets smaller than the duration of the experiment. This produced the concept of Deborah's number D_n [REI 64], ratio of relaxation time and observation time, and which is inspired from an ancient hymn of the Old Testament attributed to the prophetess Deborah (Judges 5, 5th verse): "... the mountains flowed before the Lord...". The behavior is brittle if $D_n > 1$ and ductile if not. The time constant τ, characteristic of the relaxation process, can be estimated by supposing that the kinetics of relaxation follow a linear Maxwell model. We therefore have $\tau = \eta/\mu$, where η is the coefficient of viscosity and μ the elastic shear modulus. We can also obtain precise values of τ by conducting tests of mechanical relaxation.

The BDT can be observed through relaxation tests conducted at different temperatures, through tests with imposed deformation speed by varying the speed and the temperature, or again by micro-indentation experiments [BRA 95]. In the latter case, the transition is marked by an energy dissipation (irreversible) in the load-displacement curves, which increases with decreasing speeds and/or increasing temperature. Table 8.2 gives results obtained for different materials, more often dense fine grain ceramics intended for superplastic forming operations.

8.1.5. *Experimental techniques*

8.1.5.1. *Fracture*

The measurement of breaking stress and deformation of ceramic material is particularly delicate. For taking this measurement in tension, we should have a rigid testing machine capable of providing an excellent alignment of the test specimen on the load axis, equipped with a displacement pick-up capable of reading micronic displacements. The result largely depends on the accuracy of the alignment and the surface condition of the test specimens. Typically, it is necessary to carry out more than ten measurements. A compression test will give a breaking stress which is 10 to 100 times higher than in tension. We can get close to the tensile value through 3 or 4-point-bend tests (mind the friction at support surfaces). The breaking stress σ_r, and the corresponding deformation ε_r, are derived easily from the fracture load P_r, thickness W and width B of the test specimen, and distances between lower S_1, and upper S_2 supports ($S_2 = 0$ for 3-point-bend):

$$\sigma_r = 3P_r (S_1 - S_2)/(2BW^2) \hspace{2cm} [8.24]$$

$$\varepsilon_r = \sigma_r/E \hspace{2cm} [8.25]$$

We find for example: $B = 4$ mm, $W = 3$ mm, $S_1 = 20$ mm and $S_2 = 40$ mm (French standard B41-104, or German standard DIN 51-110-1).

We can also resort to a diametric compression test, also known as the Brazilian test, for evaluating the tensile stress [CLA 70]. This test, which is particularly suitable for raw materials or for bars (non-sintered products), consists of compressing a pellet monoaxially (diameter D_n, thickness t) following a diametric direction. The breaking stress is then deduced from the rupture load using the following relationship:

$$\sigma_r = 2P_r/(\pi D_p t) \tag{8.26}$$

8.1.5.2. Toughness

a) Indentation method (IF (Indentation Fracture))

The surface of a test specimen is polished to obtain mirror finish and then is indented, generally with a Vickers probe. Knowing Young's modulus E (Pa) and the load on the probe P (N), hardness measurements H (in Pa), and length of cracks 2c (m), which radiate from the edges of the indentation, we can calculate K_{Ic} (Pa.$\sqrt{m}$). Other works are full of equations, and the following are the most frequently used ones:

$$K_{Ic} = 0.0824 \, Pc^{-3/2} \qquad \text{if } c/a > 2.5 \qquad \text{[EVA 76] [8.27]}$$

$$K_{Ic} = 0.01(E/H)^{2/3} \, Pc^{-3/2} \qquad \text{if } c/a > 2.5 \qquad \text{[LAU 85] [8.28]}$$

$$K_{Ic} = 0.016(E/H)^{1/2} \, Pc^{-3/2} \quad \text{if } c/a > 2.5 \qquad \text{[ANS 81] [8.29]}$$

$$K_{Ic} = 0.018H^{0.6}E^{0.4}al^{-1/2} \quad \text{if } c/a < 2.5 \ (l = c - a) \qquad \text{[NII 82] [8.30]}$$

This method is not self-consistent and the prefactor on the RHS is a result of an experimental calibration on test specimens with known toughness values.

The relationship proposed by Niihara is adapted to measurements obtained with relatively small loads (typically P < 5N), leading to a purely lateral system of cracks, known as Palmqvist cracks (see Figure 8.8).

b) Indentation and fracture method (IS (Indentation Strength))

An indented test specimen is subjected to a bend test, with the indentation located at the centre of the face in tension. The stress measured at fracture σ_r enables the calculation of toughness from the indentation load P, Young's modulus E and hardness H [CHA 81]:

$$K_{Ic} = 0.59(E/H)^{1/8}(\sigma_r P^{1/3})^{3/4} \tag{8.31}$$

This method helps to avoid the often inaccurate measurement, of the length of cracks resulting from the indentation.

c) Straight notch method (SENB (Single Edge Notch Beam))

This method is similar to the previous one, but it makes use of a straight notch of depth a, generally made with a diamond saw of small thickness ($\approx 0.1 - 0.2$ mm) which sets off the fracture of the specimen.

The expression for toughness in this case is:

$$K_{Ic} = \sigma_r Y \sqrt{a} \text{ with } Y = \frac{1.99 - \alpha(1-\alpha)(2.15 - 3.93\alpha + 2.7\alpha^2)}{(1+2\alpha)(1-\alpha)^{3/2}} \qquad [8.32]$$

where $\alpha = a/W$, with W: height of the specimen.

This expression is valid only for $S \approx 4W$, where S is the distance between lower supports (3-point-bend).

d) Straight precracked method (SEPB (Single Edge Precracked Beam)) [NOS 88]

This is a variation of the previous method in which the straight notch is replaced by a straight crack, that is a flaw which is more serious and closer to flaws present in the materials, by using a precracking device.

e) Chevron shaped notch method (CN (Chevron Notch))

A V shaped notch is made in a test specimen to be fractured by bending. During loading, a crack propagates from the tip of the V and the load goes through a maximum (P_{max}) corresponding to a specific theoretical value of a (or of α); the toughness is then given by [MUN 80]:

$$K_{Ic} = \frac{P_{max}}{B\sqrt{W}}(2.92 + 4.52\alpha_o + 10.14\alpha_o^2)\frac{(S_1 - S_2)}{W}\sqrt{\frac{\alpha_1 - \alpha_o}{1 - \alpha_o}} \qquad [8.33]$$

with: $0.12 < \alpha_o < 0.24$; $0.9 < \alpha_1 < 1$; $W/B \approx 1$; $S_1/W \approx 4$ and $S_1/S_2 \approx 4$.

NOTE.– there are other testing specimens that are suitable for fracture toughness or fracture energy measurements of brittle materials. These include for instance compact tension (CT) and double torsion (DT) specimens. Readers will find more details in [CHE 75].

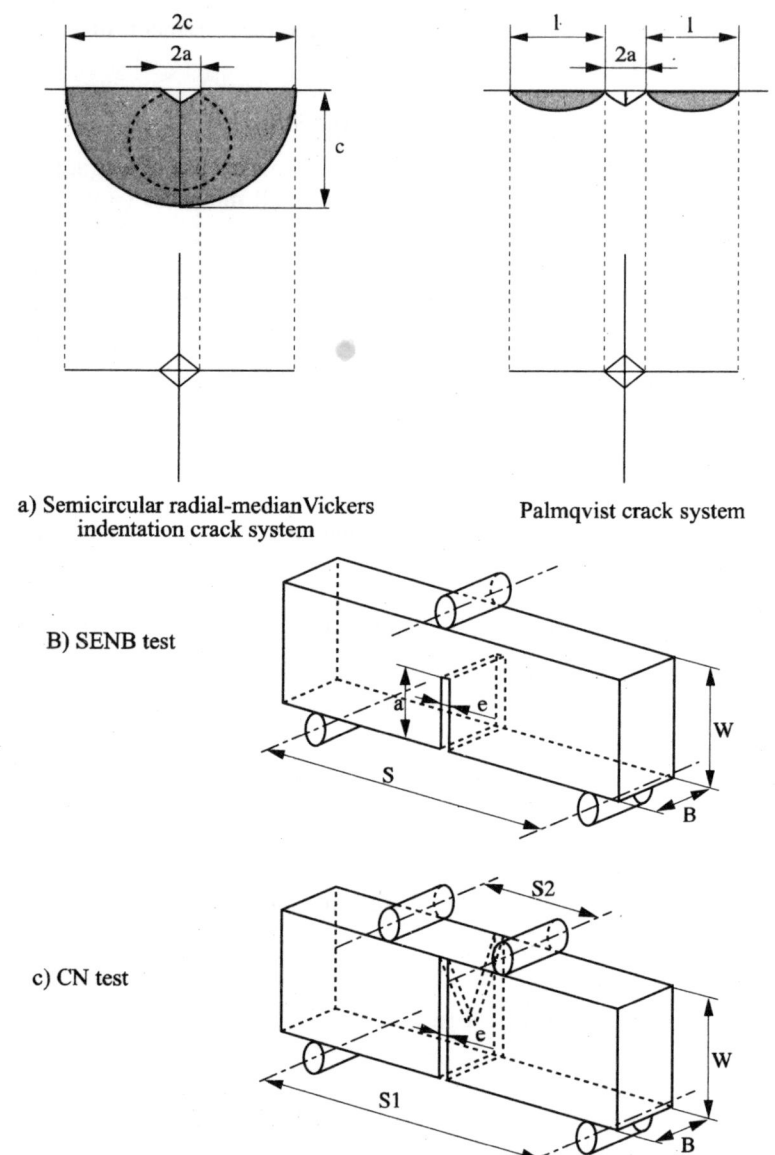

Figure 8.8. *Toughness measurement: a) method by indentation; b) SENB specimen (3-point-bend); c) CN specimen (4-point-bend)*

8.2. Friction, wear and abrasion

The relative displacement, or sliding, of two parts maintained in contact does not take place without resistance. The tangential force T needed for ensuring the movement depends on the normal force N at contact: this is the phenomenon of friction. This phenomenon is well defined by Coulomb formula $T = \mu N$, where μ is the coefficient of friction. The friction causes wear of the two parts at the level of the contact surfaces. For a couple of given materials, the wear, as well as the value of μ, depends on the temperature, the atmosphere or even on the sliding speed. This wear is generally harmful to the in-service behavior of the part and we will look for operating conditions and surface treatments which will minimize it. It is, however, desirable in machining by abrasion. Ceramics are excellent materials for applications where friction is present by virtue of their high hardness, their chemical inertia and their temperature stability. Their applications include friction parts and seal components for the automobile industry (sealing items for water pump), gears, valve heads, precision ball bearings, cutting tools, conduits for casting of refractory metals, abrasives, etc.

8.2.1. *Friction and wear*

Wear can be measured at the depth of the friction track by the volume of material removed per unit length of friction or by volume of unit length and normal force (specific wear rate).

Ceramic wear is more of the abrasive type and is characterized by the generation of debris giving rise to propagation of cracks in type I and II mixed (opening and straight sliding). It increases drastically above the critical load.

The volume of wear V is proportional to the contact load $\bar{N}$ and to the friction distance x. We may therefore account for the wear by using the specific wear rate defined by:

$$V_s = V/(xN) \qquad [8.34]$$

It is difficult to measure the quantity of wear. It can be done indirectly by measuring the decrease in the mass of the specimen, which would also include all mass changes due to physico-chemical processes or the adherence of debris, or by measuring the volume of material removed along the friction path by optical or electronic microscope. The first method gives an overall result with accuracy increasing as the specimen size gets smaller.

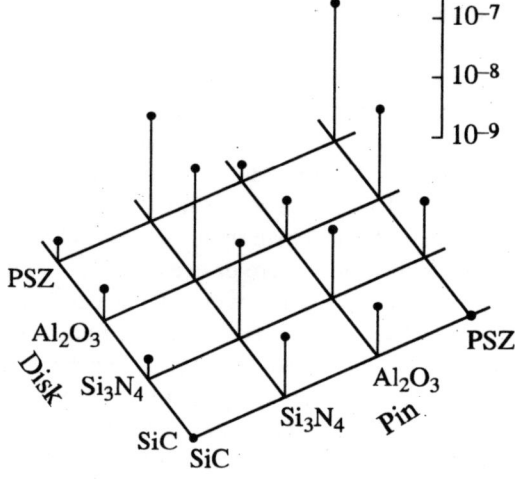

Specific wear V_s /mm²kgf⁻¹

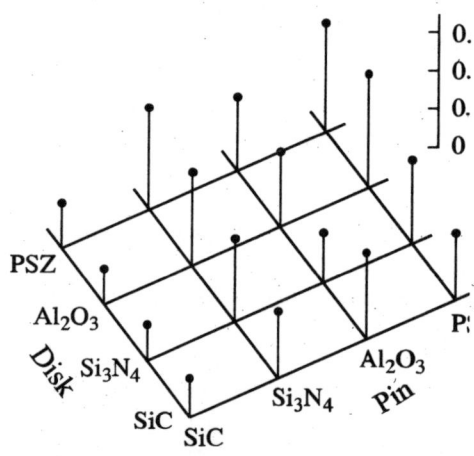

Friction coefficient μ

Figure 8.9. *Value of coefficient of friction and of specific wear rate for different ceramic couples in the pin on disk configuration [IWA 86]*

Friction and wear are complex phenomena, whose mechanisms are not well known. Following the classic model of friction and wear in the presence of adhesion, it is possible to relate V_s to the hardness and to the geometry of areas of contact. Let us suppose, for example, that the contact zones are disks of radius a and that their number is n. The force distribution on the entire surface is given by the relationship:

$$n = N/(H\pi a^2) \qquad\qquad [8.35]$$

where H is the hardness of the softer material.

We suppose that each adhesion zone disappears after a displacement over a distance of 2a, while new zones of contact get formed. The total number n_T of "points" of adhesion for a friction distance x will be:

$$n_T = nx/(2a) = Nx/(2\pi H a^3) \qquad\qquad [8.36]$$

Therefore, if k is the probability of forming a volume of wear debris at a point corresponding to a hemisphere of radius a, the total quantity of wear is given by equation:

$$V = kn_T 2\pi a^3/3 = kNx/(3H) \qquad\qquad [8.37]$$

which gives a specific wear:

$$V_s = k/(3H) \qquad\qquad [8.38]$$

The factor k is known as the coefficient of wear. For a conical indenter with half-angle at apex θ, k will be $3\cot\theta/\pi$ and $V_s = \cot\theta/(\pi H)$. This relationship confirms the observation that harder materials are more wear resistant. The wear resistance also increases with toughness and, in fact, it appears that the ratio K_{Ic}/H governs it. A brittleness index B has in fact been defined as the inverse of the ratio: $B = H/K_{Ic}$ [LAW 79]. It also varies very appreciably with the sliding speed: in most cases, the wear rate goes through a maximum for speeds between 0.1 and 10 ms^{-1}. The atmosphere also plays an important role on friction and wear. In particular, friction is higher under vacuum and humidity leads to physico-chemical modifications of the contact surfaces, which is clearly reflected in the tribological behavior (in general μ and V_s increase with humidity).

Experimental techniques commonly used for assessing the tribological behavior have resorted to pin on disk, disk on disk, ball on disk, or even cylinder on cylinder configurations. Figure 8.9 shows the results obtained for different dense ceramics couples through the pin on disk method, under a load of 1 kg, with an average relative contact speed of 0.18 ms^{-1}.

Under friction at high speed and under small load, and in ambient air, silicon nitride is presently the ceramic which offers best wear resistance [CHE 92].

8.2.2. *Abrasion*

With the increase in load or contact speed between a hard object (diamond indenter, for example) and a ceramic material, we see the transition from a relatively soft regime of scratching without removal of material to a plowing regime with removal of material (mechanical polishing) and finally to a rapid wear regime with peeling corresponding to mechanical abrasion. Materials with the highest abrasion resistance are used as abrasives and are meant for application in polishing surface finishing, or in machining (grinding). We then find that diamond, boron nitride (cBN) silicon and aluminum carbide are high on the list.

Diamond tools are widely used for shaping grinding wheels. Single tip diamond tools are used to cut Al mirror (high precision). Polycrystalline diamond may be used to cut aluminum alloys (small business though). Major technical drawbacks of diamond are its reaction with many metals and its oxidation above 600°C. Problems occur in particular with cast-iron, titanium, nickel, etc. The grinding of light metals such as aluminum and Ni-based alloys is difficult. Local burning may occur. As a matter of fact, although cBN is less hard than diamond, it is more suitable as far as ferrous alloys are concerned. The market for cBN is growing rapidly (10%/year). Three families of abrasive are defined, namely:

– coated abrasives (suitable for plastics and metals);

– bonded abrasives (suitable for hard metals and ceramics);

– superabrasives (suitable for hard ceramics and crystals).

Three classes of bonding materials are developed for the grinding wheels, namely: resin, which requires heat-treatments at 300–400°C, metal (treatment at 500°C) and vitreous bonding, with a crystallization treatment to form a glass-ceramic at 800–1,000°C. Resin (bakelite) is good for finishing; metal, mostly from powder metallurgy, is good for rough abrasion of glass; glass is good when the grinding wheel has to be reprofiled.

A remarkable success was achieved regarding the ultra-high finishing. For instance, Saint-Gobain Co. (Worcester, USA) developed a machine and a process to reach Å-quality finishing on thin silicon wafer, up to 300 mm diameter, using a special machine tool, called a Nano Slicing machine, and a very fine Chemo-Mechanical Polishing (CMP), allowing for a better than 10 Å Ra finish. The finished product can be bent without failure, demonstrating that the surface is free of large (nano-sized) defects.

In order to understand the mechanisms which govern machining by abrasion, we should analyze the different interactions between the tool and the material being machined. For a grinding operation, we refer to the following four interactions: abrasive/material, binder (grinding wheel)/material, abrasive/debris and binder/debris. This difficult problem is further complicated by physico-chemical processes linked particularly to the use of "cutting" fluids, which not only acts as a lubricant, but also condition thermodynamic machining parameters (contact temperature) and the kinetics of removal of debris. The understanding of abrasion is still at a very superficial level. All the same, technical solutions, which are satisfactory in most cases, have been found due to the ingenuity and know-how of industries confronted with this problem. For a given material and a particular surface condition to be obtained, the fabrication scheme consists of a set machine/grinding wheel/lubricant/machining parameters. Super abrasives with diamond, cubic boron nitride or other hard ceramics base, as well as new binders for grinding wheels, are in the process of development (see Chapter 14) [SUB 87].

8.3. Deformation

8.3.1. *Elasticity*

8.3.1.1. *Physical origin*

Elasticity can be seen by a reduction, at least partial, of the deformation when the load applied on a body is released. From ambient temperature (or below) and up to relatively high temperatures (T < 1,000°C), ceramics are elastic materials par excellence: their behavior under load is most often linear with a nearly full reversibility of the deformation on removal of the load and the fracture takes place during elastic loading (no plasticity), for a deformation less than 1%. Refractories are generally of mixed nature, ionic and covalent. It is difficult to deform these. In covalent crystals (for example B_4C or Si_3N_4) this is due to the directivity of the bonds, while in ionic crystals (for example NaCl or ZrO_2) it is due to deep disturbances in the electrostatic interactions resulting during the deformation of the crystal.

Let us suppose that the potential of interaction follows a binomial expression:

$$U = -\frac{A}{r^m} + \frac{B}{r^n} \qquad [8.39]$$

where the first term of the second part reflects the attraction between atoms and the second repulsion. (m) and (n) are constants such that n > m for ensuring the existence of a minimum of potential energy for an interatomic distance r_0 at equilibrium ($\partial U / \partial r = 0$ for r_0). Experimentally, n is near 9 for the ionic bond and n $\approx$ 2 for alkaline metals. This expression is not just gross approximation: the energies of attraction and repulsion can have much more complex forms (Morse potential for covalent bonds and Born-Mayer for the ionic bond) [SAK 93], where the homopolar nature of the bond and the saturation phenomena (distortion of the Fermi surface) are taken into account. It has the advantage of linking the modulus of elasticity to the bond energy and to r_0. In fact, we can show that the compressibility modulus K can be expressed as:

$$\frac{K}{V_0} = \frac{\partial^2 U}{\partial V^2}\bigg|_{V_0} \tag{8.40}$$

where V is the average volume occupied by an atom ($V = V_0$ at equilibrium).

We thus obtain from equation [8.39], with U_0 being the bond energy:

$$K = \frac{-nmU_0}{9V_0} \tag{8.41}$$

This result is known as the first rule of Grüneisen. U_0 is obtained by taking the opposite of the sum of the sublimation energy (or of cohesion) and the ionization energy. The determination of U_0 is therefore not easy and, for ceramics which have high fusion temperatures, U_0 is not always known. Typically, $U_0 \approx -200$ at $-1,000$ kJmol^{-1} for an ionic or covalent bond (-347 kJmol^{-1} for the C-C bond in the diamond carbon), against -100 to -500 kJmol^{-1} for the metallic bond (-519 kJmol^{-1} for Na) and -1 to -20 kJmol^{-1} for the Van der Waals bond (-7.8 kJmol^{-1} for the Ar-Ar bond). Furthermore, the interatomic equilibrium distances, and therefore V_0, have a tendency to decrease while the bond force increases. Ceramics therefore have high modulus of elasticity (see Table 8.3).

It appears from equation [8.41] that K is proportional to a volume density of energy. Neither the atomic volumes nor the bonding energy are accurately known and m and n depend on the chemical nature of the bonds and are likely to fluctuate with the composition. Nevertheless, the former expression gave birth to several theoretical models aimed at providing *ab initio* values for the elastic moduli of glasses. The most widely used model is the one proposed by Makishima *et al.* [MAK 73] which expresses E as a function of the volume density of energy and the atomic packing density. The energy density is calculated from the dissociation (or atomization) energies of the different oxides introduced in the starting powder mixture. A linear dependence, $K = U_0/V_0$, is observed as long as $U_0/V_0 < 50$ kJ/cm^3, which gives m = 9 and n = 1 in equation [8.41] [ROU 06].

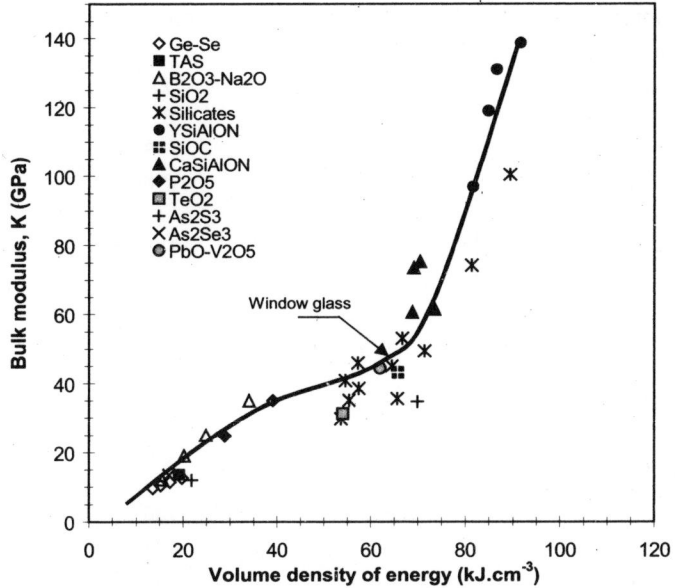

Figure 8.10. *Experimental bulk modulus as a function of the calculated volume density of energy*

8.3.1.2. Behavior laws

a) Anisotropic linear elasticity

In anisotropic linear elasticity, the stress and deformation tensors are connected to each other by 4th order tensors:

$$\tilde{\sigma} = \overset{=}{C}\,\tilde{\varepsilon} \text{ and inverted: } \tilde{\varepsilon} = \overset{=}{S}\,\tilde{\sigma} \qquad [8.42]$$

where $(\overset{=}{C})$ and $(\overset{=}{S})$ are called tensors of elasticity modulii (or stiffness tensors) and tensors of elastic compliance (or flexibility) respectively, and possess *a priori* $3^4 = 81$ independent components, expressed as C_{ijkl} and S_{ijkl} for $\overset{=}{C}$ and $\overset{=}{S}$ respectively.

In reality, because of the symmetries of $\tilde{\sigma}$ and $\tilde{\varepsilon}$ and the conditions of deformation energies, we arrived at 21 independent components for each of the two tensors. In simplified writing, the notation reduced to two indices is frequently used:

$C_{ijkl} \rightarrow C_{IJ}$ with:	(i, j) or (k, l)	1,1	2,2	3,3	2,3	3,1	1,2
	(I or J)	1	2	3	4	5	6

Materials	E (GPa)	μ (GPa)	ν
1) Ceramics			
ZrO_2 (PSZ)	200	77	0.31
SiO_2	73	31	0.17
BeO	353	147	0.2
MgO	280	110	0.27
ThO_2	240	/	/
Al_2O_3 (99.9%)	366	155	0.18
Mullite	220	87	0.27
Spinel	250	/	/
Cordierite	139	45	0.31
Si_2N_2O (HP)	220	92	0.2
SiAlON	310	122	0.27
Si_3N_4 (HPSN)	310	123	0.28
BN	350	/	/
AlN (HP)	315	126	0.25
HfB_2	500	223	0.12
TiB_2	551	249	0.11
ZrB_2	500	225	0.11
SiC (99.99%)	475	181	0.31
WC	696	/	/
B_4C	445	187	0.19
TiC	451	/	/
ZrC	348	/	/
Graphite	42-60	/	/
Porcelain	60-75	/	/
Marble	27-82	....	0.1-0.29
Granite	42-60	385	0.2
Diamond (natural)	700-1,200		
Diamond (synthetic)	925		
2) Vitroceramics			
Zerodur (Schott)	84	/	/
β-spodumen (Cor)	88	35	0.25
Type LAS	52-98	/	/
Type MAS	100-120	/	/
3) Fibers			
Carbon fiber	490	/	//
SiO_2-Al_2O_3-MgO	85	/	/

Table 8.3. *Modulus of elasticity of polycrystalline ceramic materials*

Chemical System	Glass	E (GPa)	K (GPa)	ν
	Amorphous ice (H_2O, high density)[1]	12.5	9.9	0.29
Chalcohalogenides	Se	10.3	9.6	0.322
	$Ge_{10}Se_{90}$	12.1	10.4	0.307
	$Ge_{15}Se_{85}$	13.8	11.2	0.295
	$Ge_{25}Se_{75}$	16.1	12.3	0.281
	$Ge_{30}Se_{70}$	17.9	12.6	0.264
	$Ge_{22}As_{20}Se_{58}$	18	13.6	0.28
	$Te_2As_3Se_5$ (TAS)	18	13.6	0.28
	$Ga_5Sb_{10}Ge_{25}Se_{60}$ (2S2G)	23.9	16.7	0.262
	$F_{0.746}Zr_{0.122}Ba_{0.064}La_{0.015}Al_{0.015}Na_{0.038}$ (ZBLAN - Infra. Fib. Syst. Inc.)	53.3	44.42	0.3
Leadates	$Si_{0.22}Na_{0.024}K_{0.076}Ca_{0.005}Pb_{0.09}O_{0.585}$ (type "cristal")	61	36.3	0.22
	$V_{0.27}Pb_{0.03}O_{0.7}$	42	35.0	0.3
	$V_{0.25}Pb_{0.06}O_{0.69}$	48	44.4	0.32
Phosphates	P_2O_5	31.3	24.8	0.29
	$P_2Na_2O_7$	35.9	n.d.	n.d.
	P_2CaO_6	55.3	n.d.	n.d.
	$LiPO_3$	46.3	35.1	0.28
Borosilicates	B_2O_3	17.4	12.1	0.26
	Schott BK7	81	46.6	0.21
	Schott Borofloat	63	32.8	0.18
	$Si_{0.25}Al_{0.011}B_{0.072}Na_{0.027}O_{0.64}$ (Corning Pyrex 7740)	64	35.6	0.2

Soda-Lime-Silicates	SiO_2	70	33.3	0.15
	$Si_{0.22}Na_{0.22}O_{0.56}$	65.8	45.9	0.261
	$Si_{0.2}Na_{0.27}O_{0.53}$	60.2	40.8	0.254
	$Na_{0.1}Si_{0.28}O_{0.62}$	62.4	34.2	0.196
	$Si_{0.27}Na_{0.08}Ca_{0.03}O_{0.62}$	67	37.6	0.203
	Obsidian (Greece)	71.5	38.6	0.191
	$Si_{0.25}Na_{0.092}Ca_{0.035}Mg_{0.021}O_{0.6}$ (Window glass)	72	44.4	0.23
Silico-aluminates	$Na_{0.095}Al_{0.065}Si_{0.23}O_{0.61}$	70.9	38.6	0.194
	$Na_{0.086}Al_{0.143}Si_{0.171}O_{0.16}$	77.3	47	0.226
	$Na_{0.1}Al_{0.1}Si_{0.2}O_{0.6}$	71.2	41.8	0.216
	Saint-Gobain E glass	72.3	n.d.	n.d.
	Saint-Gobain R glass	86	n.d.	n.d.
	S Glass	145	105.1	0.27
Oxynitrides	$Li_{0.6}Al_{0.1}Si_{0.6}O_{1.575}N_{0.05}$	90	55.6	0.23
	$Y_{11.9}Si_{17.8}Al_{6.8}O_{63.5}$	128	97.0	0.28
	$Y_{12.3}Si_{18.5}Al_7O_{54.7}N_{7.5}$	150	119.0	0.29
	$Y_{12.5}Si_{18.8}Al_{7.2}O_{50.3}N_{11.2}$	165	131.0	0.29
	$Y_{0.146}Si_{0.232}Al_{0.034}O_{0.31}N_{0.29}$	183	138.6	0.28
	$Y_{4.86}Mg_{6.3}Si_{16.2}Al_{11.8}O_{54.9}N_{5.92}$	134	n.d.	n.d.
Oxycarbides	$SiC_{0.33}O_{1.33}$	104	n.d.	n.d.
	$SiC_{0.375}O_{1.25}$	110	n.d.	n.d.
	$SiC_{0.5}O_{1.24}$	97.9	n.d.	n.d.
	$SiC_{0.8}O_{1.6}$	101	43	0.11

Table 8.4. *Elastic moduli of inorganic glasses from different compositional systems at 293 K. [1]at 77 K. n.d.: non determined*

	Crystal	C_{11} (GPa)	C_{12} (GPa)	C_{44} (GPa)
Cubic	LiF	111	42	63
Structures	NaCl	48.7	12.3	12.6
	KCl	39.8	6.2	6.2
	KBr	34.6	5.8	5.1
	MgO	289	87	148
	Diamond	1,079	330	578

	Crystal	C_{11} (GPa)	C_{12} (GPa)	C_{13} (GPa)	C_{33} (GPa)	C_{44} (GPa)
Hexagonal	Graphite	1,060	180	15	36.5	4.5
Structures	Si_3N_4-β	315	239	222	332	40

Crystal Trigonal Structure		C_{11} (GPa)	C_{12} (GPa)	C_{13} (GPa)	C_{14} (GPa)	C_{33} (GPa)	C_{44} (GPa)
	Quartz	85.1	6.9	14	16.8	105.3	57.1

Table 8.5. *Anisotropic modulii of elasticity of a few monocrystals (at 20°C)*

b) Isotropic linear elasticity (generalized Hooke's law)

In the case of a material showing isotropic linear behavior, stresses are easily deduced from deformations (and vice versa) with the help of the generalized Hooke's relationship:

$$\frac{1+\nu}{E}\sigma_{ij} = \varepsilon_{ij} + \frac{\nu}{1-2\nu}(\text{Trace } \tilde{\varepsilon})\delta_{ij} \qquad [8.43]$$

where (δ_{ij}) represents the Kronecker symbol and Trace $\tilde{\varepsilon}$ $= \varepsilon_{11} + \varepsilon_{22} + \varepsilon_{33}$.

Young's modulus E and the shear modulus μ are linked by the relationship:

$$E = 2(1+\nu)\mu \qquad [8.44]$$

The compression modulus is written:

$$K = \frac{\text{Trace } \tilde{\sigma}}{3\text{Trace } \tilde{\varepsilon}} = \frac{E}{3(1-2\nu)} \qquad [8.45]$$

NOTF.– for metals, the hardness is only a small fraction ($\sim 10^{-3}$) of the value of Young's modulus, while for covalent crystals, the hardness is about 10% of the value of E (the data regarding materials with basically ionic bonds cannot be interpreted so easily).

Besides, elasticity moduli are considerably higher for ionocovalent materials than for other classes of materials. Ceramics are therefore found on top of the list of hard and rigid materials. Their Poisson's ratios are between 0.2 and 0.3, and generally $\mu/E \approx 0.4$ and $K/E \approx 2/3$. By comparison, $\nu \approx 0.33$, $\mu/E \approx 3/8$ and $K \approx E$ for metals, $\nu \approx 0.5$, $\mu/E \approx 1/3$ and $K \gg E$ for polymers.

Glasses below T_g are ideal isotropic linear elastic materials. The elastic moduli of glasses depend greatly on their chemical composition. For instance, pure amorphous selenium and a glass from the YSiAlON system will have Young's moduli of 10 and 165 GPa respectively. Young's modulus (the most frequently measured elastic modulus) is plotted as a function of the glass transition temperature (T_g) in Figure 8.11. Even though the search for glasses possessing high elastic moduli is a relatively old topic, it is of paramount interest today with the need for new light and durable materials stiffer than those presently available. For instance, in order to increase both the rotating speed and the durability of computer hard disks, Al-Mg alloys are being more and more replaced by high Young's modulus (E) glasses. An enhancement of the elastic moduli also allows for a decrease of the weight of windows (for a given glass density) and thus for a significant decrease of the energy consumption in the transportation industry. Although the general tendency is an increase of E with T_g, it is worth noting that the highest values for E are not reported for the most refractory glasses (SiOC glasses). This is because SiOC glasses, such as a-SiO$_2$, are characterized by a low atomic packing density (recall from equation [8.41] that an elastic modulus reflects a volume density of energy and note that it is expressed in Pascals, with 1 Pa= 1 Jm^{-3}!). Instead, T_g seems to depend mostly on the mean bonding energy so that a-SiO$_2$ and silicon oxycarbide glasses are the most refractory glasses.

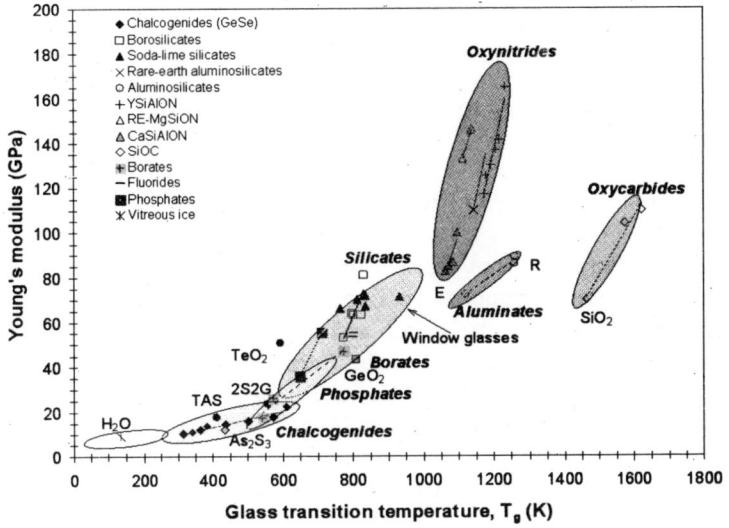

Figure 8.11. *Young's modulus at 293 K (except for amorphous ice: T=77 K is considered) and glass transition temperature of glasses. TAS and 2S2G stand for $Te_2As_3Se_5$ and $Ga_2Ge_{20}Sb_{10}S_{65}$ respectively. E and R glasses are high strength industrial aluminosilicate glasses*

Besides the essential role of elastic properties for materials selection in mechanical design, we show in this analysis that macroscopical elastic characteristics (E,ν) provide an interesting way to get insight into the short- and medium-range orders existing in glasses. In particular, ν, the packing density (C_g) and the glass network dimensionality appear to be strongly correlated. Networks consisting primarily of chains and layers units (chalcogenides, low Si-content silicate glasses) correspond to ν > 0.25 and C_g > 0.56, with maximum values observed for metallic glasses (ν ~ 0.4 and C_g > 0.7). On the contrary, ν < 0.25 is associated to a highly cross-linked network with a tri-dimensional organization resulting in a low packing density.

c) Linear thermoelasticity

The principle of superposition indicates that, in the elastic domain, the stresses and deformations of the same nature due to different systems of simultaneous external loads are respectively additive. The superposition of a stress field (associated with an elastic deformation field) and a thermal field (which generates a thermal deformation) can be summarized by the following equation: $\widetilde{\varepsilon} = \widetilde{\varepsilon}^{el} + \widetilde{\varepsilon}^{th}$, where the tensor of thermal deformations is expressed in the standard case of an isotropic material: $\varepsilon^{th} = \alpha(T - T_o)\delta_{ij}$, where α is the coefficient of thermal expansion (for example $\alpha_{Al_2O_3} = 8.7.10^{-6}°C^{-1}$, $\alpha_{Si_3N_4} = 3.2.10^{-6}°C^{-1}$, $\alpha_{SiO_2} = 0.57.10^{-6}°C^{-1}$, $\alpha_{plate\ glass} = 8.5.10^{-6}°C^{-1}$).

For an isotropic body having linear thermoelastic behavior, the behavior relationship can be written as [SAA 93]:

$$\varepsilon_{ij} = \frac{1+v}{E}\sigma_{ij} - \frac{v}{E}(\text{Trace } \tilde{\sigma})\delta_{ij} + \alpha(T - T_o)\delta_{ij} \qquad [8.46]$$

where $T - T_0$ is the difference of temperature from a reference temperature T_0.

For ceramics, the stresses of thermal origin, which are likely to appear during cooling from the sintering temperature (typically $T_{sintering} \geq 1,200°C$) or in-service at high temperature are particularly high and could lead to cracking and destruction of the component, or the pieces in contact, because the capacity of the material to accommodate these stresses is often insufficient: there is no plastic accommodation and the resistance to damage of refractories is small.

If we assume an isotropic material subjected to sudden variations of temperatures and blocked by imposed dimensions, we can examine the most unfavorable case corresponding to an inclusion in an infinitely rigid matrix; the previous equations give:

$$\sigma_{11} = \frac{E}{1+v}\varepsilon_{11} + \frac{Ev}{(1+v)(1-2v)}(\text{Trace } \tilde{\varepsilon})\delta_{ij} - \frac{3E}{1-2v}\alpha(T - T_o)\delta_{ij} \qquad [8.47]$$

and for the simple case: $\sigma_{22} = \sigma_{33} = 0$ and $\varepsilon_{11} = 0$ (blocking along the axis 1), we have:

$$\sigma_{11} = -E\alpha(T - T_0) \qquad [8.48]$$

For refractories: $\alpha E \approx 1$ MPa.°C^{-1} and with $|T - T_o| > 1,000°C$, we find $\sigma_{11} > 1,000$ MPa. It is clear that this stress is susceptible to produce a fracture.

NOTE.– in reality, for an inclusion in a matrix, the expansion (or contraction) which accompanies a variation in temperature is partially accommodated by the matrix deformation. The maximum stress σ_{rr}, which governs at equilibrium at the inclusion/matrix interface can be written as [SEL 61]:

$$\sigma_{rr} = \frac{(\alpha_m - \alpha_p)(T - T_o)}{\dfrac{1+v_m}{2E_m} + \dfrac{1-2v_p}{E_p}} \qquad [8.49]$$

where the indices (m) and (p) stand for the matrix and the particle respectively.

The damage and fracture by thermal shocks are discussed in greater detail in section 8.4.3.

8.3.2. Irreversible deformation (by heat)

Creep: phenomenology and thermodynamic parameters

In the usual sense, creep denotes the evolution undergone by the geometry of a body over a length of time, under the effect of stresses. The study of this phenomenon is of major interest because it enables the prediction of dimensional differences, which will appear in service on an element of a structure.

Creep is a thermally activated process, that is, it is faster with higher temperatures. The stationary creep regime for most of the materials is appropriately described by the relationship [MUC 69]:

$$\dot{\varepsilon}(t) = K\left(\frac{\sigma - \sigma_i}{\mu}\right)^n \left(\frac{d_o}{d}\right)^p \left(\exp\frac{-\Delta G_a}{RT}\right) \qquad [8.50]$$

where K is a constant for a given structure, $\dot{\varepsilon}$ is the speed of creep, σ and σ_i the stress and an internal threshold stress respectively (the difference $\sigma - \sigma_i$ is also called the effective stress), μ the shear modulus, n the creep exponent, d the grain size, d_o the normalization factor, p the size exponent, ΔG_a the free flow activation enthalpy, R the constant of perfect gases and T the temperature.

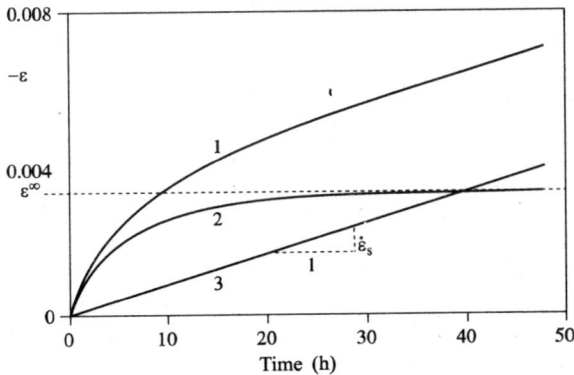

Figure 8.12. *Different elements of creep: experiment conducted on a specimen of silicon nitride under nitrogen at 1,500°C with a compressive stress of 100 MPa [TES 99]. The initial elastic deformation is negligible. The negligible damage does not show on the curve. The total deformation can be decomposed in the following manner: $\varepsilon = \varepsilon_a + \varepsilon_d$ with $\varepsilon_a = \varepsilon^\infty[1-exp-(t/\tau)^b]$ and $d\varepsilon_d/dt = constant$. $\varepsilon^\infty = 0.0032$, $\tau = 20.5.10^3 s$, $b = 0.86$ and $d\varepsilon_d/dt = 2.49.10^{-8} s^{-1}$*

The speed of deformation or creep rate increases quickly when the temperature approaches the softening temperature of the material (case of glass), or the least refractory phase forming a continuous network (intergranular phase of a ceramic), or for $T_{test} > 0.5\ T_{fusion}$ (K) for a monocrystal. This will happen, for example, at 20°C for polypropylene PP, at about 550°C for sheet glass and at about 1,200°C for a sialon with vitreous intergranular phase. For ceramic materials, creep primarily originates from sliding at grain boundaries and the mechanisms differ by their nature of accommodation (viscous flow, diffusion across the grains, etc.). The temperature sensitivity is often very high and the diminution of hot mechanical properties can have many origins, with the nature of deformation and damage being different depending on the temperature and the load applied. We realize in these conditions that the identification of a constitutive law sufficiently precise for incorporating into the structure mechanical design calculations and capable of taking into account any multidirectional load trajectories is very difficult.

While studying creep, the progress curve of deformation against time presents successively a stage of primary creep, where $d\varepsilon/dt$ decreases with time, followed by secondary creep regime (or stationary) at constant $d\varepsilon/dt$ and finally tertiary creep regime, characterized by a rapid increase in $d\varepsilon/dt$. The anelasticity is responsible for the primary stage of creep (transitory viscoelastic behavior) which precedes the stationary yield. Anelasticity manifests itself by a delayed response to stress: the return to the non-deformed state on load removal may necessitate an infinite time. As a first approximation, the anelasticity can be described with the help of linear viscoelastic models. The tertiary stage reflects the rapid progress of damage, which is often seen in the form of cavities. The total deformation ε can therefore be broken down into different parts: instantaneous elastic deformation ε_i, anelastic deformation ε_a (or delayed elasticity), a permanent often diffusional part ε_d and in extreme conditions of temperature and loading, or after a prolonged exposure, cavitational deformation ε_c:

$$\varepsilon = \varepsilon_i + \varepsilon_a + \varepsilon_d + \varepsilon_c \qquad [8.51]$$

For ceramics, the permanent deformations resulting from exposure to heat while stressed are better explained by the development of viscous and/or diffusional processes than by plasticity. Experimentally, we have access to an apparent stress exponent and activation volume, and to an activation enthalpy ΔH_a. ΔG_a is accessible only if an estimation of the activation entropy ΔS_a is possible. We can show that ΔG_a and ΔH_a are linked by:

$$\Delta G_a = \frac{(\Delta H_a + x\sigma V_a)}{1-x} \quad \text{where} \quad x = \frac{T}{\mu}\cdot\frac{d\mu}{dT} \qquad [8.52]$$

NOTE.– the confusions and errors in other works arise most often on account of wrong identification of activation variables. In particular, "ΔH_a" is often considered as free enthalpy, while the calculation of ΔG_a from ΔH_a requires knowledge of entropic contribution $T\Delta S_a$.

Physical approach

a) Ambient irreversible deformation

When sufficiently stressed, a permanent deformation described as micro-plastic appears. As per the findings of Marsh [MAR 64], the fact that the energy absorbed during the fracture process is clearly more than the cracking surface energy demonstrates the existence of "plasticity" (in the sense of a permanent deformation), even at 20°C. The threshold of plasticity τ_c is deducible from a measurement of hardness by the following relationships:

$$\frac{H(Pa)}{\tau_c(Pa)} \approx 4.3 \ [\text{HIL 47}] \qquad\qquad [\text{TAB 51}] \ [8.53]$$

$$\frac{H(Pa)}{\tau_c(Pa)} \approx 0.95\left(1+\frac{3}{3-\lambda}\ln\frac{3}{\lambda+3\xi-\lambda\xi}\right) \qquad [\text{MAR 63}] \ [8.54]$$

with: $\lambda = \dfrac{(1-2v)\tau_c}{6.89.10^6\,E}$ and $\xi = \dfrac{(1+v)\tau_c}{6.89.10^6\,E}$

Furthermore, the plastic flow stress depends on the loading time, which demonstrates the existence of the effect of a static fatigue. Typically, τ_c is in the vicinity of 0.05 E (Young's modulus).

b) Deformation mechanisms of a monocrystal

Let us recall that for metallic monocrystals, the creep speed in a permanent regimen is expressed as a power of the applied load function, with an exponent $n \approx 3 - 5$, characteristic of a deformation by dislocation mobility. We will show some of the characteristics of the deformation of iono-covalent solids. The latter are generally less plastic and the dislocations are aligned in a stable manner in potential valleys because of the existence of high Peierls forces. The propagation and multiplication of dislocations are therefore not very easy and the density of dislocations is generally low.

The critical stress of sliding is expressed as [TAK 88]:

$$\tau_p = \frac{2\mu}{1-\nu} \exp\left[\frac{-2\pi h}{(1-\nu)b}\right] \qquad\qquad [8.55]$$

with μ: shear modulus; ν: Poisson's ratio; b: Burgers vector modulus and h: distance between sliding planes.

For ceramics, μ is high and h is small. Ceramics are therefore not very deformable until a relatively high temperature is reached. It seems that in the domain where ceramics creep, the dislocation activity is very limited, so that the deformations are explained by diffusion type processes. The dislocation activity plays an important role only at very high temperatures, often unrealistic with respect to the temperature at which the material will be used. Among the covalent solids, it is the semi-conductors and in particular silicon and germanium which have been the most studied.

Material	Structure	Sliding system
Diamond, Si, Ge	Cubic	$\{111\} < \bar{1}\,10>$
$MgAl_2O_4$, Fe_2O_4	Spinel	$\{111\} < \bar{1}\,10>$
Quartz α (SiO_2)	Rhombohedra	$(0001) <11\,\bar{2}\,0>$
TiC, HfC, TaC	NaCl (rock salt)	$\{110\} < \bar{1}\,10>$ $\{111\} < \bar{1}\,10>$
α-Al_2O_3	Hexagonal	$(0001) <11\,\bar{2}\,0>$ $\{\bar{1}\,012\} <10\,\bar{1}\,1>$ $\{\bar{1}\,101\} <10\,\bar{1}\,1>$
C (graphite)	Hexagonal	$(0001) <11\,\bar{2}\,0>$
WC	Hexagonal	$\{10\,\bar{1}\,0\} [0001]$ $\{10\,\bar{1}\,0\} <11\,\bar{2}\,0>$
SiC	Hexagonal	$\{10\,\bar{1}\,0\} <11\,\bar{2}\,0>$ $\{0001\} <11\,\bar{2}\,0>$
β-Si_3N_4	Hexagonal	$\{10\,\bar{1}\,0\} [0001]$ $\{11\,\bar{2}\,0\} [0001]$ $\{1\,\bar{1}\,01\} [11\,\bar{2}\,0]$
TiO_2	Rutile	$\{101\} <10\,\bar{1}>$ $\{110\} <001>$

Table 8.6. *Sliding systems of some monocrystals*

An essential characteristic of the deformation of covalent crystals is that the speed of dislocations is controlled by the crossing of Peierls peaks until very high temperatures are reached, of the order of 0.5 T_{fusion}. In the case of the most simple ionic compounds, with two components, we have to consider two crystallographic sub-networks: the cations and the anion sub-networks. The cationic and anionic lacunae can only move in their own respective sub-networks. Thus, for example, for a pure dislocation of the structure of NaCl (structure adopted by numerous MO oxides), we have to consider two neighboring extra half-plans. The sliding does not therefore take place on the more dense planes, that is {111} for NaCl, because ions of similar charge will be brought to the immediate vicinity, which would consume a considerable amount of energy. The case of oxide M_xO_y is often very complicated.

c) Creep of a polycrystalline aggregate by diffusion in the grains [HER 50]

Herring took inspiration from the works of Nabarro [NAB 48] for studying the creep of polycrystals of different grain geometries. The problem brings us to the calculation of the deformation of a grain of a given form with different conditions at surface limits. The essential difference with the Nabarro's calculation resides in the fact that Herring used chemical potentials instead of concentrations for calculating the flux of atoms and vacancies lacunae. Yet, we must stress the fact that, while the utilization of chemical potentials is more rigorous in systems subjected to hydrostatic pressure, it poses numerous problems when the stress regime is not hydrostatic, as is the case in diffusional creep. For a monocrystalline grain in a cubic system, containing only one type of atoms and maintained at a constant temperature, the following result is obtained:

$$\dot{\varepsilon} = \frac{K_1 D^v \Omega}{kTd^2} \sigma \qquad [8.56]$$

where D^v is the coefficient of self-diffusion, Ω is the atomic volume and d is a dimensional characteristic of the grain.

When the symmetry of the grain is high, K_1 depends little on the form of the grain. K_1 is situated between 12 (cube) and 13 (sphere). The important results of this model are the following:

i) the creep by volume diffusion is expressed by a behavior law of the viscous Newtonian type;

ii) the creep rate increases with higher temperatures and with smaller grain size.

The Herring-Nabarro creep appears in ceramics at high temperatures when the grain size is large enough. At lower temperatures and for smaller grain sizes, the creep is of the Coble type.

d) Diffusion along the grain boundaries: Coble model [COB 63]

Coble has proposed a creep model for polycrystalline aggregates in which the creep speed is controlled not by the diffusion in volume across the grains, but by the diffusion along the grain boundaries. Only a fraction of the volume of material (in the form δ/d, δ: effective thickness of boundaries for diffusion) acts as diffusion paths and from the expression obtained previously by Herring, Coble gets:

$$\dot{\varepsilon} = \frac{K_2 D^j \sigma \Omega \delta}{d^3 kT} \qquad [8.57]$$

where D^j is the coefficient of atomic diffusion at grain boundaries.

The difference between the Herring-Nabarro model and that of Coble rests on the one hand on the influence of the grain size d, and on the other on the energy of self-diffusion. We expect a preponderance of Coble creep at "low temperature" where, because of the lower energy of activation, $D^j > D^v$.

For semi-spherical grains, $K_2 \sim 47$. For metals $\delta \sim 10b$ (Burgers vector); for ceramics $\delta \sim 1\text{-}10$ nm. For the latter case, the thickness of the boundary, or the intergranular film, depends largely on the nature and the quantity of sintering additives and impurities.

NOTE.– if the grain size becomes as small as the interatomic distance, we find according to Herring's expression for creep:

$$\eta = K_3 \frac{kT}{dD} \qquad [8.58]$$

where η is the viscosity; η is proportional to $\sigma/\dot{\varepsilon}$ and K_3 is constant.

We then get the relationship proposed by Gibbs for the coefficient of viscosity of glass [VAR 94]. Glass then appears to be a polycrystalline material having grain size equal to the interatomic distance.

e) Taking into account the deformation compatibilities between grains

Polycrystalline solids are made of polyhedral grains in contact with one another and it is clear that in the absence of any damage, the faces of contiguous grains should remain in contact during deformation.

Most ceramics consist of an intergranular vitreous phase and we then understand that the rheological properties of this secondary phase play a major role on the creep resistance and the damage resistance. The problem of deformation compatibility and relative movements of grains during the flow has been studied by Lifshitz [LIF 63], Raj and Ashby [RAJ 71] and re-examined, in the case of ceramics with intergranular vitreous phase, by Chadwick [CHA 92]. It has been shown that the flux of intergranular diffusion and the sliding at grain boundaries are connected. Thus, the resultant deformation can be indifferently interpreted as the consequence of creep diffusion of grains accommodated by sliding at boundaries, or conversely, as sliding of grains with accommodation by the process of dissolution-diffusion and reprecipitation of the material at grains boundaries. Finally, the total deformation in stationary creep regime can be broken up as follows:

$$\varepsilon = \varepsilon_{disloc} + \varepsilon_{slid\,or\,diff} + \varepsilon_{cavitation} \qquad [8.59]$$

where ε_{disloc} is the deformation due to the mobility of dislocations in the grains, $\varepsilon_{slid\,or\,diff}$ is the deformation by sliding along the grain joints (sliding of the Lifshitz or Rachinger type for example, or resultant from diffusion processes) and $\varepsilon_{cavitation}$ is the contribution of damage. The sliding at grain boundaries is associated to an accommodation mechanism (e.g. dissolution-migration-reprecipitation of material at grain boundaries). The deformation mechanism predominantly depends on the temperature, the stress and the microstructure of the materials.

From experimental values of n and p [CAN 88], a certain number of models have seen the light of day to define the physical behavior of materials.

For example, the mechanisms based on the mobility of dislocations correspond generally to n > 4 and the diffusion mechanisms to n = 1. n = p = 2 suggests that the sliding at grains boundaries plays an important role and characterizes the flow of the majority of superplastic materials.

f) Superplasticity

A polycrystalline material is said to be superplastic when it can support a deformation under tension of more than 100% without necking. This property, well known to metallurgists, is very widely used today for making complex architectural pieces by hot forming or diffusion welding. The revelation of superplasticity of ceramics is relatively recent: 1980 for magnesia (MgO) [CRA 80], 1985 for zircon Y-TZP [WAK 89] and for compounds with essentially ionic bonds, and 1990 for silicon nitride Si_3N_4 with highly covalent bonds [WAK 90]. The feasibility has been demonstrated for hot forming, which is the most common application of superplasticity, as well as for solid state diffusion welding, and does not limit itself to structural materials but also applies to functional ceramics for electronics ($PbTiO_3$, ZrO_2) or for medicine (hydroxyapatite) [WAK 91].

[HER 50] $\dot{\varepsilon} = 13(D_v\Omega\sigma)/(d^2kT)$ (viscous; intra-granular diffusion)

[COB 63] $\dot{\varepsilon} = 47(D_j\delta\Omega\sigma)/(d^3kT)$ (viscous; diffusion at grain boundaries)

[RAJ 71] $\dot{\varepsilon} = 14(D_v\Omega\sigma)/(d^2kT)$ (volume diffusion with compatibility)

$\dot{\varepsilon} = 13(D_j\delta\Omega\sigma)/(3d^3kT)$ (diffusion at boundaries with compatibility)

$$\dot{\varepsilon} \approx 5\frac{\sigma\Omega}{kT}\frac{1}{d^2}\left\{\frac{D_v}{R} + \frac{\delta}{d}\frac{D_j}{\sqrt{R}}\right\}$$ (strongly acicular grains)

where R is the ratio of anisotropy: R = length (L)/width (B) and d $\approx \sqrt{LB}$

[HAR 57] $\dot{\varepsilon} = K_{HD}(D_v b\sigma)/(kT)$ (viscous; edge dislocation climb)

($K_{HD} \in [2.10^{-8} - 5.10^{-8}]$ for ceramics)

[RAC 52] $\dot{\varepsilon} = 13(D_v\Omega\sigma)/(d^2kT)$ (viscous; volume diffusion; fine grain)

[ASH 73]

$$\dot{\varepsilon} = \frac{98\Omega}{d^2kT}\left(\sigma - \frac{0.72\Gamma}{d}\right)D_v\left(1 + \frac{\pi\delta Dj}{dD_v}\right)$$ (volume diffusion accommodated by

diffusion at boundaries; large deformations; fine grain) (Γ: free energy of grain boundaries)

[PAI 91] $\dot{\varepsilon} \approx 3.4\delta D_v\Omega\sigma^2/(\mu bd^2kT)$ (sliding with dislocation of grain boundaries accommodated by volume diffusion; fine grain)

[CHA 92] $\dot{\varepsilon} = \sigma\delta^3/(\eta\sqrt{3}a^3)$, where η is the viscosity of the intergranular phase and a+ the length of grain facets (viscous, intergranular phase, fine grain)

Table 8.7. *Recapitulation of equations proposed for stationary creep*

For a material to be superplastic, a certain number of conditions are required:

– medium size of grains < 1 micron for ceramics. This supposes adapted synthesis methods: hot pressing under load (HP), hot isostatic pressing (HIP), chemical vapor phase deposition (CVD);

– stability of grain size during deformation to avoid hardening resulting from an enlargement (see equation [8.50]);

– T > 0.5 fusion temperature. This is the criteria of ductility (note that many ceramic materials may decompose and volatilize in ambient atmosphere at such elevated temperatures).

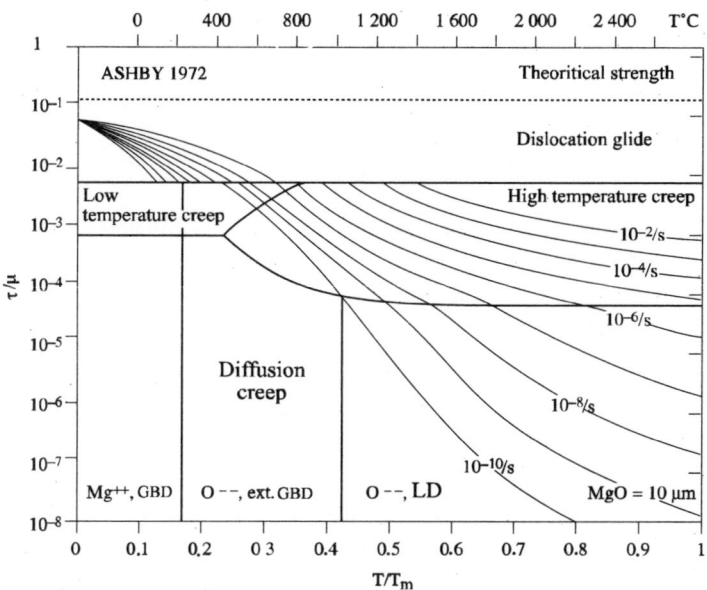

Figure 8.13. *Chart showing deformation mechanisms for MgO [ASH 83]*
(GBD: grain boundary diffusion; LD: lattice diffusion)

During deformation, due to various processes such as the alignment of grains in the main direction of stress, a phenomenon of hardening (σ grows with ε for an imposed speed $\dot{\varepsilon}$) is observed and the progressive increase in the flow stress at the rate of imposed deformation can be written as:

$$\sigma = K\dot{\varepsilon}^m \exp(\gamma\varepsilon) \tag{8.60}$$

where m is the sensitivity exponent at the speed m = 1/n and K depends only on the temperature.

According to the theory of the tension test [HAR 67], if A is the section of the specimen, there will be stability if: $d\dot{A}/dA \le 0$. After applying differential calculus on σ and ε, Hart demonstrated that the yield is stable only if $\gamma + m \ge 1$.

In the superplastic deformation regime it is the sliding at grain boundaries which is dominant and the grains move on distances close or greater than the size of grains, without too much of a change in their forms [LAN 91]. Superplasticity is characterized by a non-Newtonian flow: n ~ 2 and p ~ 2 (exponent of the grain size). The flow is generally accompanied by an enlargement of the grains. Recent observations seem to show that the deformation generates a flow super structure with relative displacements of grain groups forming rigid clusters [ZEL 93].

8.4. Damage and in-service behavior

Like deformation mechanisms, the damage mechanisms and destruction causes of a structure depend greatly on the temperature and the applied stress (or reciprocally on the speed of the imposed deformation).

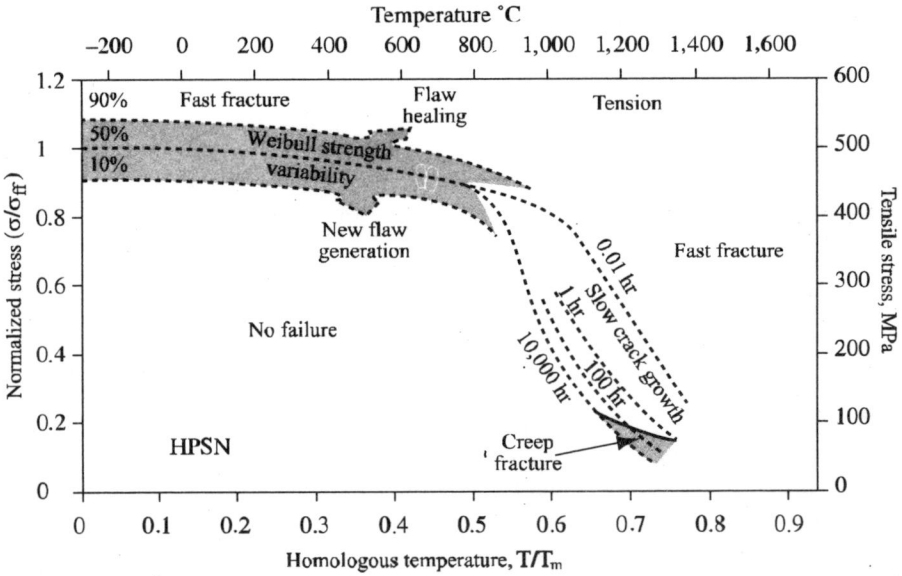

Figure 8.14. *Chart showing mechanisms of rupture of a sialon sintered in presence of MgO. Results obtained in air, in bending [QUI 90]*

Figure 8.14 illustrates the behavior of silicon nitride, with the lives observed, according to the applied stress and temperature. At ambient temperature, the behavior is elastic and near the average breaking stress of the material, the dispersion on the breaking stresses is defined by Weibull's statistics. At high temperatures, the material creeps and the damage mechanisms, generally by cavitation and/or by microcracking, appear. The intermediate zone is characterized by the predominance of a damage mechanism localized by a slow growth of sub-critical flaws.

8.4.1. *Elastic behavior and slow growth of flaws*

At relatively low temperature, i.e. for $T < 0.5\ T_f$, the behavior of materials is mainly elastic and the rupture is brittle in nature. For stresses clearly less than the average breaking stress, the probability of destruction is nearly zero. In other words, the growth of flaws is insignificant and the stress intensity factor remains, at any point on the structure, lower than its critical value. If a high deformation speed is imposed ($d\varepsilon/dt > 10s^{-1}$), we enter into the field of sudden dynamic rupture, which makes it possible, for example in the case of fibers or whiskers, to be free from all environmental influences and attain stresses approaching theoretical stress of interatomic decohesion.

When temperature increases, the fracture may appear for a stress well below the catastrophic breaking stress, following a phenomenon of gradual increase of sub-critical flaws. The fracture is then delayed (also referred to as fatigue). In the case of glass, this phenomenon should be taken into account at ambient temperature. Sub-critical cracking is a type of localized damage which is produced for values of K_I or of G inferior to critical values causing catastrophic fracture.

In a diagram representing the speed of extension of the crack face according to the stress intensity factor, we generally distinguish three regions referred to as I, II and III in Figure 8.15, which correspond to different regimes of cracking kinetics.

Region I is a region where growth, under relatively small load, is governed by the kinetic of the reaction which is produced at the level of the crack-face between the material and the environment.

In region II, the speed of cracking is governed by the kinetics of species transport present in the atmosphere towards the crack face; that is why the speed is not especially dependent on the stress.

Finally, in region III, the high stress leads to a value of K_I near K_{IC} and we approach the catastrophic fracture regime, with the speed of cracking tending towards the speed of sound in the material.

In region I, corresponding to more common in-service situations, the growth speed V increases with K_I, following the empirical relationship formulated by Paris [PAR 63], and the growth is thermally activated, as evidenced by Wiederhorn [WIE 70]:

$$V = A(K_I/K_{Io})^N exp[-Q_{clf}/(RT)] \hspace{2cm} [8.61]$$

where A is a constant, N the growth exponent (or exponent of fatigue/corrosion), K_{Io} the normalization constant and Q_{clf} the apparent energy of activation of the process of slow growth of the crack.

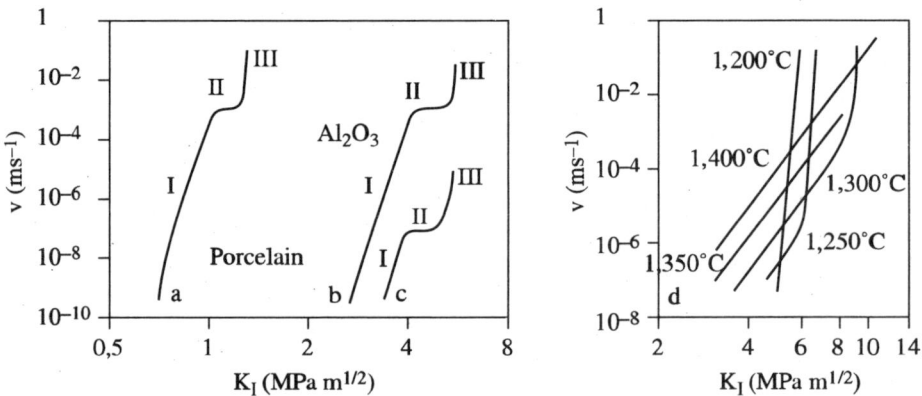

Figure 8.15. *Speed of propagation of a crack in mode I in different ceramics: a) porcelain in water at ambient temperature; b) Al_2O_3 in air at ambient temperature; c) alumina in toluene; d) Si_3N_4 at different temperatures [MEN 92]*

Typically, N falls in the range [10-300]. For example, in ambient atmosphere, N will be about 15 to 35 for soda-lime-silica glass in fiber state, 30 to 50 for alumina, 50 to 130 for silicon nitride and 110 to 130 for SiC. The exponent N tends to decrease while the temperature increases. However, the cracking speed increases with temperature. At high temperature, or under the effect of atmosphere, the crack face may become blunt, especially if a period of relaxation of stresses is observed. Thus, sometimes, the resistance to static fatigue is less than that against dynamic fatigue. In the case of a dynamic testing of cyclic fatigue, we represent da/dn, where n is the number of cycles instead of V.

Relationship [8.61], as well as the relationship for the stress intensity factor $K_I = Y\sigma_\infty \sqrt{a}$ (see section 8.1.1.2), enable easily evaluation of the life of a structure when the behavior is elastic. The rupture time t_r is the time at the end of which the condition $K_I = K_{IC}$ is satisfied near a flaw having attained the critical size a_c. In static fatigue, we also obtain the following approximate expression for the rupture time:

$$t_r = \frac{2}{(N-2)A'\sigma_\infty^N Y^N a_i^{(N-2)/2}}$$

[8.62]

where $A' = A\exp[-Q_{clf}/(RT)]/K_{Io}^N$, and a_i is the initial size of the flaw at the origin of rupture.

We can similarly express the delayed breaking stress in the case of a monotonic load ($\sigma_\infty = \dot{\sigma}_\infty t$):

$$\sigma_r = \left[\frac{2(N+1)\dot{\sigma}_\infty}{(N-2)A'Y^N a_i^{(N-2)/2}} \right]^{\frac{1}{N+1}}$$

[8.63]

NOTE.– we can expect, taking into account the presence of a population of disparate flaws, a scattering of the nominally identical lives of structures. This scattering can be characterized by employing Weibull's statistical approach. In fact, by neglecting the phenomenon of sub-critical growth under ambient temperature and rapid loading, all flaws of the initial size a_i are linked to the measured breaking stress σ_{rAT} (AT: ambient temperature) by the relationship:

$$a_i = \left(\frac{K_{Ic}}{\sigma_{rAT} Y} \right)^2$$

[8.64]

On the other hand, following the Weibull approach (see section 8.1.3) and neglecting the threshold stress, the breaking stress can be expressed according to the rupture probability P_{rV}, from the relationship:

$$\sigma_{rAT} = \sigma_{oV} \left[\frac{1}{V} \ln\left(\frac{1}{1-P_{rV}} \right) \right]^{\frac{1}{m}}$$

[8.65]

By replacing in equation [8.64] σ_{rAT} by the preceding expression, we get:

$$a_i = \left(\frac{K_{Ii}}{\sigma Y}\right)^2 = \frac{K_{Ic}^2}{\sigma_{oV}^2 \left[\frac{1}{V}\ln\left(\frac{1}{1-P_{rV}}\right)\right]^{\frac{2}{m}} Y^2} \qquad [8.66]$$

By substituting a_i by this expression in equation [8.62], we can draw up a diagram representing the variation in the lifespan according to the applied stress for a given failure probability. This type of diagram, represented more often in the logarithmic scale, bears the name of an SPT (Strength-Probability-Time) diagram.

8.4.2. *Viscoplastic behavior and creep damage*

At high temperatures and when the stress is relatively small ($< 2/3\ \sigma_r$), the structure gets deformed by creep, while the more severe flaws can be subjected to the phenomenon of slow sub-critical growth. The material therefore gets weakened, on the one hand by damaging due to creep, slow and overall, taking the forms of cavitation and/or of microcracks, and on the other by local damaging due to growth of pre-existing flaws. The cause of destruction is determined by the competition between the two damage processes. As the load increases, the mechanism of slow cracking becomes dominant. All the same, even with small loads and when the fracture deformation is significant, the fracture often starts from a pre-existing flaw.

8.4.2.1. *Slow cracking*

The phenomenon of slow cracking in the macroscopic creep regime is still not well-known and difficult to analyze, partially because the determination of the peculiarities of stresses and deformations at the crack tip in the viscoplastic domain is very delicate. However, most ceramics are affected by this mode of damaging, the present flaws (pores, impurities, etc.) serving as the beginning of an intergranular crack. This problem can be seen from two points of view: from the point of view of micromechanics, by formulating physical hypotheses on the distribution of stresses in the vicinity of the crack (models based on works of Dugdale and Barenblatt) [SHA 75], or by going on a continuum scale and using a loading parameter adapted for describing the peculiarity, specifically in the case of non-linear viscoplastic behavior, the path independent contour integral C* analogous to Rice's integral J developed in elastoplasticity. In the latter case, for a creep flow described by Norton's phenomenological law (generalized under the form of equation [8.50]),

$\dot{\epsilon} = B\sigma^n$, the asymptotic stress field valid for $r \ll a$, where r is the distance from crack tip and a the length of the crack, can be written as [RIE 87]:

$$\hat{\sigma}_{ij} = \left(\frac{C^*}{I_n Br} \right)^{1/(n+1)} f_{ij}(\theta) \qquad [8.67]$$

where I_n is a function of n and of the mode of loading, and C* is defined by $C^* = aB\sigma_{eff}^{n+1}g$ (geometry, n), where g is a function tabulated by various authors [RIE 87]. C* is replaced by J in elastoplasticity.

The growth speed of a crack can later be evaluated as soon as the damage mechanism is known. We find for example:

$$\dot{a} \propto (C^*/I_n)^{n/(n+1)} \qquad [8.68]$$

for a growth based on a criteria of critical deformation (α: proportional).

However, the analysis is often complicated due to primary creep, by the appearance, in the first stage, of a creep region confined around the crack face (transitory regime), or again by the development of softening mechanisms (blunting, multi cracking, etc.).

8.4.2.2. Generalized damage

The acceleration of the deformation speed during the tertiary stage of creep results most often from the formation and coalescence of cavities at grain boundaries. This form of generalized damage is accompanied by an appreciable decrease in the density of structures during creep and leads to an intergranular rupture.

We can therefore attribute an important fraction of the deformation to the damage.

For silicon nitride, Wiederhorn [WIE 95] came to the conclusion that in creep $f_c = \epsilon$, where f_c is the volumetric fraction of cavities.

In the case of relatively small deformations (< 0.1), the relationship of Monkman and Grant [RIE 87] helps to evaluate the lifetime t_r, according to the rate of stationary creep $\dot{\epsilon}_s$:

$$\dot{\epsilon}_s t_r = C \qquad [8.69]$$

where C is a constant near $1/n$ for materials resistant to cavitation. In the presence of cavitation, $C < 1/n$.

For ductile or superplastic ceramics, in the case of elongations of more than 10%, the fracture strain follows the empirical relationship proposed by Kim *et al.* [KIM 91].

$$\varepsilon_r = K\left[\dot{\varepsilon}\exp\left(\frac{Q}{RT}\right)\right]^f \tag{8.70}$$

where K is a constant dependent on the material and f is near −0.33.

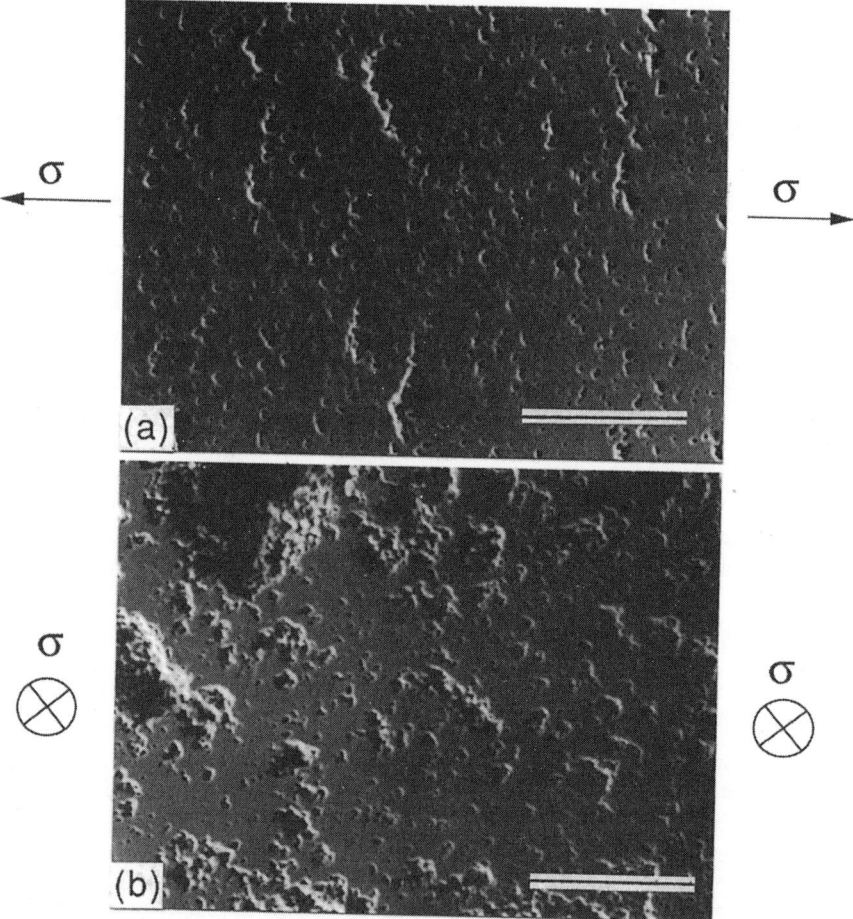

Figure 8.16. *Cavitation and microcracks resulting from the coalescence of cavities in the gauge length of a tension specimen in yttrium oxide elongated by 44% at 1,550°C. Bar = 50 μm [ROU 96]*

8.4.3. Thermal shocks

The exposure of a structure to a sudden change of temperature may cause its ruin either because locally, most often on the surface, the tension stress reaches the critical rupture value, or because a flaw of the structure propagates and finally reaches a critical size, causing a catastrophic failure. In the former case, a breaking criterion can easily be obtained from the equations which govern the thermoelastic behavior (see section 8.3.1). In the latter case, the fracture occurs only after several cycles, or at the end of a certain period of time, depending on the difference of temperature and the initial microstructure; this is the phenomenon of thermal fatigue, which is more difficult to model.

8.4.3.1. Analysis in linear thermoelasticity

The experiments generally consist of quenching identical preheated test specimens in water, or in a thermostatically controlled oil bath, or in air. The number of specimens surviving in relation to ΔT, the subjected thermal difference, helps to determine the value $\Delta T_{0.5}$, corresponding to a survival probability of 50%. This interval defines the resistance of the material to a severe thermal shock. For example, alumina and tungsten carbide are characterized by critical temperature intervals close to 200 and 400°C respectively. Adopting a hypothesis of a biaxial plane stress, a simple expression for the critical temperature interval causing destruction of a structure made of a material of elastic modulus E, with Poisson's ratio v, coefficient of expansion α and breaking strength σ_r, is given by [KIN 55]:

$$\Delta T_c = \frac{1-v}{E\alpha}\sigma_r \tag{8.71}$$

In reality, the shock is appreciably reduced by the transmission of heat, so that ΔT_c can be written as:

$$\Delta T_c = \frac{1-v}{\varphi E\alpha}\sigma_r \tag{8.72}$$

where φ is the interval [0-1] and expressed using the number of Biot $\beta = eh/k$, e is characteristic dimension of the structure (half thickness of a plate or radius of a cylinder), h is surface heat exchange coefficient and k is coefficient of thermal conductivity.

It appears that β grows with the severity of the thermal shock: $\beta = 0$ for an infinitely weak shock (k tends to infinity when e tends to zero) and β tends to infinity ($\varphi \approx 1$) for a hard or severe shock. For an infinite plate: $\varphi^{-1} = 1.5 + 3.25/\beta - 0.5\exp(-16/\beta)$.

We can see that thin structures made of materials with low coefficient of expansion and high thermal conductivity show better resistance to a sudden rupture by thermal shock.

Even though the Biot number takes into account a characteristic dimension of the object, the behavior towards thermal shock is also influenced by its geometry. The latter parameter has an influence over the distribution of stresses in the volume, which can be deciphered indirectly with the help of a function of Poisson's ratio, f(v):

$$\Delta T_c = \frac{1-v}{\varphi E \alpha} \sigma_r f(v) \qquad [8.73]$$

In practice, when $\beta > 20$ the cooling is so fast that the surface attains the final temperature before the mean temperature of the structure had time to vary, with the result that $\varphi \approx 1$. This situation is encountered when, for example, a plate of standard glass (k = 4 $Jm^{-1}sec^{-1}°C^{-1}$) is quenched in water (h = 4.10^4 $Jm^{-2}sec^{-1}°C^{-1}$) and when the thickness of the plate (or diameter of the cylinder) is equal to or greater than 4 mm. We then define the first parameter of resistance to thermal shocks by:

$$R = \frac{1-v}{E\alpha} \sigma_r \qquad [8.74]$$

R reflects the resistance of the material to a severe thermal shock. We then have $\Delta T_c = Rf(v)$. The second parameter, called R', is introduced for describing relatively weaker shocks (low Biot number) and is defined by: R' = kR. R and R' are the two characteristics of the material governing the resistance to thermal shocks, assuming that the thermoelastic analysis prevails.

The preceding equations are applicable only for isotropic structures for which physical characteristics h, k, α and E vary little with the temperature.

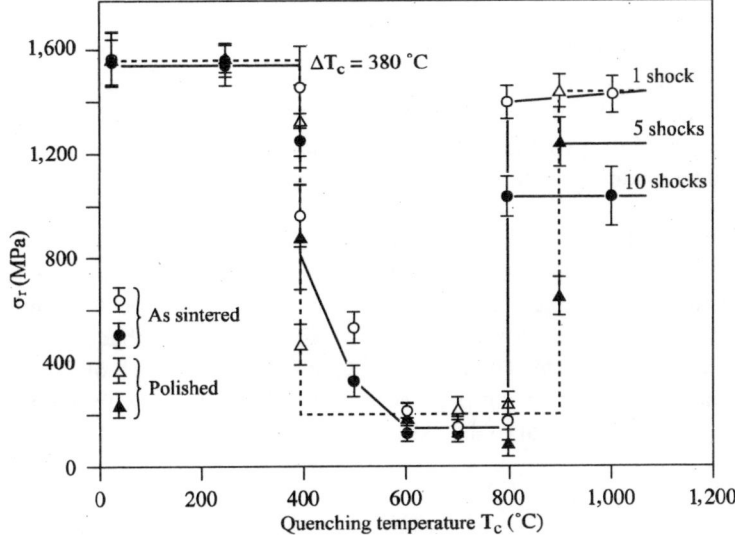

Figure 8.17. *Variation of the strength of tungsten carbide specimens according to the difference in temperature resulting from a heating in air and a rapid quenching in water at ambient temperature (transfer in less than one second) [MAI 76]*

8.4.3.2. Slow propagation of flaws and thermal fatigue.

In the great majority of cases, it is the growth of pre-existing flaws (microcracks), under the combined effects of variations in stress and temperature, which causes the destruction of an in-service structure. The degree of damage to a structure under fatigue can be evaluated by comparing the present mechanical resistance, for example after a given number of thermal cycles, to the initial strength. In developing an energy approach similar to the one conducted in linear fracture mechanics (see section 8.1.1), Hasselman [HAS 63] introduced a parameter R'''', representing the resistance to thermal shock damage:

$$R'''' = \frac{E\gamma}{(1-v)\sigma_r^2} \qquad\qquad [8.75]$$

where γ is the surface energy of cracking.

Although it is not explicit, R'''' is linked to the number and length of cracks which propagate (we use also the parameter R''' = R''''/γ).

It is therefore apparently futile, with regard to relationships [8.74] and [8.75], to try to optimize simultaneously the resistance to sudden rupture by thermal shocks (parameters R or R') and the damage resistance (parameters R''' and R''''). In other words, the optimization of the resistance to a severe shock leads to materials which are likely to be damaged rapidly by microcracking and vice versa. A choice has therefore to be made, according to the application in view.

In a series of articles, Hasselman and his team studied and modeled the phenomenon of propagation of pre-existing flaws. In one of these articles [GEB 72], an expression linking the critical difference of temperature for initiating the propagation of flaws situated at the surface of a plate, with number N_s of such flaws per surface unit, is proposed. The flaws are supposed to be parallel, of the same length of 2a and do not interact. If the plate is uniformly prestressed with an amplitude of deformation ε_o, perpendicular to the axis of flaws, then ΔT_c is given by:

$$\Delta T_c = \frac{\varepsilon_o}{\alpha} + \left(\frac{2\gamma}{\pi e \alpha^2 a} \right)^{1/2} (1 + 2\pi N_s a^2) \qquad [8.76]$$

After propagation, the crack stabilizes at a length $2a_f$, with (based on the hypothesis that $2\pi N_s a^2 \ll 1$):

$$a_f = \frac{1}{4\pi N_s a} \qquad [8.77]$$

From its extension we get a diminution of breaking strength as well as a weakening of the material, giving rise in particular to a diminution of the density and macroscopic modulii of elasticity (see Figure 8.18). It appears that the final length of flaws is inversely proportional to their initial length and to the number of flaws propagating simultaneously.

The study of a structure subjected to a thermal cycle can be made following the approach developed in the context of mechanical cycling. The service life is attained for a number of cycles N_r placing a flaw in a critical situation (in the fracture mechanical sense of the term), the stress intensity factor attaining the value K_{ic} near the edges of this flaw. The life in thermal fatigue, expressed in number of cycles, can be written as:

$$N_r = \int_0^{K_{Ic}} \frac{2K_I}{\sigma^2 Y^2 V} dK_I \qquad [8.78]$$

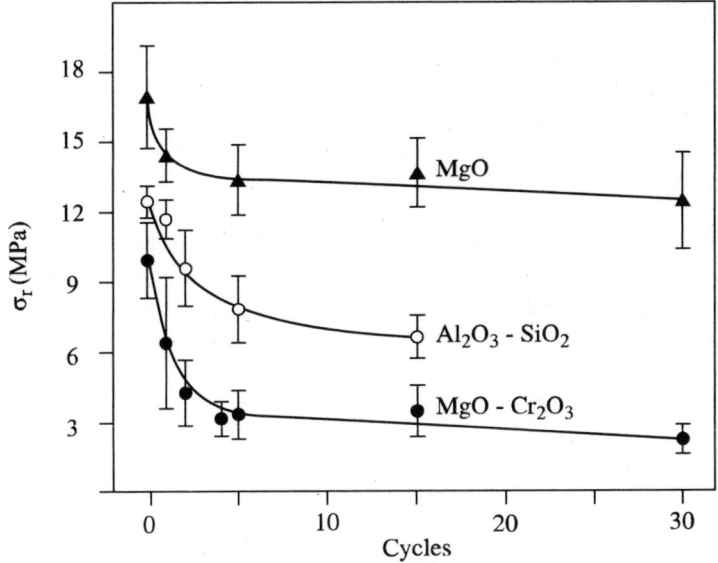

Figure 8.18. *Effects of repeated thermal shocks on the strength of different ceramic oxides [AIN 74]*

where Y is a factor dependent on the geometry of loading (see section 8.1.5) and V is the speed of propagation of flaws.

By supposing that the growth of a crack is activated thermally and follows the Paris law (equation [8.61]), we get:

$$N_r = \frac{A(1-v)^n \exp\left(\dfrac{Q_{clf}}{RT_{max}}\right)}{e^2(\alpha E \Delta T)^n (n-2) Y^n a^{(n-2)/2}}$$ [8.79]

where A is a constant, e a characteristic dimension of the structure and n the growth exponent of the crack of the Paris law (the other terms have been defined earlier).

This type of calculation nevertheless faces a major difficulty: the stress and temperature at the instant t are not fully known, thus making it difficult to incorporate in expression [8.78].

8.5. Conclusion

The "mechanical" problems which come up in ceramic material sciences are very varied. The ceramist is confronted with rheological problems – rheology of powders, law of flow of slurry for injection or molding – as well as with problems of strength – mechanical strength of unfinished parts, in-service strength of sintered parts, resistance to thermal shocks, etc. – and it is not generally possible to solve these problems with just ceramic knowledge: some basic knowledge on polymers (binders and plasticizers for molding, etc.), on metals (tooling) and on glasses are often necessary. In fact, varied knowledge on the mechanical behavior of materials is indispensable for designing, fabricating, handling and also verifying the in-service strength of a ceramic component (tile, plate, substrata for electronics, bio-medical implants, turbo charger for the automobile industry, etc.).

Today, high performance ceramic design is essentially ingenious research in architecture at nano- and microscopic scales, in order to reduce the brittleness and increase the endurance with respect to damage. During the last 20 years, ceramists have thus imagined and created some surprising materials, producing at times unexpected effects, which prove the dynamism of this science evolving at a scale often greater than the atomic or molecular one.

Composite materials, with special reinforcements, fibrous, or even with lamellar or woven architecture, have seen the light of day. Among these materials, a new promising family has emerged in the 1990s: the family of nano-composites. Under this term are grouped very different materials in which at least one phase is found to be in the nanometric scale. The higher performing varieties have exceptional breaking strengths and toughness ($\sigma_r > 1.5$ GPa and $K_{Ic} > 6$ MPa $\sqrt{m}$).

8.6. Acknowledgements

Jean-Louis Besson (ENSCI, Limoges, France) and Jean-Louis Chermant (LERMAT, ISMRA, Caen, France) are gratefully acknowledged for their careful reading of the manuscript and for their constructive comments.

8.7. Bibliography

[AIN 74] AINSWORTH J.H. and HERRON R.H., "Thermal shock damage resistance of refractories", *Ceram. Bull.*, vol. 53, no. 7, p. 533-538, 1974.

[ANS 81] ANSTIS G.R., CHANTIKUL P., LAWN B.R. and MARSHALL D.B., "A critical evaluation of indentation techniques for measuring fracture toughness: I, direct crack measurements", *J. Am. Ceram.*, vol. 64, no. 9, p. 533-538, 1981.

[ASH 73] ASHBY M.F. and VERALL R.A., "Diffusion-accommodated flow and superplasticity", *Acta Metal.*, vol. 21, no. 2, p. 149-163, 1973.

[BAR 62] BARENBLATT G.I., "The mathematical theory of equilibrium cracks in brittle fracture", *J. Advanced App. Mech.*, no. 7, p. 55-129, 1962.

[BAT 93] BATTO R.A. and SCHULSON E.M., "On the ductile-to-brittle transition in ice under compression", *Acta Metal.*, vol. 41, no. 7, p. 2219-225, 1993.

[BRA 92] BRADT R.C., HASSELMAN D.P.H., MUNZ D., SAKAI M. and SHEVCHENKO V.Y., *Fracture mechanics of ceramics*, vol. 10, *Fracture fundamentals, high temperature deformation, damage, and design*, Plenum Press, New-York and London, 1992.

[BRA 95] BRADT R.C., BROOKES C.A. and ROUTBORT J.L., *Plastic deformation of ceramics*, Plenum Press, New-York, 1995.

[BUR 73] BURESH F.E., "Crack growth resistance of ceramics", *Sci. of Ceramics*, vol. 7, p. 383-91, 1973.

[CAM 89] CAMPBELL G.H., DALGLEISH B.J. and EVANS A.G., "Brittle to ductile transition in silicon carbide", *J. Am. Ceram. Soc.*, vol. 72, no. 8, p. 1402-1408, 1989.

[CAN 88], CANNON W.R. and LANGDON T.G., "Review creep of ceramics, Part 2: an examination of flow mechanism", *J. Mat. Sci.*, vol. 23, p. 1-20, 1988.

[CHA 81] CHANTIKUL P., ANSTIS G.R., LAWN B.R. and MARSHALL D.B., "A critical evaluation of indentation technique for measuring fracture toughness: II, strength method", *J. Am. Ceram. Soc.*, vol. 64, no. 9, p. 539-543, 1981.

[CHA 92] CHADWICK M.M., WILKINSON D.S. and DRYDEN J.R., "Creep due to a non-Newtonian grain boundary phase", *J. Am. Ceram. Soc.*, vol. 75, no. 9, p. 2327-34, 1992.

[CHA 93] CHAIM R. and MAROM H., "Particle size effect on the apparent brittle-ductile transition in $YBa_2Cu_3O_{7-x}$-Sn composite superconductor", *J. Mat. Sci. Lett.*, no. 12, p. 1563-1566, 1993.

[CHE 89] CHERMANT J.L., *Les céramiques thermomécaniques*, Presses du CNRS, 1989.

[CHE 90] CHEN I.W. and XUE L.A., "Development of superplastic structural ceramics", *J. Am. Ceram. Soc.*, vol. 73, no. 9, p. 2585-2609, 1990.

[CHE 92] CHEN Y., PAVY J-C. and RIGAUT B., "Usure des céramiques à grande vitesse", *CETIM Informations*, no. 126, p. 67-71, 1992.

[CLA 70] CLAUSSEN N. and JAHN J., "Green strength of metal and ceramic compacts as determined by the indirect tensile test", *Powder Metall. Inter.*, vol. 2, no. 3, p. 87-91, 1970.

[COB 63] COBLE R.L., "A model for boundary diffusion controlled creep in polycrystalline materials", *J. App. Phys.*, vol. 34, no. 6, p. 1679-1682, 1963.

[CRA 80] CRAMPON J. and ESCAIG B., "Mechanical properties of fine-grained magnesium oxide at large compressive strains", *J. Am. Ceram. Soc.*, vol. 63, no. 11-12, 1980.

[DRY 89] DRYDEN J.R., KUCEROVSKY D., WILKINSON D.S. and WATT D.F., "Creep deformation due to a viscous grain boundary phase", *Acta Metal.*, vol. 37, no. 7, p. 2007-2015, 1989.

[DUG 60] DUGDALE D.S., "Yielding of steel sheets containing slits", *J. Mech. Phys. Solids*, vol. 8, p. 100-104, 1960.

[EVA 76] EVANS A.G. and CHARLES E.A., "Fracture toughness determination by indentation", *J. Am. Ceram. Soc.*, vol. 59, no. 7-8, p. 371-372, 1976.

[FET 93] FETT T. and MUNZ D., "Evaluation of R-curve effects in ceramics", *J. Mat. Sci.*, vol. 28, p. 742-752, 1993.

[GEB 72] GEBAUER J., KROHN D.A. and HASSELMAN D.P.H., "Thermal-stress fracture of a thermomechanically strengthened aluminosilicate ceramic", *J. Am. Ceram. Soc.*, vol. 55, no. 4, p. 198-201, 1972.

[GIL 59] GILMAN J.J., "Cleavage and ductility in crystals", in *Proceed. Int. Conf. on Atomistic Mechanisms of Fracture*, Tech. Press, MIT, Cambridge (MA), J. Wiley & Sons, NY, p. 193, 1959.

[GRI 20] GRIFFITH A.A., "The phenomena of rupture and flow in solids", *Phil. Trans. Roy. Soc. Lond.*, A221, p. 163-198, 1920.

[HAR 57] see [WANG, 1994]

[HAR 67] HART E.W., "Theory of the tensile test", *Acta Metal.*, vol. 15, p. 351-355, 1967.

[HAS 63] HASSELMAN D.P.H., "Elastic energy at fracture and surface energy as design criteria for thermal shock", *J. Am. Ceram. Soc.*, vol. 46, no. 11, p. 535-540, 1963.

[HER 50] HERRING C., "Diffusional viscosity of a polycrystalline solid", *J. App. Phys.*, vol. 21, p.437-445, 1950.

[HIL 47] HILL R., LEE E.H. and TUPPER S.J., "The theory of wedge indentation of ductile materials", *Proc. Roy. Soc. London*, vol. A 188, p. 273-289, 1947.

[HIR 91] HIRSCH P.B. and ROBERTS S.G., "The brittle-ductile transition in silicon", *Phil. Mag. A*, vol. 64, no. 1, p. 55-80, 1991.

[ING 13] INGLIS C.E., "Stresses in a plate due to the presence of cracks and sharp corners", *Proc. Inst. Naval Archit.*, vol. 55, p. 219-241, 1913.

[IRW 57] IRWIN G.R., "Analysis of stresses and strains near the end of a crack traversing a plate", *J. App. Mech.*, no. 9, p. 361-364, 1957.

[IWA 86] IWASA M., Chapter 9, in *Seramiksu no rikigakuteki tokusei hyôka* "Evaluation des propriétés mécaniques des céramiques", Nishida T and Yasuda E., Nikkan Kogyô Shinbunsha, p. 153-167, 1986 (in Japanese).

[KIM 91] KIM W.J., WOLFENSTINE J. and SHERBY O.D., "Tensile ductility of superplastic ceramics and metallic alloys", *Acta Metal.*, vol. 39, no. 2, p. 199-208, 1991.

[KIN 55] KINGERY W.D., "Factors affecting thermal shock resistance of ceramic materials", *J. Am. Ceram. Soc.*, vol. 38, no. 1, p. 3-15, 1955.

[KIN 76] KINGERY W.D., BOWEN H.K. and UHLMANN, D.R., *Introduction to ceramics*, John Wiley & Sons, 1976.

[LAN 91] LANGDON T.G., "The physics of superplastic deformation", *Mater. Sci. Eng.*, A137, p. 1-11, 1991.

[LAU 85] LAUGIER M.T., "The elastic/plastic indentation of ceramics", *J. Mat. Sci. Lett.*, no. 4, p. 1539-1541, 1985.

[LAW 79] LAWN B.R. and MARSHALL, D.B., "Hardness, toughness and brittleness: an indentation analysis", *J. Am. Ceram. Soc.*, vol. 62, no. 7-8, p. 347-350, 1979.

[LAW 83] LAWN B.R., "Physics of fracture", *J. Am. Ceram. Soc.*, vol. 66, no. 2, p. 83-91, 1983.

[LAW 93] LAWN B.R., *Fracture of brittle solids*, 2nd edition, Cambridge Solid State Science Series, Cambridge University Press, 1993.

[LIF 63] LIFSHITZ I.M., "On the theory of diffusion-viscous flow of polycrystalline bodies", *Soviet Phys. JETP*, vol. 17, no. 4, p. 909-920, 1963.

[LOU 92] LOUCHET F. and BRECHET Y., "Physics of toughness", *Phys. Stat. Sol. (a)*, vol. 131, p. 529-537, 1992.

[MAI 76] MAI Y.W., "Thermal-shock resistance and fracture-strength behavior of two tool carbides", *J. Am. Ceram. Soc.*, vol. 59, no. 11-12, 1976.

[MAI 89] MAI Y.W., "Analysis and measurement of crack resistance in engineering materials", *Mat. Res. Soc. Int'l. Meet. on Adv. Mat.*, vol. 5, p. 295-311, 1989.

[MAK 73] MAKISHIMA A. and MACKENZIE J.D., "Direct calculation of Young's modulus of glass", *J. Non-Cryst. Sol.*, vol. 12, p. 35-45, 1973.

[MAR 63] MARSH D.M., "Plastic flow in glass", *Proc. Roy. Soc. London*, vol. 279 A, p. 420-435, 1963.

[MAR 64] MARSH D.M., "Plastic flow and fracture of glass", *Proc. Roy. Soc. London*, vol. 282 A [1388], p. 33-43, 1964.

[MCM 77] MCMEEKING R.M., "Finite deformation analysis of crack-tip opening in elastic-plastic materials and implications for fracture", *J. Mech. Phys. Sol.*, vol. 25, p. 357-381, 1977.

[MEN 92] MENCIK J., *Strength and fracture of glass and ceramics*, Glass Science and Technology series, vol. 12, Elsevier, New York, Tokyo, 1992.

[MIY 95] MIYAJIMA T., YAMAUCHI Y. and OHJI T., "High temperature R-curve behavior and failure mechanisms of in-situ toughened composites", *High Temperature Ceramic-Matrix Composites I: Design, Durability and Performance, Ceramic Trans.*, Ed. Evans A.G. and Naslain R., vol. 57, p. 425-430, Amer. Ceram. Soc., Ohio, 1995.

[MUK 69] MUKHERJEE A.K., BIRD J.E. and DORN J.E., "Experimental correlations for high-temperature creep", *Trans. ASM*, vol. 62, p. 155-179, 1969.

[MUN 80] MUNZ D., SHANNON J.L. and BUBSEY R.T., "Fracture toughness calculation from maximum load in four-point-bend tests of chevron notch specimens", *Int. J. Fract.*, vol. 16, p. R137-R141, 1980.

[NAB 48] NABARRO F.R.N., "Deformation of crystals by the motion of single ions", *Rep. of Conf. on the strength of solids*, The Physical Society, London, p. 75-90, 1948.

[NII 82] NIIHARA K., MORENA R., HASSELMAN D.P.H., "Evaluation of Kic of brittle solids by the indentation method with low crack-to-indent ratios", *J. Mat. Sci. Lett.*, no. 1, p. 13-16, 1982.

[NOS 88] NOSE T. and TOSHIMITSU F., "Evaluation of fracture toughness for ceramic materials by a single-edge-precracked-beam method", *J. Am. Ceram. Soc.*, vol. 71, no. 5, p. 328-333, 1988.

[ORO 49] OROWAN E., "Fracture and strength of solids", Reports on Progress in Physics, *Phys. Soc. London*, vol. 12, p. 185, 1949.

[PAB 72] DISS PhD, Stuttgart, 1972.

[PAI 91] PAIDAR V. and TAKEUCHI S., "Grain rolling as a mechanism of superplastic deformation", *J. Phys. III*, vol. 1, p. 957-966, 1991.

[PAR 63] PARIS P. and ERDOGAN F., "A critical analysis of crack propagation laws", *Trans ASME*, vol. 79, p. 361, 1957.

[QUI 90] QUINN G.D., "Fracture mechanism maps for advanced structural ceramics, part 1: Methodology and hot-pressed silicon nitride results", *J. Mat. Sci.*, vol. 25, p. 4361-4376, 1990.

[RAC 52] RACHINGER W.A., "Relative grain translations in the plastic flow of aluminum", *J. Inst., Metals*, vol. 81, p. 33-41, 1952.

[RAJ 71] RAJ R. and ASHBY M.F., "On grain boundary sliding and diffusional creep", *Metal. Trans.*, vol. 2, p. 1113-1127, 1971.

[REI 64] REINER M., "The Deborah number", *Physics Today*, vol. 62, p. 46, 1964.

[RIC 68] RICE J.R., "A path independent integral and the approximate analysis of strain concentration by notches and cracks", *J. App. Mech., Trans. ASME*, vol. 6, p. 379-386, 1968.

[RIE 81] RIEDEL H., "Creep deformation at crack tips in elastic-viscoplastic solids", *J. Mech. Phys. Sol.*, vol. 29, p. 35-49, 1981.

[RIE 87] RIEDEL H., *Fracture at high temperatures*, Springer-Verlag, Berlin, New York, Paris, 1987.

[ROU 92] ROUXEL T., WAKAI F. and IZAKI K., "Tensile ductility of superplastic Al_2O_3-Y_2O_3-Si_2N_4/SiC composites", *J. Am. Ceram. Soc.*, vol. 75, no. 9, p. 2363-72, 1992.

[ROU 93] ROUXEL T. and WAKAI F., "The brittle to ductile transition in a Si_3N_4/SiC composite with a glassy grain boundary phase", *Acta Metal.*, vol. 41, no. 11, p. 3203-3213, 1993.

[ROU 95] ROUXEL T., BAUMARD J.F., BESSON J.L., VALIN F. and BONCŒUR M., "Fine grained yttria: processing and potentialities", *Eur. J. Solid State Inorg. Chem.*, vol. 32, p. 617-627, 1995.

[ROU 96] ROUXEL T., MURAT D., BESSON J-L. and BONCŒUR M., "Large tensile ductility of high purity polycrystalline yttria", *Acta Mater.*, vol. 44, no. 1, p. 263-278, 1996.

[ROU 99] ROUXEL T. and BUISSON M., "Physics of the brittle-ductile transition in glasses and glass-containing ceramics: time and temperature incidences", *Key Eng. Mat.*, vol. 166, p. 65-72, Trans Tech Pub., Switzerland, 1999.

[ROY 06] ROUXEL T., "Propriétés élastiques des verres", *Verre*, vol. 12, no. 4, p. 70-79, 2006.

[SAK 86] SAKAI M. and BRADT R.C., "Graphical methods for determining the non-linear fracture parameters of silica and graphite refractory composites", in *Fracture mechanics of ceramics*, vol. 7, Ed. Bradt R.C., Evans A.G., Hasselman D.P.H. and Lange F.F., Plenum Press, New York, 1986.

[SAK 87] SAKAI M., "Fracture mechanics of refractory materials", *Taikabutsu overseas*, vol. 8, no. 2, p. 4-12, 1987.

[SAK 88] SAKAI M., "The crack growth resistance curve of non-phase transforming ceramics", *Nippon Seramikkusu Kyokai Gakujutsu Ronbunshi*, vol. 96, no. 8, p. 801-809, 1988.

[SCH 91] SCHNEIDER S.J., DAVIS J.R., DAVIDSON G.M., LAMPMAN S.R., WOODS M.S., ZORC T.B., *Ceramics and glasses*, Engineered Materials Handbook, vol. 4, ASM International, USA, 1991.

[SEL 61] SELSING J., "Internal Stresses in ceramics", *J. Am. Ceram. Soc.*, p. 419, 1961.

[SHA 75] SHAPERY R.A., "A theory of crack initiation and growth in viscoelastic media", *Intern. J. of Fracture*, vol. 11, no. 3, p. 369-388, 1975.

[SRA 76] SRAWLEY J.E., "Wide range stress intensity factor expression for ASTM E399 standard fracture toughness specimens", *Int. J. Fract. Mech.*, vol. 12, p. 475-476, 1976.

[STE 92] STEINBRECH R.W., "R-curve behavior of ceramics", in *Fracture mechanics of ceramics*, vol. 9, Ed. Bradt R.C., Hasselman D.P.H., Munz D, Sakai M. and Shevchenko V.Y., Plenum Pres, New York, 1992.

[SUB 87] SUBRAMANIAN K., "Superabrasives for precision production grinding-A case for interdisciplinary effort", *Proceed. Sympo. on Interdisciplinary Issues in Materials Processing and Manufacturing*, vol. 2, p. 665-676, 1987.

[TAB 51] TABOR, *The hardness of metals*, Clarendon Press, Oxford Tabor, UK, 1951.

[TAK 88] TAKEUCHI S. and SUZUKI T., "Deformation of crystals controlled by the Peierls mechanism", in *Strength of metals and alloys*, p.161-166, Ed. Kettunen P.O., Lepistö T.K. and Lehtonen M.E., Pergamon Press, Oxford, New York, Tokyo, 1988.

[TES 99] TESTU S., "Déformation à haute température d'un nitrure de silicium à haute performance: relaxation, fluage and essais à vitesse de déformation imposée", PhD Thesis, University of Limoges (France), 25th November 1999.

[TRU 79] TRUSTRUM K. and JAYATILAKA A. de S., "On estimating the Weibull modulus for a brittle material", *J. Mat. Sci.*, vol. 14, p. 1080-1084, 1979.

[VAR 94] VARSHNEYA A.K., *Fundamentals of inorganic glasses*, Academic Press Inc., Boston, New York, Tokyo, 1994.

[WAK 89] WAKAI F., "A review of superplasticity in ZrO$_2$-toughened ceramics", *Br. Ceram. Trans. J.*, vol. 88, p. 205-208, 1989.

[WAK 91] WAKAI F., "Superplasticity of ceramics", *Ceram. Intern.*, vol. 17, p. 153-163, 1991.

[WAK 91] WAKAI F., KODAMA Y., MURAYAMA N., SAKAGUCHI S., ROUXEL T., SATO S. and NONAMI T., "Superplasticity in advanced materials", in *Superplasticity of functional ceramics*, Ed. Hori S., Tokizane M. and Furushiro N., The Japan Society for Research on Superplasticity, Osaka, Japan, 1991.

[WAN 84] WANG J.G. and RAJ R., "Mechanism of superplastic flow in a fine-grained ceramic containing some liquid phase", *J. Am. Ceram. Soc.*, vol. 67, no. 6, p. 399-409, 1984.

[WAN 94] WANG J.N. and LANGDON T.G., "An evaluation of the rate-controlling flow process in Harper-Dorn creep", *Acta Metal.*, vol. 42, no. 7, p. 2487-2492, 1994.

[WEI 51] WEIBULL W., "A statistical distribution function of wide applicability", *J. App. Mech.*, p. 293-297, September 1951.

[WIE 70] WIEDERHORN S.M. and BOLZ L.M., "Stress corrosion and static fatigue of glass", *J. Am. Ceram. Soc.*, vol. 53, no. 10, p. 543-548, 1970.

[WIE 95] WIEDERHORN S.M., LUECKE W.E. and FRENCH J.D., "Importance of cavitation to the creep of structural ceramics", in *Plastic deformation of ceramics*, ed. by Bradt R.C., Brookes C.A. and Routbort J.L., Plenum Press, New York, 1995.

[YOS 92] YOSHIZAWA Y. and SAKUMA T., "Improvement of tensile ductility in high purity alumina due to magnesia addition", *Acta Metal.*, vol. 40, no. 11, p.2943-2950, 1992.

[ZEL 93] ZELIN M.G., KRASILNIKOV N.A., VALIEV R.Z., GRABSKI M.W., YANG H.S. and MUKHERJEE A.K., "On the microstructural aspects of the non-homogenity of superplastic deformation at the level of grain groups", *Acta Metal.*, vol. 42, no. 1, p. 119-126, 1994.

Chapter 9

Materials for Cutting, Drilling and Tribology

9.1. Introduction

The productivity of the manufacturing industries has constantly improved during the 20th century. This is the result of the spectacular increase in the speed of stock removal by cutting tools, faster penetration of drills (petroleum, mines, public works), as well as the considerable improvement in the resistance to wear, fatigue and corrosion of the components of mechanical systems.

These achievements are mainly linked to the discovery, study and development of hard and superhard materials such as ceramics and cermets.

The term "cermet" denotes a composite material containing at least one ceramic phase and one metallic phase. Used here in the broader sense of the term, ceramic designates composites with highly localized bond strengths, covering not merely oxides (Al_2O_3, Cr_2O_3, ZrO_2, UO_2, etc.), but also carbides (WC, TiC, TaC, Cr_3C_2, etc.), nitrides (TiN, etc.), borides (TiB_2, CrB_2, etc.) of refractory transition metals and their combinations or solid solutions (oxycarbides, oxynitrides, carbonitrides, etc.), as well as some elements and covalent composites (diamond, BN, B_4C, SiC, Si_3N_4, AlN, etc.). The ceramic particles have an average grain size between 0.1 and 100 µm and their volumetric content in cermet is more than 15%. The metallic

Chapter written by Henri PASTOR.

binder consists of one of the transition metals (mainly Co, Ni, Fe) or their alloys (Ni/Co, Ni/Cr, Ni/Mo, Fe/Cr, etc.).

Carbide-based cermets constitute the majority of cermets produced industrially and are subdivided into four groups:

    – WC-Co, WC-Ti-Co, WC-TaC-Co, WC-TiC-TaC-Co, WC-Fe, WC-Ni, etc.;

    – TiC-Ni, TiC-Ni/Cr, TiC-Ni/Mo, TiC-Mo$_2$C-Ni/Mo, Ti(C, N)-Ni/Mo, etc.;

    – TiC-Fe, TiC- steel, etc.;

    – Cr$_3$C$_2$-Ni, Cr$_3$C$_2$-Ni/Cr, etc.

WC-based cermets are traditionally referred to as "sintered hard metals" or "hard metals".

9.2. Materials for cutting

9.2.1. *Machining and properties of tools*

The term "cutting tool" in fact covers a very large variety of tools: turning tools, drill bits, planer tools and slotting tools, scraping or shaving tools, drill spindles, milling cutters, facing cutters, recessing tools, reamers, taps and dies, files, and grinding wheels.

All machining operations involve three players: the part to be machined, the machine tool and the actual cutting tool. Productivity, with respect to the first two players, depends on:

    – characteristics (chemical nature, metallurgical state, properties) of the machined material;

    – geometry and size, accuracy requirements, finish and integrity of the surface of the machined part;

    – batch size of the parts to be machined (small batch or mass production);

    – type of cutting (rough grinding or finishing, continuous or intermittent, etc.);

    – cutting conditions (fast or slow speed);

    – condition and capacities of the available machine tools (old or new, adequacy of rigidity/power, feed speed, etc.);

    – support systems (experience and skill of the turner, sensors, controls, lubricants, inspection of chips, etc.).

As regards the cutting tool, it is not only influenced by its geometry and cutting conditions, but also by its constituent material; it should be able to sustain the rigors of the cutting action, including:

– wear, which takes place at the level of the cutting edge;

– heat, generated by the energy required for the formation and removal of chips of the material machined and thus, the temperature;

– inherent shocks in the cutting action.

To sustain these rigors, the tool should be made of a material having the following main properties:

– hardness (specially when hot) to combat the wear;

– resistance to the effects of temperature due to heat released;

– sufficient toughness to resist impact;

– resistance to thermal shocks;

– chemical stability, especially with respect to the machined material.

The tool must have a high degree of hardness and thereby a high abrasion resistance, so as to avoid any dimensional changes due to friction. The tool should maintain a high degree of hardness when hot, so that the cutting edge remains uniform and sharp at temperatures attained during machining; if not, the cutting edge degrades rapidly and the tool becomes unusable. Its toughness helps the tool to absorb the cutting forces and impact, which is particularly important during intermittent cutting. If toughness is not sufficient, the cutting edge will be shattered under the effect of vibrations induced in the tool. Resistance to thermal shocks is required to overcome the effects of continuous cycles of heating and cooling, typical of intermittent turning and milling. A tool with low resistance to thermal shocks should only be used (under certain conditions) in continuous cutting operation. Lastly, the ideal condition would be to have no chemical affinity between the materials constituting the tool and the machined part, otherwise, the rapid development of a built-up edge is inevitable. Now, this results in the modification of the geometry of the tool and consequently, a lower capacity of chip breaking. Moreover, the cutting forces increase and the surface finish of the machined part deteriorates.

9.2.2. *Processing of materials*

The performance of cutting tools depends on their chemical composition, their microstructure (grain size, etc.), their flaws (porosity, inclusions, etc.), their surface condition, their residual stresses, their microgeometry and their operating conditions.

The chemical composition has an influence on the coefficients of elasticity, thermochemical properties, fracture toughness (K_{1C}), hardness and malleability of materials. The last three properties depend, among other things, on the microstructure. It is therefore clear that the processing, that is, the preparation of the powder (composition, size and shape of particles, etc.), their strengthening and their sintering, is crucial to the performance of the tool. We therefore find it appropriate to describe very briefly the processing stages.

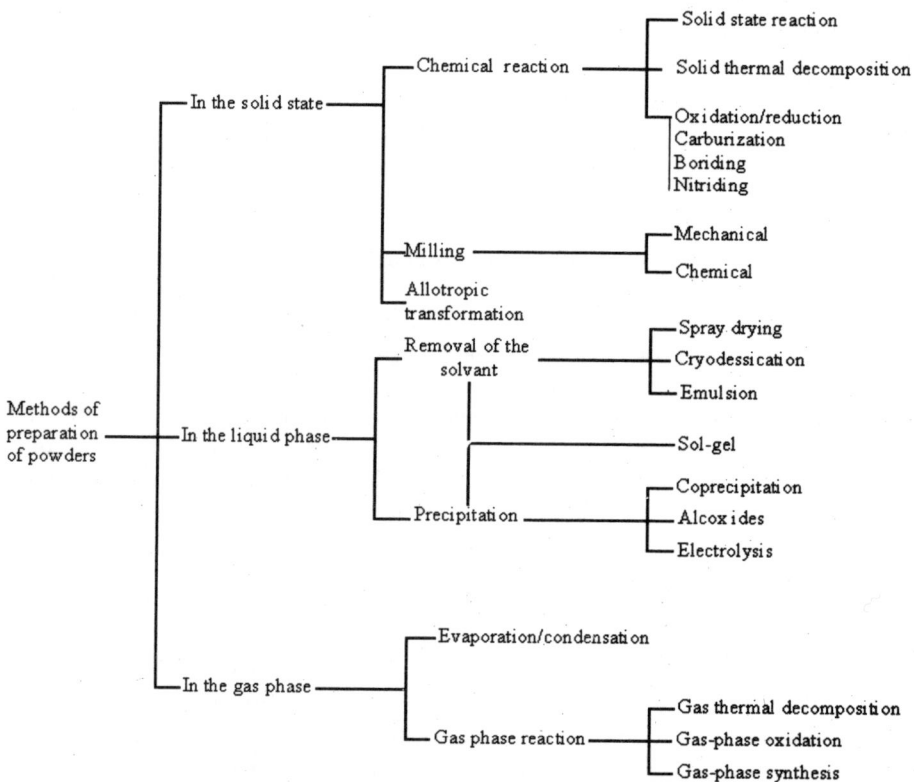

Figure 9.1. *Principal methods of preparation of powders for hard and superhard materials*

Figure 9.1 illustrates the principal methods of preparation of powders for hard and superhard materials. These materials are formed by techniques of powder ceramic-metallurgy, as per the following alternatives:

– sintering: pressure forming (in dies or isostatically), injection molding, slip-casting or extrusion casting; pre-sintering (to eliminate the binder/lubricant and sometimes, to increase the mechanical resistance when machining to dimensions); sintering (with or without liquid phase);

– hot pressing (also known as sintering under pressure): pressure forming (in dies or isostatically) or without applying pressure; hot pressing in graphite dies;

– hot isostatic pressing (HIP): i) with sheath: pressing or forming without pressure; sheathing, ii) without sheath: HIP on materials pre-sintered to more than 93% of their theoretical density, which enables attainment of residual porosity of less than 2%, iii) without sheath: sinter-HIP: pressing; conventional sintering in HIP furnace to more than 93% of theoretical density; application of a gas pressure at the end of sintering to eliminate any residual porosity.

9.2.3. *Material properties*

The range of cutting tool materials is vast and extends over three classes:

– sintered hard metals and titanium carbide based cermets (50% of global sales);

– high-speed steels (45%);

– ceramics (5%) and superhard materials (1%).

Table 9.1 gives the main properties of these different products and Figure 9.2 gives comparisons in a hardness to mechanical strength diagram.

9.2.4. *High speed steels*

High speed steels were discovered in 1868 and were developed from 1898 onwards. These derive their name from their capacity to preserve their hardness (until about 600°C) while cutting material at high speed. These are steels which can be hardened by heat treatment up to 1,000 HV (Vickers Hardness) or 65–70 HRC (Rockwell C Hardness). These steels are certainly not ceramics, but they cannot be ignored here on account of their properties, their operating techniques and the presence of reinforcing precipitates (carbides in particular).

Material	Composition or characteristics	Ø	d	α	K	H	K_{IC}	E	R
Diamond	Monocrystalline	-	3.52	1.50-4.80*	500-2,000*	K: 57-104*	6.89	1,141	91-1,168*
PCD (HP)	Polycrystal diamonds, Syndax3	-	3.43	3.80	120	K: 50	4.26	925	34
PCBN (HP)	BN cub. Polycrist. Amborite	-	3.12	4.90	100	K: 28	6.30	680	30
Al$_2$O$_3$ (S)	5% porosity	2	3.67	8.0	19.8	K: 18	3.90	350	7.1
Al$_2$O$_3$ (HP)		-	3.98	7.5	30	K: 20	5.0	400	10
Al$_2$O$_3$-TiC (HP)	70-30	-	4.26	7.6	17	K: 19	4.0	420	5.3
Al$_2$O$_3$-ZrO$_2$ (HP)	90-10	-	3.96	8.0	8.0	K: 16.6	7.3	390	2.6
Al$_2$O$_3$-SiC$_w$ (HP)	SiCw: SiC whiskers	-		6.0	27.0	K: 17.0	8.0	400	11.2
Al$_2$O$_3$-TiN (HP)	70-30	-	4.25	-	-	K: 21.0	5.0	395	-
Si$_3$N$_4$ (HP)	1% porosity	-	3.26	3.2	25.0	K: 16-22	5.0	300	26
Si$_3$N$_4$ (RB)	22% porosity	-	2.45	1.91	3-11	K: 7.0	2.3	160	10-36
SiC (HP)	< 0.1% porosity	-	3.14	4.32	95	K: 25	3.5	440	69
SiC (RB)	10% free Si	-	3.14	4.3	1.75	V: 20.3	2.75	400	1.4
B$_4$C (HP)	+ 4% Fe; 0.6% porosity	-	2.50	4.32	35	K: 40	7.8	380	37
WC (HP)		-	15.67	5.22	293	V: 17-24	8	727	25.7
TiB$_2$ (HP)		-	4.38	4.6	64.5	K: 33.7	-	540	27.3
WC-Co (S)	97-3	2.2	15.3	4.0	121	V: 20.0	8.8	641	12.3
	86.2-13.8	2.2	14.18	4.5	63	V: 12.2	13.5	485	9.2
	80.5-19.5	2.2	13.63	5.5	58	V: 0.8	17.3	447	9.5
	94-6	0.7	14.9	5.5	80	V: 15.8	9.6	630	8.5
	86.2-13.8	0.7	14.18	4.5	63	V: 14.2	11.4	505	8.8
	80.5-19.5	0.7	13.63	5.5	58	V: 12.7	13.3	447	9.5
	75-25	0.7	12.9	7.5	50	V: 7.80	14.5	470	8.2
WC-TiC-Co (S)	85-5-10	13.4		-	-	V: 14.2	9.0	540	-
WC-(TiC-TaC)-Co (S)	92.3-(1.7)-6	<1	14.9	6.2	46	V: 17.8	10.8	640	4.8
	85.3-(2.7)-12	<1	14.2	5.9	65	V: 12.9	12.7	580	7.9
	78.5-10-11.5	<1	13.0	6.4	60	V: 13.8	10.9	560	8.2
	60-31-9	<1	10.6	7.2	25	V: 8.10	15.6	520	4.5
(Ti, Mo)(C,N)-Ni/Mo (S)**	92.5-6.53/0.97	<1	6.0	6.6	12.3	V: 17.50	7.7	454	4.51
	85.0-13.05/1.95		6.15	6.8	12.4	V: 15.25	9.2	428	4.71
	70.0-26.1/3.9		6.5	7.3	12.5	V: 12.10	11.8	397	4.84
	60.0-34.8/5.2		6.8	7.8	11.5	V: 9.70	14.8	372	4.54
(Ti,Mo)(C,N)-(W,Ta)C-(Ni/Co) (S)	85.7-0.8-13.5	-	7.0	9.4	11.6	V: 16.06	7.9	450	3.50

Table 9.1. *Physical and mechanical properties of hard and superhard materials*

Notations in Table 9.1

Composition: % mass

E: Young's modulus (Gpa) at ambient temperature;

$\varnothing$: average grain size (μm);

R: thermal shock resistance coefficient: $R = \dfrac{K}{\alpha.E} . 10^9$ (W.m^{-1}GPa^{-1});

d: density (g/cm^{-3}) at ambient temperature;

*: according to orientation and/or perfection of the crystal;

α: coefficient of thermal expansion (10^{-6}K^{-1}) at ambient temperature;

S: sintered;

K: thermal conductivity (Wm^{-1}K^{-1}) at ambient temperature;

HP: hot pressed;

H: microhardness (GNm^{-12}) at ambient temperature: K: Knoop, V: Vickers;

RB: reaction bonded;

K_{IC}: critical stress intensity factor (MPa$^{1/2}$) at ambient temperature;

**: $(Ti_{0.8} Mo_{0.2})(C_{0.8} N_{0.2})$.

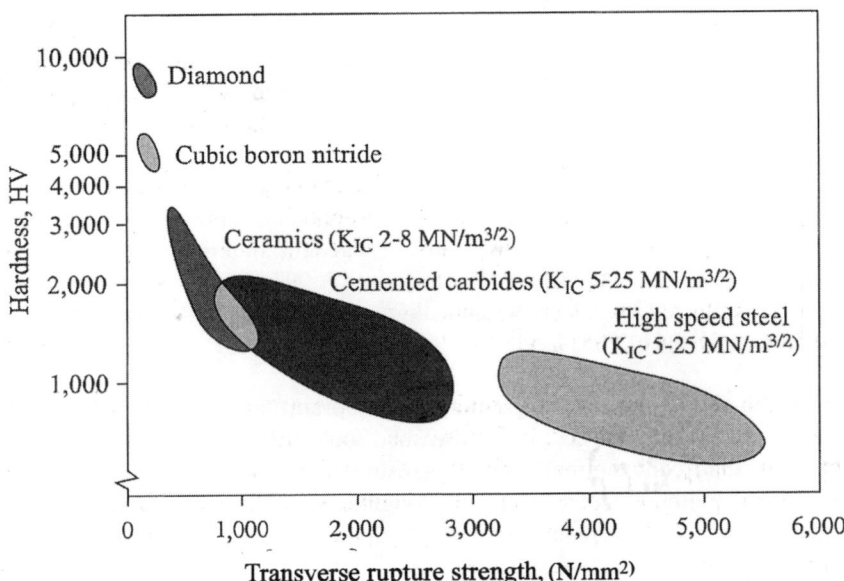

Figure 9.2. *Hardness and bend strength of materials for cutting tools (on the x-axis, feed, depth of pass and impact resistance vary in the same direction as the bend strength; on the y-axis, resistance to deformation and cutting speed vary in the same direction as hardness)*

We can classify high speed steels into two types (in this whole chapter, by content we refer to the weight content):

– type (or series) T: assaying 12% to 20% of W, with additions of Cr (~ 4%), V and Co. For example, T42 is very much used when we require an abrasion resistant cutting edge;

– type (or series) M: assaying 3.5% to 10% of Mo, with additions of Cr (~ 4%), V, W and Co. For example, M2 is very popular for making drill bits and taps, while M35 and M42 are found useful for tools requiring high temperature resistance and are particularly suitable for receiving a TiN coating.

Carbon (0.7% to 1.6%) is an essential constituent, the content of which should be strictly controlled. The introduction of cobalt increases the hot hardness and the resistance to wear, but will slightly reduce toughness.

Powder metallurgy has become the main procedure for fabrication of high-performance high-speed steels because, in comparison with high-speed fusion steels, it improves machinability, ease of grinding, dimensional control during heat treatment and cutting performances in conditions where a tough cutting edge is essential.

High-speed steel powder is obtained by water or gas atomization (N_2, Ar). The water atomized powder should be annealed in vacuum or hydrogen so as to make it ductile and suitable for die or cold isostatic pressing. The gas atomized powder is particularly suitable for hot isostatic pressing (introduced in 1970).

The conventional way (die pressed and sintered) helps to obtain tool tips "close to dimensions" with specially designed chip-breakers, optimizing the cutting geometry of the indexable throwaway inserts. Vacuum sintering, near the solidus temperature, helps to obtain total densification (> 99%). The sintered parts are subsequently subjected to successive annealing, austenitizing and tempering, which lead to hardness levels of 65% to 70% HCR.

The machinability and ease of grinding are important factors which influence the selection of the variety and cost of the finished tool. High-speed steels have better machinability than conventional products (fusion), because these tolerate a higher sulfur content, without decreasing the toughness or performances of cutting operations. There is also greater ease of grinding, because the carbide content is finer and more homogenously distributed.

It is estimated that less than 10% of turning operations (indexable throwaway inserts), but more than 80% of drilling operations (reamers, taps, bits), are carried out with high-speed steel tools. These steels are suitable for a large number of machines still in use today which lack the power and rigidity, two factors necessary for drilling with sintered hard metal tools. High-speed steels also represent 40% of the milling market (end mills, slot mills, mortice mills; hobs, etc.). By using titanium nitride coating (3 μm thick), obtained by physical vapor deposition (PVD), we can work with higher cutting speeds.

9.2.5. *Sintered hard metals*

Sintered hard metals are composite materials consisting of hard particles of carbide(s) (only or predominantly WC) bound together by a metallic binder (only or predominantly cobalt). The carbide content represents 70 to 97% of the total mass of the composite and the average grain size is between 0.3 and 20 μm.

The binary grades WC-Co, which combine the hardness of WC and the ductility of Co, form the base structure of sintered hard metals and appeared commercially in 1925. These are being successfully used for machining cast iron and non-ferrous alloys at cutting speeds much higher than those possible with high-speed steels. But it was soon realized that, when cutting ferritic steels, they were subjected to chemical attacks and wear by diffusion. The solution was found, for avoiding the resulting wear by crater forming, by adding TiC (5% to 25%), a material which is very hard and little reactive to the steel against which it forms a diffusion barrier. Later, in 1950, TaC (or a mixture (Ta, Nb) C which is less expensive) was added (5% to 10%) for improving hot hardness and preventing plastic deformation of the cutting edge. These carbides are totally soluble between themselves and can dissolve a high percentage of WC. We thus have ternary grades WC-(W, Ti, Ta, Nb)C-Co.

It is also possible to replace all or part of the cobalt by another metal like nickel (generally) and/or iron, for certain specific applications. Vanadium carbide VC or chromium carbide Cr_3C_2 are also used to inhibit grain growth of WC during sintering for fine grain and ultra fine grades of WC. Additions of molybdenum and/or chromium to the binder also help to improve corrosion resistance of these grades.

The important properties of binary nuances WC-Co are given in Figures 9.3 to 9.12. We must note here the interesting relationship between compressive strength, rigidity, hardness, low thermal expansion and high thermal conductivity offered by these binary grades. The properties of certain ternary grades are given in Table 9.1. The corrosion resistance of three grades is represented in Figure 9.13.

The international standard ISO R513 classifies sintered hard metals in three groups, according to the area of the cutting application. In each of these three groups, the type of application is denoted by a number (01, 05, 10, 20, 30, 40, 50). A low number relates to the finishing (high speed, low feed), while a high number relates to rough machining (slow speed, high feed). Thus, the fine finishing by turning of a mild steel billet, without interruption, is a P05 application; while the planing of a cast iron with sand inclusions is a K40 application. Grades recommended by two different tool manufacturers, for a given application, are not identical in composition but are reasonably close to each other, with similar properties.

The range of applications of binary grades, besides metal cutting (see sections 9.3 and 9.4) is illustrated in Figure 9.14, in a diagram representing grain size of WC versus Co content.

The present trends in research in the field of sintered hard metals are:

– the search for coarser WC powders (mining, construction) or finer, even nanometric powders (drills, end mills), at the expense of medium grades (cutting);

– the control of WC morphology (tips, lenses, spheres);

– a return to the reduction/direct carburizing of tungsten carbide oxides and the coreduction/carburization of compounds of W and Co;

– the decrease in TiC, NbC or TaC contents, which are becoming less necessary because of improvements in coatings (see below);

– the replacement of cobalt by Co/Fe/Ni (cost, resistance to thermal fatigue);

– the refinement of wet milling and granulation (in fluidized bed);

– the refinement, modeling and control of single axis pressing (near net-shape) and extrusion or injection molding;

– progress in understanding and modeling of liquid phase sintering;

– advances in studies on microwave or electric impulses sintering (resistance heating);

– the development of sintering in controlled atmosphere (methane, nitrogen, etc.).

The combined technologies of indexable throwaway inserts and coating have made coated sintered hard metal today's dominating cutting tool. Ever since its introduction in 1962, its impact on the manufacturing industry has been enormous. As of now, there are a large number of coated grades and each one of these is conceived for optimum performance in a given field of application. A modern indexable throwaway insert has a multilayer coating consisting of at least three or four refractory materials such as TiC, Ti(C, N), TiN and Al_2O_3. These four materials can be deposited by chemical vapor deposition (CVD) and the microstructure of each layer can then be regulated by controlling the chemical process. The two disadvantages of the CVD process (surface decarburization of the substrate and formation of tensile stresses and cracks due to the inadequate thermal coefficients of the substrate and the coating) can be overcome by improving the technology of the process. Although using physical vapor deposition (PVD) helps to eliminate these problems because of a lower operating temperature (500°C instead of 1,000°C) and the development of compressive stresses, only coatings of TiN and Ti(C, N) are presently made through PVD on the industrial scale.

The plasma assisted CVD (PACVD) has had, until now, very little impact, because of the problems of contamination by chlorine of Ti(C,N) or TiN coatings and because of the low crystalline nature of alumina coatings.

The CVD at medium temperature (MTCVD) helps to reduce the decarburization of the substrate. Ti(C,N) coatings could thus be obtained at 800-900°C by using acetonitrile. This MTCVD technique may also be combined very easily with the standard CVD.

The rapid development of coating technology has made it necessary to work on the substrate so as to fully utilize the effects of modern multilayer coatings. Whence the recent development of gradient coatings, so as to optimize the toughness of the cutting edge without compromising on the resistance to plastic deformation. Thus, we can increase the resistance to the propagation of cracks by creating a superficial zone enriched with a cobalt binder, which can be obtained, for example, by removing this zone of cubic carbides (TiC, TaC). The mechanisms at work are complex, which necessitate a strict control of raw materials and sintering parameters. A recent progress in the CVD technology is the development of a procedure to obtain a smooth single phase coating of α-Al_2O_3, while up to now the coatings obtained were two-phase (α/γ-Al_2O_3) and irregular. However, the most important progress is the advent of inserts coated with diamond which is used to machine aluminum alloys, fiber reinforced polymers and raw ceramics. Although the principle of diamond vapor phase deposition has been known for several years, it took a long and combined effort of specialists in the fields of coating and sintered hard metals to develop this technology to obtain a commercial product. Problems regarding low adhesion, encountered in the beginning, have been overcome by combining the PACVD process with the development of a gradient substrate, mentioned earlier.

In conclusion, sintered hard metals cover a very wide domain (50%) of machining applications. It is estimated that 70% of turning operations are carried out with carbide tools. A wide range of compositions is offered and each variety is conceived in relation to the requirements of each defined application. Coated carbides also help to reach very high levels of productivity.

9.2.6. *Cermets*

Modern cermets are the consequences of an evolution of TiC-Ni cermets created in the beginning of the 1950s. The early cutting cermets assayed 70% TiC, 12% Ni and 18% Mo_2C, with a density of 6.08 gm/cm^3, a hardness of 92 HRA and a bend strength of 860 MPa. On account of their hardness, their high mechanical strength and their low thermal conductivity, these were particularly suited to high speed finish turning operations.

Presently, the majority of cermets are of titanium carbonitride base (with possible additions of Mo_2C, WC, TaC, etc.) and Ni-Mo binder (with possible additions of Co, Al, etc.). The optimization of the hard phase and a careful control of the composition of the binder have given rise to significant improvements in the toughness and in the resistance to plastic deformation. The resultant microstructures are complex, with one or more hard phases with composite gradients (for example: core of grain in titanium carbonitride and rim enriched with molybdenum carbide).

As for sintered hard metals, the production of cermets requires an extensive control of the process, from the choice of raw materials to the conditions of sintering (and particularly the sintering atmosphere). Properties of some grades are given in Table 9.1. Finally, quite recently, coating technologies discussed in section 9.2.5 have been extended to cermets.

Cermets have performed particularly well in the Japanese market, representing about 25% of all indexable disposable inserts. In Europe, it represents for the time being only 3% to 4% and this regional difference can be explained by the following reasons:

– efforts for the replacement of tungsten carbide in the period 1950-1960 (the metal became of strategic importance at this time) accelerated the development of cermets in Japan;

– machine tools technology is very advanced in Japan: the machines are more powerful and more rigid;

– development of coated sintered hard metals has been slower in Japan than in the USA and Europe, the Japanese preferring cermets for applications where coated carbides are adopted elsewhere.

While cermets have a hardness similar to sintered hard metals, these are less sensitive to diffusion and they have a better friction and wear behavior. They are capable of machining at speeds higher than those used for sintered hard metals. They excel in the semi-finish turning and milling and finishing of steels and stainless steels. The material property deficiencies between high-speed steels and sintered hard metals is made up by the composites consisting of a dispersion of TiC or, less frequently, TiN (50% to 75%) in a matrix of steel (with chromium, martensitic Ni-Cr, or maraging), which can be thermally hardened. These materials, known as "Ferro-TiC" or "Ferro-TiN", developed for special milling and drilling, have a relatively low hardness in the annealed state (35 to 50 HRC), which allows easy machining of blanks into finished tools. Furthermore, the used tools can be annealed, re-machined and thermally treated for regaining their intended hardness (46 to 73 HRC).

9.2.7. Ceramics

The ceramic cutting tool market is shared, unequally, between aluminina (Al_2O_3)- and silicon nitride (Si_3N_4)-based ceramics.

9.2.7.1. Alumina-based ceramics

Cutting tools in sintered alumina were used from 1905 and the first patents date from 1912, but their commercial development only started in 1954. Early cutting tools were fragile, because of an insufficient densification (porosity) and a too large grain size. Improvements were brought in by hot pressing (1944-1945), or by the addition of magnesia (which inhibits, among other things, grain growth: 1948-1951) and finally by the use of submicronic powders (1968-1970).

Pure alumina is not very tough ($K_{1C} < 3.5$ MPa.m$^{1/2}$), has low mechanical strength and possesses a low thermal conductivity: it is therefore sensitive to sudden changes in stresses and temperature. Nevertheless, in the 1960s, it was used in the automobile industry for the machining of brake disks and drums in grey cast iron. It is practically no longer used today. In 1976, it was shown that the addition of 2% to 10% of zircon to alumina significantly improved the toughness and resistance to thermal shocks. This reinforcement of alumina is based on the allotropic transformation quadratic → monoclinic of zirconia, which takes place by cooling with increase in volume. If zirconia grains are ultrafine (0.3 μm) and well dispersed, these subsist in the quadratic metastable form because of the compressive forces exercised by the alumina matrix. When a crack forms and propagates, the relaxation of the compressive forces associated with the propagation allows the quadratic metastable zirconia to get transformed into a stable monoclinic, with formation of compressive forces and microcracks, if any. This results in an improvement in the overall toughness of the material. Alumina-zirconias are essentially used for interrupted cutting of cast iron and steels.

Since 1955, the addition of titanium carbide (TiC) to alumina has been recommended. We can now find in the market Al_2O_3-TiN, Al_2O_3-Ti(C,N) ceramics and particularly Al_2O_3-TiC with 60% to 70% Al_2O_3, ceramics with greater toughness, mechanical strength and thermal conductivity when compared to pure alumina, and moreover, also harder. Their strengthening can be done only by hot pressing. The important fields of application are in the machining of quenched cast iron, hardened and case-hardened steels and also finish machining of cast iron.

In 1985, it was demonstrated that the addition of 25% to 30% of whiskers of silicon carbide (SiC: diameter $\approx$ 1 μm, length $\approx$ 5 to 20 μm) to Al_2O_3-ZrO_2 and Al_2O_3-TiC ceramics helped to further improve the mechanical strength, toughness and the resistance to thermal shocks. Two mechanisms were advanced for explaining this reinforcement: the deflection (bending/torsion) of cracks and the unearthing of whiskers. Due to the large difference in coefficients of thermal expansion of Al_2O_3 ($8.6.10^{-6}$/°C) and of SiC ($4.7.10^{-6}$/°C), whiskers are in axial compression while the matrix is in axial tension. As cracks propagate perpendicularly to tensile stresses, these will be attracted by the whiskers and will propagate in parallel or perpendicularly to them. The unearthing mechanism requires whiskers, whose transverse energy of rupture is high compared to the whisker-matrix interfacial energy of rupture, such that the rupture takes place essentially along this interface. The shaping of these composites can only be done by pressure sintering or isostatic hot pressing. These were used for machining of nickel-based superalloys, as well as (in severe conditions) of alloyed and hardened steels, and quenched nodular cast iron. The problem posed by whiskers is, however, their toxicity: their short fibers appear to present (or at least are suspected of presenting) risks similar to those shown by asbestos, due to which we may have to give them up.

9.2.7.2. *Silicon nitride-based ceramics*

As indicated in Chapter 7 devoted to non-oxides, silicon nitride has a low coefficient of thermal expansion, which renders it less vulnerable to thermal shocks. It cannot be densified totally by solid phase sintering without applying pressure, but the addition of Y_2O_3, MgO, ZrO_2 or Al_2O_3 enables liquid phase sintering, the liquid being formed by reaction between the additive and the layer of silica always present on the surface of Si_3N_4 grains. During sintering, the α-Si_3N_4 powder (grains of 2 to 3 μm) dissolve in the liquid phase and re-precipitate in the form of β-Si_3N_4 grains, sintered by an amorphous phase, which should be crystallized totally or partially by slow cooling or heat treatment. By acting on the initial α-Si_3N_4 powder, on the sintering conditions and on liquid phase content (oxynitride complex), we can help the growth of β-Si_3N_4 in the form of elongated prisms (whiskers with length/diameter ratio greater than 3) which reinforce the structure. The tools in Si_3N_4, developed in the 1970s, appeared commercially only 10 years later. Their main area of application is rough cutting (turning or milling) of cast iron. Due to their reactivity with steels, these cannot be used for machining steel.

In the early 1970s, it was discovered that oxygen can be substituted for nitrogen in the β-Si_3N_4 network, on the condition of substituting in parallel aluminum for silicon so as to maintain the electric neutrality of the network. The solid solution obtained, or sialon, is symbolized by the formula $Si_{6-z}Al_zO_zN_{6-z}$ where $0 < z < 4.2$. The stability of β-sialon increases with z, but at the same time, the toughness, the mechanical strength and the resistance to thermal shocks decrease, so much so that z is fixed at less than 2 for cutting applications.

For β-sialons, sintering additives are also used, for example Y_2O_3. During sintering, the silica reacts with alumina and yttrium oxide to form a liquid phase which dissolves the silicon nitride, which re-precipitates in the form of β-sialon (equiaxed or needle-shaped grains). After sintering, the microstructure is composed of β-sialon, α-sialon $[Y_X(Si,Al)_{12}(O,N)_{16}$ where $x < 2]$ and crystalline phases formed by devitrification of the vitreous binder phase (for example $Y_3Al_{15}O_{12}$). Sialons are suitable for the machining refractory steels, hard matrix steels and cast iron (at high speed). They are also not suitable for machining of steels.

9.2.8. Superhard materials

The first reproducible synthetic diamond (C_d) dates back to 1955, and in 1957 General Electric already announced the commercial availability of powdered synthetic diamond. The same year, the same firm announced the synthesis of cubic boron nitride (BN_c), but, while the small diamond crystals and BN_C are suitable for grinding wheels and are excellent polishing agents, these cannot be used for making cutting tools (tendency for cleavage). Also, from the very beginning, researchers were trying to obtain polycrystalline diamond (PCD) and polycrystalline boron nitride (PCBN). The making of PCD dates back to the early 1970s (and was followed shortly after by PCBN) and is produced by sintering of powder, without addition of binder (or very little of binder), at very high temperature (1,500°C) and high pressure (50-60 kbar). PCD and PCBN are isotropic and have greater toughness and wear resistance than C_d or BN_c.

Since the working face of a cutting tool is basically bi-dimensional, PCD and PCBN are most often made in the form of a layer (0.2 to 1 mm thickness) on a substrate of sintered hard metal (disks of diameter 34 mm and more). This configuration gives rise to many advantages. The composite is tougher and costs less than a solid PCD or PCBN; it can be cut it into small pieces which are brazed (T < 800°C) on to the edges of an indexable throwaway insert in sintered hard metal. PCD is basically available in three grain sizes: fine, medium and coarse. The fine grain PCD is a little less resistant to shocks than the coarse grain PCD, but slightly less resistant to wear and vice versa; the medium grain PCD is a compromise between the two.

PCD is suitable for cutting non-ferrous metals at very high cutting speed. It is widely used in the automobile industry for aluminum alloys, which it machines at very high cutting speed, yielding excellent surface finish. It is also perfectly suited for the machining of bronze, copper, babbit, copper-lead, composites with metallic matrix, but also composites and laminates with base of resin, glass, fiber glass, rubber, graphite, titanium, asbestos, ceramics and even wood and derivative materials. It may also be used for the machining of sintered hard metal (for example dies) in good condition and it is economically more advantageous than grinding. PCD is not suitable for machining of cast irons and steels. PCBN is suitable for machining of high-speed steels, alloyed steels, hardened steels, quenched cast irons and stellites (cobalt based superalloys). However, it offers no advantages in the machining of mild steels, austenitic steels and nickel-based alloys.

9.2.9. Conclusion on cutting materials

On reading this brief survey about cutting materials, we understand that in each family of materials research continues to improve performances and efficiencies. It is certain that the machinist, who aims at optimizing his work, is faced with a range of cutting materials, more and more complex, with very often overlapping characteristics and fields of application: thus it is not always easy to choose. Furthermore, the tendency nowadays is to buy a machining function rather than a tool. Consequently, the suppliers need to develop a deeper partnership with the users.

9.3. Materials for drilling (petroleum, natural gas), mining, buildings and public works

The materials used, above all else, for this application are the sintered hard metals. Despite their cost, PCDs are now well established in the rotary drilling of relatively soft abrasive rocks. High speed steels are sometimes used for cutting disks for working in soft and fragile rock formations. Cermets (Ti,V)C-Ni/Mo and TiB_2-Ni/Cu have not gone beyond the stage of experimentation. PCBN and ceramics have not found, until now, any application in these fields.

9.3.1. Sintered hard metals

Binary grades WC-Co, basically used for applications considered here, represent about a third of the total production of sintered hard metals.

Figure 9.14 diagrammatically represents different fields of application of sintered hard metals. Other than the field of cutting, already described in section 9.2, and that of drilling being dealt with here, we can see the other fields of application which will be discussed in section 9.4.

9.3.1.1. *Tools for mining and drilling*

The recovery of natural resources from the Earth's crust is an important activity, in which sintered hard metals play a crucial role. It relates to both underground and open cast mining techniques for the recovery of minerals – metallic and non-metallic – like carbon (coal), potash, natron, gypsum and of course petroleum drilling (petroleum, natural gas). The methods of evacuation and the types of tools used depend on the type of stratification encountered. We can classify drilling into three types:

– rotary drilling (or "honning");

– percussion drilling (or "beating");

– subsurface drilling of flats.

The fields of application covered by rotary drilling are drilling of petroleum or gas wells and drill holes in difficult stratifications. The tool performance depends heavily on the properties of carbides and the tool design. The cutting edge should resist wear, while being at the same time sufficiently tough to avoid chipping. The important factors affecting performance are the strength and abrasiveness of the rock. The degree of resilience or plasticity which affects the cutting action affects also its drillability. A rock which caves in without fragmenting or breaking may be harder to pierce than a fragile rock which breaks easily. Rotary drilling uses rock bits fitted with insert bits (forks) or button bits (tricone bits), the latter being preferred in hard rock formations. Brazed tips on steel bodies have a problem, because tensile stresses are induced due to the large difference in the coefficients of thermal expansion of carbide and steel. The carbide in the bit should therefore be very tough and thus less resistant to wear. The solution to this problem has been the development of self-sharpening button tips (hemispherical cylinders). These buttons are mechanically embedded in the steel body. There is no metallurgic bond, thus avoiding tensile stresses. As such, we can use for these buttons a harder and therefore more wear-resistant variety of carbide, for example HV_{30}: 1,450 instead of 1,250 for a particular type of brazed bits. The speed of rotation may vary from 100 rpm (rocks) to 1,500 rpm (coal). Rotary drilling is efficient and achieves great penetration speeds in soft rocks; but in hard and abrasive rocks, there is severe wear and break.

For hard rocks, rotary-percussion drilling is preferred. In this method of drilling, when the carbide insert, in the form of a chisel, strikes the rock, the shock generates compression waves, which on reflection create tensile waves, thus pulverizing the rock. Then the insert is subjected to a rotation and starts striking the rock again, etc. The size of the crater formed under the impact is proportional to the energy of the shock. Most of the rock bits deliver some powerful energy at a frequency of 1,500 to 3,000 blows per minute. For very hard and abrasive rocks, it is recommended to operate with a high force per blow and slow speed of rotation, while for tender rocks, we work with a moderate force per blow and high speed of rotation. Rock bits which are widely used are cross bits (with 90°) or X bits (with 80° and 100°). Rock bits in X are mostly used for drilling large diameter. The main innovation in the development of percussion tools has been the replacement (towards 1960) of bits by hemispheric self-sharpening buttons, already discussed in rotary drilling. In 1986, these buttons were evolved by developing DP grades (double phase) which we will describe later.

The excavation of flat-lying (relatively horizontal seams) carbon, potash, natron, gypsum, etc., is done with the aid of an undercutting tools, where a carbide pick is brazed on a steel support. Flats usually consist of alternative soft and hard layers. The release of rock fragments in front of the cutting edge subjects the tool to a continuous series of shocks (up to 600/min). The temperature may reach 800°C at the rock-carbide contact point and the latter, as well as the brazing, are subjected to thermal fatigue.

The credit for fantastic improvement in the efficiency of drilling tools (speed of drilling in granite became 100 times faster in less than a century) goes to the tools as well as the machines. For example: the introduction of compressed air percussion hammers (1925), the introduction of removable threaded cone bits (1936), the early tool holder bits in carbide (1940-1945), popularization of diamond tipped tools (from 1945) and numerous innovations in the design of tools (buttons, DP grades) and machines (tunneling machines). The tooling materials most commonly used are the binary grades WC-Co, with a cobalt content varying from 5% to 20% (more often 6% to 15%) and an average WC grain size varying from 0.8 to 25 μm (more often 1.9 to 7 μm). For important properties of these grades, please see Figures 9.3 to 9.9. Trials with the addition of TiC and/or TaC were negative because of the harmful effect on the resistance to wear. Similarly, coated carbides had no success in this field.

The wear of the tool is the main factor which determines the energy needs and the speed of penetration, based on which the choice of the drilling method and the type of carbide variety for a given type of rock is made. The wear behavior of sintered hard metals is influenced by a certain number of factors; the most important ones are:

– spalling due to shock: wear is proportional to the force of the blow;

– spalling due to shock and fatigue.

These two effects predominate in the percussion drilling of hard rocks:

– abrasion due to sliding: the wear is proportional to the load applied, to the distance of sliding and to the resistance to grinding and the abrasiveness of the rock (quartz content). It is the most important method in rock drilling of soft and non-abrasive rocks;

– thermal fatigue: "snake skin" cracking: important in the drilling of soft and non-abrasive rocks.

All mechanisms occur at the same time, in various degrees of intensity and more or less independently from one another.

Since the surface of the drill bit is exposed to high stresses and high temperatures, the performance of the tool depends highly on the thermal conductivity and mechanical properties at high temperatures of WC-Co grades.

The resistance to wear by shock-scaling is proportional to the toughness of the carbide, and increases when the grain size of WC grows and when the mean free path (that is, the average thickness of the cobalt film between WC grains) in the cobalt increases.

The resistance to abrasive wear depends on the hardness, the cobalt content and WC grain size (and therefore also on the mean free path in cobalt). For harder grades, where wear is the result of microrupture of WC-Co interface and tearing of WC grains, fine and very fine grains (0.4 to 1 μm) basically give better wear resistance. For "softer" grades, where wear is the result of preferential removal of cobalt, the use of coarse grains (10 to 25 μm) provides better wear resistance.

The most recent innovation involves the grades known as DP (dual phase or dual properties), where it is possible to improve the wear resistance (hardness) and the toughness independently. With the help of a controlled redistribution of the cobalt binding phase, it is possible to develop components (drill bit buttons, for example) consisting of three microstructural zones with different properties. These grades of composition, associated with the resulting differences in coefficient of thermal expansion, redistribute the internal stresses. Thus, it is possible to create a superficial layer, extremely hard and wear resistant (low Co content), which is prestressed in compression, and helps in preventing the initiation and propagation of cracks. The underlying zone is on the other hand enriched in cobalt binder and therefore possesses high toughness. The core has an intermediate cobalt content corresponding to the strength of the variety before DP treatment.

Other than the quality of the carbide, the geometry of the inserts and their arrangement on the drilling head (there are sometimes more than 100 buttons on a single head), and the technique of brazing (for the drill bits) have an influence on the performance. However, these are outperformed as soon as diamonds are incorporated in the tool bits (see section 9.3.2).

9.3.1.2. *Tools for buildings, public works and woodwork.*

Sintered hard metal (binary grades WC-Co) are used exclusively for making these tools. Cobalt content may vary from 5% to 15%; the average grain size of WC may vary from 0.8 to 20 μm (more often 1.5 to 7 μm); the hardness between 1,100 and 1,600 HV_{30}.

Here are some examples in the fields of building and public works:

– drilling in special projects (Channel Tunnel, crossings in the Alps or Pyrenees);

– road works: road leveling, ground stabilization, asphalt recovery;

– vertical and horizontal drilling (excavators, drills, augers, etc.) for sewer and water collectors, trenches, wells, foundations of buildings and parking lots, tunnels for water gas or petroleum conduits;

– miscellaneous: blades of snow ploughs; anti-skid spikes for tires or golf shoes.

In the same way, for the road leveling (planing), conical carbide picks mounted on rotary drum are used to remove asphalt or concrete road surfacing, up to depths of 25 cm. Each pick is fixed in its housing, such that it can turn in its retaining sleeve, leading to self-sharpening and longer life. Concrete road pavements, harder than asphalt, require carbide grades resistant to shocks and abrasion, and yet tough: grades of 9.5% Co with coarse WC grains (10 to 20 μm) are perfectly suitable. Asphalt pavements, which are less hard, require nevertheless abrasion resistant grades (6% Co).

Sintered hard metal tools are now well established in forestry and wood working applications, where they render very high levels of productivity, closer dimensional tolerances, better surface finish and cheaper costs. For example:

– sawing (circular saws) and debarking of trunks: grades with 10%-12% Co and $\varnothing_{WC} = 1$ μm;

– planing, edge tool industry and different wood working tools for commercial or domestic use: grades at 3%-7% Co and $\varnothing_{WC}$: 0.5-2 μm; HV_{30}: 1,600-2,200.

9.3.2. *Diamonds and PCD*

Diamond is not thermodynamically stable at atmospheric pressure, but its transformation into graphite is extremely slow below 800°C. Its synthesis, achieved for the first time in 1955, takes place by treating graphite at 2,700°C, under 12 GPa pressure, in the presence of a catalyst (1% to 2% of B, Be, Si, etc.), in an equipment (anvil, punches) which besides can be made only in sintered hard metal WC-Co. Present global production reaches 400 million carats, which is more than three times the production of natural diamonds (a carat is 1/5 g).

PCD (polycrystalline diamond) is obtained by sintering synthetic diamond powder in the presence of a metal binder (Co, Ni or Fe: a low percentage by volume), at 1,350-1,500°C under 5 GPa pressure. One may also sinter a layer of diamonds (0.5 mm thick) on a sintered hard metal substrate, the cobalt of the substrate thus participating in the sintering of the diamond and the adherence of the PCD on the substrate. Inserts up to 72 mm diameter and hardness of 5,000 to 8,000 HV may thus be attained.

Petroleum drilling is a special field of application for diamonds and PCD. A petroleum well is drilled (up to depths of over 5,000 m) by a drill bit fixed at the end of a drill pipe string, the length of which may grow shaft by shaft while the bit drives into the rock. The drill bit, the diameter of which may vary from 2 to 40 cm, rests on the rock under the weight of the drill pipe string and is set to rotate, around the longitudinal axis, by a couple transmitted from the surface or produced by a turbine mounted just above the bit and driven by the flow of a cooling agent. This agent, whose function is to also clear the rock debris, is injected in the axis of the drill pipe, passes along the inlet channels in the drill bit and rises up in the annular space formed by the drill pipe and the well wall. The fluid is essentially water or a clayey suspension (bentonite).

Drill bits in steel and in sintered hard metal have to face strong competition from diamond drill bits, which have two advantages. Firstly, on account of its high hardness, a diamond drill bit can be used for drilling very hard rocks, at adequate penetration speeds. Secondly, on account of its higher wear resistance, a diamond drill bit, drilling a soft but abrasive rock, has a much longer life resulting in lower replacement costs (lifting the drilling string) of the used drill bit. In a drill bit, whole stones are arranged individually in regular geometric patterns on the faces of a matrix in WC-Co. Small lateral bars (barrettes), carrying very small diamonds, prevent an excessive wear of the sides of the drill. The centre of the drill bit is hollowed, because the diamonds nearer the axis of rotation have a reduced linear speed and would suffer considerable damages if these came in direct contact with the intact rock. Channels are laid between diamond bands to ensure the flow of the cooling fluid. The diamonds on the crown may weigh from 4 to 20 carats, according to the type of rock involved.

Trials to replace monocrystaline diamonds with PCDs of the same shape (quasi-spherical) were not successful. On the other hand, drills in steel fitted with bits in sintered PCD (1 mm thick) on sintered hard metal (located on steel studs) have been very successful for drilling relatively soft rocks. The bits act practically as cutting tools with negative clearance; they create small splinters of rock which coalesce under high ambient pressure and form long "chips". As carbides wear faster than PCD, there is the effect of self-sharpening, which tends to preserve the sharpness of the cutting edge of the PCD.

9.4. Materials for tools and components resisting wear, fatigue and corrosion

Wear, fatigue and corrosion represent the three plagues of any mechanical system, because these entail higher maintenance and replacement costs of the structural components. Due to the special and balanced combination of their properties (compressive resistance, good abrasion resistance, high modulus of elasticity, impact resistance, ability to assume and retain an excellent surface finish), sintered hard metals replace here the majority of other materials and from day to day new applications are discovered. Table 9.2 shows the market share of "wear parts" in sintered hard metals.

Items	Market share
Cold working tools	
– forming tools: dies, rolls, roller dies, punches and dies for powder pressing	17%
– cutting up and related tools: guillotine shears, pre-piercing punches, cutting disks and wheels	7%
Hot working tools	
– rolls and guides of rolling mills; high temperature anvils and punches (diamond synthesis)	3%
Others	
– gaskets and bearings	4%
– miscellaneous	32%
Semi-finished products	
– inserts for sawing of wood and metals	3%
– extruded rods and bars	9%
– drilling, brazing tools	12%
– miscellaneous	13%
Total	100%

Table 9.2. *Market distribution of wear parts in sintered hard metal*

High performance ceramics (Al_2O_3, SiC, Si_3N_4), pure or reinforced by fiber, have a lower density, a higher chemical inertia and often greater hot hardness than sintered hard metals. However, in the majority of cases, a minimum of toughness must be guaranteed and it is always the sintered hard metal that is finally chosen.

9.4.1. *Cold working tools*

9.4.1.1. *Forming tools*

The WC-Co grades having 14% to 26% Co, with an average grain size of WC $\varnothing_{WC}$ = 0.7 to 3.5 µm and hardness of HV_{30} = 850 to 1,300, are used for making forging dies for nuts, bolts, rivets and coins, as well as artistic die punches for cutlery, medals, buttons, etc. The productivity of a carbide tool is ten times higher than its counterpart in steel.

For cold rolling (as well as hot), mill rolls of diameter ranging from 6 to 400 mm are made in binary grades, having 6% to 30% Co, grain size $\varnothing_{WC}$ = 3 to 5 µm and hardness of HV_{30} = 600 to 1,400. Compared to steel rolls, a carbide roll for cold sheet rolling achieves a more uniform thickness for the full width of the sheet.

Compression dies used in powder metallurgy have 6% to 16% Co, with $\varnothing_{WC}$ = 2 to 8 µm and serve, for example, in the pharmaceutical industry (tablets, pills, etc.). Drawing dies generally have 3% to 16% Co, with $\varnothing_{WC}$ = 0.4 to 1.8 µm and HV_{30} = 1,350 to 2,300.

Diamond and PCD are greatly used in the drawing of wires of small diameters. Drawing dies are very often made of a core of PCD and a ring in WC-Co. Alumina and stabilized zirconia find applications in the industry for manufacture of capstans, pulleys, rollers and wire guides.

9.4.1.2. *Cutting tools and similar items*

Irrespective of the material (except paper and non-reinforced plastics) to cut or punch (up to 1.3 mm thick), binary grades with 11%-2% Co have largely replaced steel and high speed steels, with the tool being capable of ensuring a production of at least ten times more (1 to 10 million parts per tool). Knives of scissors or blades of paper cutters have 6% to 10% Co, with $\varnothing_{WC}$ = 0.5 to 2 µm. In the stamping of food cans, we can see the appearance of some competing ceramic materials: stabilized zirconia, Al_2O_3-SiC_W, Si_3N_4-SiC_W (the index w standing for whiskers). Cold extrusion uses dies (12%-20% Co; $\varnothing_{WC}$ = 1-3 µm) and punches (11%-12% Co; $\varnothing_{WC}$ = 1 µm) in WC-Co, whose performance cannot be surpassed.

9.4.2. *Hot working tools*

Rolls of plate mills, mentioned in section 9.4.1.1, can attain, during hot lamination, working speeds of more than 500 m/s. In a particularly corrosive medium, cobalt can be replaced by a Ni/Co/Cr (15% to 20%) binder.

Without sintered hard metals, the synthesis of diamond and CBN, and the sintering of PCD and PCBN would not have been possible. High temperature (500°C on the carbide wall, 1,500°C at the core of the chamber) and high pressure (12 to 15 GPa) chambers are made up of a die (10% to 12% Co; $\varnothing_{WC}$ = 2-3 μm) and punches (6% Co) in binary grades WC-Co.

9.4.3. Components

9.4.3.1. Swing joints

The usage of swing joints in sintered hard metals has become very common, because of their reliability in severe operating conditions combining corrosion, abrasion, high speed of rotation, high pressures and high temperatures. In very corrosive media, we prefer the use of grades with binders Co/Ni, Co/Ni/Cr/Mo, Ni, Ni/Cr/Mo, Ni/Cr/Mo/Al, Ni/Al or Ni/Fe, Co/Ni/Fe, or even grades without binders (WC with less than 0.1% Co).

In applications requiring swing joints of single form, standard size and without any requirements of toughness, silicon carbide can replace sintered hard metals.

Cermet 85 $(Ti_{0.8} Mo_{0.2})(C_{0.8} N_{0.2})$-13 Ni-2Mo, whose properties are given in Table 9.1, appears very promising. At room temperature, a pin disk friction test on itself, gives a wear coefficient varying from 10^{-7} to 10^{-6} mm^{3}N^{-1}m^{-1} when the sliding speed is more than 0.03 to 3 m/sec; which is ten times less than that of ceramics SiC, SiC/Si, SiC-TiC (50-50) and SiC-TiB$_2$. Under the same conditions, at 100°C, the wear coefficient decreases when speed increases from 4.10^{-6} to 4.10^{-7} mm^{3}N^{-1}m^{-1}, sign of a tribological oxidation (formation of Magneli phases).

9.4.3.2. Other components

Other components in sintered hard metals are diverse and varied and their mere enumeration would take up many pages. We will therefore cite here only a few: bearing bushes, shanks and valve seats, sand blasting nozzles, agricultural powdering nozzles, atomization nozzles (food, paints, enamels industry, etc.), parts of worm screws of centrifugal separators of turbid liquids, screws of industrial coffee grinders, knives for the tobacco industry, plungers for the petroleum industry (fabrication of polyethylene), ejectors for sand crushers, emergency jet fuel pumps for fighter aircrafts, control ball joints of helicopters, ammunition, armor plating, ball points for pens, and master gauges.

9.5. Conclusion

As we have seen in this chapter, on various occasions, the excellent combination of properties on account of their composite nature explains the almost unequalled success of sintered hard metals in a large number of fields of technology. They are still growing in importance despite having been around for 75 years. The appearance of new materials (ceramics, superhard materials) has not dampened the momentum of fundamental and applied research on sintered hard metals. The accent is now placed on technoeconomic gains and the increase in duration of the life of tools, combined with effective methods of recycling.

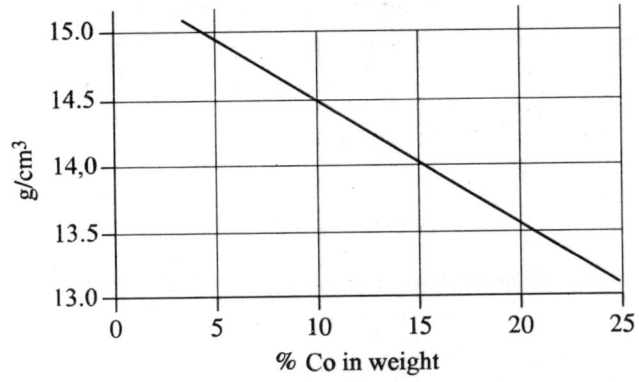

Figure 9.3. *Density of carbides WC-Co according to cobalt content*

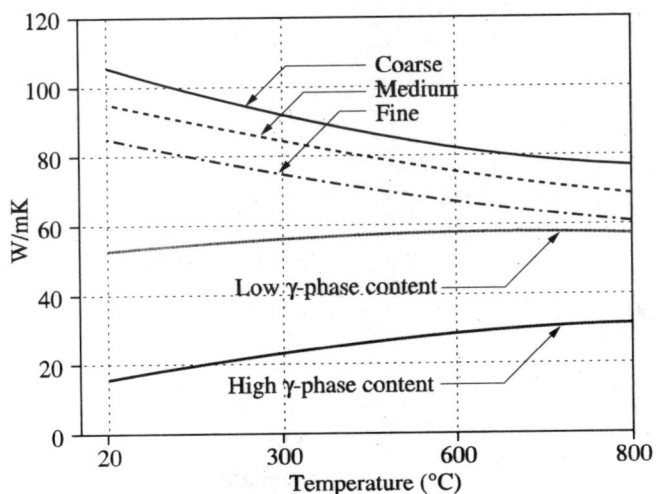

Figure 9.4. *Thermal conductivity as a function of the temperature for different microstructures and WC grain sizes*

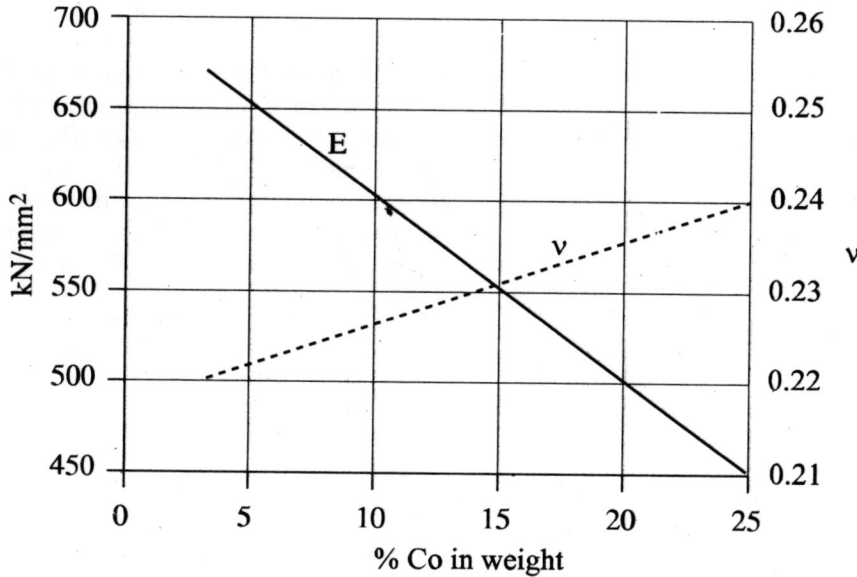

Figure 9.5. *Young's modulus and Poisson's ratio according to the Co content*

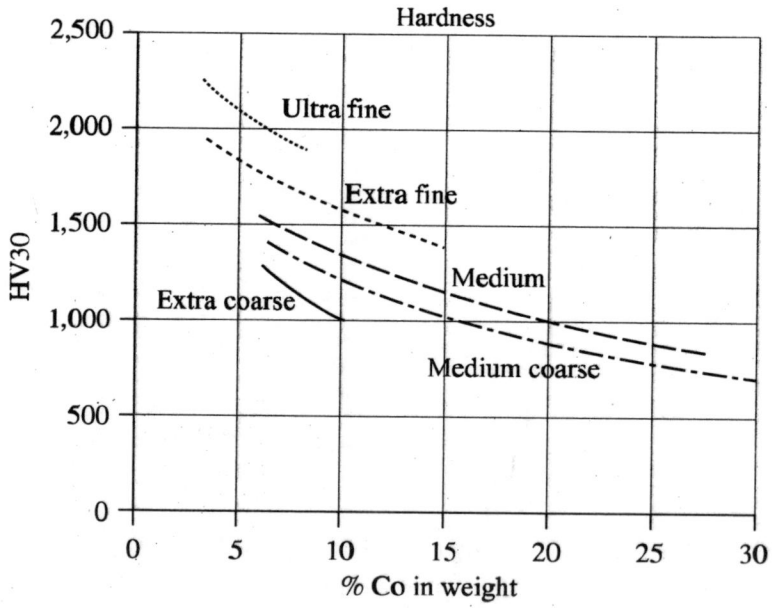

Figure 9.6. *Hardness as a function of Co content for different WC grain sizes*

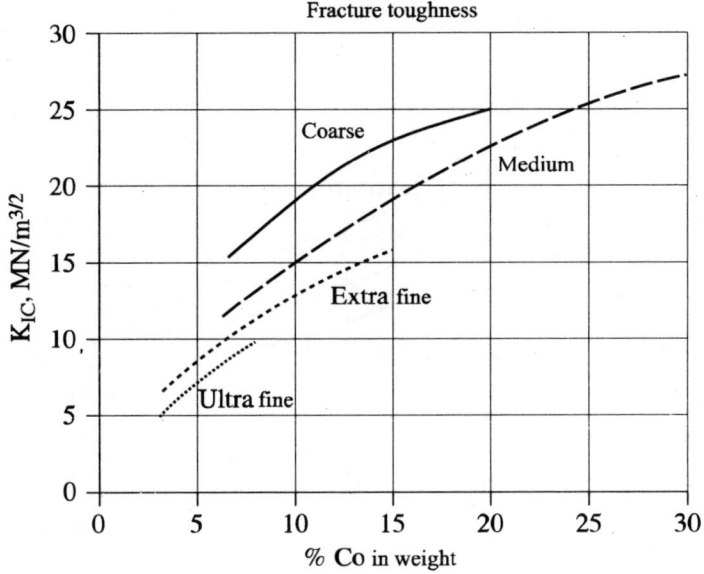

Figure 9.7. *Fracture toughness as a function of Co content for different WC grain sizes*

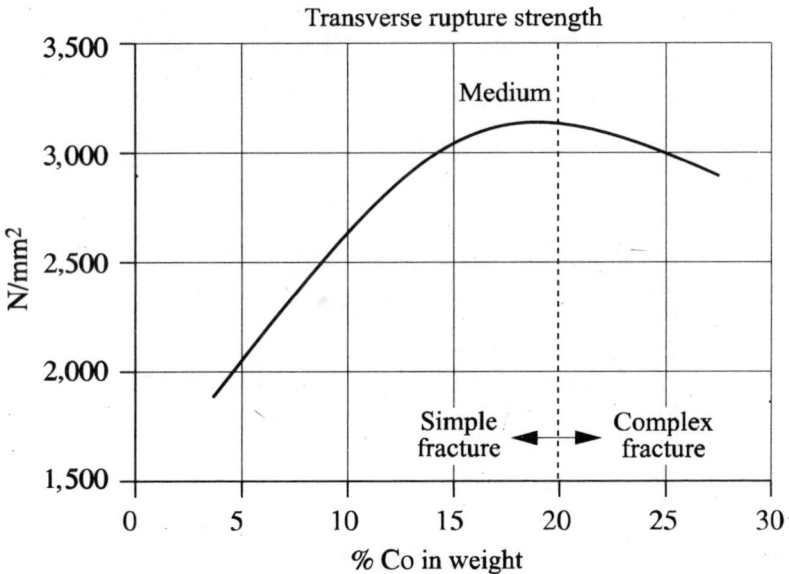

Figure 9.8. *Transverse rupture strength as a function of Co content. Beyond 20% Co, the relation is interfered by other rupture mechanisms*

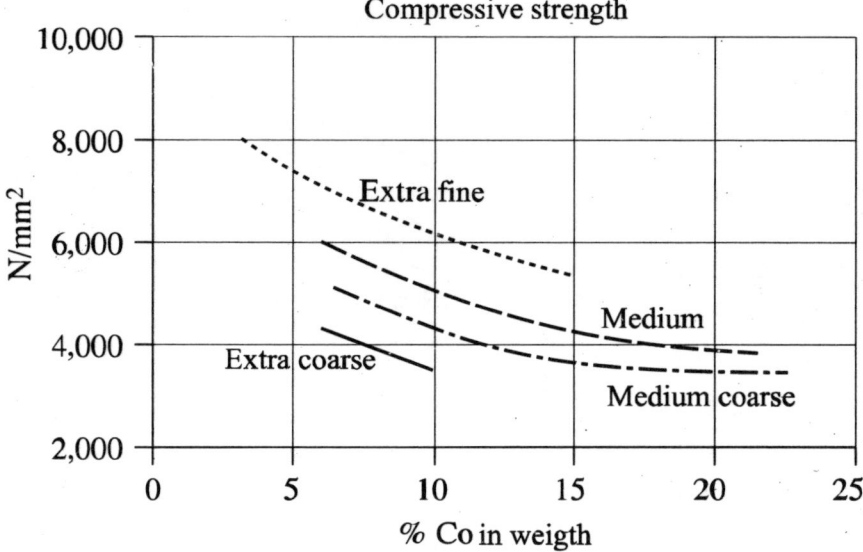

Figure 9.9. *Compressive strength as a function of Co content for different WC grain sizes*

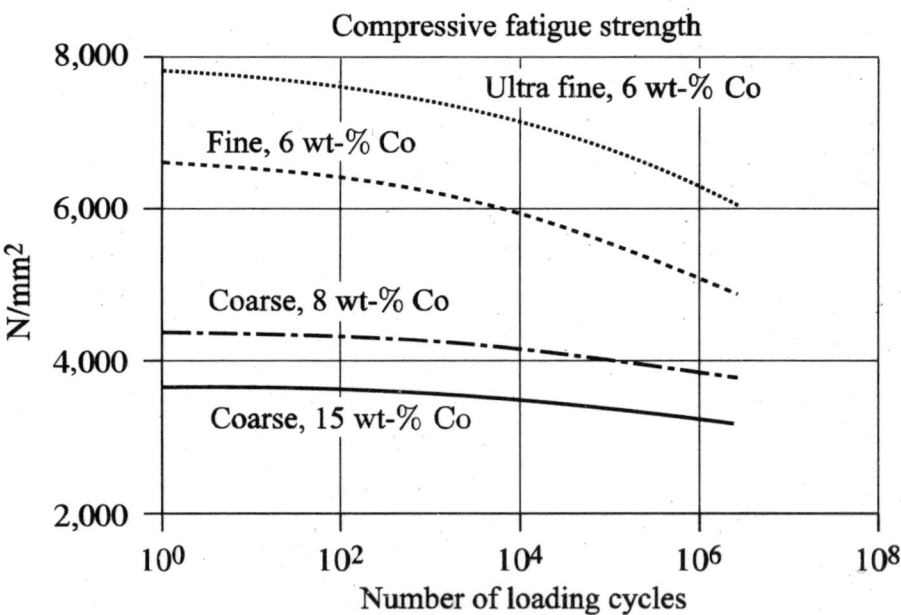

Figure 9.10. *Wohler curves related to fatigue tests in compression of different carbides. The lower limit of loading is 250 Mpa*

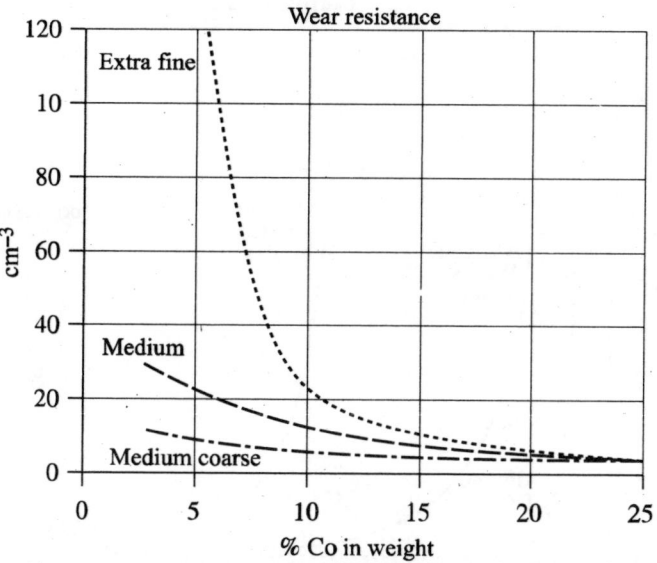

Figure 9.11. *Wear resistance as a function of Co content at different WC grain sizes according to the testing method ASTM-B611-85*

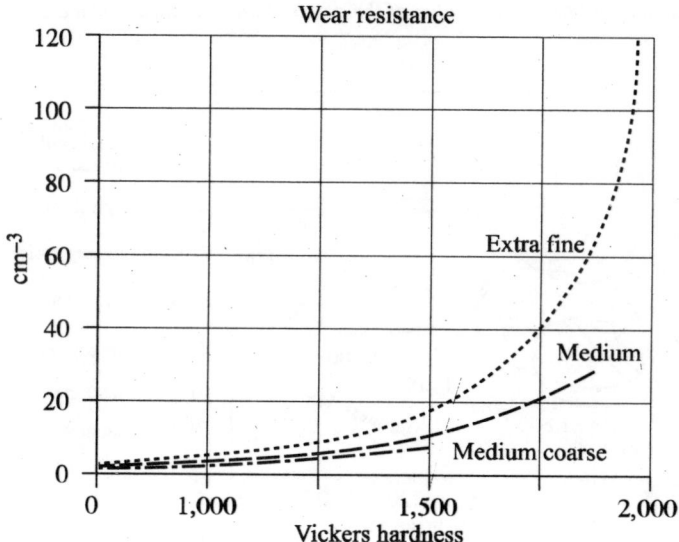

Figure 9.12. *Wear resistance as a function of hardness (ASTM B611-85)*

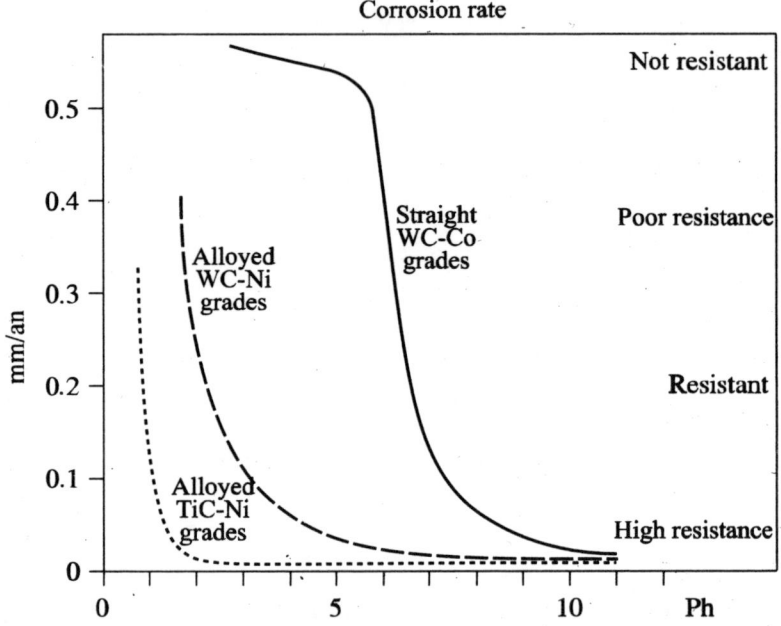

Figure 9.13. *Corrosion rate as a function of pH for different cemented carbides tested in buffered solutions. These tests include a final surface wear treatment by tumbling, so as to obtain a real value of the depth of the corroded surface zone.*

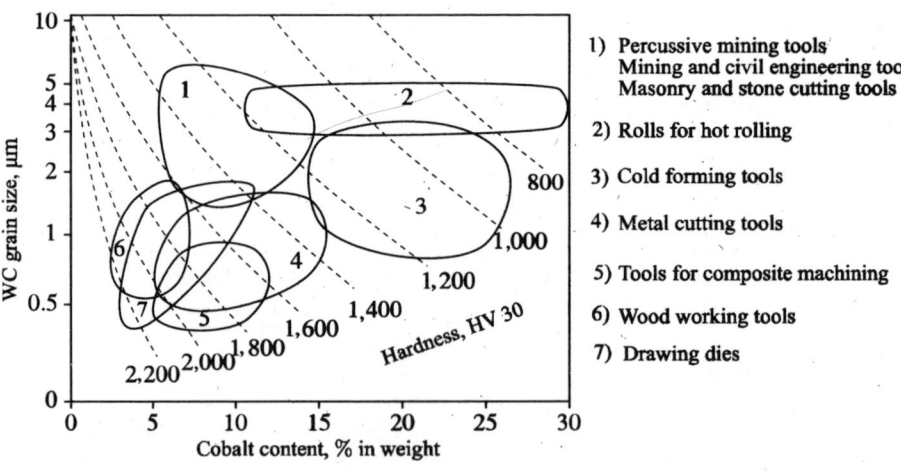

Figure 9.14. *The application range of straight grade cemented carvide*

9.6. Bibliography

[ASM 98] ASM Handbook vol. 7: Powder Metal Technologies and Applications, ASM International, Materials Park, OH, USA, 1998.

[BCE 96] L'évolution des matériaux pour outils de coupe (ENS des Mines de Saint-Etienne, 13-14 Nov. 1996), *Bulletin du Cercle d'Etudes des métaux*, 16, (13), 1996.

[DAV 95] DAVIS J.R. (Ed.), *Tool Materials*, ASM International, Materials Park, OH, USA, 1995.

[DOW 94] DOW WHITNEY E., *Ceramic Cutting Tools Materials, Development and Performance*, Noyes Publications, Westwood, NJ, USA, 1994.

[EDW 93] EDWARDS R., *Cutting Tools*, The Institute of Materials, London, UK, 1993.

[EPM 96] European Powder Metallurgy Association, *Advances in Hard Materials Production*, (*Proceedings of the 1996 European Conference, Stockholm, Sweden, May 27-29, 1996*), Shrewsbury, UK, 1996.

[EPM 99a] European Powder Metallurgy Association, *Advances in Hard Materials Production (Proceedings of the 1999 European Conference, Torino, Italia, Nov. 8-10, 1999)*, Shrewsbury, UK, 1999.

[EPM 99b] European Powder Metallurgy Association, *International Workshops on Diamond Tool Production* (Proceedings, Torino, Italy, Nov. 8-10, 1999), Shrewsbury, UK, 1999.

[GOR 81] GORSIER F.W., *Cutting Tool Materials (Proceedings of an International Conference Ft. Mitchell, KY, USA, Sept. 1980)*, American Society for Metals (ASM), Metals Park, OH, USA, 1981.

[HOY 88] HOYLE G., *High Speed Steels,* Butterworths & Co (Publishers), London, UK, 1988.

[JUN 87] JUNEJA B.L., SEKHON G.S., *Fundamentals of Metal Cutting and Machine Tools*, Wiley Eastern Ltd, New-Delhi, India, 1987.

[KOL 92] KOLASKA H., *Pulvermetallurgie der Hartmetallo Fachverband Pulvermetallurgie*, Hagen, Germany, 1992.

[KOM 76] KOMANDURI R. (Ed.), *Advances in Hard Material Tool Technology, Proceedings of the 1976 International Conference on Hard Material Tool Technology*, Carnegie Press, Pittsburgh, PA, USA, 1976.

[LI 94] LI X.S., LOW I.M. (Ed.), "Advanced Ceramic Tools for Machining Applications, I", *Key Engineering Materials*, 96, 1994.

[LOW 96] LOW I.M., LI X.S. (Ed.), "Advanced Ceramic Tools for Machining Applications, II", *Key Engineering Materials*, 114, 1996.

[LOW 98] LOW I.M. (Ed.), "Advanced Ceramic Tools for Machining Applications, III", *Key Engineering Materials*, 138/140, 1998.

[MET 86] METZGER J.L., *Superabrasive Grinding*, Butterworths, London, UK, 1986.

[MPR 88] Metal Power Report, *Advances in Hard Materials Production* (Conference Proceedings, London, UK, Apr. 11-13, 1988), Metal Powder Report Publishing Services Ltd, Shrewsbury, UK, 1988.

[MPR 92] Metal Power Report, *Advances in Hard Materials Production* (Conference Proceedings, Bonn, Germany, May 4-6, 1992), Metal Powder Report Publishing Services Ltd, Shrewsbury, UK, 1992.

[SCH 88] SCHEDLER W., *Hartmetall für den Praktiiker,* VDI-Verlag Gmbh, Düsseldorf, Germany, 1988.

[TRE 91] TRENT E.M., *Metal Cutting*, 3rd edition, Butterworths-Heinemann Ltd., Oxford, UK, 1991.

[UPA 98] UPADHYAYA G.S., *Cemented Tungsten Carbides Production, Properties and Testing*, Noyes Publications, Westwood, NJ, USA, 1998.

[WIL 91] WILKS J., WILKS E., *Properties and Applications of Diamond*, Butterworths-Heinemann Ltd, Oxford, UK, 1991.

Chapter 10

Refractory Materials

10.1. Introduction

Refractory materials are materials which resist high temperatures. They are basically used in the "firing industries" and first and foremost in the steel industry, which by itself represents two-thirds of the applications of these materials. We also use refractory materials in foundries, in the fabrication of non-ferrous metals [HOU 99], cement, glass and ceramics and, to lesser extent, in the petrochemical and the chemical industries, in the incineration of waste and energy production.

Theses materials are of considerable economic and strategic importance and striving for better performances is a major challenge to these industries [POI 96]:

– the direct consumption cost of these refractory materials is very high. For example, they represent more than 10% of the transformation costs in steelworks;

– these materials play an important role in ensuring the reliability of fabrication units and the security of personnel. Failures in refractory linings are often the cause of major incidents: breaches or breaks leading to production stoppages, damages that are often significant and risks for the people concerned;

– they contribute strongly in manufacturing processes, particularly in metallurgy, due to their influence on the quality of the finished product, including composition and inclusionary cleanliness.

Chapter written by Jacques POIRIER.

Other than the high level of temperature and hence infusibility, which is the main characteristic of all refractory ceramics, these materials must possess a large number of additional properties in order to resist operating stresses. Insofar as their behavior is mainly governed by the phenomenon of corrosion, their basic characteristics are their chemical composition, mineralogy, microstructure and porous network. We should also consider the thermomechanical properties of materials and the stresses of the pieces and lining (thermal shocks, erosion, expansion absorption, etc.) subjected to high temperatures. All refractory materials possess a common characteristic: their in-service properties frequently depend on their organization at the mesoscopic level: the microscopic level typically ranging from the micrometer to the millimeter. This intermediate field between microscopic and macroscopic is the key parameter to understand most of the properties and mechanisms of degradation of refractory materials. A few examples are given below:

– thermomechanical behavior of these materials: non-linear with a smoother phase (decrease in mechanical strength with damage);

– unbalanced physicochemical mechanisms, which govern the impregnation of refractories by slags, under thermal gradient: transport of liquid oxides in the capillary network, liquid/refractory chemical reactions during this progression (dissolution, precipitation and separation of new solids);

– coupling mechanisms between corrosion and thermomechanics. For "basic" refractories, impregnating liquid oxides which crystallize create a solid skeleton in the porous network at the operational temperature. Sintering of the refractory grains takes place, favoring brittle behavior. For "acid" refractories, the slow migration of liquids in the impregnated refractory favors ductile behavior by activating sliding and diffusion at grain boundaries.

We will first give a summary of the refractory materials and their operational properties and then deal with the wear and degradation factors, particularly those of a thermochemical and thermomechanical nature.

10.2. Characteristics and service properties of refractory materials

10.2.1. *Definition of a refractory product*

Etymologically, "refractory" is derived from the Latin word "refractarius", meaning to break, resist, or refuse to subjugate. When applied to a material, it means "which resists high temperatures" [CAR 92]. But what exactly may be considered as high temperature for a refractory material? Pyrex glass, for example, is a material which can withstand temperatures greater than 350°C but is not, however, considered to be a refractory material. We can thus see that the temperatures which the material must "resist" (that is, retain its physical properties and its integrity) must be found above a certain threshold for it to be classified as refractory material. The answer is provided through standardization in this area:

> *"Refractory materials are materials and products other than metals and alloys (without excluding those containing a metal constituent), whose pyroscopic resistance[1] is equivalent to at least 1,500°C."*

[AFNOR standard NF B 40-001]

This definition provides two important clarifications:

– the exclusion of metals and alloys from the materials used as main constituent;

– refractories are made up in most cases of mixtures of crystallized and glassy components, their solidus and liquidus temperatures being different.

10.2.2. *Role and function of refractory products*

The basic functions of refractory materials are given below:

– to ensure the physical safety of personnel and installations between the hot material (the processed product) and the outer shell of the processing tool: let us note that the hot material may be found under any of its usual forms like:

- *liquid*: for example, steel, cast iron, aluminum, molten glass,

- *solid*: for example, loads of coal, minerals or metal ingots fed into furnaces (coke ovens, cement works, blast furnaces, reheating or annealing furnaces),

- *gas*: for example, hot-air flue, cowpers, petrochemical industry;

– to reduce heat loss.

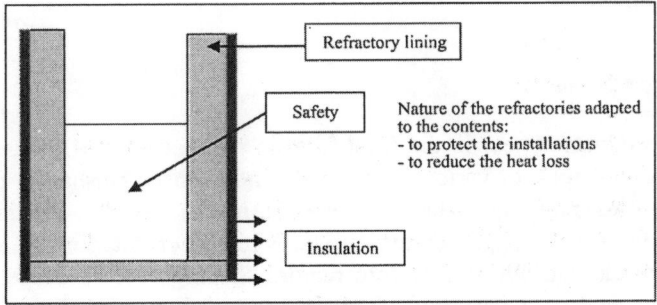

Figure 10.1. *Role and function of refractory products*

---

1 The pyroscopic resistance of a material is the temperature at which a sample of the product, subjected to a gradual increase in temperature under normal conditions, softens and sags under its own weight.

10.2.3. *Classification of refractory products*

Due to the large variety of available refractory materials, it is possible to approach its classification from different angles.

We will describe below the four principal methods of classification providing a quick presentation of a refractory product.

By its chemical and mineralogical nature

There are basically three main families of refractories:

– acid refractories of the silica-alumina system;

– basic refractories: magnesia, dolomite, chromite;

– special refractories: carbon, carbides, nitrides, spinels, zirconia, etc.

By its weight and compactness

We can distinguish:

– dense products;

– light products (hence, heat insulating materials).

By definition, light insulating products have a total porosity greater than 45% in volume. Below this value, a refractory is considered to be dense.

By its form

There are:

– shaped products;

– unshaped products;

– semi-rigid products.

Shaped products appear in the final form in which they will be used: they are bricks or special shapes of various dimensions and formats. Unshaped products are made of castables, ramming materials, cements, mortars, pastes, plastics, coatings. These generally consist of pulverulent mixtures ready (or not) for use, delivered in bags or in barrels and which therefore require shaping before usage. Semi-rigid products form part of a category of products shaped separately, which can be "deformed" during processing: it generally refers to ceramic fiber-based materials delivered in the form of plates or panels.

By its manufacturing technology

We can distinguish two main manufacturing methods:

– the fusion of raw materials, making the production of electrofused blocks or special shapes possible;

– the transformation of a pulverulent product made up of non-cohesive particles into a consolidated material. The cohesion of the refractory material takes place through sintering or chemical reaction (with the help of an organic, mineral or hydraulic binder). In practice, shaping is the result of very varied processes: dry pressing, vacuum extrusion, injection molding, isostatic pressing, casting/vibrating or gunning.

10.2.4. *Design and components of electrofused refractory products*

The essential property of an electrofused refractory is its excellent corrosion resistance due to its extreme compactness (its porosity is less than 5% and often even 3%) and high refractoriness of its constituent muddled crystals. On the other hand, it is very sensitive to thermal shocks, thus limiting its main utilization to glass furnaces.

The electrofusion technique consists of melting extremely pure raw materials in arc furnaces, followed by cooling of the liquid in moulds where it solidifies. This process is very complex, particularly in the solidification phase:

– the passage from liquid to solid takes place with considerable contraction and a high cavity forming tendency;

– too quick a cooling or temperature gradients can generate stress and provoke fractures at the edges and corners of the casting.

After drawing, the shapes show a complex form and must be ground.

Different types of electrofused refractories [TAR 98]

The first electrofused mullite materials were developed by G. Fulcher from Corning Glass Work in 1923. Nowadays, the electrofused can be sub-divided into five groups according to their chemical composition: silica-alumina, alumina, alumina-zirconia-silica (AZS), zirconia and chromium (alumina-chromium or chromium-magnesia) based refractories. Table 10.1 gives the principal characteristics of a few products [TAR 98].

Electrofused products of the alumina-silica system

Obtained by the fusion of raw materials of bauxite and clay, these refractories are made up of crystals of mullite, corundum and a glassy phase. They have limited usage.

Properties	$Al_2O_3 - SiO_2$	Al_2O_3 β	AZS 32-36% rO₂	AZS 39-41%ZrO2	$Al_2O_3 - Cr_2O_3$
Composition (%)					
SiO_2	16–20	0.1–0.2	11–17	10–13	1–3
Al_2O_3	73–79	93–94.5	48–53	45–48	56–60
Fe_2O_3	1.5–2.0	0.02 >	0.15 >	0.15 >	5–7
Cr_2O_3	–	–	–	–	26–28
ZrO_2	0–3.5	–	32–36	39–41	–
MgO	–	0–0.1	–	–	5–7
Na_2O	–	5.0–6.5	1.1–2.0	1.0–1.3	–
Mullite (%)	55–70	–	–	–	–
Alumina α	20–30	–	46–50	42–45	–
Alumina β	–	99–100	–	–	–
Baddeleyite	–	–	31–36	39–44	–
Spinel	–	–	–	–	50–55
Al_2O_3-Cr_2O_3	–	–	–	–	60–70
Glasses	10–15	0–1	17–21	15–17	0–3
Density (g/cm3)	2.95–3.10	2.90–3.10	3.40–3.55	3.6–3.7	–
Apparent porosity (%)	0.5–4	3–5	0–1	0–1	3–6
Cold compressive strength (MPa)	250–300	20–70	200–500	200–500	150 <
Thermal conductivity at 1,000°C (W/m K)	4.1–4.4	3.1–3.7	3.7–4.3	3.7–4.3	3–3.5
Thermal expansion between 25 and 1,000°C (%)	0.5–0.6	0.65–0.75	0.6–0.85	0.6–0.9	0.7–0.9

Table 10.1. *Main characteristics of electrofused refractories*

Electrofused alumina β

Consisting totally of alumina β crystals, this type of refractory has an excellent resistance to corrosion by alkaline vapors. It is used in superstructures of glass furnaces.

Electrofused products of the alumina-zirconia-silica (AZS) system

AZS products are made up of baddeleyite (ZrO_2) and corundum (Al_2O_3) crystals associated with a glassy phase indispensable to restrict internal stresses generated by the polymorphism of ZrO_2. The zirconium oxide imparts to the product an excellent resistance to corrosion by molten glass. We can distinguish two types of electrofused AZS: products whose ZrO_2 content is found between 32% and 35% and those containing 41% of ZrO_2.

Electrofused Al_2O_3-Cr_2O_3

Since the aggressiveness of the "insulating" glass is higher than that of the sodio-calcic glass, alumina chromium refractories deviating from the AZS formula have been developed. These have a very compact structure made up of spinel crystals and Al_2O_3-Cr_2O_3 solid solutions.

10.2.5. Design and components of "cohesive particles" refractory products

These refractory products are heterogenous materials whose composition broadly consists of one or more minerals making up the refractory medium as such, with or without additives which provide or reinforce specific properties, whose cohesion is ensured, before or during the installation of refractory linings, by an appropriate binder. Added to these different constituents is an additional phase constituted by porosity (open or closed) of the product which greatly contributes to fixing the properties of the material. This design definition can be considered as "universal" with the exception of electrofused refractories (see section 10.2.4).

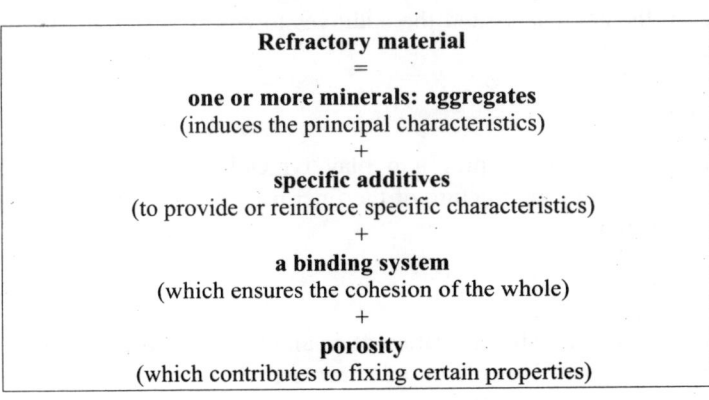

Refractory material
=

one or more minerals: aggregates
(induces the principal characteristics)
+

specific additives
(to provide or reinforce specific characteristics)
+

a binding system
(which ensures the cohesion of the whole)
+

porosity
(which contributes to fixing certain properties)

Figure 10.2. *Designing of a refractory product*

Each manufacturer uses his technical know-how to get the desired result in assembling these constituents.

10.2.5.1. *Aggregates*

These are in general made up of one or more minerals, basically oxides in an amorphous or crystalline form, which imparts the main quality (refractoriness and complementary properties) to the finished product. These minerals constitute between 70 and 100% of the product composition by very often occupying the whole granulometric distribution and always the coarser part. We can distinguish between different types of raw materials based on their origin and method of production:

– natural raw materials: these are almost always impure and must therefore be processed most often;

– synthetic raw materials: these exhibit very high purity and are processed either by heat treatment from an already transformed natural raw material, or by physicochemical reactions from synthetic products. Table 10.2 shows the main raw materials used for the production of refractory products.

10.2.5.2. *Different binding systems*

Binding elements are the essential means by which we can permanently ensure the cohesion of the set of constituents of refractory products, "permanently" because this cohesion must be effective from the ambient temperature to the service temperature and during the whole life of the material. A part of the material added as binder does not always possess refractory qualities: in general these are added in small quantities and always belong to the finer fraction of grading of the finished products. This is obviously logical, since it refers to that part of the material which gets coated on the aggregates and the additives to ensure material cohesion. There are four binding systems.

Ceramic bond

This is a bond which comes into play typically at high temperatures. The physical and mineralogical structure of the finer part of the materials is assigned by ceramization reactions.

Hydraulic bond

The bond is ensured by the hydration of an aluminous refractory cement added to the product. This cement is made up of a mixture of calcium aluminates which are hydrated in the presence of water at low temperatures, by provoking hardening of the material. It therefore refers to a cold binding method. Irrespective of its content: high (around 30%), low (around 2% to 10%) or ultra low (less than 2%), this cement is considered to be the typical binder for refractory castables.

Group	Raw materials	Chemical classification		Type
Silica-alumina system		Al_2O_3	SiO_2	
Alumina	Tabular alumina (sintered)	99		Synth
	Calcinated alumina	99		Synth
	White electrofused alumina	99		Synth
	Brown electrofused alumina	94–96		Synth
	• Refractory bauxite	82–89		Nat
	Sintered and electrofused mullite	72–78		Synth
	Kyanité	55–59		Nat`
	Andalusite	52–59		Nat
Silica-Alumina	Clay chamottes	35–60		Nat
	Refractory clays	30–45		Nat
	Bentonite-pyrophyllites	20–35		Nat
	Perlite-vermiculite	9–12	40–75	Nat
Silica	Quartzites		99	Nat
	Vitreous silica		99	Nat
	Volatile silica		90–95	Synth

Group	Raw materials	MgO	CaO	SiO₂	Cr₂O₃	Al₂O₃	Type
Basic		MgO	CaO	SiO_2	Cr_2O_3	Al_2O_3	
Magnesia	Electrofused magnesia	93–99					Synth
	Natural magnesia	70–97	2–20	inf 20			Synth
Forsterite	Olivine	39–44		41–50			Nat
Dolomite	Sintered dolomite	38	59				Synth
Spinel	Electrofused spinel	28–31				67–72	Synth
	Sintered spinel	10–28				67–90	Synth
Chrome	Chromite	16–20			32–45	15–28	Nat
	Co-clinker magnesia/chromite	55–60			15–20		Synth
	Chrome oxide green				99		Synth

Group	Raw materials	C	SiC	ZrO₂	CaO	Al₂O₃	Type
Others		C	SiC	ZrO_2	CaO	Al_2O_3	
Carbon	Graphite	77–98					Nat
	Carbon black	99					Synth
Carbide	Silicon carbide		87–99				Synth
Zirconium	Zircon sand			64–67			Nat
	Dense zirconia			80–95			Synth
	Electrofused zirconia/mullite			35–37		46–52	Synth

Synth = synthetic; Nat = natural

Table 10.2. *List of main aggregates*

Mineral chemical bonds

There are various mineral chemical binders used in refractory products. These react in a complex manner i.e. by formation of gels which modify their specific structures at low temperatures and sometimes through direct chemical reaction with the finer part of the aggregate. In the latter case, these ensure hardening around a mean temperature, sometimes even after the service temperature. Generally rather low contents (less than 5%) are added, on account of the not very refractory nature of these binders (most contain alkalis and are fusible at low temperatures).

Organic bonds

These refer to "low or average temperature" bonds. We will mainly consider carbonaceous binders: coal pitch, tar and thermoset resins. These bonds leave behind, during the rise in temperature, very fine residual carbon in the porosity of the finished products, thereby improving the texture and corrosion resistance. Their hardening method only involves the temperature, which acts through polymerization (resin) or polycondensation and cross-linking in the case of pitch or tar.

Nature of bond	Binder	For	
		SP	UP
Ceramic	Clay	x	x
	Pulverulent oxides	x	x
	Mullite, Si_3N_4, SiAlON, AlN	x	x
Chemical inorganic	Sodium silicate	–	x
	Silica gel	x	–
	Ethyle silicate	x	–
	Phosphoric acid	x	–
	Sodium, aluminum phosphates	x	x
		(x)	x
	Boric acid	x	x
	Magnesia salts	–	–
Chemical organic	Cellulose derivatives	x	x
	Tar, pitch	x	x
Hydraulic	Aluminous cements	–	x

(SP and UP: shaped/unshaped products)

Table 10.3. *Binders used in refractories*

10.2.5.3. *Additives*

Additives are added in the composition in order to impart or reinforce specific characteristics of the finished products [TAF 94]. We will find in Table 10.4 a few non-exhaustive examples of properties provided by additives.

Additives	Associated properties
Chromite, sintered and electrofused mullite, spinel	Increase in the refractoriness
Aluminum (at mean temperatures) Silicon metal (at high temperatures) combined with carbon	Increase in hot mechanical strength
Silicon carbide Graphite chromium oxide Zirconia Spinel Carbon black, pitch, resin	Increase in the corrosion resistance
Silicon carbide Electrofused alumina	Increase in the resistance to abrasion and erosion
Mullite Graphite Vitreous silica	Increase in the resistance to thermal shocks
Silicon carbide Graphite	Increase on thermal conductivity
Pyrolisable products (creation of porosity through different additions)	Decrease in thermal conductivity
Impregnation by pitch and tar	Increase in compactness and decrease in porosity
Clay (shrinkage) Kyanite, quartz sand (expansion)	Permanent change
Clay, bentonite, various organic products (plasticizers) Miscellaneous organic products (defloculants, dispersants)	Rheological modifications

Table 10.4. *Examples of additives and their associated properties*

10.2.5.4. Characteristics of refractory products composed of "cohesive particles" -

There is a great variety of refractory materials. Without being exhaustive, we will consider those refractories belonging to the silica-alumina family, basic products, carbons, carbides, without forgetting ultra-refractory materials like thorium oxide for special purposes. The combination of aggregates/binders/additives aims at adapting the characteristics of the material to the service conditions [ALI 79].

Thus, a refractory product will not only resist "high temperatures", but also other aggressions like:

– chemical corrosion (by liquid oxides, fused metals, gases);

– hot mechanical erosion;

– thermal shocks;

– mechanical stresses.

Today, it is not rare to come across products containing more than a dozen constituents at the time of production. This shows the extent of improvement in the know-how and technical skill of manufacturers, in order to better satisfy the specific needs of the users and to increase the service life and the safety of the installations. Examples of types of refractory materials used in different industrial sectors like cement works, electrometallurgy, pyrometallurgy, steel industry, chemical industry will be given later.

Silica-alumina refractories

This refers to an extended family made up of:

– silica refractories ($SiO_2 > 93\%$) whose basic characteristics are an exceptionally high resistance to thermal shocks for temperatures greater than 800°C and a low creep. These are used as roofs for glass furnaces, cowpers cupola, coke ovens;

– clay refractories ($20\% < Al_2O_3 < 45\%$). These are characterized by low expansion and thermal conductivity. The main applications concern the construction of furnaces: anode furnaces, coke ovens, blast furnaces, cowpers, safety lining of steel ladles, cement works, glass furnace insulation;

– refractories with high alumina content ($Al_2O_3 > 45\%$). The corrosion resistance and the refractoriness increase with the alumina content. Their applications include blast furnaces (mullite, corundum), cowpers (andalusite), cement works (bauxite), chemical industry (mullite, corundum), incineration (corundum), etc.

Product type	Clay	HTA Group 1	Alumina/Carbon
Binding	Ceramic	Ceramic	Carbon
Processing	Firing	Firing	Polymerization
Raw materials			
Aggregates and additives	Chamotte	Bauxite	Electrofused alumina Graphite Silicon
Binder	Clay	Clay	Resin
Chemical analysis %			
Al_2O_3	42	79.5	85
SiO_2	54	14	0.9
Fe_2O_3	1.5	1.6	0.7
TiO_2	1.4	2.9	0.8
Si	–	–	2
C	–	–	9
Others	1.1	2	1.6
Mineralogical phases	Mullite Cristoballite Corundum Glassy phase	Corundum Mullite Glassy phase	Corundum Graphite Silicon
Physical and thermal properties			
Bulk density (g/cm^3)	2.25	2.75	3.09
Open porosity (%)	16.5	20	10.5
Compressive strength (MPa)	65	90	59
Refractoriness under load (°C) *(under 0.2 MPa)*	1,450	1,460	–
Permanent linear change at 1,450°C/5h (%)	+ 0.1	+ 3	0
Thermal expansion at 1,000°C (%)	0.65	0.71	0.7
Thermal conductivity at 1,000°C (W/ m.K)	1.4	2.2	4
Specific heat between 20–1,000°C (kJ/kg K)	1.06	1.10	1.11

Table 10.5. *Characteristics of a few SiO_2-Al_2O_3 refractory materials*

Table 10.5 gives the main characteristics of a few SiO_2-Al_2O_3 refractory products.

Basic refractories

These are characterized by a high refractoriness and a very high corrosion resistance to liquid oxides (slag). They are used in converters, electric furnaces (magnesia), steel ladles (magnesia, dolomite), fusion of non-ferrous metals Cu and Pb (magnesite-chrome), rotary furnaces in cement works (magnesite-chrome, spinel magnesia, dolomite). By way of illustration, Figure 10.3 presents the main characteristics of two basic refractory ceramics: a magnesia carbon used in converters and a magnesite-chrome refractory used in steel degassers under vacuum (RH/OB).

Carbon refractories

Carbon refractories are used mainly in the aluminum industry to line alumina electrolytic tanks, as blast furnace hearths in the steel industry and in the ferro-alloy and pure silicon industry as furnace lining and electrodes [DUM 84]. These materials are used on account of the following unique and exceptional properties of carbon:

− it does not melt. It goes directly from the solid state to the gaseous state by sublimation at a temperature of 3,500°C;

− it is does not get wet by most of the molten metals and by slag;

− it has a low coefficient of expansion (clearly lower than traditional refractories) and possesses an excellent resistance to thermal shocks;

− it exhibits a significant thermal conductivity range;

− it has an excellent chemical resistance to most acids and bases.

However, carbon materials are oxidizable starting from 450°C. This is their greatest limitation as far as using them as refractories is concerned.

The raw materials used consists of aggregates (anthracite, natural and artificial graphite, coke) binders (pitch, tar or resin), whose role is to agglomerate the solid particles and if necessary the additives (silicon, silicon carbide, fine alumina). The characteristics of a few industrial refractory materials are given in Table 10.6.

Chromium magnesia refractory

– Main constituents: magnesia, chromite
co-clinker electrofused chromite
magnesia

– Ceramic bond: "direct bonded"

– Chemical composition (as a %)

MgO	Cr_2O_3	Fe_2O_3	Al_2O_3	CaO	SiO_2
61	18	13	6	1.5	0.5

– Physical properties
Bulk density: 3.35 g/cm^3
Open porosity: 16%
Cold compressive strength: 70 MPa

– Shaping: pressing and firing at 1,700°C

⊢———⊣ 400 μm

Carbon magnesia refractory

– Main constituents: magnesia, graphite

– Organic bond: phenolic resin

– Chemical composition (as a %)

MgO	C	CaO	SiO_2	Al
85	10	1.5	0.7	2.8

– Physical properties
Bulk density*: 3.02 g/cm^3
Open porosity*: 8.5%
Cold compressive strength*:
30 MPa

– Shaping: pressing followed by tempering
at 150°C

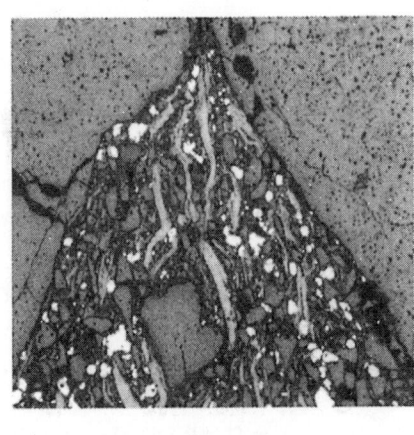

⊢———⊣ 400 μm

Figure 10.3. *Characteristics and microstructures of two ceramic refractories*

Characteristics	C.1	C.2	C.3	C.4
Bulk density (g/cm^3)	1.59	1.71	1.61	1.65
Open porosity (%)	14	16	15.6	20
Ash content (%)	5.4	23	3.8	1.6
Cold compressive strength (MPa)	32	65	28	31
Thermal expansion between 20 and 600°C (in 10^{-6}/K)	3.2	3.5	3.3	3.5
Thermal conductivity at 20°C (W/m K)	6.5	20	10.2	27

C.1: anthracite-based for lining of furnaces and blast furnaces;

C.2: graphite and anthracite-based with silicon, silicon carbide and fine alumina additives for hearths of blast furnaces;

C.3: electrodes for silicon or phosphorous furnaces, it refers to a quality for which the raw materials are anthracite and graphite;

C.4: for this product, the raw material is only graphite; it is used for the cathode lining of tanks for aluminum manufacturing.

Table 10.6. *Characteristics of industrial carbon refractories*

10.2.6. *Manufacturing principles*

Manufacturing processes of refractory products are very varied: ranging from a simple mixture of powders starting from aggregates to elaborate processes which could include reactive firing in a controlled atmosphere:

– under reducing atmosphere refers for example to alumina-carbon products which instead develop a bond of type SiC by a high temperature reaction between silicon and carbon;

– under nitrogen atmosphere refers to forming nitrides: Si_3N_4, AlN or SiAlON.

The flow chart [LAP 86] shown in Figure 10.4 illustrates the broad outline of operations in the manufacturing of electrofused refractory products, shaped and fired (with ceramic bond, chemical bond or with carbon bond).

Figure 10.4. *Refractory product fabrication flow chart*

10.2.7. *Service properties of refractory products*

The evaluation of the usability of a refractory product in a defined industrial context is a difficult technical procedure. The characteristics of the material must be known and its effect on the in-service behavior must be evaluated. Acquiring knowledge about the service properties of the material starts with its raw materials: in fact, a refractory retains through the very nature of its manufacturing process, the memory of its raw materials. Insofar as the behavior of the refractory is mainly governed by phenomena of corrosion and capillary impregnation, the chemical composition, open porosity (volume and distribution into pore sizes) and microstructure are essential parameters. The arrangement of the different constituents needs to be taken into account:

– distribution of mineralogical phases;

– nature and distribution of impurities;

– particle size distribution;

– size of crystals in the sintered aggregates;

– distribution of internal faults, microfissures and porosity;

– local variations in the chemical composition;

– development of bonds;

– location of certain additions.

We shall remember that the raw materials are considered to be opponents of corrosion, porosity is its progress, the secondary phases are its brakes or accelerators depending on their arrangement, nature and distribution. These characteristics have to be supplemented by thermal and thermomechanical properties, which help to evaluate the resistance of materials to thermomechanical stresses. We will mention as examples:

– expansion coefficient;

– thermal conductivity and specific heat;

– elastic modulus;

– bending strength, tensile strength, crushing strength;

– creep and, more generally, laws of mechanical behavior which integrate the phenomena of viscoplasticity and damage.

These properties, which vary with temperature, imply a sustained activity in the field of mechanical testing, taking into account more and more complex stresses and the integration of microstructures.

10.3. Wear and degradation factors

Refractory wear has two main sources: corrosion and thermomechanical degradation. In practice, in industrial installations, these two types of degradation are very often associated. For example, the impregnation of slag in the porosity of a refractory leads to chemical corrosion which deeply transforms the nature and the arrangement of the phases and is followed by a modification of the thermomechanical properties of the material. Users then notice that the impregnated layer gets frequently removed by a fracture known as "structural scaling" which limits the life of the lining. Corrosion and thermomechanical degradation may be accelerated or retarded, depending on various parameters regarding the material, the lining and the service conditions, the importance of which must be evaluated. Figure 10.5 shows the different parameters influencing refractory wear.

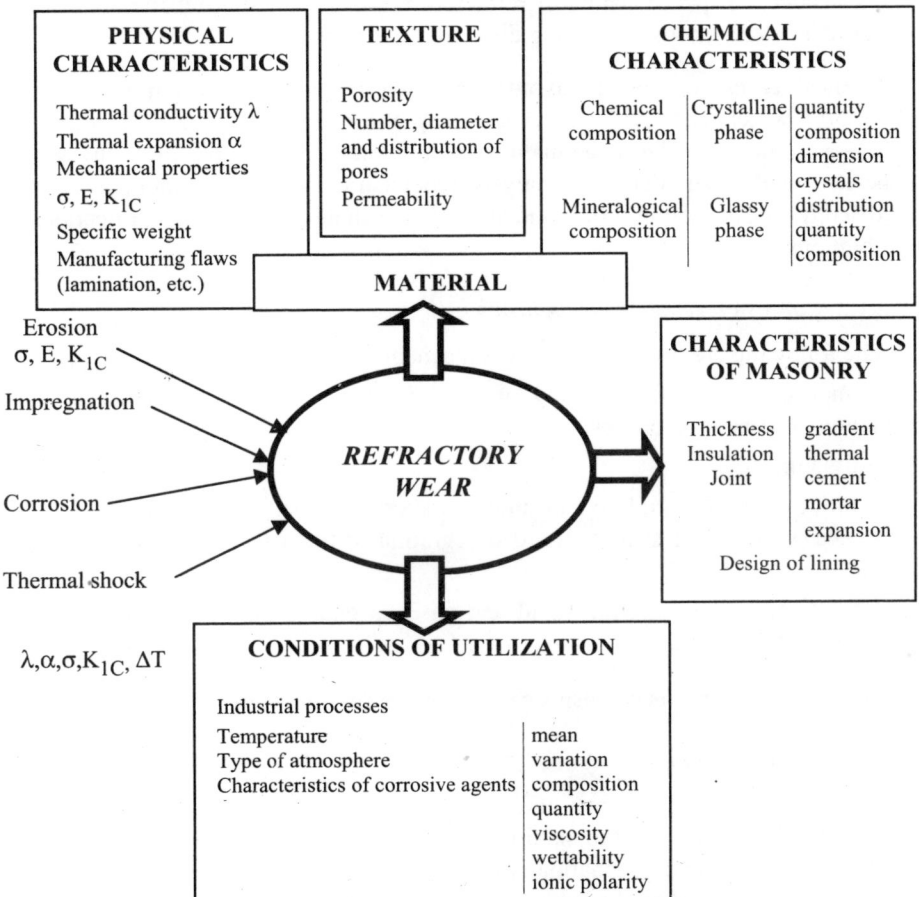

Figure 10.5. *Different parameters influencing refractory wear*

10.3.1. *Thermochemical behavior of refractories*

10.3.1.1. *Corrosion tools*

In order to explain the phenomena related to corrosion we need to [PEC 96]:

– resort to the basic notions of thermodynamics [POI 91]. We cannot escape the laws of chemistry and particularly the laws of thermodynamics. We have to use them in order to avoid very common errors. It refers to the prediction of the state of equilibrium between the aggressive agents (gas/metal/slag/solid loads) and the refractory material;

– integrate the kinetic factors associated with the transitional nature of the reactions which come into play while considering the microstructure of the material and viscosity parameters, surface tension and wettability of slags;

– consider the special nature of refractory products which are materials that are porous and heterogenous combining different constituents.

In most cases, we cannot avoid simulation methods (corrosion tests in the laboratory) which can guide us in choosing the refractory. However, industrial reality is complex. The examination of refractories used on different scales associated with techniques of analysis (chemistry, x-ray diffraction, electronic microscopy) is essential in order to identify the mechanisms and to give directions to the solutions.

10.3.1.2. *Corrosion reactive mechanisms*

The following are the main corrosion mechanisms:

– impregnation in the porosity of the refractory by liquid oxides (slag) and formation of weathered phases;

– dissolution;

– infiltration of liquid oxides in the material "matrix" made up of a bond associated with the finest particles, disintegration of the affected area and carrying over of aggregates;

– reactions between solid, liquid and gaseous phases like oxidation-reduction, oxidation and dissociation.

In practice, several mechanisms can take place simultaneously.

Capillary impregnation

The refractories exhibit open porosity which allows impregnation by slags. This impregnation depends on the nature of the materials. It occurs for carbonless refractories, however, impregnation is very limited and even non-existent for carbon refractories (since carbon has an anti-wetting effect). For a given porosity and porosimetry, the amplitude of the phenomenon depends on the "capillary motor force", which is a function of the slag surface tension and its connecting angle on the refractory, but also of the viscosity. Penetration, by increasing the slag-refractory interfacial area, facilitates the evolution of slag composition, and thus of viscosity and wetting parameters [ONI 80]. Figure 10.6 shows the microstructure of a refractory impregnated by a slag.

Impregnation phenomena [BLU 95] differ following the basicity of the slag-refractory system, depending on the extent of polymerization and viscosity of liquid oxides (see Figure 10.7):

– in basic refractories, highly basic liquids, lowly polymerized and fluid, migrate rapidly at speeds of the order of 1 cm/hr. They also progress rapidly towards compositions in overall chemical equilibrium with the refractory due to a high speed of diffusion;

– in aluminous and silico-aluminous refractories, liquids with low basicity are, however, strongly polymerized and viscous. They migrate slowly, at a speed less than 1 mm/hr and react preferentially with the fine grains of the refractory binder.

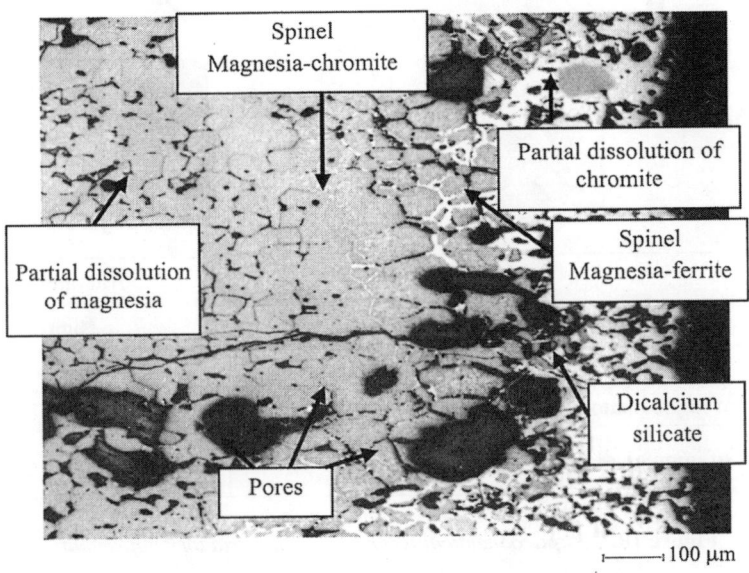

Figure 10.6. *Capillary impregnation of a magnesia chromite refractory by a calcic slag*

The consequences on the assemblies of phases present in the impregnated layers and on their physical state at the service temperature are clear:

– in basic refractories, the liquids migrate towards the back that is the coldest of the refractory wall until they reach their solidification isotherm. On the way, they precipitate on the porous network of solids thermodynamically compatible with the phase assembly of the new refractory;

– in acidic and aluminous refractories, liquids do not reach their end of crystallization isotherms. Impregnation does not exceed the depth where liquid migration is not faster than surface dissolution. In these conditions, a liquid fraction persists in the whole impregnated layer, which gets vitrified while cooling.

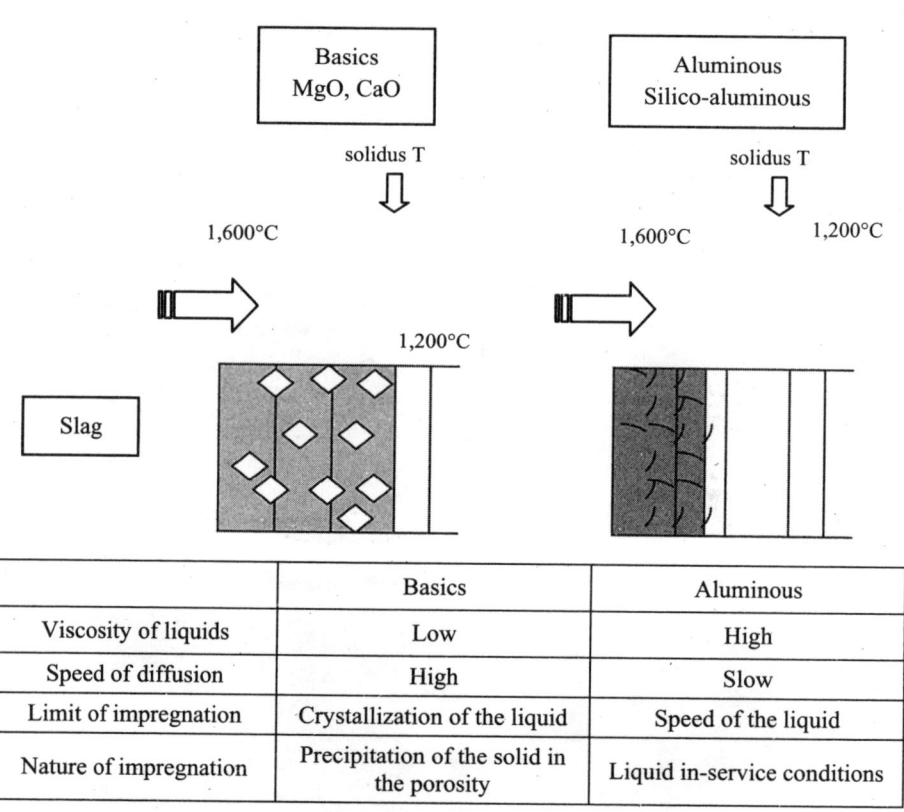

	Basics	Aluminous
Viscosity of liquids	Low	High
Speed of diffusion	High	Slow
Limit of impregnation	Crystallization of the liquid	Speed of the liquid
Nature of impregnation	Precipitation of the solid in the porosity	Liquid in-service conditions

Figure 10.7. *Physicochemical mechanisms of refractory impregnation*

Dissolution

Since industrial slags are practically never saturated by refractory constituents, dissolution takes place and is controlled basically by diffusion and the difference between the chemical potential of oxides in the refractory and slag (see Figures 10.8 and 10.9).

However, the enrichment of refractory constituents resulting from this dissolution could lead to precipitation, at the refractory-slag interface, of new phases less soluble than the initial oxides, and these could slow down any subsequent dissolution. Thus, the process of corrosion by dissolution is dependent on precipitation which can be interpreted only based on the knowledge of phase diagrams with more than three constituents.

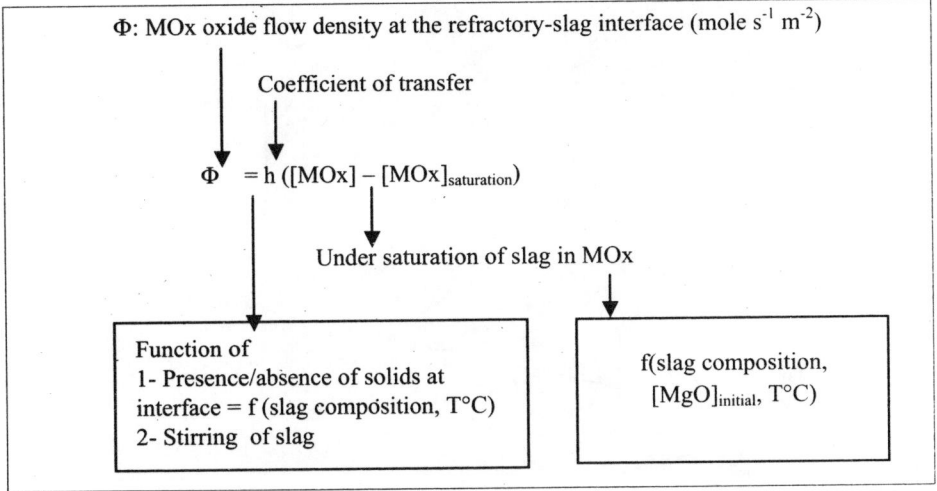

Figure 10.8. *Dissolution kinetics and mechanism of a refractory oxide in a slag*

Infiltration of liquid oxides in the material matrix

This corrosion mechanism, which concerns carbon refractories, is similar to erosion due to disintegration and carry-over in the slag of solid oxide grains without notable dissolution, accompanied by carbon particles. It is induced by the infiltration of the slag in the matrix. The corrosion resistance will essentially depend on the constitution and the structure of the fine material matrix: strength and size of graphite, additions and ultra-fine particles.

Reactions between solid, liquid and gaseous phases

These often heterogenous reactions take place between different types of phases: solid-liquid-gas. There is a whole list of possible reactions: dissociation, volatilization, reduction, notably by carbon, oxidation-reduction, formation of new compounds.

As an illustration, Table 10.7 shows the numerous reduction reactions which take place in carbon magnesia refractories at high temperatures. These reactions bring into play carbon (originating from the bond and the graphite), periclase grains, lime silicates which constitute the bonds between the magnesia crystals and the metal additions (which act as carbon antioxidants).

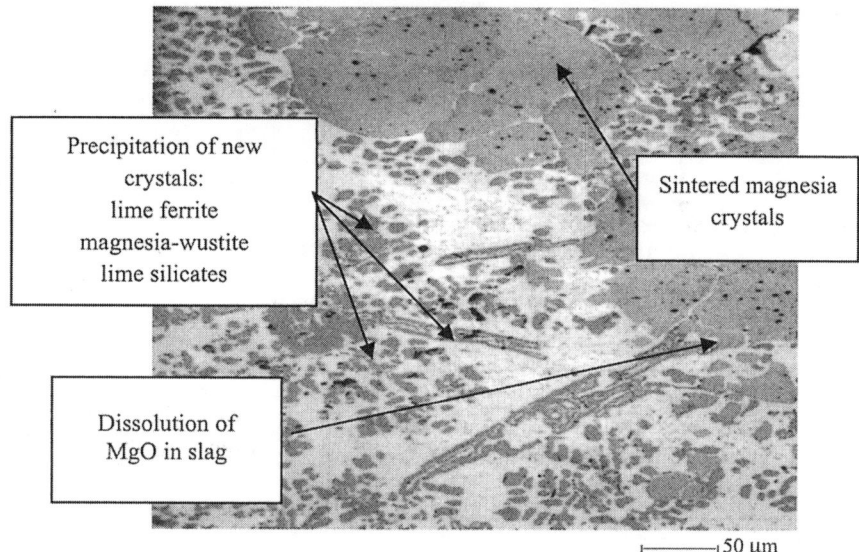

Precipitation of new crystals:
lime ferrite
magnesia-wustite
lime silicates

Sintered magnesia crystals

Dissolution of MgO in slag

├───────┤50 µm

Figure 10.9. *Dissolution of magnesia by a converter slag*

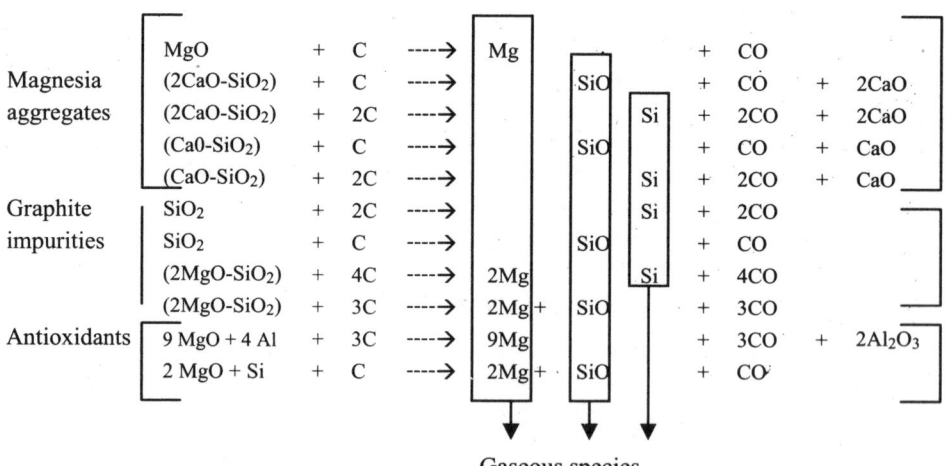

	Reactant			Mg	SiO	Si		Products		
Magnesia aggregates	MgO	+	C	----→	Mg			+	CO	
	(2CaO-SiO$_2$)	+	C	----→		SiO		+	CO	+ 2CaO
	(2CaO-SiO$_2$)	+	2C	----→			Si	+	2CO	+ 2CaO
	(Ca0-SiO$_2$)	+	C	----→		SiO		+	CO	+ CaO
	(CaO-SiO$_2$)	+	2C	----→			Si	+	2CO	+ CaO
Graphite impurities	SiO$_2$	+	2C	----→			Si	+	2CO	
	SiO$_2$	+	C	----→		SiO		+	CO	
	(2MgO-SiO$_2$)	+	4C	----→	2Mg		Si	+	4CO	
	(2MgO-SiO$_2$)	+	3C	----→	2Mg+	SiO		+	3CO	
Antioxidants	9 MgO + 4 Al	+	3C	----→	9Mg			+	3CO	+ 2Al$_2$O$_3$
	2 MgO + Si	+	C	----→	2Mg+	SiO		+	CO	

Gaseous species

Table 10.7. *Reduction reactions inside MgO-C bricks*

10.3.2. *Thermomechanical behavior of refractories*

10.3.2.1. *Thermomechanical tools*

Thermomechanics consists of:

– studying the status of stresses and their consequences on the integrity of linings and refractory parts, in order to predict in-service behavior;

– designing technical solutions to ensure control of mechanical phenomena.

The quantification of thermomechanical stresses therefore plays a central role. This is done through the tool of modeling by finite elements. Modeling has developed considerably, due to the arrival of powerful computers on the market accessible to laboratories and to the wide distribution of suitable software.

Observation, a second tool, more traditional in nature, retains its importance. It is in fact observation associated with measurements, if necessary, which helps to identify thermomechanical degradations and to guide further action.

10.3.2.2. *Thermomechanical degradation*

Thermomechanical degradation of refractories can be largely classified into three categories [THE 94].

Damage

Damage takes place on the microstructural scale of the material, at the level of the aggregates and the bond. It brings about a loss in rigidity and by a fragilization of the microstructure. The resulting consequences will be a high sensitivity of the material to erosion or mechanical shock type of stresses. This type of degradation cannot be seen by the naked eye, but can be detected through experimental non-destructive techniques (measurement of the speed of ultrasound propagation) or destructive ones (curve of force and deformation until fracture).

Spalling

This refers to a surface degradation of the refractories which affects the thickness from a few millimeters to a few centimeters. Scaling results in a loss of material caused by the propagation of a crack parallel to the brick surface. The volume of the material lost is significant, when compared to the size of the microstructure, but remains unimportant with respect to the volume of the refractory lining. Spalling is most often observed during the establishment of thermal regimes with strong temperature variations at the hot face of the bricks. By using simple fracture criteria, modeling helps predict spalling in many cases.

Fracture

Fracture results from the propagation of an in-depth fissure in a part of the lining or in a brick. In practice, insofar as the shaped pieces dissociate from each other, fracture translates in the loss of material, of thickness ranging from several to around 10 cm. It therefore refers to a catastrophic phenomenon.

These different degradations originate from three factors: the stresses imposed by the processes and the service conditions, the laws of material behavior and the design of the masonry. They affect the refractory material on different scales: at the material level, piece or brick level and lining level. Figure 10.10 illustrates this scale effect.

10.3.2.3. *Stresses imposed by service conditions*

Thermomechanical stresses imposed by service conditions are of two types:

– those due to hot mechanical stresses: impact of solid loads, the weight of tools, erosion through moving loads, and tool movement provoking new load distribution;

– those due to thermal conditions during the operating cycle of the tool. These can be quantified by temperatures, heating speeds (more violent or less violent thermal shocks), convection and radiation thermal flux. They constitute the thermal boundary conditions imposed on refractory linings at the interfaces with the hot material and the ambient air.

10.3.2.4. *Material behavioral laws and properties*

Temperature distributions within the material are governed by the following characteristics: density p, specific heat c, thermal conductivity λ and thermal diffusivity $a = \lambda/p\ c$. All these characteristics depend on the chemical composition of the refractories. Conductivity and diffusibility also depend on mineralogy and morphology of phases. The stress fields resulting from thermal distribution and mechanical boundary conditions are governed by the thermal expansion and the behavioral laws. Thermal expansion depends on the chemical and crystallographic nature of the material.

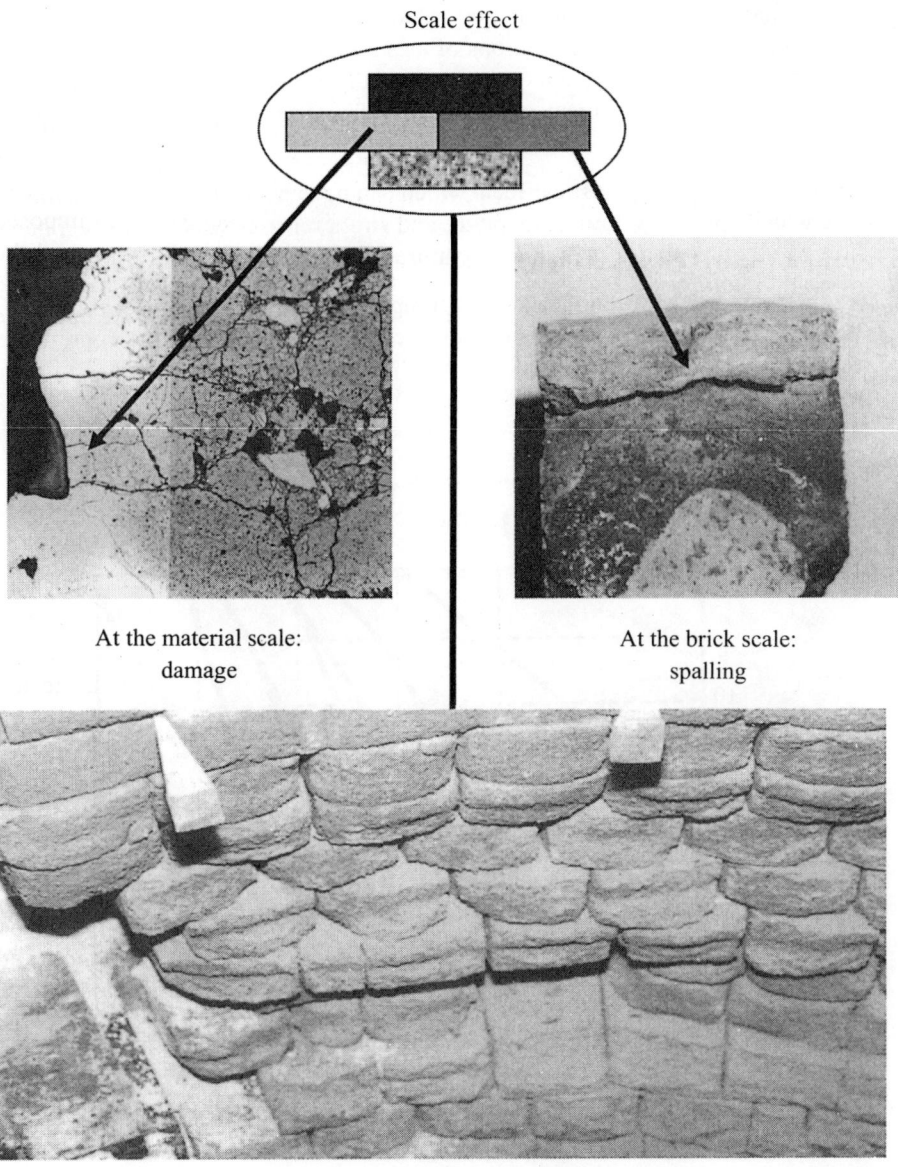

At the material scale: damage

At the brick scale: spalling

At the masonry scale: fracture of bricks

Figure 10.10. *Example of thermomechanical degradation.*
Magnesite-chrome brick lining of a metallurgical reactor in vacuum

The behavioral laws express the relationship between the stress σ and deformation ε, as well as the evolutions of ε as a function of time. These behavioral laws strongly depend on the microstructure of the material and its manufacturing process. The thermomechanical behavior of refractories is characterized by two major effects:

– non-linear and viscoplastic effects which can be explained by two mechanisms: damage which appears at low temperature and stress releases which are often due to the flow of vitreous phases at high temperatures [THE 93];

– a dissymmetry between tension and compression. In fact, there is a zone where the behavior of refractories is tension-compression symmetrical. This zone, which corresponds to the elastic field, is of very small amplitude [ROB 98].

As an example, Figure 10.11 shows a stress-deformation experimental curve of a magnesia carbon refractory.

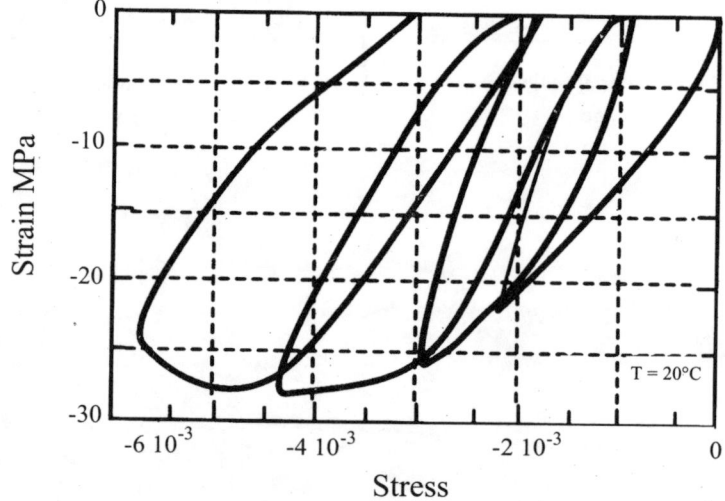

Figure 10.11. *Stress-deformation experimental curve of a magnesia carbon refractory (at ambient temperature)*

We understand that the traditional mechanical models (law of linear elasticity described by an elastic modulus whose value depends on the temperature associated with a creep law) do not allow the satisfactory description of the behavioral laws of such materials by taking into account the damage and viscoplastic effects. To take these effects into account, it is necessary to use behavior models using thermodynamic concepts of irreversible processes and consisting of laws of state, load surface and laws of evolution.

10.3.2.5. *Masonry design*

Refractory linings are broadly classified into two families:

– cylindrical masonry: these have a revolutionary axis and are subjected to a radial thermal flux;

– wall type masonries: these are flat and subjected to a thermal flux perpendicular to their surface.

Most installations consist of both types of masonry: cylindrical for the wall and wall type for the bottom.

Designing a masonry involves defining the format and the geometric arrangement of bricks or special shapes, the joint dimensions and the nature of the jointing material, with the aim of controlling the thermomechanical behavior of the assembly. The principal mechanism which governs the thermomechanical behavior of a refractory masonry is the absorption of expansion. This refers to the limitation of movements due to interactions between the different elements of the masonry [SCH 95]. These interactions take place through the joints and the interfaces. Masonries react to transitional thermal regimes in two stages:

– firstly, each brick or element gets deformed in an independent manner without interaction with its neighbors. For bricks, and more generally for parallelepipedic parts, this is reflected in the form of stresses parallel to the thermal flux;

– secondly, the interactions become significant and we can observe assembly specific behavior like fracture, peeling, buckling followed by compression in the perpendicular plane in the direction of thermal flux.

The thermomechanical behavior also shows the characteristic of being non-linear: there is no proportionality between stress and strain at any point. This results from joints and interfaces. Most of the thermomechanical problems concern excessively blocked masonries; it is therefore desirable to evolve towards less constrictive and thus more stable masonries. The design of such masonries must reconcile the reduction of thermal stresses and compressions induced by blocking with requirements for maintaining the lining during tool handling and limitations of joints to which are associated risks of infiltration and corrosion.

10.4. Conclusion

Refractories are part of an extended family of ceramics. A certain number of key words characterizing in some way their complexity can be associated with these materials. These key words are: heterogeneity, diversity of aggregate/binder/additive assemblies, capillary network, corrosion through unbalanced physicochemical processes, non-linearity of the thermomechanical behavior and scale effect.

Appreciating the aptitude of a refractory product in an industrial context requires a pluridisciplinary technical approach which requires basic knowledge in material and process engineering, thermics, thermomechanics and physicochemistry of high temperatures. The improvement in the life of refractories goes through the following progress paths:

– criteria of material selection based on better knowledge of stresses;

– constant improvement in service properties;

– mastery of thermomechanical stresses in linings;

– careful processing of refractory linings installations.

10.5. Bibliography

[ALI 79] ALIPRANDI G., *Matériaux Réfractaires et Céramiques Techniques*, Editions Septima, Paris, 1979.

[BLU 95] BLUMENFELD P., Comportement des réfractaires sans carbone à l'acierie: imprégnation par les slags et effet sur les propriétés mécaniques, Rapport RI 95015, IRSID 1995.

[CAR 92] CARNIGLIA S.C., BARNA G.L., *Handbook of Industrial Refractories Technology: Principles, Types, Properties and Applications*, Noyes Publications, 1992.

[DUM 84] DUMAS D., PARISOT C., BUSCAILHON A., Carbones et graphites, A 7 400, Techniques de l'ingénieur, 1984.

[HOU 99] HOUSSA C.E., "A review of the non-ferrous refractories market", *Industrial Minerals*, July 1999.

[LAP 86] LAPOUJADE P., LE MAT Y., DEVIDTS P. et al., *Traité pratique sur l'utilisation des produits réfractaires,* Editions H. Vial, 1986.

[ONI 80] ONILLON N., PERRIN J., "Hydrodynamique et physico-chimie de l'attaque des réfractaires industriels par les slags sidérurgiques", *Bull. Soc. Fr. Céram.*, 128, 1980.

[PEC 96] PECORARO A., MARRA C., WINZEL T., "Corrosion of Materials by Molten Glass", *Ceramic Transaction*, vol. 78, 1996.

[POI 91] POIRIER J., PROVOST G., RIGAUD M., "Applications pratiques des données thermochimiques au comportement des produits réfractaires dans quelques outils sidérurgiques", *La Revue de Métallurgie – CIT*, February 1991.

[POI 96] POIRIER J., "Recent tendencies in refractories in relation with service conditions in the steel industry", *Proceedings of the 30th Colloquium on Refractories*, Aix La Chapelle, 1996.

[ROB 98] ROBIN J.M., BERTHAUD Y., SCHMITT N., POIRIER J., THEMINES D., "Thermomechanical behaviour of magnesia-carbon refractories", *British Ceramic Transactions*, vol. 97, no. 1, 1998.

[SCH 95] SCHACHT C.A., *Refractory Linings: Thermomechanical Design and Applications*, Marcel Dekker Inc., New York, 1995.

[TAF 94] TAFFIN C., POIRIER J., "The behaviour of metal additives in MgO-C and Al_2O_3-C refractories", *Interceram,* vol. 43, 1994.

[TAR 98] The Technical Association of Refractories, *Refractories Handbook*, TARJ, Japan, 1998.

[THE 93] THEMINES D., POIRIER J., PROVOST G., BLUMENFELD P., "Modeling of the thermomechanical behaviour of refractory shapes and linings – experimental validation", *Unitecr'93*, Sao Paulo, 1993.

[THE 94] THEMINES D., Séminaire "Réfractaires", Centre d'Etudes Supérieures de la Sidérurgie Française, Maizière-Lès-Metz, 1994.

Chapter 11

Ceramics for Electronics

11.1. Conductors and insulators

In a crystalline solid, electrons and ions are possible charge carriers. However, the conductance is the product of charge concentration and mobility. Electrons, because of their low mass are much more mobile and conducting materials are most often electronic conductors. The majority of this section will therefore be dedicated to the electronic properties of the solid, whilst ionic conductivity will be more specifically dealt with in section 11.1.6.

The distinction between insulators and conductors comes naturally within the framework of the band theory developed by Bloch and Wilson (1929). Please refer to other works for a detailed presentation, [KIT 98] for example, as only the basic elements will be mentioned here.

A solid is an assembly of charged particles, electrons and atomic nuclei. Because of the difference in mass, atomic nuclei are very slow when compared to electrons and their positions can be considered merely as parameters and not as unknown functions of time, solutions of the Schrödinger equation. Furthermore, Coulomb repulsive interactions of one electron with others are averaged over the electronic population. These are the two basic assumptions of the band theory, assumptions which are often not entirely justified in the case of ceramic materials (this will be discussed later on). It is therefore sufficient to consider the motion of a single electron placed in an effective potential and to determine its energy levels and states

---

Chapter written by Pierre ABÉLARD.

by applying the equations of quantum mechanics. Despite the approximations made, it remains a difficult problem. In the case of complex solids like oxides, silicates, etc., the wave functions, $\varphi_{\vec{k}}(\vec{r})$, are most often built as linear combinations of atomic orbitals $\varphi_{at,j}(\vec{r})$ (or LCAO theory, [MAR 79]), i.e.:

$$\varphi_{\vec{k}}(\vec{r}) = \sum_j \sum_i c_{ij} \cdot \varphi_{at,j}\left(\vec{r} - \vec{R}_i\right) \exp\left[j\vec{k}\left(\vec{r} - \vec{R}_i\right)\right] \qquad [11.1]$$

the summation referring on the one hand to the positions of the nuclei $\vec{R}_i$, and on the other to the atomic orbitals retained. The condition of energy minimization allows the determination of the coefficients c_{ij}. The wave vector $\vec{k}$, identified in the reciprocal space, helps to index the electronic states. The overlap of orbitals centered on different atomic nuclei is responsible for the degeneracy removal of the considered atomic levels and the distribution of energies of the electronic states into different energy bands. The larger the overlap, the larger the band. Very often, it is not necessary to acquire detailed knowledge of the variations of the electron energy according to the wave vector, but a determination of the state density N(E) is sufficient (see Figure 11.1).

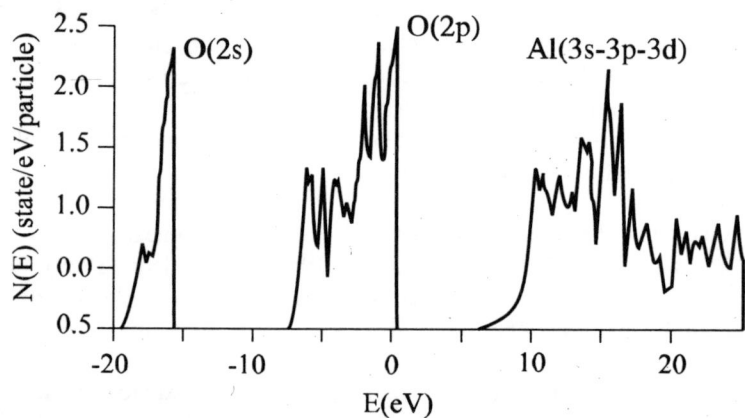

Figure 11.1. *Electronic state density N(E). Example of α-Al$_2$O$_3$*

These states are later filled with the available electrons by following the Pauli principle: only two electrons, of opposite spin (± 1/2), can each occupy the states. The energy of the last electron placed is called Fermi level. The band where the Fermi level is found is called the valence band. If this is fully occupied, the solid is an insulator, otherwise it is a conductor (see Figure 11.2).

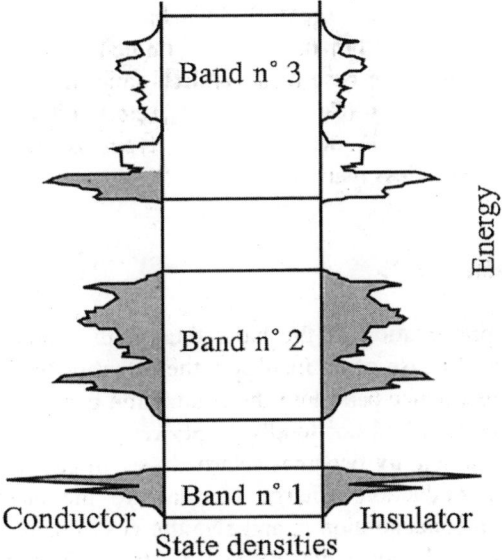

Figure 11.2. *Filling of electronic states. The occupied states are shaded: case of a conductor (left-hand side) and an insulator (right-hand side)*

11.1.1. *Insulators*

Several ceramic materials are oxides. The valence band, essentially built from the 2p atomic orbitals of oxygen, is 2/3 filled with oxygen electrons (atomic configuration $2p^4$) and is completed by the electrons provided by cations. In case of MgO, for example, the band 2pO is saturated, the immediate higher energy band 3sMg is empty and we can legitimately speak of Mg^{2+} and O^{2-} ions. This is an insulator.

All oxides are not as simple. Let us consider, for example, quartz SiO_2. The Si-O bonds are essentially of covalent type, which can be described with the help of molecular orbitals. Atomic orbitals s and (p_x, p_y, p_z) of silicon combine linearly to form hybridizations sp^3, which are four in number. These interfere with the six orbitals (p_x, p_y, p_z) of the two oxygen atoms to form the ten molecular orbitals of the SiO_2 group. They are used in equation [11.1] instead of atomic orbitals to determine the quantum states of an electron in the crystal. These are grouped into two distinct energy bands, corresponding to six and four molecular orbitals per SiO_2 group respectively. The 12 electrons of such a group totally fill the first lowest energy band. Quartz is thus an insulator.

In these two examples a large forbidden band, 7.6 and 8.4 eV[1] respectively, separate the valence band from the band of immediate higher energy, called the conduction band. This is the case of several ceramic materials, like oxides (MgO, BeO, Al_2O_3, etc.), silicates (SiO_2, Mg_2SiO_4, etc.), nitrides (AlN, Si_3N_4, etc.). Conductivity, both of ionic and electronic type, is basically determined by impurities. It is typically less than 10^{-10} Ω^{-1} cm^{-1}.

11.1.2. Semi-conductors

In the above presentation of the band theory, the temperature was implicitly assumed to be 0 K. In case of an insulator, the effect of temperature is to promote the electrons of the valence band into the conduction band. These two energy bands are then neither totally filled nor totally empty respectively, and both contribute to conductivity. If conductivity becomes relatively significant ($\cong 10^{-6}$–10^3 Ω^{-1} cm^{-1}), we refer to semi-conduction. Charge carriers are the electrons present in the conduction band, in concentration n, and also the electrons of the valence band. For the latter, their contribution with respect to the different physical properties, conductivity, specific heat, etc., is the same as positively charged particles called electron holes, with concentration p, the fraction of unoccupied states. The concentrations n and p are fixed by the following relationships[2]:

$$n = N_c \exp\{(\mu - E_c)/kT\} \quad \text{and} \quad p = N_v \exp\{(E_v - \mu)/kT\} \qquad [11.2]$$

if μ is the chemical potential of the electrons, still called by extension as Fermi level. E_c and E_v are energies at the bottom of the conduction band (effective state density N_c) and at the top of the valence band (effective state density N_v) respectively:

$$N_c = 2\left(\frac{2\pi m_e^* kT}{h^2}\right)^{3/2} \quad \text{and} \quad N_v = 2\left(\frac{2\pi m_t^* kT}{h^2}\right)^{3/2}$$

m_e^*, m_t^* are the effective masses of conduction electrons and electron holes (defined below). Or even:

$$n\,p = N_c N_v \exp\left(-E_g/kT\right) \qquad [11.3]$$

---

1 1 eV = 1.6 10^{-19} J.
2 At least for non-degenerated semi-conductors, i.e. when the double condition $E_v + kT \ll \mu \ll E_c - kT$ is satisfied.

where E_g (= $E_c - E_v$) is the width of the forbidden band separating the valence band and the conduction band. This law of mass action is associated with a chemical equilibrium:

$$e'(valence\ band) \Leftrightarrow e' + h^\bullet\ (\Delta H = E_g)$$

corresponding to the transfer of an electron of the valence band to the conduction band (e'), with creation of an electron hole ($h^\bullet$) in the valence band. The required energy is equal to E_g.

11.1.2.1. *Intrinsic semi-conductivity*

Under effect of temperature only:

$$n = p = \sqrt{N_c N_v}\ \exp\left(-E_g/2kT\right)\ \text{and}\ \mu \cong E_g/2$$

We refer to the intrinsic semi-conductivity. On closer examination of the values of E_g for a certain number of semi-conductor oxides or carbides (see Table 11.1) we can see that the concentrations of electrons and holes may be significant only at very high temperatures, unlike what we see for more traditional semi-conductors like silicon.

Material	E_g (eV)
Si	1.2
Cu_2O	2.0
Cubic SiC	2.2
WO_3	2.6
SnO_2	2.7
In_2O_3	2.9
Hexagonal SiC	2.9
TiO_2 rutile	3.2
ZnO	3.3
$SrTiO_3$	3.4
CeO_2	5.5

Table 11.1. *Width of the forbidden band E_g, for a few semi-conductors, determined by measurements of optical absorption (T = 300 K)*

11.1.2.2. *Extrinsic semi-conductivity*

Doping, i.e. the introduction of aliovalent foreign elements, still remains a method for inducing semi-conductivity, referred to as extrinsic. Thus, is niobium likely to be substituted for titanium in the oxide $SrTiO_3$? This element has an electronic configuration ($5s^2\ 4d^3$) to be compared with titanium ($4s^2\ 3d^2$). In the solid, four electrons/atoms are transferred in the valence band $2pO$. The fifth electron remains in the vicinity of the ion Nb^{5+} as a consequence of the electrostatic attraction between the two opposite charges. Ionization of this electron, which enables its delocalization on all the cationic sites, requires an energy E_d. From the point of view of the band theory, these electrons are within electronic states present in the forbidden band, an energy E_d below the conduction band. We refer to them as electron donor bands. This process can be described through a chemical equilibrium:

$$Nb_{Ti}^x \Leftrightarrow Nb_{Ti}^{\bullet} + e'\ (\Delta H = E_d)$$

$$\frac{\left[Nb_{Ti}^{\bullet} \right].n}{\left[Nb_{Ti}^x \right]} = K_d = N_c \exp(-E_d/kT) \tag{11.4}$$

where the brackets [...] denote a concentration, with the Kröger-Vink notation.

The Kröger-Vink notation is written in this manner: A_{site}^{charge}:

– A is the element, A = V indicates a vacancy site;

– the lower index, the concerned site;

– the upper exponent, the charge, evaluated with the ideal crystal as reference.

The symbol (x) indicates an effective charge of 0, (n•) of +nq, (m') of –mq.

In the absence of source or sink for this element, a law of conservation has to be verified:

$$\left[Nb_{Ti}^x \right] + \left[Nb_{Ti}^{\bullet} \right] = N_d \tag{11.5}$$

if N_d is the initial concentration of the dopant. Finally, electrical neutrality must be adhered to:

$$n = \left[Nb_{Ti}^{\bullet} \right] \tag{11.6}$$

All these equations ([11.4]–[11.6]) help to calculate the concentration of electrons. If the condition $N_d << K_d$ is satisfied, then all the donor levels are ionized: $n = N_d$. Quite a small fraction of dopant ($\cong 100$ ppm) is enough to induce a sizeable semi-conductivity. Equation [11.2] helps to calculate the chemical potential and to place the Fermi level in the forbidden band. The greater the electronic concentration, the closer it is to the conduction band.

In a crystalline solid, *point defects* thermodynamically stable at any temperature T other than 0 K, are also effective to induce an extrinsic semi-conductivity. Thus, in oxides [KOF 72], we notice quite frequently the presence of oxygen vacancies, i.e. unoccupied anionic sites, following the exchange of oxygen between the solid (at least its surface) and the gaseous atmosphere which surrounds it. At the thermodynamic equilibrium, this exchange is represented by a chemical equilibrium:

$$O_O^x \Leftrightarrow V_O^x + 1/2 O_2(gas) \quad \Delta H_f$$

$$\left[V_O^x\right] P_{O_2}^{1/2} = K_{ox}(T) = K_o \exp(-\Delta H_f / kT) \tag{11.7}$$

where P_{O_2} is the partial pressure of oxygen in the atmosphere and ΔH_f the enthalpy of formation of an oxygen vacancy. During oxygen transfer from the solid (made up of O^{2-} ions) into the gaseous phase (oxygen molecules), the two electrons of the O^{2-} ion remain on the oxygen site. They are strongly attracted by the nearby cations and easily delocalized on all the cationic sites of the solid. We must however provide for this a certain energy of ionization E_d, of a few 0.01–0.1 eV which differs depending on whether it refers to the first or second ionization.

Material	ΔH_f(eV/vacancy)
V_2O_5	1.3
WO_3	3
Nb_2O_5	4.4
TiO_2	4.8
$SrTiO_3$, $BaTiO_3$	6
ZrO_2	8.7
Y_2O_3	10

Table 11.2. *Enthalpy of formation of oxygen vacancies ionized twice*

As an example, let us consider the twice ionized oxygen vacancies. At thermodynamic equilibrium, we can express the chemical equilibrium as:

$$V_O^x \Leftrightarrow V_O^{\bullet\bullet} + 2e' \text{ and } \frac{[V_O^{\bullet\bullet}] \cdot n^2}{[V_O^x]} = K_{d2} = N_c^2 \exp(-E_{d2}/kT) \qquad [11.8]$$

We must also take into consideration the condition of electroneutrality, i.e.:

$$2[V_O^{\bullet\bullet}] = n \qquad\qquad [11.9]$$

Combining equations [11.7]–[11.9], we get:

$$n = \left(2N_c^2 K_o\right)^{1/3} P_{O_2}^{-1/6} \exp\left\{-(\Delta H_f + E_{d2})/(3kT)\right\}$$

The concentration of electrons is a temperature sensitive function. Thus, in the case of $SrTiO_3$, sintered in reducing atmosphere ($P_{O2} = 10^{-25}$ bar) the concentration of electrons at 1,200 K is of $4.2.10^{19}/cm^3$, but only of $3.5.10^2/cm^3$ at 400 K! It also depends on the partial oxygen pressure, which must therefore be controlled during heat treatment.

We have assumed on several occasions to have realized the thermodynamic equilibrium state. It is true that the kinetics of the exchange of oxygen between the surface and the gaseous atmosphere, or even the kinetics of ionization of oxygen vacancies, can be generally considered as fast. This is not always true concerning the process of diffusion of vacancies from the surface towards the core of the material, which alone is capable of regularizing the oxygen activity. Since this process is activated thermally, it becomes very slow at low temperature. Below a certain temperature, called the quenching temperature T_t, the oxygen vacancies diffuse so slowly that the total concentration of defects can be considered to be fixed, i.e. $C_o(T_t)$. Relationship [11.7] must be replaced by a law of conservation:

$$[V_O^x] + [V_O^{\bullet\bullet}] = C_o(T_t)$$

The equations which govern the different concentrations, laws of mass action, electroneutrality, balance equations, are very similar whether they be dopants or defects, leading to similar calculations. Let us consider once again the earlier example of $SrTiO_3$, the concentration of electrons at 400 K after quenching at 1,000 K is $8.7.10^{17}/cm^3$ instead of $3.50.10^2/cm^3$, value calculated at equilibrium.

The cooling rate, which determines T_t, is an important parameter of the thermal treatment. Thus, a fall in the quenching temperature of 100 K divides the electron concentration at 400 K by a factor of 10. It is clearly seen that quenching is responsible for the semi-conductivity observed at ambient temperature for several oxides.

The law of mass action (equation [11.3]) shows that the electron holes are in the minority if the electrons are in the majority. This is why their contribution has not been considered in the previous development. Some dopants or defects are associated with levels close to the valence band and are susceptible to acquire electrons, creating electron holes in the valence band. These are referred to as acceptor levels. If the electron holes are in the majority, we refer to them as p-type semi-conductors.

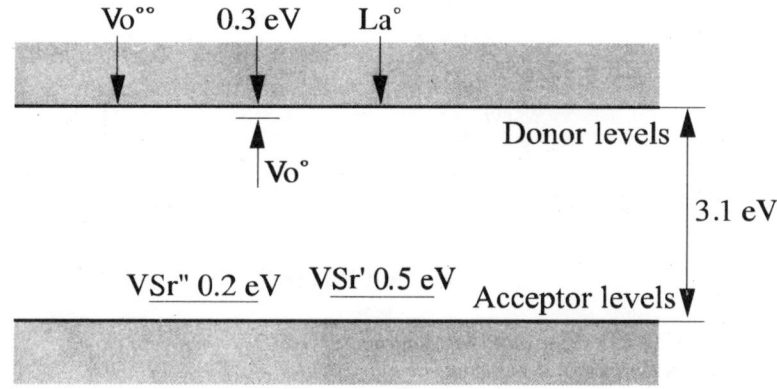

Figure 11.3. *Donor and acceptor levels in the forbidden band of SrTiO₃*

Donors and acceptors can still compensate their effects if present simultaneously, while electrons and electron holes destroy each other according to the equilibrium [11.3].

The nature of the majority defects depends on the temperature and the partial pressure of oxygen. An example is given in Figure 11.4 in the case of strontium titanate doped with niobium. In reducing atmosphere ($p_{O2} < 10^{-18}$ bar), oxygen vacancies ionized twice and the electrons are in majority. By increasing the partial pressure of oxygen (10^{-18} bar $< p_{O2} < 10^{-12}$ bar), the concentration of oxygen vacancies strongly decreases and the electrons mainly come from the ionized donors. Lastly, in oxidizing atmosphere ($p_{O2} > 10^{-10}$ bar), cationic vacancies, now numerous, capture the electrons released. Donors and acceptors compensate their effects.

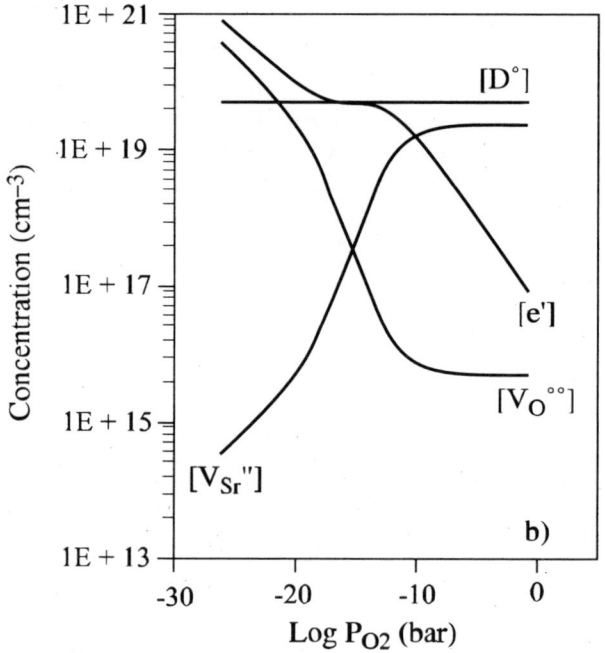

Figure 11.4. *Kröger and Vink diagram for SrTiO$_3$ doped with Nb$_2$O$_5$. Variations in defect concentrations with P$_{O2}$ (T = 1,273 K)*

11.1.2.3. *Mobility of charge carriers*

Band theory also helps to understand the dynamics of electrons. In an ideal crystal, i.e. strictly periodic, the motion of an electron is not affected by the presence of nuclei and is therefore rectilinear and uniform. The velocity depends on the state occupied and in the case of semi-conductors, it is in the order of $\sqrt{(3kT/m*)}$, where m* is the effective mass. This may be smaller or greater than the real mass of the electron, since it includes the effect of the interaction of the electron with its environment. It is small in the case of large bands built from the "s-p" states and large in case of narrow bands "d" or "f" (see Table 11.3).

Material	Si	MgO	SnO$_2$	ZnO	SrTiO$_3$	TiO$_2$	NbO$_2$
Electrons	0.26	0.4	0.16	0.24	6–8	8–10	13
Electron holes	0.4	0.6		1.72			
α		2.8		0.85			3.9

Table 11.3. *Effective mass of charge carriers, related to the mass of the electron $m_o = 0.9 \ 10^{-30}$ kg, for a few materials. α is the electron-network coupling constant*

Deviation from the ideal crystal, lattice vibrations, point defects and impurities are examples of what modifies the electron's trajectory. The interaction with such an imperfection is local, intense and of very short duration. We speak of collision and we call τ the mean time between two collisions. In the absence of an applied electric or magnetic field, to any electron of velocity $\vec{v}$ corresponds an electron of velocity $-\vec{v}$ and overall, the charge flow generated by all the electrons of an energy band is nil. In the presence of an applied electric field $\vec{E}_{app}$, any electron is accelerated, according to Newton's laws:

$$m * \frac{d\vec{v}}{dt} = -q\vec{E}_{app}$$

i.e. an increment in velocity between two collisions:

$$\Delta \vec{v} = -(q\tau / m^*)\vec{E}_{app}$$

the same for all electrons, proportional to the applied electric field. The proportionality factor is called electronic mobility:

$$\Lambda_e = -q\tau / m^*$$

The applied electric field is therefore at the origin of the charge flow:

$$\vec{j} = -nq\Delta\vec{v} = (nq^2\tau / m^*)\vec{E}_{app} = \sigma_o \vec{E}_{app}$$

thus demonstrating Ohm's law. Conductivity is the product of the charge concentration nq, and mobility Λ_e[3]. It can be deduced by the measurement of

---

3 $\sigma_o = pq\Lambda_h$ in the case of holes, $\sigma_o = nq\Lambda_e + pq\Lambda_h$ in the general case.

resistance, the velocity increment obtained by measurement of the Hall effect and the charge concentration by measurement of the Seebeck effect [TAL 74]. However, experimental difficulties and the assumptions required for the interpretation of experimental data explain why the mobility values published are quite disparate.

In Table 11.4, we can see that electronic mobility is low for several ceramic materials. The knowledge of Λ helps to calculate the time τ and the distance traveled $l = v\tau$ between two collisions. A mobility of around 0.1 cm²/(V s) corresponds roughly to an interatomic distance, which makes us think that there exists in these materials a very strong interaction of the electron with the lattice. This is not surprising, considering their strong polar nature, but it contradicts one of the basic assumptions of the band theory, i.e. the decoupling of the motion of electrons and atomic nuclei. If the electron-lattice interaction, quantified by the dimensionless parameter α (see Table 11.3), is sufficiently small, $1 < \alpha < 6$, the band theory is broadly preserved and the main consequence is an increase of the effective mass:

$$m_p^* = \frac{m^*}{1-\alpha/6}$$

Material	Mobility (cm²/(V s)	E_Λ (eV)
Si: (e'/hole e')	1,600/400	0
ZnO (e')	180	0
SnO₂ (e')	100	0
MgO: (e'/hole e')	25/8 (1,700 K)	0
TiO₂ (e'/hole e')	0.17/1 10⁻⁵ (1,200 K)	0/0.4
SrTiO₃ (e')	6	0
BaTiO₃(e')	0.2	0.02?
NbO₂ (e')	0.5 (600 K)	0
V₂O₅ (e')	0.1	0.2
NiO(Li) (e'/hole e')	0.14/0.30	0

Table 11.4. *Some values of mobility at 300 K (unless otherwise mentioned)*

The electron is referred to as a large polaron. In the reverse case, the electron is trapped by its environment and becomes localized: we then refer to small polarons.

Conduction takes place through hopping and the mobility of the charge carriers is lower, even much lower than 0.1 cm²/(V s).

Generally speaking, mobility depends on the temperature according to the relationship:

$$\Lambda = \frac{A}{T^a}\exp(-E_\Lambda / kT) \qquad\qquad [11.10]$$

where the value of "a" (generally found between 0 and 2 at high temperature) indicates the type of collision. A non-zero activation energy E_Λ is a characteristic of conduction by small polarons.

11.1.3. Metal conductors

A partially filled valence band denotes a conductor. Conductivity is high ($> 10^4\ \Omega^{-1}\ cm^{-1}$) and decreases slightly with increasing temperature. Let us mention oxides like ReO_3, RuO_2, CrO_2 and non-oxides like $MoSi_2$, LaB_6, which display this metallic behavior. The tungsten bronze family with formula Na_xWO_3 ($x > 0.4$) is particularly interesting because it is possible, by adjusting the quantity of sodium, to vary the electronic concentration.

Electrons of the valence band, more exactly electrons in an energy interval kT near the Fermi level, bring about a specific contribution to several physical properties like electrical conductivity, specific heat, magnetic susceptibility (Pauli paramagnetism) etc.; contributions which can be measured experimentally.

Property	Relation
– Electrical conductivity v_F, velocity, τ_F time between collision at Fermi level	$\sigma = \dfrac{1}{3}q^2 v_F^2 \tau_F N(E_F)$
– Magnetic susceptibility $\mu_0 = 4\pi10^{-7}$, μ_B Bohr magneton ($= 9.274\ 10^{-24}$ A m²) – Specific heat	$\chi_p = \mu_0 \mu_B N(E_F)$ $\delta C_p = \dfrac{2}{3}\pi^2 k^2 T N(E_F)$

Table 11.5. *Free electron contributions to various physical properties*

The different estimates in the state density at the Fermi level $N(E_F)$, which can be deduced, are generally not concordant. Sometimes, it is enough to just question the simplicity of the free electron model (see [KIT 98]) used to establish the relations in Table 11.5 and to introduce corrective terms, but most often it seems necessary to reconsider the hypothesis of independent electrons.

The principle of Pauli exclusion introduces a coupling between the spin variables and the space coordinates. In fact, two electrons with antiparallel spin can occupy the same state (same wave function, same electronic density, maximum repulsion), which is no longer the case if their spins are parallel. The difference in repulsive energies for these two configurations, K, is called the exchange energy. It stabilizes the electron pairs with same spin. Magnetic susceptibility increases, and according to Stoner:

$$\chi_m = \frac{\chi_p}{1 - KN(E_F)/2}$$

an effect seen in several conductor compounds, for example, $LaNiO_3$. This relationship also shows that if the interaction is sufficiently strong (K $N(E_F) > 2$), the paramagnetic state is unstable. This is how the existence of ferromagnetism is explained in the oxide CrO_2 below 450 K.

These interactions therefore seem to question the band theory. It is however possible to adapt it by separately considering electron populations of spin + 1/2 and −1/2, introducing in the effective potential an exchange term [CHI 90].

11.1.4. Transition metal oxides

With respect to the band theory, the electron is delocalized, which makes conduction possible. It is again this delocalization which, through the mechanism of screening, minimizes the repulsive interactions between electrons. This mechanism is more effective when the orbital overlap is significant and the energy band is large. We have already noted the small expansion of the orbitals "d" or "f" leading to the formation of narrow bands. The question of the applicability of the band theory in the case of transition metal oxides arose as early as 1937 by Boer and Verwey [COX 95].

Transition elements have an electronic configuration ($4s^2 3d^m$). The (m + 2) electrons are first of all placed in the vacant states of the band "2pO" until saturation. If there are no more 3d electrons available, configuration d^0, the band 3d is empty and the oxide is an insulator, as is the case of TiO_2, CrO_3 oxides or even WO_3. The same is applicable if the band 3d is saturated, configuration d^{10} which we find in Cu_2O. In all other cases, the band 3d should be partially filled and the oxide should be a metallic conductor. This is true of the oxides TiO or VO for example. However, these oxides are very defective (vacancy concentration of 15% to 20% on both the sub-lattices, cationic and anionic, in stochiometric compounds!), which causes a decrease in the cell parameter, an increase in the orbital overlap and greater stability of the metallic state. The oxide immediately next in the series of oxides MO, CrO does not exist (CrO_2 is a metallic conductor, Cr_2O_3 an insulator). The other oxides of the series: MnO, FeO, CoO, NiO, CuO are all insulators.

The first effect to be taken into consideration is the influence of the crystal field. In the MO oxide, each metal cation is surrounded by six oxygen ions forming an environment with octahedral[4] symmetry. The electrical field generated by anions, called the crystal field, removes the degeneracy of the 3d level and gives rise to two distinct levels, denoted by $t_{2g}(d_{xy}, d_{xz}, d_{yz})$ and $e_g(d_{x2-y2}, d_{xyz})$ respectively (see Figure 11.5) and thus two energy bands (which will eventually overlap). The saturation of levels t_{2g} can explain an insulating state as in the case of $LaCoO_3$, with the ion Co^{3+} having the configuration $3d^6$.

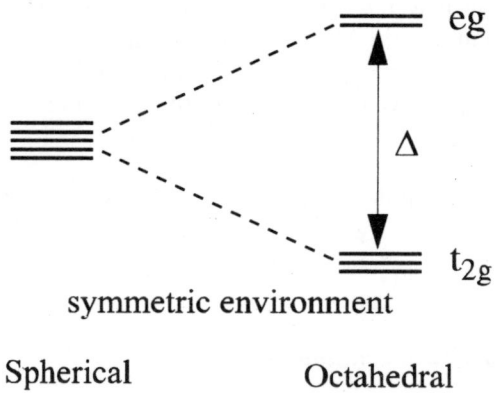

symmetric environment

Spherical   Octahedral

Figure 11.5. *Influence of the crystal field on the energy levels 3d of the cation for an octahedral anionic environment*

However, if the decomposition of level 3d by the crystal field is always present, it cannot by itself justify the insulating nature of the aforementioned MO oxides. According to Hubbard, it is a result of the localization of electrons on distinct sites,

---

4 For structures other than those of the MO oxides, the configuration of anions could be of tetrahedral symmetry, square plane, etc., but always of symmetry lower than the free ion and leading in any case to the removal of degeneracy.

the only way to minimize in an optimal manner Coulomb's repulsive interactions within the limit of a band of nil width. Conduction is then possible only through electron hopping from one site to a neighboring site. This is based on the prior assumption that the crystal, considered as a whole, has moved from the fundamental state $(...d^n, d^n, d^n, d^n...)$, where each of the metal cations bear the same number of electrons 3d, to an excited state $(...d^n, d^{(n-1)}, d^{(n+1)}, d^n...)$. The transition energy is equal to the additional repulsive energy U of the configuration $d^{(n+1)}$. This is the equivalent of E_g, the width of the forbidden band in the band theory, while $(d^n, d^n) \to (d^{(n-1)}, d^{(n+1)})$ represents the creation of a pair of electron-hole. U decreases when the orbital overlap d, quantified by the parameter b, increases [GOO 73]. The band theory is applied when b is greater than U.

As in a traditional semi-conductor, it is possible to induce an extrinsic semi-conductivity by doping with aliovalent elements, for example lithium in the oxide NiO. Charge compensation takes place through a change in valence $(d^n \to d^{(n-1)})$ of a nickel ion. Charge transport, which can be outlined as $d^n d^{n-1} d^n d^n \to d^n d^n d^{n-1} d^n$, does not require the creation of an excited state as described in the Hubbard model.

Point defects can play the same role as doping elements. Metal vacancies are effectively present in several MO transition metal oxides. The FeO compound is particularly impressive because under atmospheric pressure it is never stoichiometric, irrespective of conditions of temperature and partial oxygen pressure.

Let us note that the polarization of the network induced by this localization ensures an additional energy gain by the formation of a small polaron (see section 11.1.2).

11.1.5. *Localization or delocalization?*

The band theory and the Hubbard model are borderline descriptions, which take into account only one of the factors: either a delocalization resulting from the orbital overlap, or a localization induced by strong electron-electron interactions. The exchange interaction assumes in fact both delocalization and Coulomb interactions. Thus, in the case of metallic conductors, it favors an ordering of spins (CrO_2 ferromagnetism). In the case of insulating oxides of transition metals, described in the previous section, it is a super-exchange interaction which helps to understand the existence of an antiferromagnetic order at low temperatures (see section 11.3.1). It is called superexchange since Coulomb's interactions take place between electrons localized on distinct cationic sites, separated by an oxygen ion. It therefore assumes an overlap of cationic and anionic orbitals.

Magnetic properties therefore particularly reveal the importance of different factors. The diversity of the encountered situations, para-, ferro-, antiferro-, ferrimagnetic, spin waves [KIT 98], shows the complexity of the problem. Small modifications in composition, temperature or pressure are susceptible to modify the state of conduction. Hence, at low temperatures, NbO_2 is a semi-conductor. Up to 500-600 K, the electrons behave like small polarons, and above 600 K, like large polarons. The increase in the orbital overlap with temperature is accompanied by the decrease in the width of the forbidden band, i.e. 0.25 eV to 300 K. This disappears at 1,350 K, with conduction becoming metallic. Another example is V_2O_3 which can be an insulator or a conductor, antiferromagnetic or paramagnetic depending on thermodynamic conditions (see Figure 11.6).

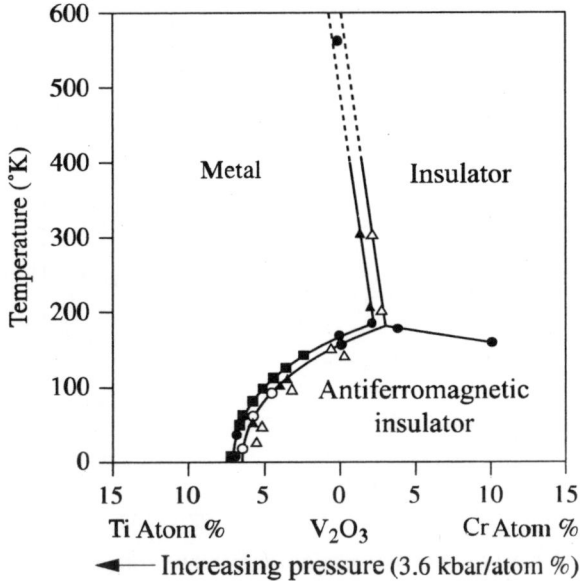

Figure 11.6. *Phase diagram (temperature-pressure) showing the different states possible for V_2O_3. Doping has an effect which is equivalent to the effect of pressure*

Dimensionality is one of the parameters to be taken into consideration. Peierls has demonstrated that the state of the metal conductor is unstable for a one-dimensional solid. In fact, a structural modification, like the doubling of the cell parameter, is effective to introduce a forbidden band at the Fermi level, ensuring an energy gain. Such an effect can be observed for compounds having a strong anisotropy of properties, like for example molybdenum bronzes with composition $A_{0.3}MoO_3$ (A = K,Rb). For two-dimensional compounds, there is competition between a conducting state and an insulating state. An example is La_2CuO_4, antiferromagnetic insulator of orthorhombic structure. By substituting a fraction of lanthanum with barium (or strontium), the material becomes a conductor in the planes (001), a transformation which is accompanied by an orthorhombic-quadratic transition, which could be interpreted as a Peierls distortion. Such an effect cannot be observed for three-dimensional compounds.

A certain number of oxides become superconductors at low temperature [SHI 95]. The fact that the transition temperature T_c (see Table 11.6) could be higher than the temperature of liquid nitrogen (77 K), thus making numerous applications possible, explains the large number of studies undertaken in the 1980s (35,000 publications in seven years!).

Material	T_c
$SrTiO_{3-x}$	< 1 K
NbO	11 K
$BaPb_{0.75}Bi_{0.25}O_3$	13 K
$(La,Ba)_2CuO_4$ (214)	35 K
$YBa_2Cu_3O_{7-x}$ (123)	93 K
Bi-Al-Ca-Sr-Cu-O (2,212)	114 K
Tl-Ca-Ba-Cu-O (2,223)	120 K

Table 11.6. *Temperature of superconductor metal transition for a few materials (with usual denomination in brackets)*

Let us recall the basic characteristics of the superconductor state [KIT 98]:

– zero resistivity, which implies the possibility of transporting current without dissipation through the Joule effect;

– expulsion of induction lines outside the sample volume during the application of a magnetic field, made possible by the appearance of a surface current. In other words, the material has a perfect diamagnetic behavior.

According to the BCS (Bardeen, Cooper and Schrieffer) theory, electrons form pairs below the critical temperature T_c, stabilized by a decrease in energy $\Delta \approx kT_c$. Lattice vibrations, more precisely acoustic phonons, are mediators of the interaction. This interpretation is confirmed through different experimental observations when the transition temperature is low ($T_c < 30$–40 K). For higher temperatures, and hence stronger interactions, the mechanism of pair formation stated by BCS must be re-examined. Several proposals have been made but to date none have been impressive. The phase diagram in Figure 11.7 shows how superconductivity appears under extreme conditions.

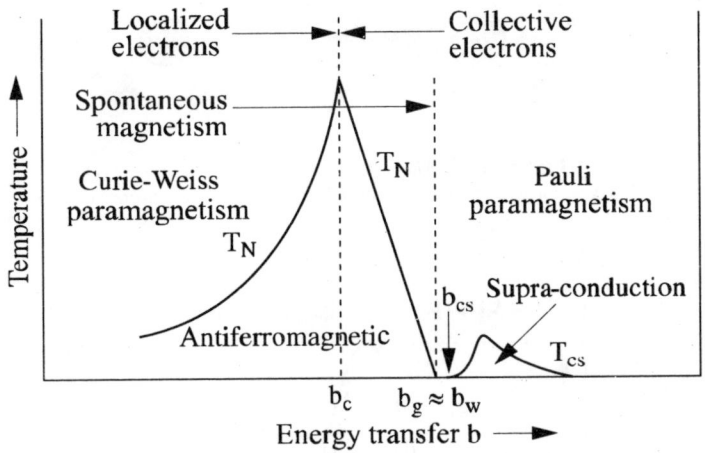

Figure 11.7. *Diagram (Temperature-b) showing conditions for the appearance of superconductivity. b is an energy parameter, function of orbital overlap;*
$$b_g \cong U \text{ Hubbard energy}$$

11.1.6. *Contribution of ions to conductivity*

Previously in this discussion, the contribution of ions to electrical conductivity has been neglected. However, it must be taken into account under two circumstances, either in the case of insulating material, when the electronic contribution is small and of the same magnitude as ionic contribution, or when it is very significant and greatly dominating, as in the case of solid electrolytes.

The transport of ions takes place via point defects, always present in a crystalline solid, mainly cationic or anionic vacancies or interstitials (see Figure 11.8).

The transport mechanism is a diffusion process which will be described briefly. In the solid state, the ions are localized on sites of minimum free energy G. Localization is ensured by the potential barriers which separate one site from the neighboring sites (see Figure 11.9).

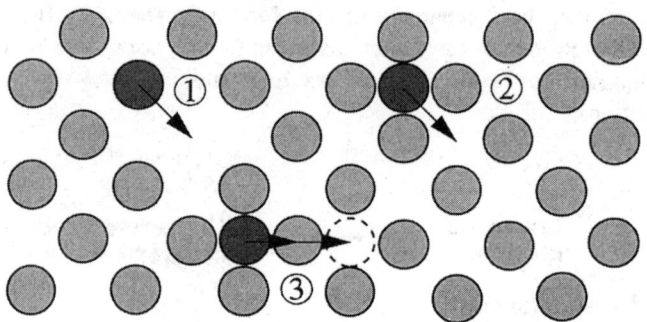

Figure 11.8. *Main mechanisms of diffusion.*
1 = vacancy, 2 = direct interstitial, 3 = indirect interstitial

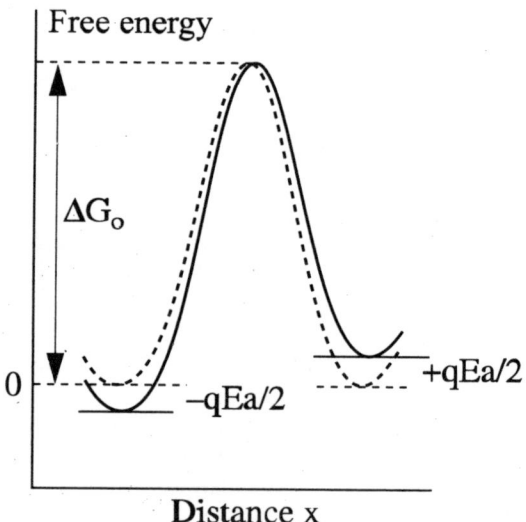

Figure 11.9. *Ion hopping a) in the absence and b) in the presence of an electric field (one-dimensional model)*

More precisely, the ions oscillate around these equilibrium positions, a harmonic motion of frequency v_o, typically 10^{13} Hz, in the case of low amplitude displacements.

High amplitude displacements are not excluded, but these are rare because the probability of occurrence is proportional to:

$$\exp\left[-\frac{G(x)}{kT}\right] = \exp\left(\frac{S}{k}\right)\exp\left[-\frac{U(x)}{kT}\right]$$

if S and U are entropic and enthalpic contributions of free energy respectively. Crossing the potential barrier, and therefore ion hopping from one site to a neighboring site, requires a minimal energy ΔG_o, on the condition nevertheless that the neighboring site is not occupied. This is the case if it refers to a vacancy site or an interstitial site. In the absence of an applied electrical field, the direct and reverse hoppings take place with the same frequency:

$$v_{\rightarrow} = v_{\leftarrow} = v_o \exp\left(\frac{\Delta S_o}{k}\right)\exp\left(-\frac{U_o}{kT}\right) = v_o^* \exp\left(-\frac{U_o}{kT}\right) \quad 5$$

and the macroscopic charge flow is zero. In the presence of a uniform electric field E_{app}, the system is no longer symmetric. In fact, an additional contribution to the free energy, $\pm q.E_{app} x$, the sign following the hopping direction, has to be taken into account, i.e.:

$$v_{\rightarrow} = v_o^* \exp\left(-\frac{U_o - qE_{app}a/2}{kT}\right) \cong v_o^* \exp\left(-\frac{U_o}{kT}\right)\left(1 - \frac{qE_{app}a}{2kT}\right)$$

A Taylor expansion to first order has been made, justified by the usual values of the electric fields applied. Similarly:

$$v_{\leftarrow} \cong v_o^* \exp\left(-\frac{U_o}{kT}\right)\left(1 + \frac{qE_{app}a}{2kT}\right)$$

The charge flow is equal to the number of hops per unit of time across a unit surface perpendicular to the applied electrical field, i.c. v [D] a, if [D] is the volume concentration of defects. The total current density J is equal to:

$$J = J_{\rightarrow} - J_{\leftarrow} = q(v_{\rightarrow} - v_{\leftarrow})[D]a = [D]\frac{q^2 v_o^* a^2}{kT}\exp\left(-\frac{U_o}{kT}\right)E_{app}$$

---

5 As an example, in the case of migration of sodium in NaCl, $\Delta S_o/k = 1.75$ and $v_o = 4.9. 10^{12}$ Hz.

Ohm's law is proved, which allows the definition of an ionic conductivity σ_{ion}:

$$\sigma_{ion} = [D]\frac{q^2 v_o^* a^2}{kT} \exp\left(-\frac{U_o}{kT}\right)$$

and a defect mobility Λ_D:

$$\Lambda_D = \frac{q}{kT} v_o^* a^2 \exp\left(-\frac{U_o}{kT}\right)$$

Mobility is therefore a very sensitive function of the potential barrier U_o and most often only one type of ion is indeed mobile.

Material	Mobile ion	Defect	$U_0(eV)$
$SrTiO_3$	O^{2-}	Vacancy	0.98
$0.88ZrO_2 + 0.12CaO$	O^{2-}	Vacancy	1.13
NaCl	Cl^-	Vacancy	0.86
	Na^+	Vacancy	0.65
CaF_2	F^-	Interstitial	1.04
$\alpha\text{-PbF}_2$	F^-	Interstitial	0.45
BiCuVOx	O^{2-}	Vacancy	0.37
Nasicon	Na^+	Interstitial	0.25
Alumina β	Na^+	Interstitial	0.24
α AgI	Ag^+	Interstitial	0.14
$RbAg_4I_5$	Ag^+	Interstitial	0.08

Table 11.7. *A few values of the activation energy U_o*

Most often, U_o is of the order of 0.5 to 1.5 eV (see a few examples in Table 11.7). Consequently, ion mobility is low, very low compared to the electrons, but it is a highly temperature sensitive function. Thus for an activation energy Uo = 0.8 eV, mobility is approximately 10^{-15} cm²/(V s) at 300 K, 10^{-6} cm²/(V s) at 1,200 K, values which can be compared to the data in Table 11.4. Among the factors promoting low values of U_o, let us mention:

– the nature of the ion: a small valence (F^- rather than O^{2-}, Na^+, Ag^+ rather than Ca^{2+}), a small ionic radius (Li^+, H^+);

– but also the characteristics of the environment: the coordination of the ion must be small (tetrahedral rather than octahedral), the arrangement of polyhedron coordination (tetrahedral combinations by faces rather than vertices), the compactness of the crystalline sub-lattice (the migration of oxygen in an oxide of compact structure is difficult and much easier in the open structure of perovskite or fluorite). In a compound M_aX_b, the ease of migration of one of the ions will increase with the polarizable nature of the other (for example Bi, Pb if conduction is anionic).

The concentrations of defects is controlled by a set of coupled equations, mass action laws and laws of conservation, as illustrated in section 11.1.2, and [D] is most often activated thermally. Consequently, the activation energy of ionic conductivity is the sum of two contributions, a transport term and a term reflecting the influence of temperature on the concentration of defects. A change of slope in a diagram Log (σT)–$1/T$ (see Figure 11.10) indicates a change in the defect regime.

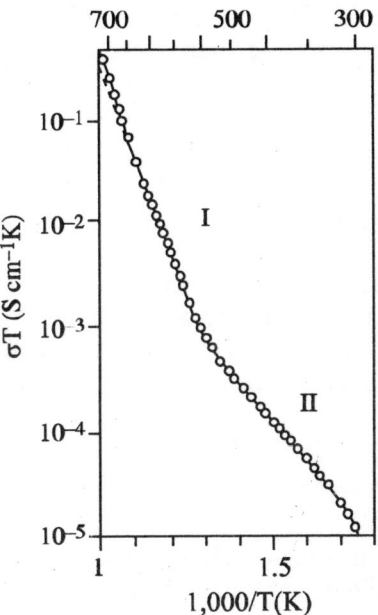

Figure 11.10. *Variation in the ionic conductivity of a monocrystal of NaCl doped with CaCl$_2$ according to temperature. I = intrinsic regime, II = the concentration of defects is fixed by doping (extrinsic regime)*

Let us note that it is sometimes difficult to know whether there is really a change of slope or continuous variation of the activation energy with temperature. In the latter case, U_o is not the slope of $Log(\sigma T)$ versus $x = 1/T$. In fact:

$$\frac{dLn(\sigma T)}{dx} = -U_o - x\frac{dU_o}{dx}$$

is the differential equation which helps to calculate $U_o(T)$ if the curve becomes refined in a wide enough temperature range.

It is possible to fix the concentration of defects by doping. Thus, zirconia ZrO_2 is susceptible to form with lime CaO a solid solution quite extended, with calcium getting substituted to zirconium:

$$CaO \Rightarrow Ca_{Zr}'' + V_O^{\bullet\bullet} \text{ and } \left|V_O^{\bullet\bullet}\right| = \left|Ca_{Zr}''\right| = N_D$$

The concentration of oxygen vacancies thus created is high, around $10^{23}/cm^3$ for a doping of 10 moles %, which explains the high anionic conductivity of stabilized zirconia[6]. This condition of electroneutrality in which only two terms have been retained ceases to be valid in highly reducing atmosphere, the oxygen valencies having a double origin, extrinsic and intrinsic, resulting from the chemical equilibrium:

$$O_O^x \Leftrightarrow V_O^{\bullet\bullet} + 2e' + 1/2O_2(gas) \quad \Delta H_f + E_{d2}$$

(see section 11.1.2). Conduction becomes mainly electronic.

Conductivity as deduced from resistance measurement include the contributions of all charge carriers. It is necessary to make use of additional experimental techniques, [DES 94], in order to evaluate the respective contributions of ions σ_i, and electrons σ_e. We therefore define ionic transport number t_i:

$$t_i = \frac{\sigma_i}{\sigma_i + \sigma_e}$$

and electronic t_e, the sum being equal to 1. These depend on the thermodynamic parameters which fix the thermodynamic state, i.e. T and P_{O2} in the case of binary

---

6 The zirconia structure is monoclinic at ordinary temperature, then under the effect of increasing temperature it becomes quadratic and finally cubic. The addition of a dopant like Ca (but also Mg, Y, a lanthanide) stabilizes the cubic phase.

oxides (the total pressure is fixed at 1 atm.). We can define a domain of ionic conductivity by arbitrarily fixing a value t_i equal to 0.99 (see Figure 11.11).

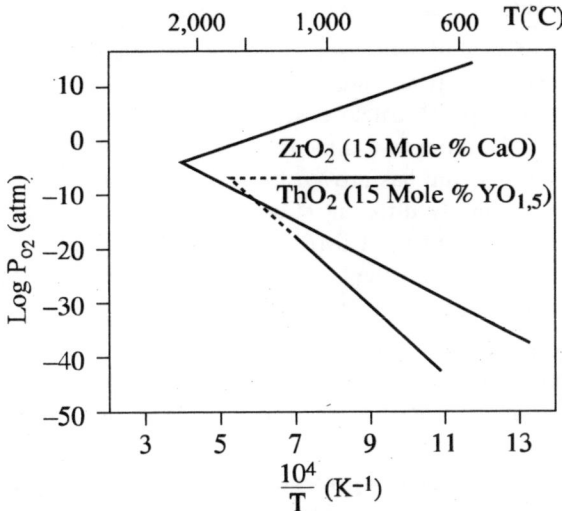

Figure 11.11. *Ionic conductivity domain of two solid electrolyte conductors of oxygen*

Inside this domain, the material is described as a solid electrolyte. In the range $t_i = (0.01-0.90)$, materials are called mixed conductors, with ionic and electronic conductivities of the same order of magnitude. These are often materials showing electronic conduction by small polarons (see section 11.1.2). If the ionic transport number is less than 0.01, the material is an electronic conductor.

Table 11.7 shows that for some materials, U_o is very much smaller than 0.5 eV. The ionic conductivity can exceed 0.01 Ω^{-1} cm^{-1}. These refer to ionic superconductors and several works have been dedicated to this subject [DES 94]. These materials are very diverse. Let us mention:

– silver iodide AgI. With hexagonal structure at ordinary temperature, it assumes a cubic structure (α phase) above 147°C. The entropy variation during transition is close to the fusion entropy of a normal ionic conductor. This corresponds to the appearance of a significant disorder on the cationic sub-lattice at the transition temperature. Interstial sites and normal sites have very close free energies and the concentration of available sites for conduction becomes very high. It is possible to shift the transition temperature below ambient temperature by cationic or anionic substitution, like for example RbAg$_4$I$_5$. The behavior of bismuth oxide Bi$_2$O$_3$ is similar: transformation of the low temperature α phase to a high temperature δ form

729°C, with a significant variation in entropy, a quarter of the anionic sites being unoccupied in the new structure. Conductivity is then $1\ \Omega^{-1}cm^{-1}$. However, the fusion temperature of 830°C restricts the potential uses of this material. For a certain number of compounds, like some fluorides, transition is diffused;

– the existence of a significant disorder on the cationic sub-lattice of sodium ions is also a characteristic of β alumina (11 Al_2O_3, Na_2O). It is made up of Al_2O_3 constituent blocks with spinel structure, separated by planes containing sodium ions. Ionic conduction is therefore two-dimensional. Sodium can be replaced by other monovalent ions, but sodium has proved to be the most mobile. The variant β" (5-7Al_2O_3, Na_2O), with slightly different structure is more conducting. Other similar compounds are $Bi_4V_2O_{11}$ and the BIMEVOX family obtained by substitution of a fraction of vanadium by divalent ions. Copper gives the best results with conductivity as high as $0.01\ \Omega^{-1}cm^{-1}$ at 350°C;

– very open structures promote high ionic conductivity. The displacement of ions takes place in interconnected channels. This is the case of the Nasicon family and the typical example of $Na_3Zr_2PSi_2O_{12}$, with sodium being again the mobile ion.

To be able to speak of ionic superconductors, and there is a certain haziness on this subject, we need at the same time a high concentration of mobile charges ($\cong 10^{22}/cm^3$) and a small value of U_o (thus stabilized zirconia is not an ionic superconductor). The residence time of an ion on its site is comparable to the hopping time and the simple model given here can no longer be applied. The lattice dynamics of the mobile ions is very complex.

11.2. Dielectrics

What are dielectrics? They are first of all insulators, but this is not their full definition: polarization of the material in the presence of an electrical field should be the basic electrical property. Hence, cordierite will be categorized as insulator if it is used for the fabrication of a substrate and as dielectric in the case of a capacitor.

11.2.1. Basic concepts

In a solid, the nuclei and the electrons in perpetual motion are the sources of an intense electromagnetic field, very rapidly varying in space as well as in time. The calculation of this field is all the more difficult since at the atomic level only a quantum description is appropriate. The traditional approach developed by Maxwell consists of dividing the sample into regions, small with respect to the macroscopic scale, but large on the atomic scale, of a few hundreds of cells approximately. The Maxwell field is an average carried out on such a region and also in time. We can then use the electrostatic laws by considering the solid like an assembly of immobile ions.

The potential created at a point $\vec{R}$ by one of these regions is given by:

$$\delta V = \frac{1}{4\pi\varepsilon_o} \sum_i \frac{q_i}{\left|\vec{r}_i - \vec{R}\right|}$$

the summation applying to all the ions in the considered region; ε_o is a universal constant equal to $8.85.10^{-12}$ F/m. At a large distance ($R \gg r_i$), a Taylor expansion in $1/R$ is possible:

$$(4\pi\varepsilon_o)\ \delta V = \frac{\sum_i q_i}{R} + \frac{\left(\sum_i q_i \vec{r}_i\right) \bullet \vec{R}}{R^3} + O(\frac{1}{R^3})$$

Imposing each region to be electrically neutral, the basic contribution is dipolar in nature. Polarization density $\vec{P} = \sum_i q_i \vec{r}_i$ is the dipolar moment per unit volume.

Considering the fact that a region can be assimilated to a point on the macroscopic scale, polarization density may very in space. In the following, it will be assumed to be uniform. In the absence of an applied electric field, polarization density is most often zero because of the cancellation of the contributions of all electric charges. If not, polarization density is said to be spontaneous. With application of an external electric field, an induced polarization density is established.

11.2.1.1. Spontaneous polarization

Spontaneous polarization $\vec{P}_s$ must be invariant for all the operations of symmetry of the crystal. A non-zero polarization density exists only for 10 of the 32 groups of point symmetry. Crystals are then referred to as pyroelectric (see Table 11.8) since polarization density is a function of the temperature.

$p = \dfrac{\partial P_s}{\partial T}$ is known as the pyroelectric coefficient. A few values are given in Table 11.9. Polarization induces charges on the surface of the sample, with a density $q_s = \vec{P}_s\, \vec{n}$, if $\vec{n}$ is the unit vector perpendicular to the surface at the considered point. A variation in temperature therefore gives rise to a current in an external circuit. Pyroelectric materials are used in the detection of infrared radiation (see section 11.6.6).

Bravais network	Piezoelectric crystals	Pyroelectric crystals
Triclinic	1	1
Monoclinic	2, m	2 m
Orthorhombic	2 mm, 222	2 mm
Quadratic	4, 422, 4 mm, $\overline{4}$, $\overline{42}$ m	4, 4 mm
Rhomboedric	3, 32, 3 m	3, 3 m
Hexagonal	6, 622, 6mm, $\overline{6}$, $\overline{6}$ m 2	6, 6 mm
Cubic	23, $\overline{43}$ m	

Table 11.8. *Influence of crystal symmetry on their properties*

Material	p $(10^{-6}$ C/(m^2 K))
Tourmaline	4
$Pb_5Ge_3O_{11}$	95
$LiTaO_3$	190
PZT ceramic	380
$(0.5Sr, 0.5Ba)Nb_2O_6$	600

Table 11.9. *Pyroelectric coefficient p of a few materials*

Ferroelectric materials are pyroelectric materials whose polarization can be reoriented by applying an electric field. On account of their significance, a specific section is dedicated to them. If reorientation takes place under the effect of stress, we speak of ferroelasticity.

11.2.1.2. *Induced polarization density*

Materials which do not display spontaneous polarization are called paraelectric (antiferroelectric materials, which will be dealt with in·section 11.2.2.4 should however be excluded). It is possible to induce polarization in these materials by applying an external electric field or even stress, which is the piezoelectric effect, resulting in the separation of barycenters of positive and negative charges. The increase in free energy per unit volume, at constant temperature and pressure in usual conditions, is equal to:

$$dG = \varepsilon_o \sum_i E_i \ dE_i + \sum_i E_i \ dP_i + \sum_{i,j} X_{ij} \ dx_{ij} \qquad i,j = (1,2,3)$$

if X_{ij} is the symmetrical tensor of stresses and x_{ij} of the deformations. The first term represents the increase in the energy density of the electrical field, the second the necessary work for polarization and the third the mechanical energy generated by material deformation. By defining the electric displacement vector as $\vec{D} = \varepsilon_o \vec{E} + \vec{P}$, this equation can be expressed in the form:

$$dG = \sum_i E_i \ dD_i + \sum_{i,j} X_{ij} \ dx_{ij}$$

If the chosen variables are not electric displacement and deformation, but electrical field and stress, which are easier to control, the generalized free energy:

$$\overline{G} = G - \sum_i E_i D_i - \sum_{i,j} x_{ij} X_{ij}$$

is a minimum at equilibrium, or $d\overline{G} = \sum_i D_i \ dE_i + \sum_{i,j} x_{ij} \ dX_{ij}$

11.2.1.2.1. Polarization density induced by an external electric field

A Taylor expression to the first order of the relation $\vec{P}(\vec{E})$ is generally sufficient, i.e.:

$$P_i = \varepsilon_o \sum_j \chi_{ij} E_j \qquad i,j = \{1,2,3\}$$

χ_{ij}[7] is the symmetrical tensor of dielectric susceptibility and:

$$D_i = \sum_j \varepsilon_o (\varepsilon_r)_{ij} E_j \qquad i,j = (1,2,3)$$

---

[7] χ_{ij} and $(\varepsilon_r)_{ij}$ are symmetrical tensors with 6 distinct components. We often use a notation with a single index, m = 1.6 with the correspondence: $11 = 1$, $22 = 2$, $33 = 3$, $32 = 23 = 4$, $13 = 31 = 5$, $21 = 12 = 6$.

$(\varepsilon_r)_{ij} = 1 + \chi_{ij}$ of relative permittivity. In the case of a piezoelectric material the measuring conditions must be specified: stress ("free sample"[8]) or constant deformation ("clamped sample"), since both the values could be quite different (see equation [11.12]).

The crystal symmetry imposes conditions on the eigenvectors and eigenvalues of the tensors of susceptibility and permittivity. Thus, for a crystal with cubic symmetry, the three eigenvalues are identical, and the susceptibility/permittivity of the material and the eigenvectors can be chosen at will. For a crystal of tetragonal symmetry, the eigenvectors coincide with the crystal axes and two eigenvalues are identical. The relative permittivities vary depending on whether the electric field is applied along the "c" axis i.e. $\varepsilon_{r//}$, or perpendicular to this axis the doubly degenerate value ε_r. A few examples are given in Table 11.10 (for crystals of lower symmetry; see [NYE 87]).

Material	ε_r	$\varepsilon_{r//}$
SiO_2	4.51	4.64
$Pb_5Ge_3O_{11}$	22	40
$LiNbO_3$	30	84
TiO_2	86	170

Table 11.10. *Anisotropy of relative permittivity (tetragonal symmetry)*

Different mechanisms contribute to the induced polarization, for example the deformation of the electronic cloud with respect to the nucleus (polarizability known as electronic) or even the displacement in the opposite direction of cations and anions (ionic polarizability). In the latter case, each normal mode of vibration of the lattice has a specific contribution. It is often considered that the different mechanisms add up their effects, omitting the tensor indices: $\varepsilon_r = 1 + \Sigma \, \delta\chi$.

Matter does not react instantaneously to the application of an external electric field. Each of the mechanisms responds with a specific characteristic time. If the applied electric field is harmonic of angular frequency ω, the response is also

---

8 We have given in brackets the more commonly used expression.

harmonic with the same frequency but with a phase lag between polarization and electric field, well described by a complex susceptibility $\delta\overline{\chi}(\omega)$. Thus, for the two mechanisms mentioned previously, $\delta\overline{\chi}(\omega)$ takes the form:

$$\delta\overline{\chi}(\omega) = \frac{\delta\chi_o}{\omega^2 - \omega_o^2 + j\Gamma\omega}$$

reflecting a damped vibration, solution of a second order differential equation. For other mechanisms, displacement of domain walls in ferroelastic materials, relaxation of dipoles formed by defects of opposite charges, interfacial polarization for example, the response is described by a first order differential equation. The frequency response is defined by the Debye relationship:

$$\delta\overline{\chi}(\omega) = \frac{\delta\chi_o}{1 + j\omega\tau}$$

These two responses, which are most commonly stated, are not the only possible responses. Figure 11.12 gives a diagrammatic representation of the variation of the real part of ε_r with the frequency.

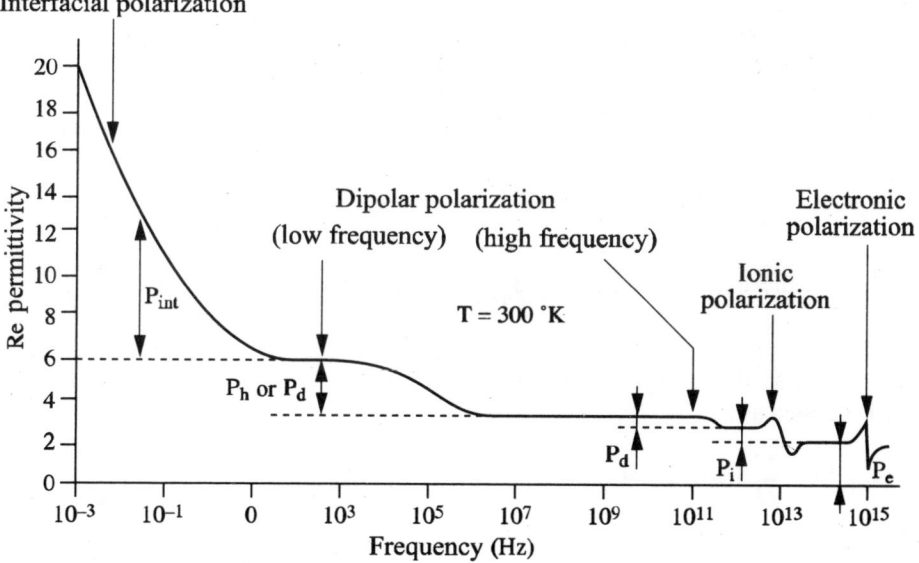

Figure 11.12. *Variations of Re ε_r with frequency (diagrammatic representation)*

The real and imaginary parts of $\varepsilon_r(\omega)$ are not independent, since they obey the integral Kramers-Kronig relations.

According to Maxwell, an alternating current $\vec{J}$ is the sum of two contributions, one due to the displacement of mobile charges (Ohm's law), and the other generated by the reorientation of polarization:

$$\vec{J} = \sigma \vec{E} + \frac{\partial \vec{D}}{\partial t} = (\sigma + j\varepsilon_o \varepsilon_r \omega)\vec{E}$$

i.e. a generalized conductivity:

$$\sigma_t = \sigma - \varepsilon_o \omega \operatorname{Im} \varepsilon_r(\omega) + j\varepsilon_o \omega Re\varepsilon_r(\omega) \qquad [11.11]$$

experimentally obtained from admittance measurements. The imaginary negative part of relative permittivity thus induces a dissipation of energy, described as dielectric losses, very much like conductivity. They increase with the frequency and become essential in the microwave domain. The loss factor $tg(\delta)$ is defined as:

$$tg(\delta) = \frac{\sigma - \varepsilon_o \omega \operatorname{Im} \varepsilon_r}{\varepsilon_o \omega Re\varepsilon_r}$$

δ being the phase lag angle between current and voltage.

11.2.1.2.2. Polarization induced by applying a stress

A Taylor expansion of the characteristic $\vec{P}(X_{ij})$ at constant field (short-circuit):

$$P_i = \sum_{j,k} d_{ijk} X_{jk} \quad i,j,k = \{1,2,3\}$$

defines the piezoelectric coefficient d_{ijk}[9], third order tensor.

If the sample is not connected to a voltage source, the surface charge of the sample and so the electrical displacement is constant. The piezoelectric coefficient g_{ijk} is defined as:

$$E_i = \sum_{j,k} g_{ijk} X_{jk} \quad i,j,k = \{1,2,3\}$$

---

9 $d_{ijk} = d_{ikj}$. There are therefore only 18 distinct constituents and a two-index notation is normally used: d_{im} i = (1.3), m = (1.6).

Both the tensors are not independent:

$$d_{ijk} = \varepsilon_o \sum_l (\varepsilon_r)_{il} g_{ljk}$$

Of the 32 crystal classes, 21 are non-centrosymmetric (not having a center of symmetry), and of these, 20 exhibit direct piezoelectricity (the 21st is the cubic class 432). Ten of these are polar (i.e. they spontaneously polarize), having a dipole in their unit cell, and exhibit pyroelectricity. If this dipole can be reversed by the application of an electric field, the material is said to be ferroelectric (see Table 11.8). Then, the ferroelectric materials are not the only ones to exhibit this effect, as shown by quartz (symmetry group 32), zinc oxide (6 mm) or aluminum nitride (6 mm).

The double derivation of the free energy $\overline{G}$ shows that:

$$d_{ijk} = \left(\frac{\partial D_i}{\partial X_{jk}}\right)_E = \left(\frac{\partial x_{ij}}{\partial E_k}\right)_X$$

predicting a reverse effect, i.e. a deformation of the sample under electric field. When the piezoelectric effect is not allowed for reasons of symmetry, it is necessary to carry out a Taylor expansion of $x_{ij}(E_k)$ up to the next order by introducing a quadratic term with respect to the electric field. This is referred to as electrostriction, characterized by a fourth order tensor. This effect exists all the time irrespective of the symmetry. It is normally very small, with the notable exception of ferroelectric relaxors.

Due to the electromechanical coupling made by the piezoelectric effect, elastic waves can be excited in a piezoelectric element by the application of an alternating voltage. Mechanical resonance occurs if the condition $2L = n\lambda$ is verified, λ the wavelength, L a characteristic length of the element (the diameter or length in the case of a cylinder) and n an integer identifying the excited mode of resonance. The electromechanical coupling made by the piezoelectric effect helps to excite such modes by applying an alternating electric field. The frequency f of the alternating voltage must be equal to v/λ if v is the velocity of the wave, i.e.:

$$f = \sqrt{\frac{\rho K}{\lambda^2}}$$

if ρ is the density and K the appropriate elastic constant for the considered mode. It is also necessary to specify whether it refers to the constant electric field, K^E, or constant electric displacement, K^D, elastic constant. There are therefore two frequencies known as resonance (f_r) and antiresonance (f_a) for each mode. With a few approximations [ROY 99], the simple equivalent circuit of Figure 11.13 is a good representation of the frequency behavior of the element. It consists of two branches in parallel, corresponding to the dielectric (C_0) and mechanical (L_1, C_1, R_1) properties respectively. The inductance L_1 represents the inertial effect of the mass of the element, the capacitance C_1 the elastic restoring force, the resistance R_1 a dissipation of energy.

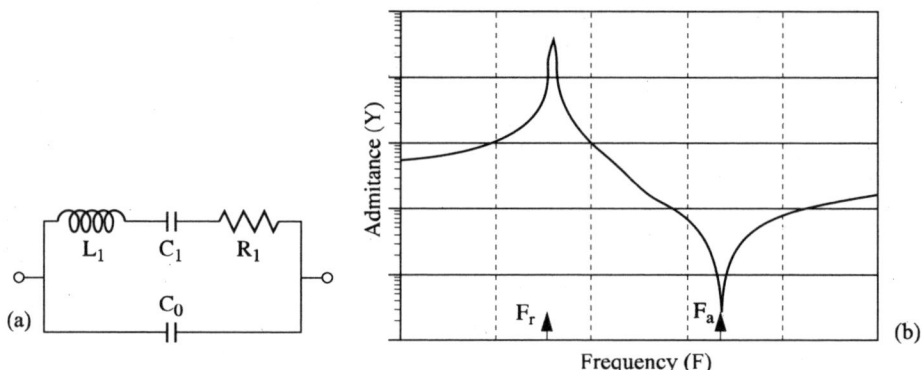

Figure 11.13. *a) Equivalent circuit for a piezoelectric material subjected to an alternating electrical field and b) variations of the admittance with frequency*

The presence of electrical resonances and anti-resonances make the piezoelectric impedance unique (see Figure 11.13). They are given by:

$$L_1 C_1 \omega_r^2 = 1 \text{ and } L_1 \frac{C_o C_1}{C_o + C_1} \omega_a^2 = 1 \quad \omega_{a,r} = 2\pi f_{a,r}$$

At resonance, the impedance Z_1 is minimal ($= R1$) and the same applies to the overall impedance Z. At antiresonance, $Z_o + Z_1$ is minimal and Z is maximal.

The piezoelectric effect helps to convert the electrical energy into mechanical energy. The conversion factor is commonly noted as k:

$$k^2 = \frac{W_{meca}}{W_{elec} + W_{meca}} = \frac{K^E - K^D}{K^E} = \frac{\varepsilon^X - \varepsilon^x}{\varepsilon^X} \qquad [11.12]$$

It depends on the chosen configuration of the element (see Table 11.11).

Vibratory mode			
longitudinal	**transversal**	**planar**	**shear**
$k_{33}^2 = \dfrac{d_{33}^2}{K_{33}^E \varepsilon_{33}^X}$	$k_{31}^2 = \dfrac{d_{31}^2}{K_{11}^E \varepsilon_{33}^X}$	$k_p^2 = \dfrac{2d_{31}^2}{\left(K_{11}^E + K_{12}^E\right)\varepsilon_{33}^X}$	$k_{31}^2 = \dfrac{d_{15}^2}{K_{55}^E \varepsilon_{11}^X}$

Table 11.11. *Some vibration modes of piezoelectric ceramics*

11.2.2. *Ferroelectric materials*

11.2.2.1. *Para-ferroelectric transition*

For some paraelectric materials, the decrease of the temperature induces a spontaneous polarization for a well-defined temperature T_c, called the Curie temperature. This transformation displays all the characteristics of a phase transition, $\vec{P}_s$ being the order parameter. Below T_c, $\vec{P}_s$ varies with the temperature giving rise to the pyroelectric effect. The low temperature phase is called ferroelectric phase. The transition is generally accompanied by a change in structure. A typical example is barium titanate $BaTiO_3$, of perovskite structure, cubic above T_c (=120°C) and tetragonal below (see Figure 11.14). In the ferroelectric state, polarization results from displacements (of the order of 0.01 nm) of cations and anions in opposite directions). A further decrease in the temperature promotes successive changes in the crystal structure, but the material remains ferroelectric (different ferroelectric states). The driving force is the resulting decrease in the free energy.

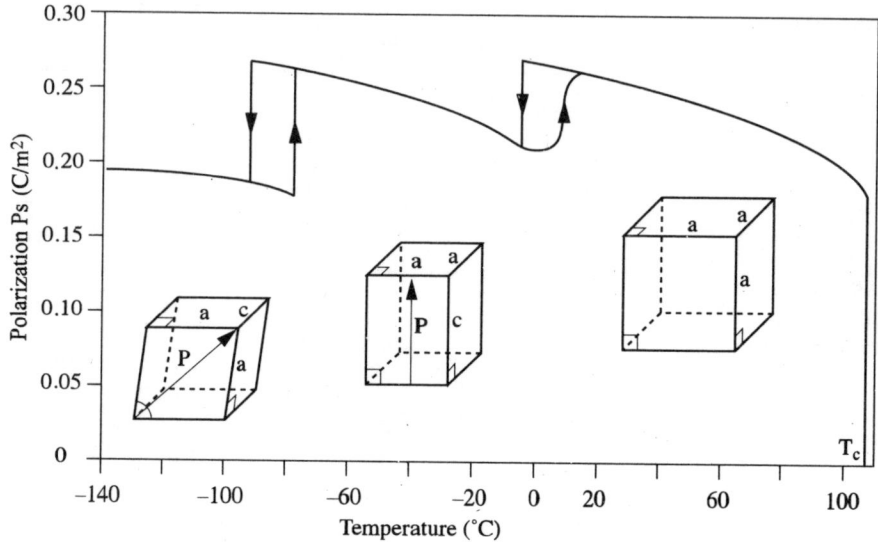

Figure 11.14. *Variation of spontaneous polarization with temperature for the compound BaTiO$_3$*

11.2.2.2. *Discussion on energy balance*

The interaction energy of two dipoles ($\vec{p}_1, \vec{p}_2$) is equal to $-\vec{p}_1.\vec{E}(\vec{p}_2)$ if $\vec{E}(\vec{p}_2)$ is the electric field created at dipole 1 by dipole 2. The calculation shows that this energy, proportional to the product $\vec{p}_1, \vec{p}_2$, is negative and minimal if the dipoles are parallel and both perpendicular to the line that joins them. It is this interaction which tends to align the dipoles along a common direction and which explains the stability of the ferroelectric state. However, creating permanent dipoles necessitates the displacement of ions and this increases the free energy of a quantity U of elastic origin, i.e.:

$$G = G_o(T) + U_{elastic} - U_{dipolar}$$

$G_o(T)$ being the free energy of the paraelectric state. For small deformations, "*u*", the interaction potential between the ions, can be considered as harmonic:

$$U_{elastic} \approx \frac{1}{2}ku^2 \approx AP^2$$

with polarization being proportional to deformation. Both the terms are therefore quadratic in P. Assuming a linear temperature dependence of the stiffness k, the free energy G is written again as:

$$G = G_o(T) + \alpha \ (T - T_c)P^2 = G_o(T) + \frac{1}{2}k_{app}u^2$$

The ferroelectric state is therefore stable for temperatures less than T_c. Let us note that everything takes place as though there was an apparent stiffness k_{app}. In the paraelectric phase, the frequency of one of the vibration modes of the lattice, referred to as soft mode:

$$\Omega^2 \cong \frac{k_{app}}{m} \approx (T - T_c)$$

gets canceled at the transition temperature. A permanent deformation, like the one appearing in the ferroelectric state, can in fact be considered as a singular vibration mode of zero frequency.

Spontaneous polarization is the polarization that minimizes the free energy. For $T = T_c$, $G = G_0$, which shows the inadequacy of a Taylor expansion limited to the 2nd order. We must take into account the inharmonic contribution to the elastic energy, keeping only to even order terms, the polarizations P_s and $-P_s$ being equally possible, i.e.:

$$G = G_o(T) + \alpha(T - T_c)P^2 + \beta P^4$$

Minimizing G, we get:

$$0 = \alpha(T - T_c) + 2\beta P_s^2 \text{ i.e. } P_s(T) \approx \sqrt{(T_c - T)}$$

In the paraelectric phase, in the presence of an external field:

$$E = \left(\frac{\partial G}{\partial P}\right)_T \text{ i.e. } E = 2\alpha(T - T_c)P$$

Susceptibility therefore varies with temperature according to the Curie-Weiss law:

$$\chi = \frac{C}{T - T_c}$$

It can also be proved that transition is of the second order. To explain a first order transition, as experimentally observed for barium titanate and several other oxides [XU 91], the free energy has to be expanded up to sixth order term. The previous relations still hold good on the condition of replacing T_c by T_o (slightly lower than T_c).

11.2.2.3. *Ferroelectric domains*

In the ferroelectric state, polarization can be oriented along several equivalent crystallographic axes (which is not the case for non-ferroelectric pyroelectric material), which explains after cooling below T_c the formation of domains polarized uniformly but in different directions (see Figure 11.15). Globally, the material appears as non-polar.

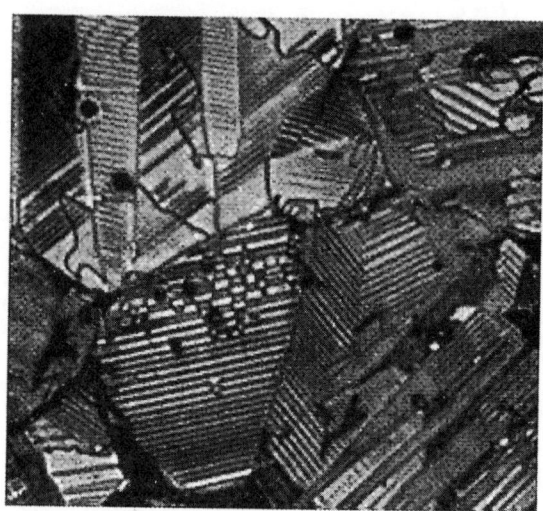

Figure 11.15. *Observation of domains in a BaTiO$_3$ ceramic*

To induce a macroscopic polarization, an external electric field has to be applied, which promotes the displacement of the domain walls. For low electric fields, polarization is proportional to the field, which helps to define an apparent relative permittivity, function of the temperature (see Figure 11.16). This depends on the orientation of the electric field with respect to the crystallographic axes, as already stated earlier. For a ceramic sample, the relative permittivity measured is an average of all possible orientations.

For higher values of the applied field, polarization is no longer a linear function of the field and tends towards a saturation value (see Figure 11.17). By decreasing the amplitude of the electric field, hysteresis is commonly observed. The field which must be applied to remove polarization is called the coercive field. The cause of hysteresis is the presence of imperfections, like point defects, impurities, dislocations, grain boundaries and pores in the case of ceramic samples, which impede the domain walls displacements.

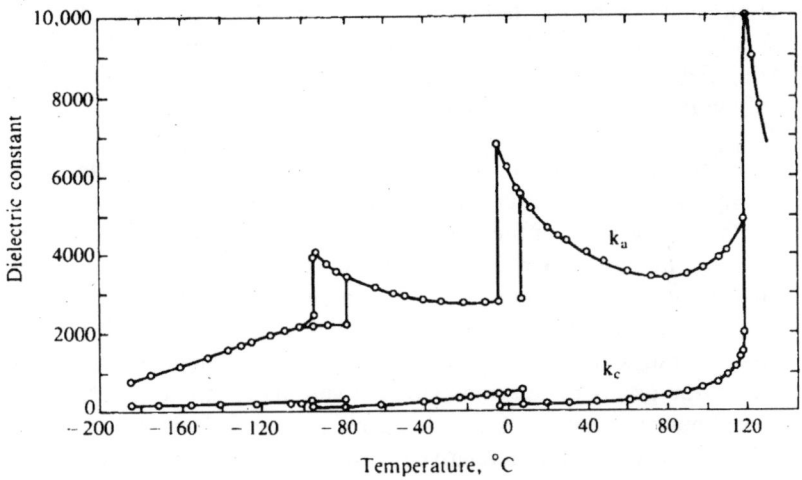

Figure 11.16. *Variation of apparent relative permittivity of BaTiO₃ with temperature. Case of a single crystal*

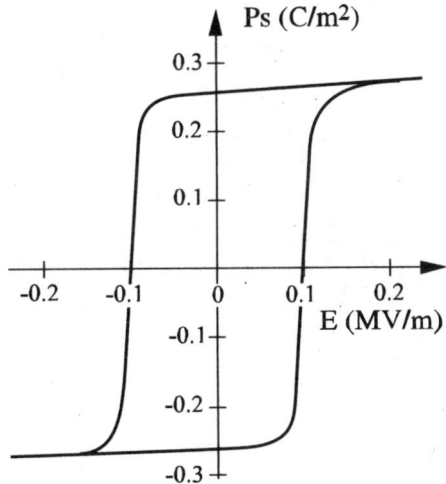

Figure 11.17. *Hysteresis loop for a BaTiO₃ single crystal*

The surface of the cycle $\oint E.dP$ represents the dissipated energy.

On both sides of a domain wall, ions displacements are in opposite directions. The existence of such a wall is therefore associated with an additional energy of elastic origin. We have already mentioned the fact that a uniform polarization induces a charge at surfaces, a source of an electrical field that is referred to as depolarizing, since it opposes polarization [KIT 98]. An appropriate distribution of domains helps to decrease this surface charge and thus the contribution of the depolarizing field to the total energy. The multi-domain configuration at equilibrium is the configuration that optimizes the net gain in free energy.

In a polycrystalline sample of barium titanate, the cubic-quadratic transformation of differently oriented grains develops considerable internal mechanical stresses. Nucleation and the growth of the domains at 90° help to relax these stresses and to reduce the induced elastic energy. The energy for wall creation varies like D^2 if D is the average diameter of the grains, while the energy gain due to the relaxation of the stresses grows like D^3. It therefore appears that a multi-domain configuration is stable only for grain sizes greater than a critical size D_c of approximately 1 μm. For smaller sizes, the grains remain monodomain and under stress, the c/a ratio of the quadratic structure becomes closer to 1. In this manner it is possible to understand the influence of the grain size on relative permittivity (see Figure 11.18).

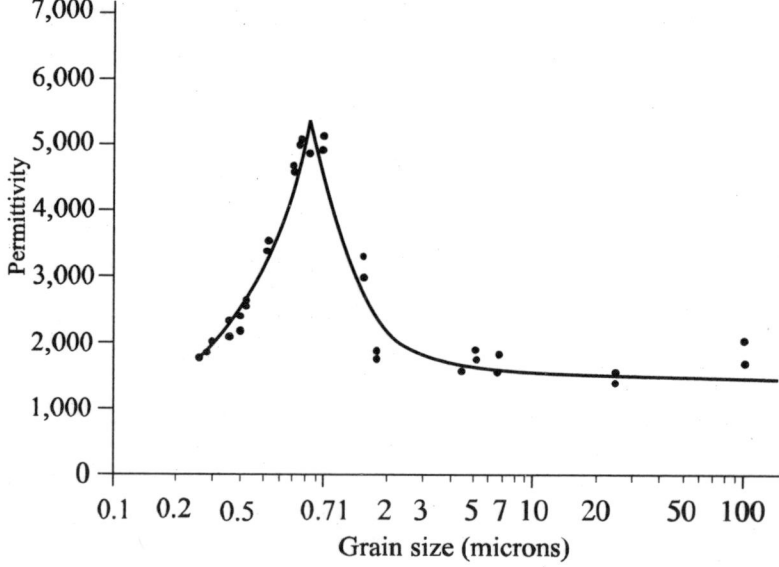

Figure 11.18. *Influence of the grain size on the relative permittivity of a ceramic of $BaTiO_3$ ($T = T_{amb}$)*

A ferroelectric material is also piezoelectric. The variations in deformation during a cycle, i.e. x(E), are shown in Figure 11.19. For piezoelectric applications, the material is previously polarized by the application of a high field (several kV/mm). The apparent piezoelectric coefficient is defined as the local slope of the curve x(E) close to E = 0.

$$d_{jk} = \left(\frac{\partial x_{ij}}{\partial E_k} \right)_X \ (E \cong 0)$$

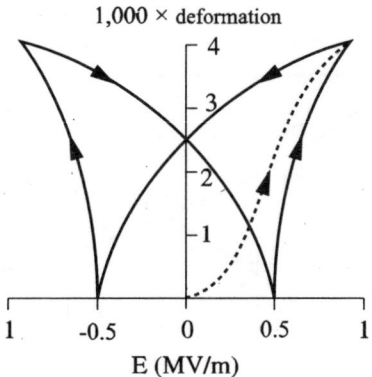

Figure 11.19. *Variations in deformation with the electric field on cycling (the dotted curve is the first polarization curve)*

Density: $\rho = 6.01$ g/cm^3
Relative dielectric permittivities: $\varepsilon_{33}^T = 168$, $\varepsilon_{11}^T = 2,920$
Piezoelectric coefficients (in 10^{-12}C/N): $d_{33} = 85.6$; $d_{31} = -34.5$; $d_{15} = 392$
Coupling coefficients: $k_{33} = 0.56$; $k_{31} = 0.32$; $k_{15} = 0.57$
Elasticity constants (in 10^{-12} m^2/N): $K_{11}^E = 8.05$; $K_{12}^E = -2.35$; $K_{13}^E = -5.24$ $K_{33}^E = 15.7$; $K_{44}^E = 18.4$; $K_{16}^E = 8.84$

Table 11.12. *Characteristics of single crystalline BaTiO$_3$*

NOTE.– the tensor of the elasticity modulus is of the 4th order since it links deformation and stress: $x_{ij} = \sum K_{ijkl} X_{kl}$. It is doubly symmetrical with 36 distinct components, hence the use of 2 indices m and n varying from 1 to 6, i.e. K_{mn}.

The domains in the grains of a ceramic are generally in a metastable state. The evolution towards a state of equilibrium is known as aging. Empirically, the variations in characteristics are quite slow following a logarithmic scale (see Figure 11.20). In view of the utilization, the components are aged prior to marketing.

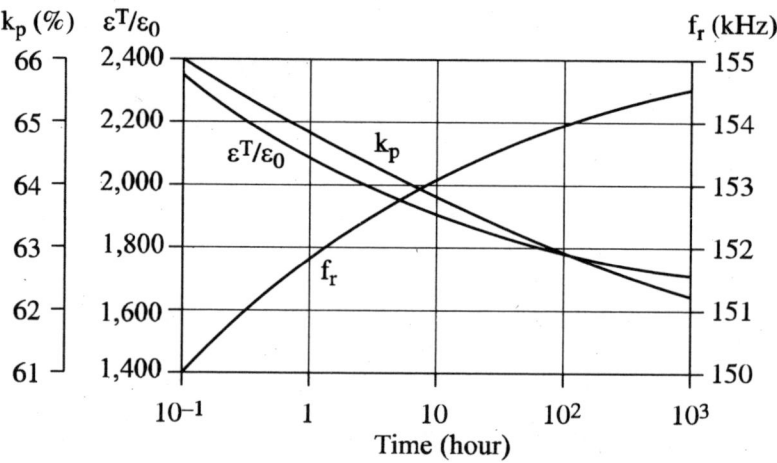

Figure 11.20. *Evolution in time of the electrical characteristics (relative permittivity ε_r, coupling factor k_p, resonance frequency f_r) of a PZT ceramic*

11.2.2.4. *Antiferroelectric materials*

In certain cases, ion displacements minimize the free energy below a temperature T_c, but without the appearance of a spontaneous polarization. However, a maximum of relative permittivity is observed, associated with a structural change. The low temperature state is referred to as antiferroelectric [XU 91]. This applies to $PbZrO_3$, for example. The substitution of zirconium by a few percent of titanium is enough to stabilize a ferroelectric state, which proves that these states have neighboring free energies.

11.2.3. *Relaxors*

The perovskite structure of formula ABO3 may accommodate many elements in A and B sites, leading to complex compositions with particularly interesting properties. Let us cite $Pb(Zr,Ti)O_3$ (PZT) (see section 11.6.3), $Pb(Mg_{1/3}, Nb_{2/3})O_3$ (PMN), among others, with lead in the A sites. Sometimes an adequate heat treatment is sufficient to order cations on B sites, for example, scandium and tantalum in the compound PST $\{Pb(Sc_{1/2}, Ta_{1/2})O_3\}$. The material then displays an ordinary paraferroelectric transition. The ordering kinetics may be particularly slow, as in the case of compounds $Pb(B'_{1/3}, B''_{2/3})O_3$. These perovskites, in disorder and metastable, exhibit an unusual dielectric behavior known as relaxors, which we will now be described.

Varying temperature, the permittivity goes through a maximum for some temperature T_m, as in the case of an ordinary ferro-paraeelectric transition, but the curve $\varepsilon_r(T)$ is considerably larger and furthermore T_m varies with the frequency of the measuring signal (see Figure 11.21). The transition is said to be diffuse.

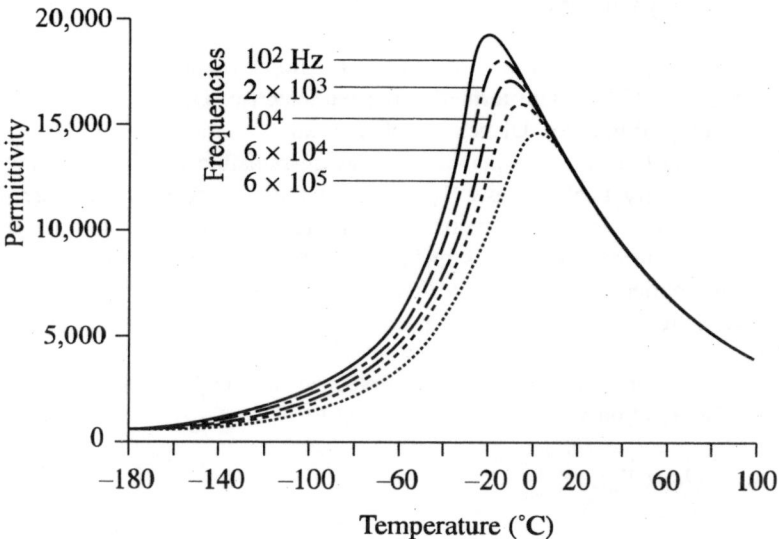

Figure 11.21. *Variation of apparent relative permittivity of PMN as a function of temperature for different measurement frequencies*

For temperatures lower than T_m, a pyroelectric effect is seen, implying the presence of a spontaneous polarization which disappears at T_m. The same applies to the hysteresis loop.

As deduced from thermal expansion measurements, a polarization is still present for temperatures greater than T_m, but fluctuating in time and in space with a zero mean: no pyroelectric effect can be detected. These fluctuations persist for temperatures much above T_m, for example in the case of PMN up to 200–300°C ($T_m \approx -15$°C). The relationship P(E) is strongly non-linear. Dielectric susceptibility measured shows significant differences with respect to the Curie-Weiss law.

The structure remains cubic at any temperature. However, for temperatures less than T_m, diffusion halos in the x-ray diffraction spectra show the existence of a disorder on the atomic scale. Furthermore, no domain structure is seen and no soft mode is detected by infrared spectroscopy.

The disorder on sites is clearly at the origin of this particular behavior. It generates not only variations in composition, but also differences with respect to electrical neutrality, cations B', B'' having different charges, Mg^{2+} and Nb^{5+} in the case of PMN. The presence of point defects, oxygen and lead vacancies is susceptible to ensure the local electroneutrality and to stabilize the composition disorder. Thus, the elimination of lead vacancies in a PST ceramic is sufficient to induce an ordinary paraferroelectric transition.

It is usually admitted that polar domains of nanometric size exist at temperatures much higher than T_m, a state termed super-paraelectric. During cooling, they will increase in size and number. Dispersion of sizes and variations in composition are at the origin of the frequency dispersion and are responsible for the widening of the relative permittivity peak. Below T_m, the polar phase occupies a large volume fraction of the sample, even attains the percolation threshold, without however invading the whole sample, like is the case for any ordinary ferroelectric material. However, an applied electric field is susceptible to provoke the transition towards a ferroelectric state.

Under the effect of a stress (X_{ij}) or an electric field (E_k), a relaxor material gets deformed. The relation $x(X_{ij}, P_k)$ is quadratic in P:

$$x_{ij} = Q_{ijkl}(X_{mn})P_k P_l$$

a behavior referred to as electrostrictive. Experiments show that the relation P(E) is non-linear but reversible, unlike the ferroelectric materials. For isotropic ceramics, it is well represented by the relation:

$$P = A\ th(\gamma E)$$

in agreement with a thermodynamic model of independent dipoles. The combination of the two previous equations then allows the calculation of x(E,X) (see Figure

11.22). Local values of the compliance, given by the local slope of the curve x(X), $s = \dfrac{\partial x}{\partial X}$, of the piezoelectric coefficient given by the local slope of the curve x(E), $d = \dfrac{\partial x}{\partial E}$, and the coefficient of electrostriction (local curvature of x(E)), $M = \dfrac{1}{2}\dfrac{\partial^2 x}{\partial E^2}$ can be defined. They are functions of electric field and stress, which makes them adaptative. The absence of hysteresis is a sure advantage in the applications.

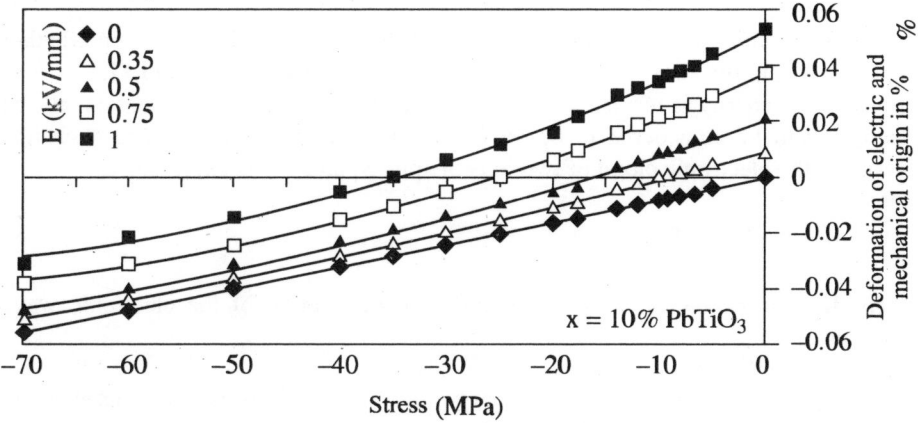

Figure 11.22. *Characteristic deformation versus stress for different values of the electric field in the case of a relaxor ferroelectric material (PMN-PT)*

11.3. Magnetic materials

All materials placed in an applied magnetic field $\vec{H}_{app}$ display a diamagnetic behavior. This is a reaction of the material which tends to screen the applied field by a modification of spatial distribution of electrons. The induced magnetization is proportional to $\vec{H}_{app}$, but in the opposite direction. This effect, of low amplitude, can be neglected when some ions of the solid have an incomplete electronic shell and consequently a total non-zero spin. They carry a magnetic moment equal to n_B μ_B, if μ_B is the Bohr magneton [KIT 98]. Some values are given in Table 11.13. This is the case of transition metal ions and lanthanides.

Ion	n_B
Fe^{3+}, Mn^{2+}	5
Fe^{2+}	4
Co^{2+}	3
Ni^{2+}	2

Table 11.13. *Theoretical number of Bohr magnetons of transition metal ions*

The Coulomb's repulsive energy of two electrons differs depending on whether their spin are in the same or opposite direction (see sections 11.1.3 and 11.1.5). The difference in energy is termed as exchange energy. In the case of insulators, all the electrons can be considered to be localized on an atomic nucleus. The exchange energy of the electronic cloud of two ions of spin $\vec{S}_1$ and $\vec{S}_2$ is expressed as:

$$\delta U = -2J_{12}\vec{S}_1 \ \vec{S}_2$$

where J_{12}, the exchange integral, depends on the overlap of the electronic wave functions.

The free energy F, minimal in the equilibrium state, is the difference in an energetic contribution U and an entropic contribution TS. There is therefore competition between a spin ordering process, lowering U, and a tendency for disorder which increases S. At high temperatures, the entropic contribution is preponderant and the material is paramagnetic. During cooling, the appearance of a magnetic order is possible, for a critical temperature T_c.

11.3.1. *Paramagnetism*

In the paramagnetic state, the application of a magnetic field $\vec{H}_{app}$ induces a magnetization $\vec{M}$ proportional to the magnetic field and in the same direction:

$$\vec{M} = \chi_m \vec{H}$$

thus defining magnetic susceptibility χ_m[10]. Magnetic induction $\vec{B}$ [11] is equal to:

$$\vec{B} = \mu_o \vec{H} + \vec{M} = \mu_o(1 + \chi_m)H = \mu_o \mu_r \vec{H}$$

μ_r is the relative permeability of the material. χ_m varies with the temperature following the Curie-Weiss law:

$$\chi_m = \frac{C}{T - T_o}$$

the parameter T_0 could be positive if the material becomes ferro-(or ferri-) magnetic at low temperature, negative if it becomes antiferromagnetic, or zero in the absence of the magnetic order at low temperature.

11.3.2. Antiferromagnetism

In the oxides, each cation is surrounded by oxygen ions of zero spin and the exchange interaction between cations is indirect: it is the interaction of superexchange already discussed in section 11.1.5. It is negative, promoting the antiparallel alignment of spins when there is only one kind of cation occupying the same type of site. These materials are antiferromagnetic below a critical temperature, called Néel temperature T_N. There is no spontaneous magnetization, the cationic and anionic sub-lattices compensating each other. However, the magnetic order can be proved by neutron diffraction. This applies to binary oxides of transition metals (see Table 11.14) or even zinc ferrite, $ZnFe_2O_4$.

11.3.3. Ferrimagnetism

Occupation of different sites with different cations induce a complex behavior.

---

10 Strictly speaking, magnetic susceptibility is a second order tensor, but ferrites are materials of cubic symmetry and the tensor gets reduced to a scalar.
11 Let us remember the units in the international system: the induction in Tesla (T) or even Weber/m^2, the magnetic field in A/m $\mu_0 = 4\pi 10^{-7}$ SI.

Oxide	T_N (K)	T_0 (K)
MnO	118	−417
FeO	198	−500
CoO	292	−300
NiO	523	−2,000

Table 11.14. *Examples of antiferromagnetic oxides* $(kT_N \cong J_{12})$

This is the case for ferrites MFe_2O_4 with spinel structure (cubic symmetry), whose magnetite Fe_3O_4 is a well-known example. It consists of eight tetrahedral sites A and 16 octahedral sites B per cell. The exchange integrals J_{AA}, J_{BB} and J_{AB} are all negative, the interactions A-B being predominant. Consequently, spins of cations A are parallel with each other as well as those of cations B but in opposite directions. Thus, in the case of $NiFe_2O_4$, the ions Fe^{3+} get divided in half between the sites A and B and the ions Ni^{2+} occupy the remaining sites B. The resulting moment is 16 μ_B/cell or even 2 μ_B/group MFe_2O_4. The small value of the ionic radius of the ion Zn^{2+} explains the preference for a tetrahedral coordination. Replacement of nickel by zinc is reflected by a transfer of Fe^{3+} ions from the tetrahedral sites to the octahedral sites. There is no longer compensation of Fe^{3+} ions moments on both the sub-lattices A and B. The addition of a non-magnetic element like zinc is reflected by an increase of the spontaneous magnetization (see Figure 11.23). For a Zn percentage greater than 40%, an effect of dilution of magnetic ions and a weakening of the exchange interaction A-B explains the decrease of the magnetization.

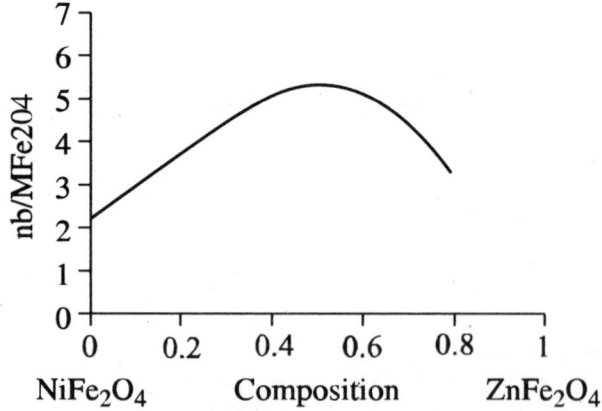

Figure 11.23. *Effect of replacement of nickel by zinc on magnetization at saturation*

When substitution is complete, Fe^{3+} ions in $ZnFe_2O_4$ occupy octahedral sites only. The material is antiferromagnetic as mentioned earlier.

The magnetic properties of iron garnets can be interpreted in the same manner. They are oxides of general formula $M_3Fe_5O_{12}$ or $(3M)_c(2Fe)_a(3Fe)_dO_{12}$, showing the dodecahedral (c), octahedral (a) and tetrahedral (d) cationic sites of the garnet structure (cubic symmetry). M is a trivalent non-magnetic metallic ion, most often yttrium (yttrium iron garnet (YIG)). The resulting magnetic moment is of $5 \mu_B$/group $M_3Fe_5O_{12}$. If the ions M are lanthanides, they also contribute to the overall magnetization. Since both the contributions change differently with temperature (the exchange interactions are different for the ions M^{3+} and Fe^{3+}), a peculiar variation of the spontaneous magnetization with temperature can be seen in Figure 11.24.

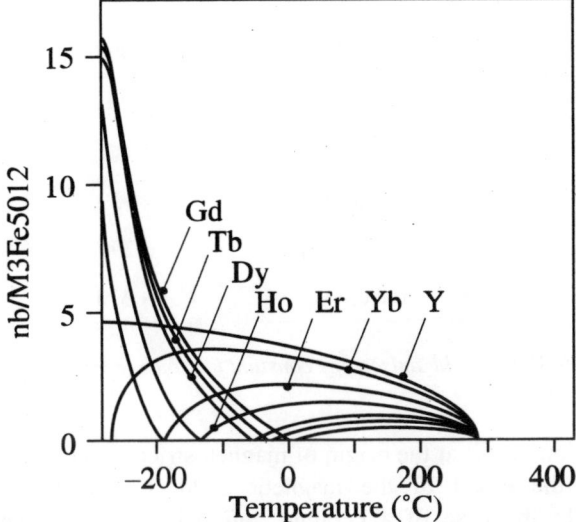

Figure 11.24. *Influence of temperature on spontaneous magnetization of iron garnets*

We must also mention the family of hexaferrites with magnetoplumbite structure (hexagonal symmetry) of composition $MFe_{12}O_{19}$, where M(= Ba, Sr, Pb, etc.) is a divalent cation. It is possible to substitute other trivalent ions (= Al^{3+}, Ga^{3+}, etc.) for the Fe^{3+} ions.

In all these compounds, it is crucial to control the distribution of cations in the different sites. It depends on the temperature and partial pressure of oxygen, which controls the degree of oxidation and even conditions of cooling if the equilibrium state is not achieved. We refer to ferrimagnetism when there is partial compensation of different sub-lattices.

The common direction of spin is not arbitrary, but is determined by an interaction spin-lattice via the orbital motion of electrons. In the case of cubic symmetry, the interaction energy as a first approximation has the following form:

$$E = K\left\{\cos^2\left(\theta_1\right)\cos^2\left(\theta_2\right)+\cos^2\left(\theta_1\right)\cos^2\left(\theta_3\right)+\cos^2\left(\theta_2\right)\cos^2\left(\theta_3\right)\right\}$$

if θ_1, θ_2, θ_3 are the angles defining the orientation of spontaneous magnetization with respect to the three axes of the crystallographic structure. If K is a positive constant, the direction of easy magnetization is [100], in the reverse case it is [111] (see examples in Table 11.15).

Ferrite	K in kJ/m^3
Fe_3O_4	−11
$NiFe_2O_4$	−6.2
$Ni_{0.5}Zn_{0.5}Fe_2O_4$	−3
$Co_{0.8}Fe_{2.2}O_4$	+290
$BaFe_{12}O_{19}$	+ 330

Table 11.15. *Value of the anisotropy constant for some ferromagnetic materials*

This interaction is still at the origin of magnetostriction, i.e. a deformation of the material under the effect of the magnetic field. Conversely, a stress induces magnetization. In the case of a ceramic, internal stresses which appear during sintering are likely to decrease the effective permeability of the material.

On cooling, the material gets divided into uniform magnetization domains, called Weiss domains. The reorientation of domains under the effect of a magnetic field is the source of a hysteresis loop B(H$_{app}$) (see Figure 11.25) similar to the hysterisis cycle P(E) of a ferroelectric material. Ferroelectricity and ferro(or ferri)-magnetism formally show several analogies.

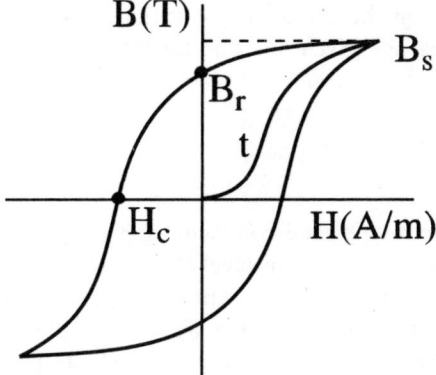

Figure 11.25. *Hysteresis loop B(H_{app}). B_r (B/μ_0) remanent induction, H_c coercive field*

11.4. Electronic properties of surfaces and interfaces in semi-conductor ceramic materials

Surfaces and interfaces break the periodicity of the ideal crystal. This can be taken into account by introducing surface (interface) electronic levels in the forbidden gap, of density of states $N_s(E)$. The filling of these specific states defines a Fermi $E_{F,s}$ level different from the bulk Fermi level E_F.

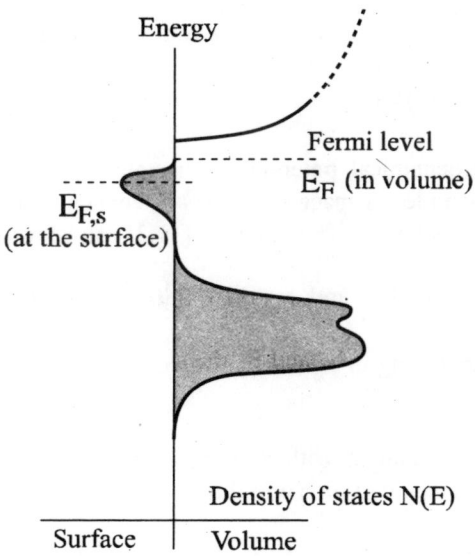

Figure 11.26. *Density of electronic states at the surface of a semi-conductor*

For more details on the origin of the surface states, see [NOW 88], for metal semi-conductor states, see [SHA 84], and for interfaces in polycrystalline materials, see [HAR 85].

11.4.1. *Surface*

Let us consider an n-type semi-conductor doped with donors of concentration N_d and let us assume the surface Fermi level ($E_{F,s}$) to be lower in energy than the bulk Fermi level (E_F) (see Figure 11.26). The difference in Fermi levels, or chemical potential, gives rise to an electron flows from the highest chemical potential to the lowest potential. The electrons get accumulated at the surface, leaving behind ionized doping elements. This non-uniform charge distribution, negative at the surface, positive in the depletion layer (also known as space charge), creates an electric field. Quantitatively, the electron flow is written as:

$$\vec{J}_{e'} = qD_{e'}\vec{grad}(n) + nq\Lambda_e\vec{E} \qquad [11.13]$$

if n is the electronic concentration, and D_e the coefficient of diffusion associated with mobility Λ_e using the Nernst-Einstein relation:

$$\frac{qD_{e'}}{kT} = \Lambda_e$$

Introducing the chemical potential μ (see section 11.1.2, equation [11.2]), we get:

$$\vec{J}_{e'} = n.\Lambda_e \; \vec{grad}(\mu - qV)$$

$\eta = \mu - qV$ is the electrochemical potential of the electrons. At equilibrium, it is uniform in the whole sample, as space variations of concentration accompany space variations of the electric potential. The relations [11.2] can be written in the form:

$$n = N_c \exp\{(E_c - \eta)/kT\} \text{ and } p = N_v \exp\{(\eta - E_v)/kT\} \qquad [11.14]$$

with the condition of including in E_c and E_v the contribution ($-qV$). These energies then vary in space.

The variations in potential V with x, if x is the distance at the surface, are determined by Poisson's equation. In a one-dimensional model:

$$-\frac{d^2V}{dx^2} = \frac{qN_d}{\varepsilon_o\varepsilon_r}$$

if N_d is the dopant concentration, ionized in the depletion layer [0, w] and ε_r the relative dielectric permittivity of the material. The electric field being zero at the depletion layer boundary, the solution of this differential equation is:

$$V(x) = -\frac{qN_d}{2\varepsilon_o\varepsilon_r}(x-w)^2$$

thus justifying the parabolic appearance of the band diagram according to distance (see Figure 11.27). This increase in energy at the interface is a potential barrier ϕ_B, called a Schottky barrier, opposing the transfer of electrons across the interface.

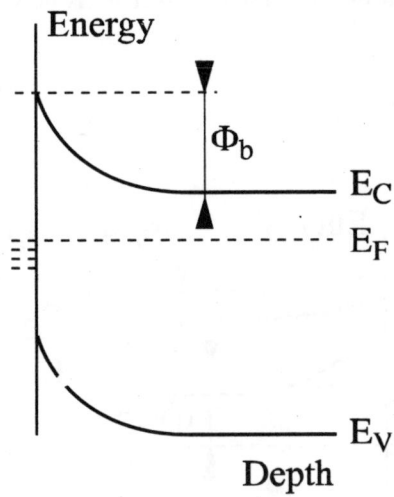

Figure 11.27. *Schottky barrier at equilibrium*

The depth of the depletion layer w is determined by the surface charge Q:

$$Q = -qN_dw \cong -q \int_{E_{F,s}}^{E_F} N_s(E)\,dE$$

The barrier height $\phi_b = -qV(0)$ is equal to:

$$\phi_b = \frac{q^2N_d}{2\varepsilon_o\varepsilon_r}w^2 \qquad\qquad [11\ 15]$$

We will see that it can be deduced from conductance or capacitance measurements, i.e. typically a few tenths of eV. The semi-conductor is depleted on a depth of a few nm to a few μm.

A device with two Ohmic metallic contacts on either side of the junction has a non-linear current-voltage characteristic. In fact, conduction is possible only for electrons having sufficient energy to cross the potential barrier, i.e. a proportion $\exp\{-\phi_b/kT\}$. In the absence of an external applied voltage, the charge flows in both directions are equal and the total current is zero. An applied voltage increases or decreases the potential barrier, following the V_{app} sign, in the semi-conductor-metal direction, i.e. $(\phi_b + qV_{app})$. It remains unchanged in the opposite direction (see Figure 11.28). The depth of the depletion layer w' is now the solution of equation [11.16]:

$$\phi_b + qV_{app} = \frac{q^2 N_d}{2\varepsilon_o \varepsilon_r} w'^2$$

[11.16]

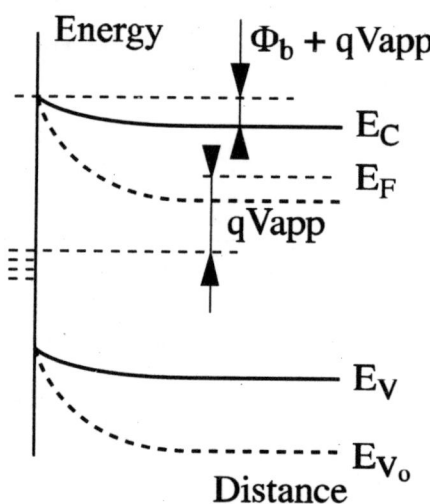

Figure 11.28. *Schottky barrier in the presence of an applied electric field*

The charge flows are no longer balanced and the total current is equal to:

$$J = J_o \exp\{-\phi_b / kT\}\left[1 - \exp(-qV_{app} / kT)\right]$$

For low applied voltages, a first order Taylor expansion lead to an Ohmic behavior with a thermally activated conductivity:

$$\sigma \approx \exp\left(-\phi_b \, / \, kT\right)$$

The Schottky barrier is associated with a high electrical resistance. Measurements of resistance at different temperatures help to determine ϕ_b.

A variation of applied potential δV_{app}, induces a charge variation:

$$\delta Q = -qN_d \delta w$$

Differentiating equation [11.16], we get:

$$\delta V_{app} = \frac{qN_d w \delta w}{\varepsilon_o \varepsilon_r}$$

It is possible to define a dynamic capacitance C_{dyn}:

$$C_{dyn} = \frac{|\delta Q|}{\delta V_{app}} = \frac{\varepsilon_o \varepsilon_r}{w}$$

This is the formula giving the capacitance of a flat capacitor of thickness w and of unit surface, but only describing the dynamic behavior of the Schottky barrier. This capacitance varies with the applied voltage according to:

$$\frac{1}{C^2} = \frac{1}{C_o^2} + \frac{2}{q\varepsilon_o \varepsilon_r N_d} V_{app} \qquad [11.17]$$

obtained by taking into account equation [11.16]. The experimental determination of the capacitance-voltage relationship helps to access the doping concentration in the depletion layer.

In this theoretical approach, the state density and the Fermi level at the surface contain the useful information regarding electrical properties. These are adjustable parameters, which should be associated with other observations. Research has been focused in two directions: on the one hand a refined characterization, structural as well as analytical of the surface or interface, and on the other, atomic modeling, taking advantage of the rapid development of computers. These studies have highlighted a reconstruction of the surface by low amplitude displacements, a few 0.001 nm, and of cations and anions of the surface in opposite directions. A dipolar layer is therefore created, which appreciably affects the local electrical field and various properties like surface energy, defects formation energy etc. The clean surfaces, obtained by the cleavage of monocrystals, under ultrahigh vacuum, do not show appreciable modification in the band diagram near the surface, in agreement with the sketch in Figure 11.26. The oxygen vacancy is the most often encountered defect at the surface of oxide samples, associated with specific levels in the forbidden gap, for example in the case of S_rTiO_3, at mid-distance of valence and conduction bands. These defects help in the absorption of molecules at the surface, firstly oxygen, but also H_2O, CO, CO_2, etc., explaining the catalytic properties of these oxides.

11.4.2. Metal semi-conductor interface

Band structures of a metal and a semi-conductor are shown in Figure 11.29 with a common origin in energy, that of an electron at rest in vacuum. Electrons in both materials have different Fermi levels or, equivalently, different electrochemical potential:

$$\eta\left(semi\text{-}cond.\right)-\eta\left(metal\right)=q\left\{\phi_m-\chi_s-\Delta\eta\left(semi\text{-}cond.\right)\right\}=\phi_b \qquad [11.18]$$

ϕ_m is the output work of an electron in the metal and χ_s the electronic affinity of the semi-conductor. $\Delta\eta$ is fixed by the dopant concentration according to equation [11.2]:

$$\Delta\eta(semi\text{-}cond.)=E_c-\eta(semi\text{-}cond.)=kTLog\left(\frac{N_c}{N_d}\right) \qquad [11.19]$$

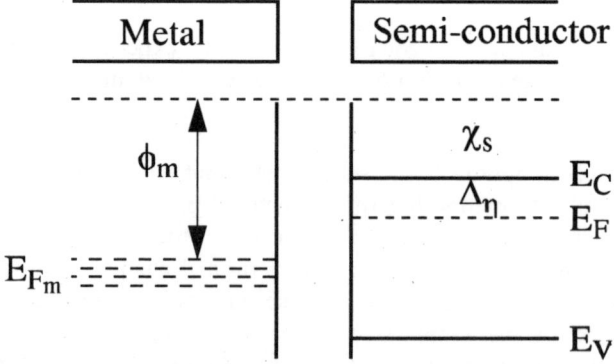

Figure 11.29. *Compared band diagrams of a metal and a semi-conductor*

If this difference in the electrochemical potential is positive, a Schottky barrier is formed through electron exchange between the two materials, which is most frequent for semi-conductor oxides (in the reverse case, the contact is Ohmic).

Interface states are always present and participate to the exchange of electrons. Equilibrium is realized, then the Fermi levels are equal in the metal, in the semi-conductor and at the interface. If the density of these states is very low, the contact is said to be "ideal" and the Schottky barrier is fixed by equations [11.18] and [11.19]. If it is high, the Schottky barrier is almost insensitive to the nature of the metal and obeys equation [11.15]. This is the case for silicon, for example.

11.4.3. *Polycrystalline materials*

In polycrystalline materials, interface states are present at grain boundaries. They are associated with the atomic rearrangement induced by the difference in crystallographic orientation, as well as the segregation of impurities. Local observations are difficult, since they require prior fracture of the material following the grain boundary plane. Published results are sometimes contradictory, highlighting not only the difficulty of these studies, but also the easily metastable state of the grain boundaries. In what follows, we will consider only one of these boundaries.

The electrons, by occupying the interface conditions, deplete the adjacent crystals on a depth of $w/2$. At equilibrium, a symmetrical potential barrier is formed with height:

$$\phi_o = \frac{q^2 N_d}{8\varepsilon_o \varepsilon_r} w^2$$

When an applied voltage is imposed on both sides of the junction, the system becomes asymmetrical. The depleted depths are the solutions of the equations:

$$\phi_1 = \frac{q^2 N_d}{2\varepsilon_o \varepsilon_r} w_1^2 \; ; \; \phi_2 = \frac{q^2 N_d}{2\varepsilon_o \varepsilon_r} w_2^2 \text{ and } \phi_2 - \phi_1 = q V_{app}$$

The charge at the interface is equal to: $Q = -q N_d (w_1 + w_2)$, i.e.:

$$w_1 = + \frac{\varepsilon_o \varepsilon_r V_{app}}{Q} - \frac{Q}{2 q N_d} \; ; \; w_2 = - \frac{\varepsilon_o \varepsilon_r V_{app}}{Q} - \frac{Q}{2 q N_d}$$

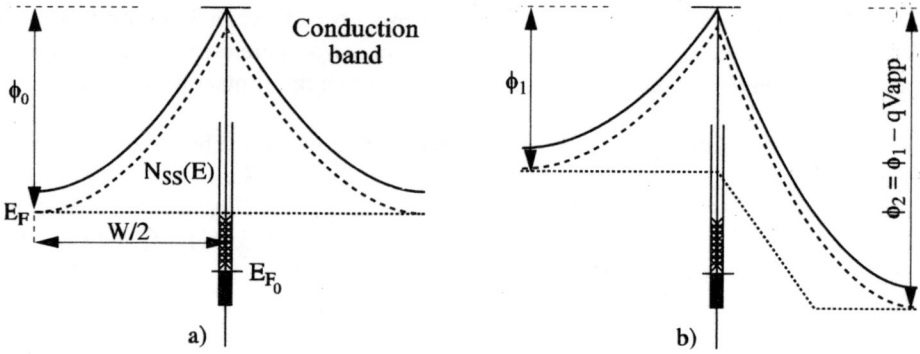

Figure 11.30. *Band diagram of a bicrystal, a) at equilibrium, b) under electrical field*

When the positive applied voltage V_{app} increases, the depleted layer of depth w_2 expands. Conversely, w_1 diminishes, and the same applies to the potential barrier ϕ_1.

A reasoning analogous to that elaborated above shows that current is given by:

$$J = J_o \exp\left\{-q\phi_1\left(V_{app}\right)/kT\right\}\left[1 - \exp(-q\left|V_{app}\right|/kT)\right] \qquad [11.20]$$

The current-voltage characteristic is symmetrical with respect to the voltage. For low applied voltages ($qV_{app}/kT << 1$), a first order Taylor expansion leads to an Ohmic behavior with a thermally activated conductivity, with the activation energy being equal to ϕ_0. For higher voltages, current increases with the applied voltage in a non-linear fashion. The appearance of the characteristics reflects the evolution of ϕ_1 with the applied voltage:

$$\phi_1 \cong -\frac{kT}{q} Log\left(J/J_o\right)$$

Equation [11.20] is obtained by assuming that the crossing of the interface takes place through a thermolectronic mechanism, i.e. the transfer of electrons over the potential barrier. Other mechanisms may also contribute to the current, like the transfer of charge by tunneling. In this case, when the applied voltage is slightly above the value of the width of the forbidden band E_g, an electron having crossed the barrier has acquired enough energy to create an electron-hole pair through a collision in the depleted region. The internal electric field dissociates the pair and accelerates the electronic holes towards the interface enabling their recombination with the electrons. The interface charge and the potential barrier are strongly reduced: the grain boundary opposes no more resistance to the current. A certain number of conditions must be fulfilled for this mechanism to be effective and are indeed operative in the case of zinc oxide based varistors.

Potential barriers can be commonly seen in semi-conductor ceramics. The applications include: the use of SiC, thus insulated, as substrate (see section 11.6.1), grain boundary layer capacitors (section 11.6.2), PTC thermistors (section 11.6.6), varistors (section 11.6.7), ferrites (section 11.6.8) and gas detectors [WOL 91].

11.5. Influence of microstructure on electrical properties

A ceramic material is heterogenous: it is polycrystalline and contains pores; it is often multiphase, the phases being distributed at grain boundaries, as a result of a liquid phase sintering, or forming inclusions. We may ask ourselves in what measure it is still possible to speak of electrical conductivity, relative dielectric permittivity, piezoelectric constant, etc. of the material. Or, to put it plainly, does the notion of material have meaning? The answer is experimental, as we will see in section 11.5.1.

By mastering the microstructure, new properties can be obtained. We then enter the domain of composite materials. A few examples will be given in section 11.5.2.

Finally, in section 11.5.3, we will see how impedance measurements can bring forward answers, with respect to the electrical heterogenity of ceramic materials.

11.5.1. *Modeling of apparent conductivity*

Experimentally, a current I is measured in response to an applied voltage V. It is always possible to calculate the ratio V/I *a priori*, but giving it the significance of a resistance R is already assuming a linear response, i.e. a current proportional to the applied voltage. If that is not the case, it is possible to get round the difficulty by superimposing a continuous polarization voltage and an alternating component δv of frequency $\omega/2\pi$. The current is also the superimposition of a continuous current and an alternating component δi, proportional to δv if the amplitude of δv is sufficiently small. The ratio $Y = \delta i/\delta v$, varies with polarization voltage and frequency[12].

---

12 The frequency is, however, sufficiently low to be able to treat separately the electrical and magnetic properties. In this case, the electric field is derived from a scalar potential.

An apparent conductivity σ_{app} may be calculated according to the equation:

$$Y = \frac{\sigma_{app} S}{e}$$

where S is the electrode area and e the thickness of the disk sample. However, this does not insure *a priori* that the apparent conductivity is a material property, independent of the sample shape and dimensions.

A necessary condition to define material properties in the case of heterogenous materials is that heterogenities are uniformly distributed at a microscopic level. Is it therefore possible to link this apparent conductivity to the intrinsic properties of the different phases? This is a difficult problem, which does not have a simple answer, as we are going to see now. In what follows, it will be assumed that each phase is characterized by a generalized conductivity σ_i (see equation [11.11]).

11.5.1.1. *Series and parallel associations*

These two types of associations are the simplest ones that could be imagined and the association laws are well known.

Series association: the current density J goes through the different phases and voltage drops get added, i.e.:

$$\delta v = \sum_i \delta v_i = -\sum_i E_i\, e_i$$

if E_i is the uniform electric field in the phase i of thickness e_i, determined by the current density J:

$$I = J\,S = \sigma_i E_i S$$

By definition, the apparent conductivity is such that:

$$\sigma_{app,\perp} = \frac{\sum_i e_i}{S} \frac{\delta i}{\delta v}$$

or:

$$\sigma_{app,\perp}^{-1} = \sum_i p_i\, \sigma_i^{-1} \qquad\qquad [11.21]$$

if p_i is the volume fraction of the phase i ($\sum_i p_i = 1$).

Parallel association: we leave it to the reader to check for himself that in this case:

$$\sigma_{app,//} = \sum_i p_i \ \sigma_i \ . \tag{11.22}$$

This modeling may seem very simple, even simplistic; however, its significance is twofold:

– theoretical, since it is possible to show that the values thus obtained encompass all values obtained by a more realistic model:

$$\sigma_{app,\perp} \leq \sigma_{app} \leq \sigma_{app,//}$$

– practical, if we admit that the microstructure of a ceramic could be considered like an assembly of cubic grains of size "a" and of average grain boundary thickness "d" (see Figure 11.31).

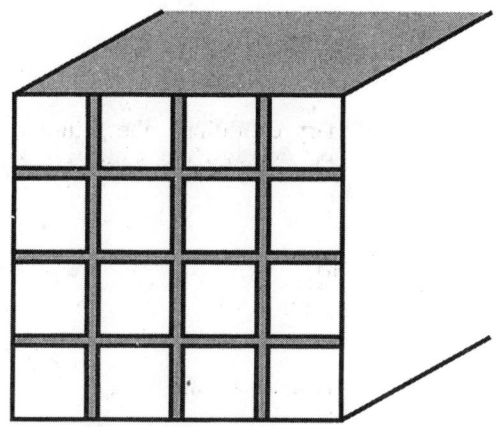

Figure 11.31. *Cubic assembly of grains*

The laws obtained (equations [11.21]–[11.22]) differ only by the value of the exponent, $n = \pm 1$. Assuming that in a ceramic sample, there are as many series and parallel associations, we can make the assumption of an average exponent not very different from zero:

$$\sigma^n_{app,//} = \sum_i p_i \ \sigma^n_i \ \ n \approx 0$$

A first order Taylor expansion of $\sigma^n = \exp[n\ Log(\sigma)]$ ($\approx 1 + n\ Log(\sigma)$) leads to the new relation:

$$Log(\sigma_{app}) = \sum p_i\ Log(\sigma_i) \qquad [11.23]$$

which is nothing but the empirical Lichtenecker relation.

11.5.1.2. *Theories of effective medium*

Let us consider a disk sample with two electrodes; the material is homogenous with a conductivity σ_o. The application of a ddp δv creates a uniform electrical field $E_{app} = \delta v/e$ if e is the distance between electrodes. Let us assume now that there is a very small spherical inclusion in the center, of radius R (R $<<$ e), of a material of different conductivity σ_i. In each part, the electrical potential obeys the Laplace equation:

$$div(\vec{J}) = 0 \; ; \; \vec{J} = \sigma\ \vec{E} = -\sigma\ \overline{grad}(V) \; \Rightarrow \; \Delta V = 0$$

Taking into account the boundary conditions, the solution is unique and well known. In polar coordinates $(r,\ \theta)$, θ being the angle $(\vec{r}, \vec{E}_{app})$, the electrical potential in the inclusion is equal to:

$$V(r,\theta) = -E_{app}r\cos(\theta)\frac{3\sigma_o}{\sigma_i + 2\sigma_o}$$

The electrical field and the polarization are uniform.

Outside the inclusion:

$$V(r,\theta) = -E_{app}.r.\cos(\theta)\left\{ 1 - \frac{R^3}{r^3}\frac{\sigma_i - \sigma_o}{\sigma_i + 2\sigma_o} \right\}$$

The inclusion is responsible for a dipolar contribution which opposes the applied electric field. It is termed a depolarizing field. At the boundary, it is equal to:

$$E_{dép.}^i = \frac{\sigma_o - \sigma_i}{2\sigma_o + \sigma_i}\vec{E}_{app}.$$

Let us use this result to define an *effective* medium. An effective medium is a homogenous medium which "at best" represents the properties of the real material. In the theory developed by Brüggeman, the apparent conductivity of this medium is adjusted, so that the depolarizing fields created by the different inclusions compensate each other:

$$\sum_i p_i E^i_{dép.} = \sum_i p_i \cdot \left(\frac{\sigma_{app} - \sigma_i}{2\sigma_{app} + \sigma_i} \right) = 0 \qquad [11.24]$$

p_i being the volume fraction of the phase "i". The average power dissipated by the Joule effect is then correctly predicted. For a biphased medium, equation [11.24] becomes a quadratic equation in p_i:

$$2\,\sigma_{app}^2 + \sigma_{app}\{\sigma_1(3p_2 - 2) + \sigma_2(3p_1 - 2)\} - \sigma_1\,\sigma_2 = 0 \qquad [11.25]$$

In Figure 11.32, the apparent conductivity is depicted versus p_1, the volume fraction of the most conducting phase. Beyond a critical volume fraction of 33%, the material is a conductor. Below this value, the apparent conductivity varies almost linearly with the volume fraction:

$$\sigma_{app} \cong \sigma_1(3p_1 - 1)/2$$

The theory of percolation helps to interpret this result. Let us consider a uniform network of points in space and let us introduce, progressively, bonds between nodes. At first isolated clusters of different sizes, of connected points are formed. When approaching the critical value p_c, called percolation threshold, one of these clusters grows very rapidly through the conglomeration of isolated clusters. For values greater than p_c, this cluster is infinite. The theory of percolation, applied to the study of conductivity, reproduces the equations of the effective medium, except at the immediate vicinity of the percolation threshold, where apparent conductivity then varies like:

$$\sigma_{app} \cong (3p_1 - 1)^n \sigma_1$$

n is called the critical exponent ($1 < n < 2$).

In Figure 11.32, we have also represented the Lichtenecker prediction. It is very different from the effective medium since it does not show any percolation. To understand this apparent contradiction, it is necessary to come back to the association of elements in series or in parallel. When the conductances σ_1 and σ_2 are very different, equations [11.21]–[11.22] indicate that the current follows a very small number of paths, those essentially made up of high conductance elements, and the percolation effect is maximal. In fact, the percolation model described earlier corresponds to the extreme case of a mixture of zero and infinite conductance. In the case where they are of the same magnitude, the percolation effect disappears. When the conductivity is the main contribution to the generalized conductance, a percolation effect is likely to be seen because conductivities may vary by several orders of magnitude. Conversely, it is not surprising that the Lichtenecker relation explains the experimental observations in the case of dielectrics, since relative permittivities of paraelectric materials do not differ much (a factor 10 at the most), and the same applies to their generalized conductance ($\sigma = j\varepsilon_o\varepsilon_r\omega$).

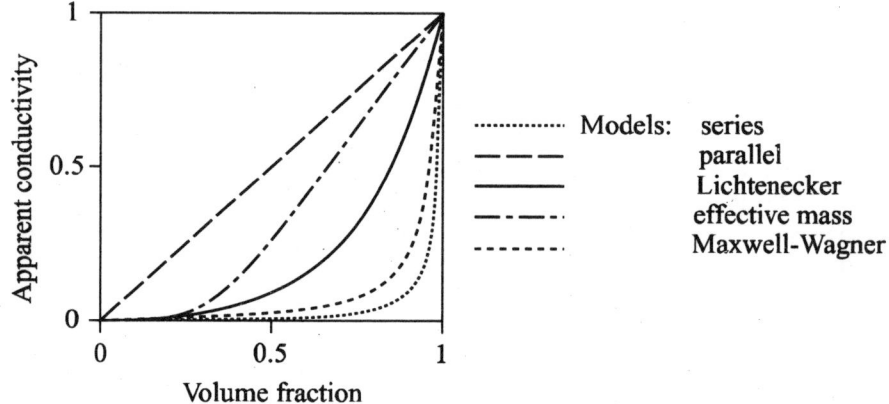

Figure 11.32. *Apparent conductivity of a biphasic material versus the volume fraction of the conducting phase for different models*

11.5.1.3. *Influence of topology*

A closer look at relationship [11.24] shows an invariance of apparent conductivity by the permutation of indices. Both the materials must therefore play equivalent roles. This is the case of a biphasic material, made of grains of similar size but of different kinds. The same does not apply if one of the phases is made of inclusions in a matrix. This is, however, a microstructure frequently observed when as a result of liquid phase sintering. It is also possible, however, to define and calculate an apparent conductivity. Let us consider a spherical inclusion, consisting of a nucleus of the dispersed phase "i" and a shell of the continuous phase "o" in adequate proportions. Similar developments to those of the previous section lead to the following expression, proved by Wagner:

$$\sigma_{app} = \sigma_o \frac{\sigma_i - (1-D)p_o(\sigma_i - \sigma_o)}{\sigma_o + Dp_o(\sigma_i - \sigma_o)} \qquad [11.26]$$

The D factor is known as the depolarization factor. It equals $1/3$ in the case of spherical inclusions and can take on any value between 0 and 1 for different shapes, like needles, rods, etc.

This model helps, for example, to estimate the influence of porosity ($\sigma_i = 0$, $p_i = P$, $D = 1/3$):

$$\sigma_{app} = \sigma_o \frac{1-P}{1+P/2}$$

or even to evaluate the influence of an insulating intergranular phase ($p_o \ll 1$, $\sigma_o \ll \sigma_i$, $D = 1/3$):

$$\sigma_{app} = \frac{3\sigma_o \sigma_i}{3\sigma_o + p_o \sigma_i}$$

The material is insulating if the condition ($3\sigma_o \ll p_o \sigma_i$) is met, showing the basic series nature of such a topology.

Applied to the case of dielectrics, it is also possible to show that a ferroelectric material experiences a strong decrease of relative permittivity if a non-ferroelectric phase is present at grain boundaries.

The different models are compared in Figure 11.32. As already stated, the series and parallel associations limit the range of possible values, but this range can be wide.

11.5.2. Composite materials

The development of a diversified range of processing techniques has brought us towards a real material engineering [NEW 86]. The microstructure is then designed in view of a specific application. The elementary bricks are the grains, fibers or filaments, plates or bands, dispersed in a matrix. Dimensionality is a parameter of great importance. From this point of view, the pores are considered to be specific grains.

11.5.2.1. *Composites 3-0*

Let us have isolated grains (phase 1) in a matrix (phase 2). One of the phases is continuous, the other discontinuous, which can be assimilated to points: the connectivity is 3-0. This is the topology of the Wagner model given in the previous section. A liquid phase sintering helps to preserve this topology, even when the volume fraction of the matrix is small. However, most often, connectivity evolves with the volume fraction of phase 1. In fact when it exceeds approximately 10%, isolated clusters are formed which can no longer be assimilated, strictly speaking, to points. Beyond the percolation threshold, both the phases are continuous and connectivity is 3-3. It is then the theory of the effective medium which is most appropriate. It is possible to modify the percolation threshold by varying the respective sizes of the grains of the two phases. If the dissymmetry is very large, the small grains will get distributed at the surface of the larger grains. Connectivity becomes 3-2 and the percolation threshold becomes very small, as shown in Figure 11.33. This microstructure is adopted for the fabrication of thick film resistors (see section 11.6.1) prepared from a mixture of glass frit (grain size of a few µm) and a metal oxide like RuO_2, finely divided (grain size of a few nm).

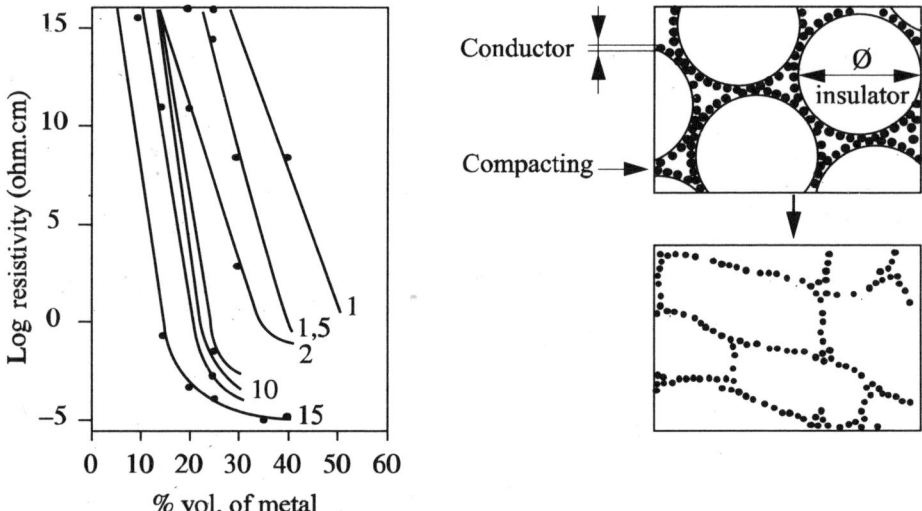

Figure 11.33. *Variations in the apparent resistivity of a insulator-metal composite as a function of the volume % of metal. Influence of the particle size ratio on the final microstructure and on resistivity ($x = D_{glass}/D_{metal}$)*

The grain shape must also be taken into account and we have already noted' its influence on the properties (see equation [11.26]). Finally, the interfaces, where segregation takes place, have special properties (see section 11.4.3).

11.5.2.2. Composites 3-1: fiber reinforced composites

Different symmetry groups can be obtained, depending on the arrangement of fibers, and this in turn influences the material properties. Examples are given in Figure 11.34.

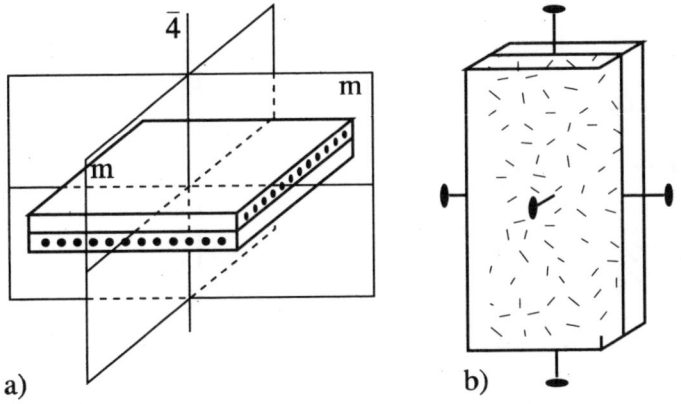

Figure 11.34. *Different symmetries, a)* $\overline{4}2$ *m,*
b) 222 for different fiber configurations

We could still mention the multilayered structure, of connectivity 2-2, obtained by stacking the ceramic bands and more complex structures prepared with more than two phases.

The different models, developed in section 11.5.1, have shown the influence of the microstructure topology on properties. Conductivity is in fact particularly simple, since only one property comes into play. Irrespective of the complexity of the model, apparent conductivity cannot be larger than the conducting phase or smaller than the insulating phase. The preparation of a composite is therefore interesting:

– adjust it to a precise value. Thus, there is no dielectric which shows no variation in relative permittivity with temperature. Such a property can be obtained by a composite, prepared by mixing two dielectrics with opposite thermal behavior (see section 11.6.2);

– in the vicinity of the percolation threshold. Then, the conductivity varies greatly with composition. As an example, PTC thermistors have been developed by mixing an insulating polymer (polyethylene) and a very divided conductive phase (carbon black) in proportions corresponding to the percolation threshold. With the help of thermal expansion, the number of contacts between carbon particles varies largely with temperature, which explains the variations in resistance.

Combining different properties is even more interesting, like for example the piezoelectric effect. Let us consider a piezoelectric multilayer in two configurations, different by the orientation of polarization (see Figure 11.35).

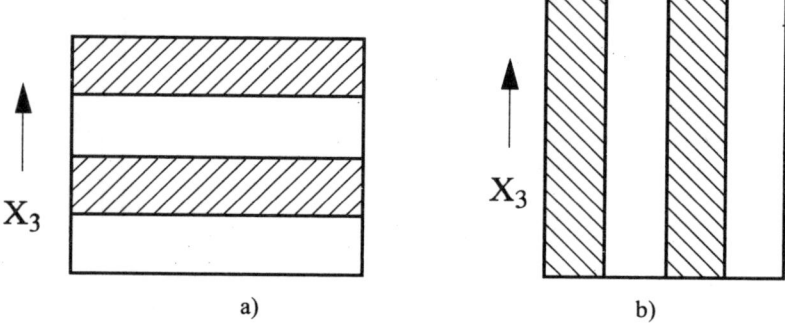

Figure 11.35. *Effect of direction of polarization with respect to the orientation of layers on properties of a piezoelectric composite: a) series arrangement, b) parallel arrangement*

In case a), the system is almost unidirectional. The apparent piezoelectric coefficients d and g are given by the relations:

$$d_{33}^{app} = \frac{p_1 d_1 \varepsilon_2 + p_2 d_2 \varepsilon_1}{p_1 \varepsilon_2 + p_2 \varepsilon_1} \,, \quad g_{33}^{app} = p_1 g_1 + p_2 g_2 \,.$$

The indices 1 and 2 refer to the values of d_{33} and g_{33} in both the phases. We notice that d_{33} varies between d_1 and d_2, and g_{33} between g_1 and g_2. The same no longer applies to case b):

$$d_{33}^{app} = \frac{p_1 d_1 K_2 + p_2 d_2 K_1}{p_1 K_2 + p_2 K_1} \,, \quad g_{33}^{app} = \frac{d_{33}^{app}}{\left(p_1 \varepsilon_1 + p_2 \varepsilon_2 \right)} \,.$$

The coupling between the phases is no longer electrical but mechanical via the elastic constants K_1 and K_2. Thus, combining a phase 1, which is non-piezoelectric and flexible and a phase 2, which is piezoelectric and rigid, the composite is still piezoelectric ($d^{app} \cong d_2$), but the coefficient g^{app} can be specifically increased by minimizing the denominator of g^{app}. These special properties of piezoelectric composites, prepared from PZT and a polymer, have been taken advantage of in various applications (medical and sonar imaging, for example).

A third type of utilization of composites is even more interesting, particularly when the properties involved in each of the phases are different. A new property appears, which none of the two phases possess. Let us mention as an example a composite, intimate mixture of two phases: a ferromagnetic phase (ferrite), and a piezoelectric phase ($BaTiO_3$). The application of a magnetic field produces the deformation of the magnetic phase by magnetostriction, which in turn generates a polarization of the piezoelectric material. We have therefore created a magnetoelectric effect via the deformation of the composite.

11.5.3. Complex impedance measurements: a technique for studying heterogenous systems

In accordance with the generalized Ohm's law, the equivalent circuit of homogenous material is a cell R-C, made up of a resistance and a capacitance in parallel:

$$Z = \frac{R}{1 + j\omega\tau} \quad \text{with: } \tau = RC$$

It is now possible to measure the real and imaginary parts of the impedance with the help of a vectorial analyzer in a large range of frequencies. The experimental data are plotted in the complex plane (ReZ, -ImZ) and they are expected to draw out a semi-circle centered on the x-axis of diameter equal to R. At low frequency, the behavior is basically resistive; at high frequency it is capacitive and the top of the semi-circle corresponds to the condition $\omega\tau = 1$. Experimentally, we often observe a deformation of this ideal curve, a flattening reflecting the fact that the capacitance (or equivalently the relative permittivity) is a function of the frequency (see section 11.2.1).

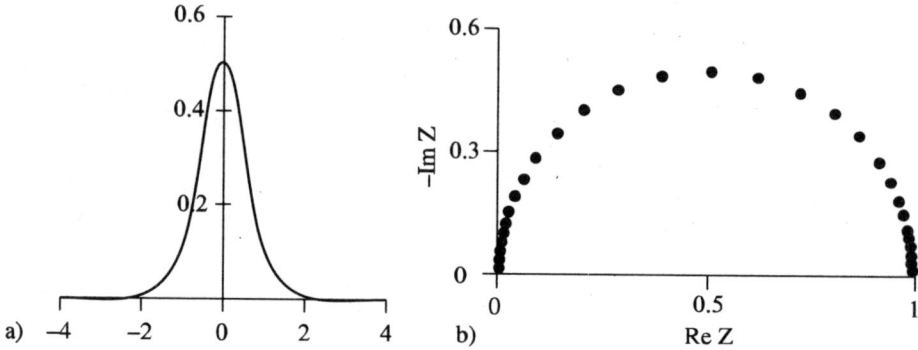

Figure 11.36. *Variations of impedance with frequency for a cell R-C*
a) –Im(Z)/ Log(f), b) –Im Z/Re Z

The representation in the complex impedance plane is not the only possibility. Thus, the imaginary part of Z, depicted as a function of the frequency on a logarithmic scale shows a maximum for $\omega\tau = 1$ (see Figure 11.36).

To carry out electrical measurements, the component, commonly a disk sample, is provided with two electrodes. We have seen that a metal-semi-conductor interface has a specific capacitance and conductance (see section 11.4). It is therefore possible to take into account the contribution of the electrodes to the total impedance by considering a second cell R_{el}-C_{el} in series with the first. We thus expect to have two semi-circles in the complex impedance plane, at least if the time constants differ by a minimum of two orders of magnitude. This is generally the case, since the electrical thickness of the metal-ceramic interface is much smaller (1 nm to 1 μm) than that of the sample (typically 1 mm).

In a ceramic sample, grains and grain boundaries are in series. There is thus a third arc of a circle, grain boundary specific. Considering the microstructure as a stack of cubic grains (see Figure 11.31), the thicknesses of the different interfaces between grains, in the direction of the applied electrical field, get added. We may therefore feel that the apparent capacity of grains boundaries is greater than that of the grains, but smaller than that of the electrodes. This order applies also to the time constants, and the grain boundary semi-circle comes in between the other two. We can frequently see considerable flattening of this arc. The main cause is the heterogenity of the ceramic.

From the complex impedance spectrum, depicted in the complex impedance plane, it is thus easy to separate the different contributions of the bulk material, the grain boundaries and the electrodes, and to determine their respective electrical characteristics, i.e. resistance and capacitance. The last, but not the least, is that complex impedance spectroscopy is non-destructive. For more details, see [ROS 87].

11.6. Ceramic components in electronics

The global market for technical ceramics was estimated in 1999 at approximately $15 billion, with Japan as the leader, the USA in second place and Europe in third position. The growth rate is high (7% to 10%), with active research and development. We must note that the industrial markets encompassed by the ceramic components are twice as large as the market for the components themselves.

The components are extremely varied. Some form part of the economic niche while others, like packaging or capacitors (several billions of pieces/year), are produced in large quantities. Here we will not attempt to be exhaustive and we will restrict ourselves to describe discrete components, despite the fact that deposited components in the form of films, integrated in complex systems, are currently experiencing tremendous growth.

11.6.1. Substrates, interconnection circuits and hybrid circuits

11.6.1.1. Substrates

The purpose of the substrates is to support discrete active components, integrated and passive circuits, capacitors, resistors, varistors, thermistors etc. connected by conducting tracks. The coefficients of expansion of the different materials must be as close as possible in order to avoid the appearance of stresses of thermal origin, susceptible to induce failure of the circuit. Roughness is also an important parameter. It therefore appears that the choice of a material for making a substrate must take into account not only electronic requirements, but also mechanical, thermal and even chemical requirements.

On reading Table 11.16, we will see that none of the materials mentioned show at the same time a small relative permittivity (minimizing the capacitive coupling between tracks), a volume expansion close to silicon (3 ppm/K) and a high thermal conductivity for removal of extensive dissipated power. Boron nitride is a good candidate, but is difficult to prepare. Aluminum nitride is interesting for its thermal conductivity and is less toxic than beryllium oxide, largely used in power applications. Cordierite has a low relative permittivity, important for high frequency operations. In very specific cases, materials can be used as single crystals, mainly quartz and sapphire.

Material	Properties				
	Electrical		Thermal		Expansion
	Re ε_r 1MHz	Breakdown kV/mm	K_{th}(25°C) W/(m.K)	Impact resistance	α_T 25–300°C ppm/K
Alumina				Good	7
96%	8.9–10.2	14–24	20–26		
99.5%	9.7–10.5	15–36	29–37		
AlN	8.0–9.2	14–27	110–260	Good	3.8–4.4
Glycine (BeO)	6.5–7.0	10–43	260–290	Excellent	7.2–8.0
Cordierite	4.1–5.4	5.5–10	1–3.2		1–3
SiC + 3% BeO	40	1	270		3.7
BN	4.1		60		4.3
Si	12		90		2.6

Table 11.16. *Properties of some good candidates for substrates*

NOTE.– all these materials have an electrical resistivity greater than 10^{14} Ω cm, except SiC.

Silicon carbide is known for its good thermal conductivity, but is a semi-conductor. The addition of boron or beryllium oxide in low quantities promotes potential barriers at grain boundaries (see section 11.4.3) and confers the necessary insulating character.

Most often, alumina is chosen, which is a good compromise, particularly on account of its low cost. There are several varieties depending on the amount of additives, alumina at 96% being the most commonly used. Even though we are restricting ourselves to crystalline materials, the possible utilization of glass-ceramics and enameled sheets should be mentioned.

11.6.1.2. *Hybrid circuits and interconnection*

We can see, ever since electronics has arrived, a double evolution:

– *the deposition of conducting tracks and passive components on the substrate by a screen printing process.* The assembly after curing between 750°C and 1,000°C (a complete cycle of less than an hour) constitutes a passive circuit, which after depositing the active components is called a hybrid circuit [SCH 90].

This technology, referred to as thick layer, has several advantages:

– any circuit made with conventional technology is easily convertible into hybrid technology;

– the density of interconnection and integration is very high, accompanied by a significant improvement in electrical performances;

– an excellent thermal dissipation;

– a good reliability;

– a better mechanical behavior due to the decrease in weight.

Screen printing makes use of a screen, a steel or nylon cloth, on which a motive is printed by the photoresist technique. Some ink is deposited. The ink is transferred on the substrate through the sieve mesh, at least on the parts not printed on the cloth, by passing a squeegee, which induces a high shear stress (see Figure 11.37).

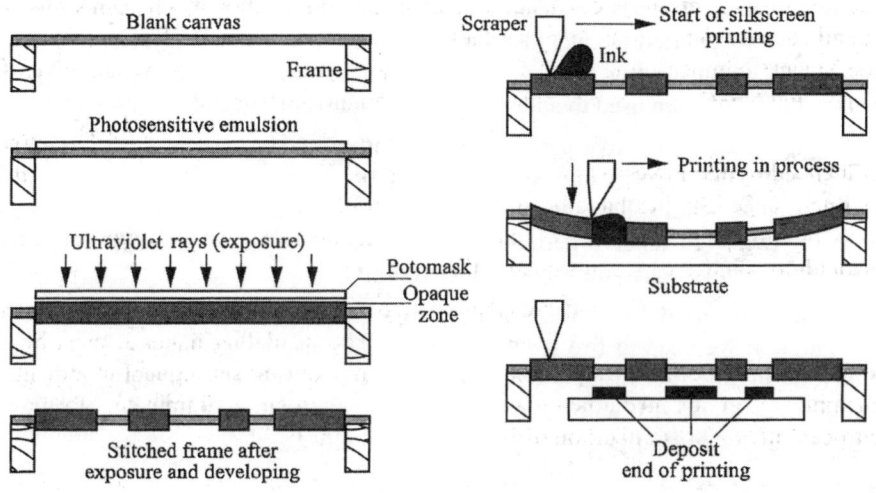

Figure 11.37. *Screen printing technique*

The ink is a mixture of a solid fraction, the active material and an adhesive material, and an organic vehicle which controls rheology. This liquid combines binders and plasticizers, polymers or resins of high molecular weight like methacrylate, surface active agents like ethylene oxide, and a solvent with low vapor pressure, for example terpineol or butylcarbitol acetate. Apparent viscosity must be high at rest and decrease strongly while the squeegee is passed.

The active material is mixed with glass frit, which must have a low relative permittivity and a high resistivity, excluding the existence of alkaline elements. It most often consists of a lead borosilicate (typically 63% mPbO-25% mB$_2$O$_3$-12% mSiO$_2$). The deposited elements are [VES 91]:

– conducting tracks: the active material is a metal. Selection is based on weldability, resistivity and the price. There is no ideal candidate. Gold is soluble in Sn-Pb weld, the most commonly used, silver migrates easily under an electric field in the presence of humidity, copper gets oxidized;

– resistors: the active material is a metal oxide like RuO$_2$ with small grains, approximately 10 nm. The volume fraction is close to the percolation threshold (see section 11.5.2), which helps to cover a wide range of resistances [10–10^6] Ω. Special care is taken with respect to the temperature coefficient which should not exceed + 100 ppm/K in the temperature range [–55, +125]°C, a value very much smaller than the temperature coefficient of only RuO$_2$, + 5,670 ppm/K. In fact, the microstructure is very complex and contacts between grains of oxides play an important role. There is a large variety of grain boundaries depending on the degree of sintering and the quantity of glass at interfaces. The same applies to conduction mechanisms involved, but all have in common the fact that they induce an NTC effect (negative temperature coefficient) compensating the PTC contribution of the oxide. Some additives like MnO$_2$ and Nb$_2$O$_5$ are used to adjust this temperature coefficient;

– capacitors: the relative permittivity of a mixture (see section 11.5.1) is greatly influenced by the lowest value, specifically glass. We will therefore try to minimize its importance. Studies have been undertaken to develop new compositions of glass for promoting high relative permittivities of certain dielectric materials. It remains difficult to achieve capacities greater than 100 pF;

– insulators, used for the crossing of two conducting tracks: to obtain such a configuration we need to fire it several times. The insulating material must be able to melt during the first firing and no longer change during subsequent treatment. An example is a devitrifiable glass, a barium borosilico-aluminate, forming a vitroceramic by crystallization of BaAl$_2$Si$_2$O$_8$ at 800°C.

Let us also recall the possibility of making sensors, thermistors, varistors in thick layer technology. The presence of a vitreous phase generally degrades the properties.

The constant evolution towards a miniaturization and higher integration of electronic circuits has rapidly shown the limits imposed by the two-dimensional nature of the substrate. The multilayer technology helps to make use of a third dimension. On each layer, prepared by tape casting, conducting tracks are deposited, holes are reserved for the passage of via, ensuring the connection between various layers. The number of layers could be several dozens with the length of conducting tracks exceeding 100 m.

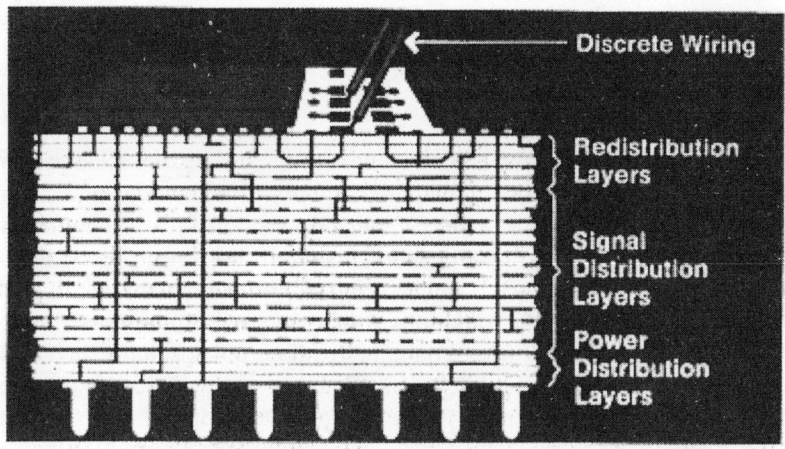

Figure 11.38. *Example of an interconnection circuit*

Most often, the assembly is cofired at a temperature around 850°C, which imposes the replacement of alumina by a mixture of alumina and a vitreous phase. Several studies are undertaken to insert resistors and capacitors in different levels, thus achieving a truly three-dimensional circuit.

11.6.2. *Capacitors*

A capacitor is an electronic component ensuring a specific function, defined by the relation:

$$q = Cv \ \text{ or: } \ i = C\frac{dv}{dt} \qquad [11.27]$$

with q the charge, and i the current going through the component under the effect of the applied voltage v. The applications are derived directly from these two equations [11.27], either the charge storage, the capacitor then being a reservoir, or signal processing, for example the filtering action.

In its simplest form, the capacitor is a disk of small thickness, fitted with two electrodes. The linearity of relationship [11.27] supposes the linearity of the relationship P(E). The material must be paraelectric or, if it is ferroelectric, the electric field must be sufficiently weak. The capacitance is then given by the relation:

$$C = \varepsilon_o \varepsilon_r \frac{S}{e}$$

if S is the electrode surface and e the thickness of the disk, ε_r the relative permittivity of the material. Let us note the implicit assumption of the homogenous nature of the material, which is rarely the case for ceramics. In section 11.5.1, the question of the apparent permittivity of a heterogenous material has been discussed. Ideally, capacitance is a constant. Since it can be achieved through a material, it is a function of frequency and temperature, variations which we strive to minimize or, at least to control. Furthermore, several uses require high values of capacitance and therefore of relative permittivity. These multiple requirements are contradictory and compromises must be made. For this reason, different classes of capacitors have been defined on the basis of variations of capacitance with temperature (see Figure 11.39).

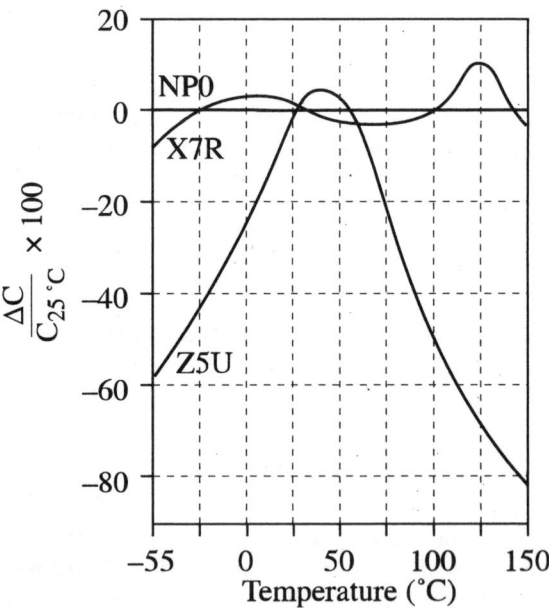

Figure 11.39. *The main classes of dielectric materials*

Another classification is based on the type of material used:

– *Type I.* Type I materials are paraelectric. The dielectric losses (tg δ) are less than 0.003, with the temperature coefficient $CT\varepsilon = \dfrac{1}{\varepsilon_r}\dfrac{\partial \varepsilon_r}{\partial T}$ perfectly controlled and which can be modulated within a range of [–2,000, + 100] ppm/K. A negative coefficient $CT\varepsilon$ can for example compensate the effects of temperature on other components of the circuit, like ferrite based inductance, materials whose magnetic permeability increases with the temperature. The NPO capacitors, which use this type of materials, are mainly used in electronic circuits. They represent around 10% of capacitors. The materials available have low or average relative permittivity, between 15 and 150, and a coefficient $CT\varepsilon$ which becomes more negative with higher relative permittivity (see Table 11.17). The preparation of composites helps to adjust it to the required value. In the absence of reactions between phases, the empirical Lichtenecker law (see equation [11.23]) proved to be useful. It predicts that the temperature coefficient of the composite is very simply the linear combination of temperature coefficients of the different constituents. Thus, a mixture of sphene ($CaSiTiO_5$, 40%), rutile (TiO_2, 49%) and strontium titanate ($SrTiO_3$, 11%) has a $CT\varepsilon$ of 0 and a relative permittivity of 108 at 1 MHz.

– *Type II* [NIE 94]. The objective is to build high capacitance capacitors (1 nF-10 μF), which is possible by optimizing the relative permittivity as well as the form factor. The requirement will be less demanding as far as the effect of temperature is concerned. This refers to capacitors of class X7R, Z5U, mainly used as decoupling capacitors.

Material	Relative permittivity ε_r (1MHz–25°C)	$CT\varepsilon$ in ppm/K
$CaSiTiO_5$	45	+ 1,200
$MgTiO_3$	12–15	+ 80
TiO_2	80–90	–750
$CaTiO_3$	130	–1,500

Table 11.17. *Some type I dielectric materials*

Optimizing relative permittivity: from this point of view, ferroelectric materials (see section 11.2.2) seem particularly interesting, with very high values of relative permittivity, 10^3–10^4, close to the Curie temperature T_c. The structure of perovskite is susceptible to accommodate several elements forming solid solutions. It is therefore possible to adjust T_c by a judicious choice of the composition (see Figure 11.40).

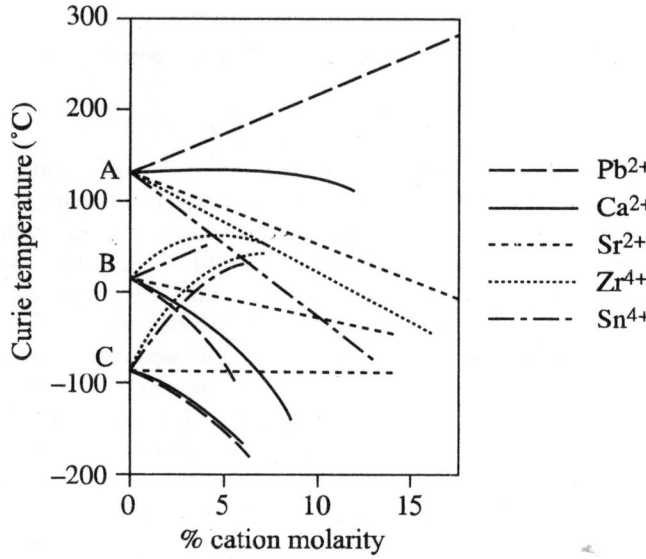

Figure 11.40. *Modification of transition temperatures A, B, C of BaTiO₃ by substitution of barium or titanium by a different cation*

Thus, the Curie temperature of the material (0.75Ba, 0.25Sr)TiO$_3$ is 25°C. This is not the only condition to fulfill with respect to the application, since relative permittivity is a highly sensitive function of the temperature around T$_c$. A heterogenity of composition induces a statistical distribution of Curie temperatures and therefore a flattening of the curve ε$_r$(T). This heterogenity can be introduced on different scales:

– at the microstructure level: for example, by using mixtures of BaTiO$_3$ and CaZrO$_3$, which form a partial solid solution. The composition is chosen in the two-phases region of the phase diagram, one rich in titanium and the other in zirconium. With the addition of SrTiO$_3$ the Curie temperature can be adjusted. Sintering conditions are selected so as to limit the interphase reactions and to prepare heterogenous and unbalanced materials;

– at the grain level: the addition of low temperature melting oxides (lead and bismuth borosilicate) promotes grain growth. Controlling the sintering temperatures (1,100–1,150°C), the grains can be made heterogenous with a BaTiO$_3$ core and a shell enriched in bismuth;

– at the atomic level: this is the case of relaxors (see section 11.2.3). The value of ε$_r$ is very high at its maximum ($\cong$ 2–3 10^4) and the curve ε$_r$(T) very expanded. The use of solid solutions helps to adjust the value of T$_m$, the temperature of diffuse transition, 40°C in the case of PMN–10%PT.

We can also make use of the influence of the grain size on relative permittivity. For a grain size smaller than 1 μm, the grains are monodomain (see section 11.2.3) and submitted to high internal stresses. This results in the flattening of the curve $\varepsilon_r(T)$ (see Figure 11.41). Additives can be used to limit grain growth during sintering.

Optimizing the form factor: it is possible to increase the surface S or decrease the thickness e, or both. This is can be achieved through a multilayer configuration, which consists of the alternative stacking of dielectric layers and internal electrodes (see Figure 11.42). The elementary capacitors of individual layers constitute are connected in parallel via the terminals:

$$C = n^2 C_o$$

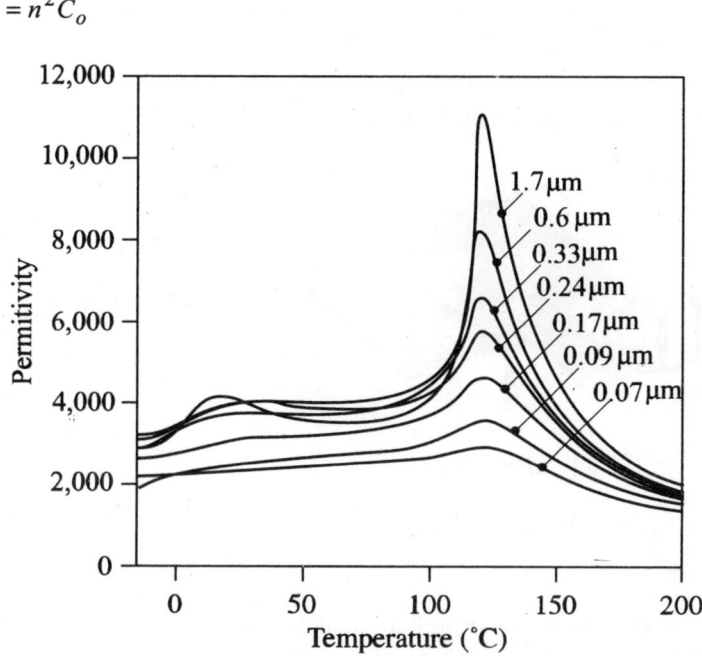

Figure 11.41. *Variations of relative permittivity with temperature.*
Influence of grain size

with n as the number of layers, and C_o the capacitance of a monolitic capacitor with the same external dimensions. The dielectric foil, in which the different layers are cut, is prepared by tape casting. The internal electrodes are deposited by screen printing. After stacking and thermocompression, the chips are sintered at temperatures varying between 1,100 and 1,350°C, depending on the composition.

The metal used for the internal electrodes is most often an alloy (Ag, Pd). By lowering the sintering temperature, compositions rich in silver, (70Ag, 30Pd), which are less expensive, can be used. A lot of work has been devoted to the development of dielectrics compatible with the more common metals, like nickel. A second heat treatment is required to fire the external electrodes, most often a silver paint, a mixture of silver and a glass frit, at a temperature of approximately 600-900°C. Let us also mention a similar technology for preparing electrodes by infiltration of molten metal in the sintered ceramic. It helps to reduce costs and minimizes the problems of metal-ceramic interaction during co-sintering. Multilayer capacitors have been designed for automatic transfer on surface of electronic boards.

– *Type III*. The capacitance of a conducting ceramic with insulating grain boundaries is generally high on account of the small thickness of grain boundaries. Such a material can therefore be used for making capacitors [GOO 91].

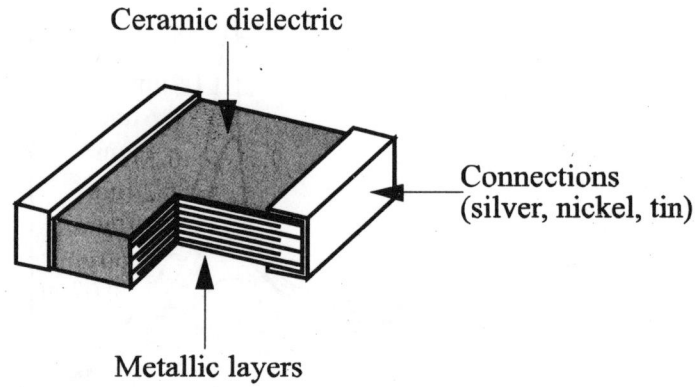

Figure 11.42. *Multilayer capacitor*

The material retained is strontium titanate, made into a semi-conductor by doping with niobium and sintered in reducing atmosphere (see section 11.1.2.2 and Figure 11.4). Additives like Al_2O_3, SiO_2 ensure a liquid phase sintering, promoting grain growth. Low temperature melting such as lead borosilicate, during a second treatment in air, insulates the grain boundaries. The simplest equivalent circuit consists of a resistance of the semi-conductors grains, in series with a capacitance of the grain boundaries. The admittance is expressed as:

$$Y = \left(R + \frac{1}{jC\omega} \right)^{-1} = jC_{app}\omega$$

thus defining the apparent capacitance of the sample, i.e.:

$$C_{app} = \frac{C}{1 + jRC\omega}$$

A real capacitance behavior is observed only for frequencies less than $1/(2\pi RC)$. A wide frequency range is obtained by increasing the grain conductivity as much as possible. This is the reason why strontium titanate was preferred to barium titanate, with the mobility of electrons being greater (see Table 11.4). Typically, conductivity is 10^3 Ω^{-1} cm^{-1} and the cut-off frequency 1 GHz. The apparent relative permittivity is of the order of 20,000.

11.6.3. *Electromechanical transducers*

The piezoelectric effect enables the transformation of a force or a deformation into electrical signals and vice versa. Applications are numerous in electroacoustics, like ultrasonic microphones and transducers, in telecommunications, like filters and timers, in metrology like measurements of forces and displacements and in the field of electrical appliances, the standard example being the gas lighter.

11.6.3.1. *Materials*

In the applications, barium titanate was rapidly replaced by the PZT family of composition $Pb(Zr_xTi_{1-x})O_3$. These are paraelectric and cubic at high temperatures, ferroelectric below the Curie temperature T_c, at least for values of x smaller than 95% ($PbZrO_3$ is antiferroelectric). For temperatures less than T_c, the compounds rich in titanium are of quadratic structure, polarization being oriented along [100], while compounds rich in zirconium have a rhombohedral structure, with polarization being oriented along [111]. The composition line separating these two regions of the diagram is called the morphotropic boundary (see Figure 11.43). At ambient temperature the limit corresponds to a ratio Zr/Ti of 52/48. The piezoelectric coefficients and the relative permittivities then goes through a maximum. Doping by acceptors, like alkaline elements (in site A of perovskite), trivalent elements (in site B) introduces oxygen vacancies which have the effect of reducing relative permittivity and piezoelectric coefficients, which increase the depolarization resistance. They are called "hard" PZT. Reverse effects are obtained by doping with donors.

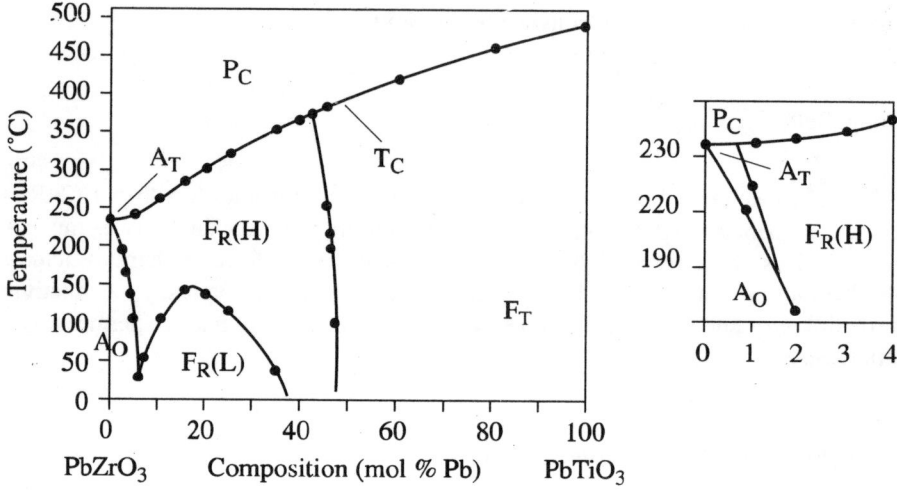

Figure 11.43. *Phase diagram of PZT*

The fabrication process is standard: mixing of raw materials, solid state reaction, grinding, shaping by pressing or casting, sintering between 1,200°C and 1,350°C. The forms produced are very varied: sheets, disks, rings, cylinders, tubes, blocks, caps, etc. The sintered parts are adjusted and polished. The electrodes are most often in silver, deposited by screen printing or by cathodic spraying and fired at 600°C. The ceramics, dipped in an oil bath, are then polarized by the application of a high electrical field, 4 kV/mm. It is also possible to operate in air by using the corona effect.

The material properties required for a given application depend on the amplitude of the deformations involved and the frequency bandpass. We can distinguish between:

– standard ceramics for electromechanics and electroacoustics, characterized by a high coupling factor. These are suitable for all quasi-static applications for which the amplitude is small. The relative permittivity could be medium (1,000–2,000) or high (2,000–5,000) depending on the composition, making the adaptation of the impedance possible. Materials of high relative permittivity have a low Curie temperature and a weak coercive field. The ranges of temperature and amplitude of deformation are thus reduced;

– ceramics with a wide frequency bandpass. The mechanical resonance quality factor is low. These are used for making ultrasonic transducers for non-destructive testing of materials and medical sonography. Ceramic-polymer composites (see section 11.5.2), provide materials with low relative permittivity, imposed by impedance matching; see [GUR 87];

– power ceramics. In normal ceramics, losses become very significant with strong fields or under high stress, due to the displacement of the domain walls. Heating can lead to the destruction of the ceramic. For details regarding mechanical behavior of PZT, see [POH 87].

11.6.3.2. Applications

There are several hundreds of applications and it is not possible to be exhaustive: [TAN 82]. We will now specifically deal with positioning actuators, whose importance is increasing in fields like microelectronics, micromechanics, mecatronics, optics, etc. The operation may be static or dynamic.

Quasi-static operation

In the longitudinal configuration (see Table 11.11):

$$x = \frac{\delta l}{l} = d_{33} E = d_{33} \frac{U}{l}$$

i.e. $\delta l = d_{33} U$, if U is the applied voltage. For normal values of d_{33}, the generated displacement is very small and does not depend on the thickness. It is possible to amplify it by using a multilayer assembly, already described in section 11.6.2.

A transverse configuration (see Table 11.11), introduces a dimensional amplification factor (l/d). Deformation:

$$\delta l = d_{31} \frac{l}{d} U$$

is limited by the value of the field (U/d) which might induce a dielectric breakdown.

Other devices, like the moonie transducer for example, have an even larger amplification factor, but the largest displacements could probably be obtained with the bimorphs (see Figure 11.44), to the detriment of the generated force (see Table 11.18):

$$\delta l = \frac{3}{2} d_{31} \left(\frac{l}{d} \right)^2 U$$

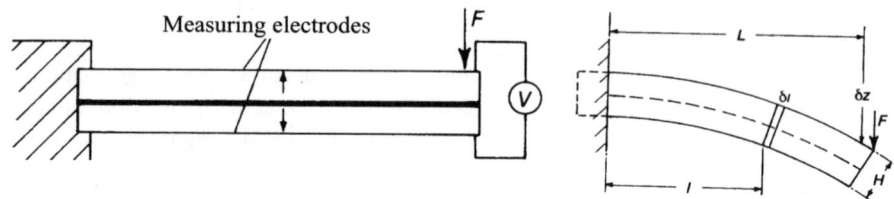

Figure 11.44. *Bimorph actuator*

Piezoelectric materials show a hysteresis which degrades the positioning reproducibility (3% to 15%). From this point of view, electrostrictive materials show greater performance.

Configuration	Displacement	Time response	Force
Disk	< 1 μm	+ +	4,000 N
Tube	< 10 μm	+ +	2,500 N
Multilayer	10–100 μm	+	1,000 N
Moonie	< 200 μm	0	1 N
Bimorphic	< 1 mm	0	0.1 N

Table 11.18. *Displacement actuators*

Piezoelectric motors

Lord Rayleigh has shown that elastic waves, combining both longitudinal and transverse deformation, can move across a surface within a thickness of approximately 1 μm. The surface particle moves in a circle or ellipse in the direction of propagation, inducing a horizontal displacement t. The wave is progressive in nature and may be seen as the combination of two stationary waves of the same amplitude and a phase difference of 90°:

$$u = u_o \cos(\omega t - n\theta) = u_o \{\cos(\omega t)\cos(n\theta) + \sin(\omega t)\sin(n\theta)\}$$

if n is the index of the excited mode. The stator is a disk divided into two sectors, separated by a space between $\lambda/4$ and $3\lambda/4$ in order to create the 90° phase difference. Each of them is made up of a sequence of segments, polarized in an alternate manner and fitted with electrodes for the application of an alternating voltage of angular frequency ω, creating the required stationary wave ($n\lambda = 2\pi R$). The rotor is a disk driven by friction (see Figure 11.45).

11.6.4. *Resonators*

A resonator is characterized by a transfer function showing a narrow frequency bandwidth. In its simplest expression, it can be represented by a circuit consisting of three elements in series, an inductor L, a capacitor C and a resistor R, the latter reflecting the damping of the system. The important parameters are the resonance frequency f_o in the absence of damping and the quality factor Q:

$$4\pi^2 L C f_o^2 = 1 \text{ and } Q = \frac{f_o}{f_2 - f_1} = \frac{1}{R}\sqrt{\frac{L}{C}}$$

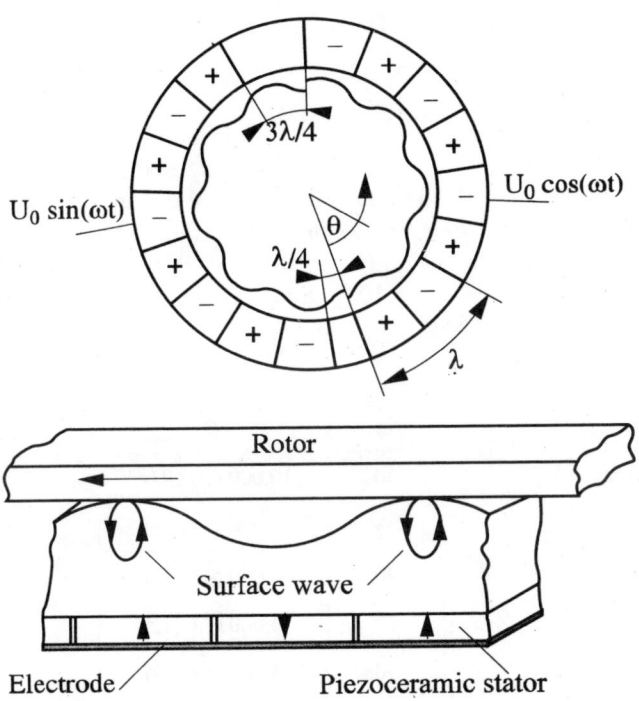

Figure 11.45. *Constituent elements of a piezoelectric motor*

if f_1, f_2 are the frequencies corresponding to the maximum amplitude divided by two. Q is still the ratio of the energy stored to the energy dissipated during a period. The system is actually resonant only if the quality factor is greater than 1.

A piezoelectric material is naturally a resonator (see Figure 11.13). Monocrystalline quartz has very high quality factors ($> 10^5$) and is used for timing purposes. The resonance frequency is very stable in temperature and time. The shape of the crystal and the crystalline orientation of the cutting planes depend on the application [VAN 61]. The PZT, though of lower performances, can also be used.

A dielectric cylinder acts like a resonant cavity for electromagnetic waves whose wavelength is equal to the diameter of the cylinder D, i.e. a frequency:

$$f_o \approx \frac{c}{D\sqrt{\varepsilon_r}} \text{ and } Q = \frac{1}{tg(\delta)}$$

if c is the speed of light in vacuum. Deriving with respect to the temperature, we get:

$$CT_f = \frac{1}{f_o}\frac{\partial f_o}{\partial T} = -\left(\frac{1}{2}CT\varepsilon + \alpha_T\right)$$

if α_T is the coefficient of thermal expansion. We can derive the following specifications: a high value of ε_r (between 30 and 100); a negative $CT\varepsilon$ in order to minimize the temperature deviation ($CT_f < 3$ ppm/°C); minimal dielectric losses to obtain a good selectivity ($tg\ \delta < 5.10^{-4}$). Some examples of materials are given in Table 11.19 (see also [OUC 85]).

Material	ε_r	Q	$CT\varepsilon$ (ppm/°C)
Ba(Zn,Zr,Ta)O$_3$	30	10,000 (10 GHz)	0
(Zr,Sn)TiO$_4$	38	7,000 (7 GHz)	0
Ba$_2$Ti$_9$O$_{20}$	40	8,000 (4 GHz)	2
BaO Sm$_2$O$_3$.5TiO$_2$	77	4,000 (2 GHz)	15
BaO-PbO-Nd$_2$O$_3$-TiO$_2$	90	5,000 (1 GHz)	0

Table 11.19. *Some materials used as dielectric resonators*

11.6.5. Heating elements

Most metals melt or get oxidized in air above 1,100°C. Conducting ceramics, on account of their refractoriness and their corrosion resistance, are all recommended for manufacturing high temperature heating elements:

– SiC: there are two crystallographic forms, α (hexagonal) and β (cubic). Resistivity is 0.1 Ω cm at 1,000°C. Unlike what is normally thought of a semi-conductor, resistivity increases with temperature. It doubles approximately between 1,000°C and 1,600°C. The heating elements are rods or tubes and are largely used in electrical industrial furnaces. The maximum service temperature is approximately 1,550°C;

– $MoSi_2$: this is a metallic conductor with a resistivity of 4.10^{-4} Ω cm at 1,800°C. The maximum service temperature is 1 900°C. The heating elements are rods or pins in Super Khantal, a composite of $MoSi_2$ (80%) and a vitreous phase (20%) acting as a binder. During the first heating, a silica film is formed at the surface of the element, which protects it from subsequent oxidation;

– $LaCrO_3$: this is a p-type semi-conductor, conduction takes place by the hopping of electrons between Cr^{3+} and Cr^{4+} ions. The maximum service temperature is 1,650°C. Resistivity is 1 Ω cm at 1,400°C. We should note the tendency for chromium to volatilize at high temperatures;

– SnO_2 and more generally ITO (indium-tin oxide) solid solution, with resistivity of a few 10^2 Ω cm at 25°C. A large forbidden band explains the transparency of these materials in the visible spectrum. Deposited in the form of films on a vitreous medium, they make effective heating possible without changing the vision (electrically heated windscreen).

11.6.6. Temperature sensors

11.6.6.1. NTC thermistors

These materials are oxides of transition metals, with conduction taking place through hopping (see section 11.1.4). The concentration of carriers is fixed by the doping element concentration. The temperature dependency of the resistance is therefore that of mobility (equation [11.10]):

$$R(T) = A\exp(B/T) \tag{11.28}$$

i.e. a negative temperature coefficient (NTC):

$$NTC = \frac{1}{R}\frac{dR}{dT} = -\frac{B}{T^2}$$

Mainly two families of materials are used: solid solutions MO (M = Mn, Ni, Co, etc.), with structure NaCl doped with lithium, and M_3O_4 with spinel structure. Depending on the composition, the constant B varies from 2,000 K to 5,000 K, i.e. a coefficient NTC of several % at ambient temperature. There is a strong correlation between the value of B and the value of the pre-exponential factor A (see Table 11.20).

Mn (%)	Co (%)	Ni (%)	Cu (%)	Resistivity at 25°C (Ω cm)	A (Ω cm)	B (K)	NTC (%/K)
56	8	16	20	10	1.10^{-4}	2,580	−2.9
70	10	20	0	10^3	$3.4.10^{-6}$	3,600	−4.0
85	0	15	0	10^4	$3.9.10^{-7}$	4,250	−4.7

Table 11.20. *Some materials used for making NTC thermistors*

We can distinguish two types of applications depending on whether the temperature of the thermistor is the temperature of the environment, or a result of its heating by the Joule effect [HIL 91]. In the first case, thermistors are used as temperature sensors; they provide excellent stability in time. In the second case, let T be the temperature of the thermistor in service conditions. In a stationary regime, the electrical power is dissipated in the form of heat in the environment where temperature is T_{amb}:

$$P = R(T)I^2 = k(T - T_{amb})$$

k is a heat dissipation factor which depends on the shape of the thermistors and the nature of connections, but also on the characteristics of the environment, liquid or gaseous for example. This equation, coupled with equation [11.28] defines, for a given electrical power, the voltage, the current and the operating temperature. The current-voltage characteristic is non-linear (see Figure 11.46). We should note the existence of a part with negative dynamic resistance.

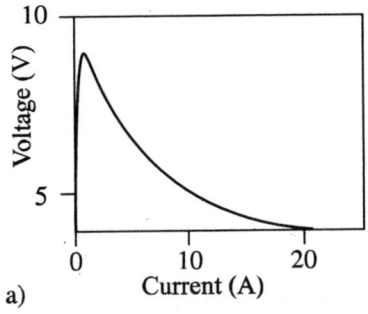
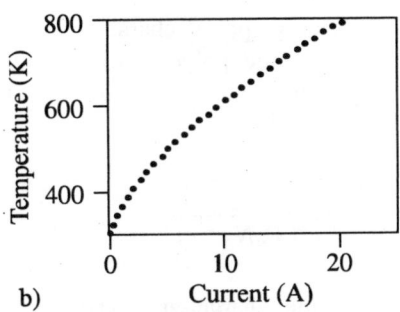

a)

b)

Figure 11.46. *a) Operating V(I) characteristic for a thermistor NTC and b) corresponding temperature variations*

11.6.6.2. *PTC thermistors (positive temperature coefficient)*

The basic material is barium titanate doped with yttrium (in substitution of barium) or niobium (in substitution of titanium). The Kröger and Vink diagram (see Figure 11.4) tells us that at high temperatures, in reducing atmosphere, it is a semi-conductor, but in air the material becomes an insulator, with the ionized dopants being compensated by cationic vacancies. In fact, the diffusion of cations in perovskite is known to be very slow and the state of equilibrium is never achieved at the level of the grains when these are large enough (> a few μm). Such a material, sintered in air, displays at ambient temperature a resistivity of around 50 Ω cm. This decreases slightly when the temperature increases and then grows rapidly, several orders of magnitude, above 120°C, the Curie temperature of the material. This part of the characteristic with a strongly positive temperature coefficient (PTC ≅ 20–30%/K!) is interesting for applications. Above 200°C, the resistance decreases again (see Figure 11.47).

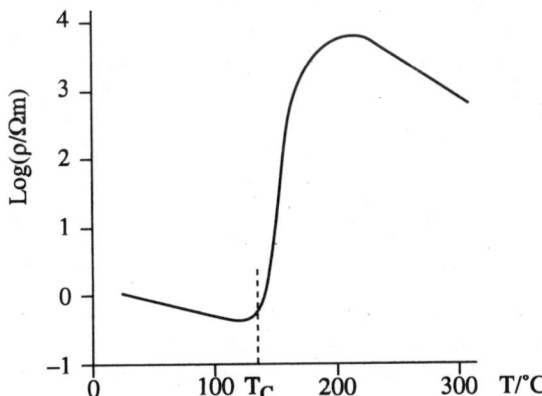

Figure 11.47. *R(T) characteristic of a PTC thermistor*

This very special characteristic is not seen for single crystals and is therefore induced by the polycrystalline microstructure of the material. All experimental observations suggest the existence of potential barriers at grain boundaries ϕ_o (see section 11.4.3):

$$\phi_o = \frac{Q^2}{8qN_d\varepsilon_o\varepsilon_r}$$

if Q is the charge at interface, q the charge of the electron, N_d the dopant concentration, ε_r the relative permittivity. For low applied voltages, the grain boundary has a resistance R:

$$R = R_o \exp\left(\frac{\phi_o}{kT}\right)$$

In the paraelectric state, $T < T_c$, the relative permittivity obeys the Curie-Weiss law (see section 11.2.2) and:

$$\phi_o \approx (T - T_c)$$

if Q is independent of temperature. This model predicts a linear variation of Log R with temperature, which is verified experimentally. This condition is satisfied as long as the interface levels are located below the Fermi level (see Figure 11.30). It is no longer satisfied when the potential barrier, or similarly the temperature, exceeds a certain value. The decrease in resistance is thus attributed to the depopulation of interface levels.

The isothermal current-voltage characteristic is non-linear, well fitted by the empirical relationship:

$$I = \frac{V}{R(T)}\exp\left(\frac{V}{V_o}\right)$$

V_o being a parameter of non-linearity which depends on the composition and grain size.

The material is prepared from rutile and barium carbonate with a slight excess of TiO_2 ($\cong 1$ mol %) and low dopant concentration ($\cong 0.2$–0.3 mol %). The addition of some transition elements in small quantities considerably increases the resistance jump. They induce specific interface levels, with manganese being more effective. Liquid phase sintering takes place in air at around 1,300°C, which enables the integration of the dopant by a mechanism of dissolution-recrystallization of grains. It is also possible to modulate the temperature T_c by using solid solutions: (Ba, Sr)TiO$_3$ (decrease of T_c) or even (Ba, Pb)TiO$_3$ (increase of T_c) (see Figure 11.40).

The PTC thermistors are not often used as sensors, since they are very sensitive but are not very accurate. In most applications, the thermistor gets heated under the effect of power dissipated by the Joule effect, like the NTC thermistors. The heat balance equation:

$$P = \frac{V^2}{R(T)} \exp\left(\frac{V}{V_o}\right) = k(T - T_{amb}) \qquad [11.29]$$

enables the calculation of the non-isothermic current-voltage characteristic (see Figure 11.48). Current decreases when the temperature of the thermistor exceeds the switching temperature T_c. When the temperature coefficient becomes once again negative, the operating state becomes unstable. The heat burst leads to the destruction of the component.

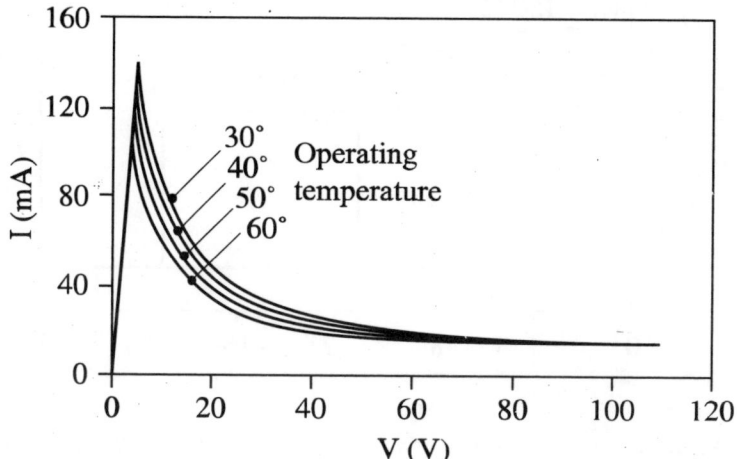

Figure 11.48. *Operating I(V) characteristic of a PTC thermistor*

The uses are very diverse and we will give only three examples.

EXAMPLE 1.– the heating of the thermistor is not instantaneous and requires a characteristic time τ. In fact, in dynamic regime, equation [11.29] becomes:

$$MC_p \frac{\partial T}{\partial t} + k(T - T_{amb}) = P \text{ or } \tau = \frac{MC_p}{k}$$

if M is the mass of the thermistor, C_p the specific heat of the material. This is a first order differential equation, with the approach of the stationary state being exponential. Figure 11.49 shows the delay induced by the heat regime for the establishment of a stationary state. A PTC thermistor thus works like an automatic switch. In practice, the delay depends on the resistance introduced in series with the thermistor.

EXAMPLE 2.– the dissipated power in a PTC thermistor could be significant, hence it is used as a heating element. The fact that the operating current-voltage characteristic displays a negative range of dynamic resistance confers it noteworthy advantages, like for example low variations of dissipated power in response to significant variations of the supply voltage. Furthermore, a decrease in the ambient temperature is reflected by a modification of the operating current-voltage characteristic and an increase in the delivered power. This is a self-regulated heating element.

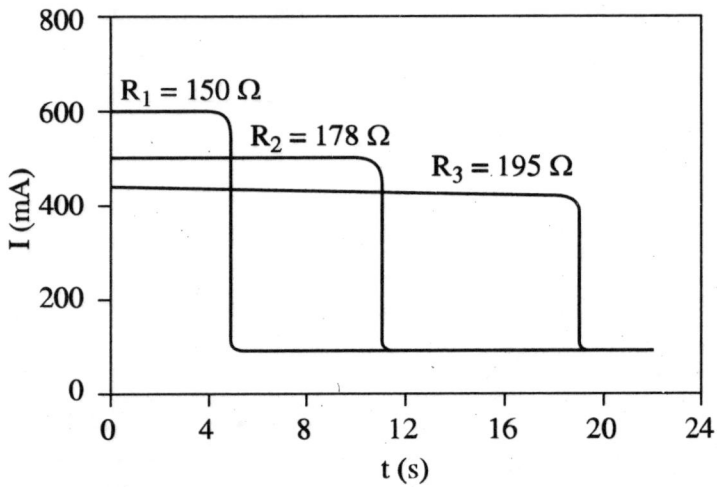

Figure 11.49. *Characteristic of a PTC thermistor used as a time delay component for different values of the load resistance*

EXAMPLE 3.– a PTC thermistor is naturally a current limiter, as seen from the current-voltage characteristic (see Figure 11.48), which makes it complementary to the varistor (see section 11.6.7).

11.6.6.3. *Pyroelectric bolometer*

This is basically a capacitor whose dielectric material is pyroelectric [MOU 90]. The absorption of infrared radiation provokes the heating of the material and thus output a variation of charge and voltage, i.e.:

$$\Delta V = pS \frac{\Delta T}{C}$$

if p is the pyroelectric coefficient, S the electrode surface area (typically 1 mm $\times$ 1 mm), C the capacitance.

Two operating modes are possible (see Figure 11.50):

– the conventional pyroelectric mode, expressed as P, which makes use of the variation of spontaneous polarization with temperature;

– the dielectric mode, expressed as D. In this case, the ferroelectric material is polarized under high voltage and the sensor makes use of the variations of the dielectric constant with temperature.

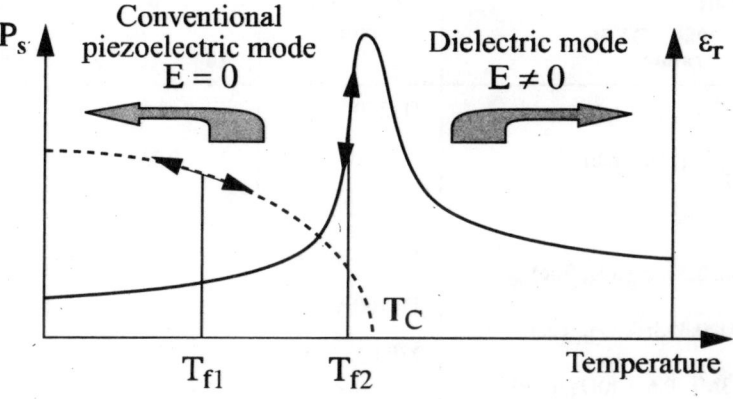

Figure 11.50. *Operating modes of a pyroelectric sensor*

This leads us to the definition of a generalized pyroelectric coefficient:

$$p = \frac{\partial D}{\partial T} = \frac{\partial}{\partial T}\left\{ P_s + \varepsilon_o \int_0^E \varepsilon_r(x)\, dx \right\}_E$$

D being the electric displacement, ε_r the dynamic relative permittivity:

$$\varepsilon_0\varepsilon_r(E) = \left(\frac{\partial D}{\partial E}\right)_T$$

i.e. the slope of the first polarization curve. Materials with high performance are not many (see Table 11.21). The figure of merit F_d:

$$F_d = \frac{p}{C_p\sqrt{\varepsilon_r tg\,(\delta)}}$$

must be greater than 10^{-4}

The generated signal is low and must be amplified by a field effect transistor. For reasons of impedance matching, the capacitance must be of low value, i.e. a material of low relative permittivity. The sensitivity of the device will be greatly affected by the different noise sources.

Materials (mc) = single crystal, (cer) = ceramic	Operating mode	p 10^{-6} C/(cm^2 K)	F_d Pa$-1/2$
K(0.67Ta, 0.33Nb), (mc)	D (250 V/mm)	8.0	4.6 10-4
Pb$_5$Ge$_3$O$_{11}$: Ba, (mc)	P	0.032	1.9 10-4
LiTaO$_3$, (mc)	P	0.18	5.1 10-5
(0.65Ba, 0.35Sr)TiO$_3$, (mc)	D (500 V/mm)	0.3	3.4 10-5
(0.67Ba, 0.33Sr)TiO$_3$, (cer)	D (100 V/mm)	23	8.4 10-4
Pb(0.33Mn, 0.67Nb)O$_3$, (cer)	D (9,000 V/mm)	0.085	1.0 10-4

Table 11.21. *Materials suitable for pyroelectric bolometers*

Furthermore, as in the case of thermistors and for the same reasons, the response time to temperature variations is not instantaneous, but requires a characteristic time:

$$\tau = \frac{MC_p}{k}$$

Reasonable response times (typically 0.1 to 10 s) require a small thickness of the dielectric (approximately 0.1 mm).

Let us list a few applications:

– measurement of infrared radiation. Since the sensor is sensitive only to temperature variations, a strobe has to be used using a frequency f such that $2\pi f \tau = 1$;

– chemical analysis of gas which absorbs in the infrared such as CO or NO;

– detection of moving people;

– infrared imaging. Sensors of 256×128 pixels have been developed, using thin films deposited by the sol-gel method.

11.6.7. *Varistors*

A varistor is a component whose current-voltage characteristic (given on Figure 11.51) is strongly non-linear. The non-linearity coefficient *CNv* is the slope of the characteristic LogI (LogV), i.e.:

$$CNv = \frac{dLnI}{dLnV} = \frac{R_{stat}}{R_{dyn}}$$

with R_{dyn} the slope of the graph V(I) and R_{stat} the ratio V/I. In Ohmic regime, $CNv = 1$, the dynamic and static resistances are equal. This happens to be the case at low voltage and the resistance is thermally activated. Then the coefficient *CNv* increases, passes through a maximum, approximately 30 to 40 for commercial varistors, close to a threshold voltage V_s, before decreasing. V_s corresponds roughly to a current density of 1 mA/cm^2. The consequence of this strong non-linearity is the very significant variations of current, by several tens of units, for small variations in voltage.

The natural function of this component is the protection against transient surges. The nominal operating voltage is chosen in the bend of the current-voltage characteristic, the dissipated power then being negligible. In case of a surge, the varistor gets short-circuited, the voltage being limited to V_s. It acts like a non-destructive fuse and shows a very short response time (< 1 ns), among other advantages. The threshold-voltage can assume very different values, between 10 V and 100,000 V. They are therefore used to protect windscreen wiper motors (12 V) or motors (220 V) or even high tension lines. In the latter case, they are referred to as lightning arresters.

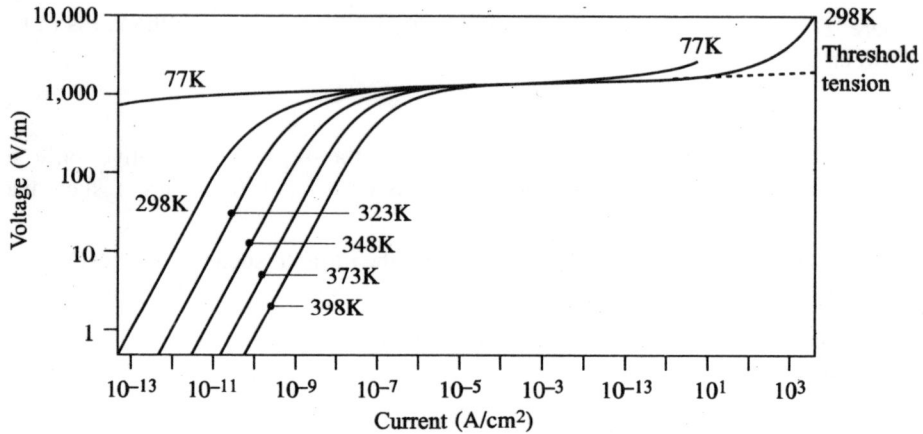

Figure 11.51. *I(V) characteristic of a varistor*

The material is zinc oxide, a natural semi-conductor of n-type as a consequence of a non-stoichiometric composition (Zn/O > 1, $\rho \cong 1 \ \Omega$ cm). A ZnO ceramic displays Ohmic behavior. A certain non-linearity appears, if it is blended with bismuth and/or antimony oxides (approximately 0.5 mole %), but the covalent CNv is small ($\cong$ 3–10). Microscopic observations show that these elements are located at grain boundaries. Then, this non-linearity is a grain boundary effect. CNv increases considerably with the addition of manganese, which is therefore an essential ingredient. Micromeasurements, performed on individual grain boundaries, show a wide diversity of electrical characteristics, but only the interfaces, i.e. grain boundaries without intergranular phase, display the varistor effect. The threshold voltage is then 3.4 V with a low statistical dispersion. See section 11.4.3 for a theoretical understanding of the mechanisms involved.

Processing is conventional. The liquid phase sintering takes place at a temperature varying between 1,150°C and 1,350°C. Annealing at a lower temperature (600–800°C) improves performances. The sequence of reactions is complex and has been the subject of several detailed studies. Microscopic observations show grains of ZnO, often twinned, of size between 5 and 20 µm, an intergranular phase rich in bismuth and secondary phases, of spinel structure ($Zn_7Sb_2O_{12}$) or pyrochlore with variable composition ($Bi_{1.4}Mn_{0.4}Zn_{0.7}Sb_{1.5}O_7$ for example). It is this heterogenity of the microstructure which explains that the average threshold voltage per boundary:

$$V_{threshold} / jdg = V_{threshold} / (1mA/cm^2)\frac{\bar{D}}{e}$$

where $\overline{D}$ is the average grain size and e the thickness of the sample is generally between 2.5 and 3 V, thus clearly lower than 3.4 V.

11.6.8. *Ceramic magnets*

These are developed from magnetic oxides, the ferrites (see section 11.3 and [GOL 87]). They are called hard or soft ferrites depending on the values of the parameters B_r and H_c (see Table 11.22).

	μ_i	H_c (A/m)	B_r (mT)
Hard ferrites	1–10		
Isotropic		100,000	200
Oriented		250,000	400
Soft ferrites			
Mn-Zn	500–10,000	5–100	350–500
Ni-Zn	10–2,000	10–2,000	100–400

Table 11.22. *Hard ferrites and soft ferrites (see Figure 11.25 for definition of headings)*

11.6.8.1. *Hard ferrites*

Hard ferrites have high remanence B_r and coercive field H_c. The hysteresis cycle is quasi-rectangular, characterized by the product $B_r H_c$ which has the dimension of an energy per volume unit. The materials are, most often, hexaferrites of barium and strontium. They are used as permanent magnets, for example in speakers and direct-current motors. The grains are of small size ($\cong 1 \ \mu m$) and monodomain, which eliminates all demagnetization by migration of domain walls. To increase the remanence, the magnets are shaped under high magnetic field, whose purpose is to orient the grains along the direction of easy magnetization, the c-axis in the case of the magneto-plumbite structure.

Let us also mention the oxides of iron γ-Fe_2O_3 or chromium CrO_2 used for magnetic recording. For better performance, the grains are acicular.

11.6.8.2. Soft ferrites

Soft ferrites, with spinel structure, display an S-shaped narrow loop with low dissipated energy and easy magnetization reversal. They exhibit high initial permeability. These characteristics make them suitable for alternating current applications. Requirements differ following the magnetic induction value.

In low level applications (inductors, filters, etc.), the initial permeability must be high, stable in time and with temperature. To that purpose, grain boundaries, pores and second phases, which oppose the motion of domain walls, have to be minimized; the microstructure must be uniform with large grains and almost no intragranular porosity. In alternating regime, permeability has real and imaginary parts, $\mu_{r,i} = \mu'_{r,i} - j\mu''_{r,i}$, the imaginary part characterizing the power dissipated by the Joule effect P per unit of volume, during a period, i.e.:

$$P = \pi H_o^2 f \mu''_{r,i} = \pi H_o^2 f \mu'_{r,i} tg(\delta)$$

where H_o is the amplitude of the magnetic field and f the frequency. $tg(\delta) = \mu'_{r,i} / \mu''_{r,i}$ is the loss factor. The ratio $tg(\delta) / \mu'_{r,i}$ (typically 10^{-6}) is the figure of merit of the material for the applications. It must be as small as possible. The permeability depends on the frequency and there is a cutoff (of the order of 100 to 300 kHz), beyond which it decreases greatly.

In power applications (transformer, switch mode power supply, etc.), the magnetic induction has the saturation value. The output voltage is given by the relation:

$$V_s = -N \frac{d\Phi}{dt} = 4.44.10^{-8} NSBf$$

where N is the number of turns and S the cross-section. High frequency operations help to minimize the size and therefore the weight of the devices, but considerably increase the dissipated power which varies f^2. This stems from two sources: losses by hysteresis and losses by the Joule effect due to induced current. The latter are inversely proportional to resistivity, which must therefore be at its highest. In ceramics, the insulating grain boundaries confine these currents in the grains. The losses will be smaller with smaller grain sizes. The insulating nature of grain boundaries can be reinforced by some additives, like lime or silica, introduced in small quantities. From this point of view, Ni-Zn ferrites have higher performances than Mn-Zn ferrites ($\rho > 10^6$ Ω cm against 100–1,000 Ω cm).

Soft ferrites are also used in microwave devices (f > 1 GHz) like circulators, phase shifters, etc. Requirements with regard to the composition, microstructure and homogenity are particularly stringent. Ni-Zn ferrites compete with garnets whose resistivity is even higher ($\rho > 10^{12}$ Ω cm).

11.6.9. *Gas sensors and fuel batteries*

Let us consider two compartments where we have different chemical potentials of oxygen, separated by a solid electrolyte membrane, an oxygen conductor. The chemical potentials may be fixed with the help of either a gas mixture (Ar + O_2, CO + CO_2, H_2 + H_2O for example):

$$\mu_{O2} = cte + RT \ Ln(P_{O2})$$

or a mixture of solid phases, for example a metal and its oxide. Local thermodynamic equilibrium is assumed to be achieved, ensuring a reversible working of the cell [DES 94]. The two sides of the membrane may be provided with two permeable electrodes. There are several applications stemming from such a set-up like oxygen probe, fuel batteries, electrochemical pumps, etc. The material most often used is stabilized zirconia (see section 11.1.6) and the operating temperature is higher than 800°C.

11.6.9.1. *Gas sensors*

At thermodynamic equilibrium, in the absence of current (open circuit), the chemical potential gradient is compensated by an electrical field. The resulting electric potential across the membrane, called the Nernst potential, is given by:

$$E^o = \frac{1}{4F} \int_{\mu_{O2}^a}^{\mu_{O2}^c} t_i d\mu_{O2} = \bar{t}_i \frac{\mu_{O2}^c - \mu_{O2}^a}{4F}$$

F is the Faraday, μ_{O2} the chemical potential of molecular oxygen ($= 2\mu_O$), t_i the ionic transport number (see section 11.1.6) function of μ_{O2} and T, $\bar{t}_i$ the average transport in the membrane thickness (not very different from 1 for a solid electrolyte). The measure of this electric potential is a direct measure of the difference in oxygen chemical potential. If one of the two is a reference, then the cell acts as an oxygen probe. The solid electrolyte is zirconia stabilized with lime or yttrium. The addition of magnesium reinforces the resistance to thermal shocks. The reference could be either air ($P_{O2} = 0.21$ bar), or the combination of a metal and its oxide (Pd-PdO for example). Let us state a few examples of uses: controlling combustion in industrial burners, measuring chemical potential of oxygen in a molten metal bath or even the regulation of air-fuel mixture in spark ignition engines (see Figure 11.52).

11.6.9.2. *Fuel cells*

The cell works in a closed circuit in stationary regime. A current I flows from one electrode to the other as oxygen ions in the solid electrolyte and as electrons in the external circuit. Redox reactions at the electrodes ensure the conversion:

– at the cathode (electrode +): $O_2 + 4e' \Rightarrow 2O^{2-}$;

– at the anode (electrode –): $2O^{2-} \Rightarrow O_2 + 4e'$.

In a fuel battery, the cathode is in equilibrium with continuously renewed air flow. At the anode, the oxygen reacts with a fuel H_2, CO, CH_4 for example, but also more complex mixtures like natural gas.

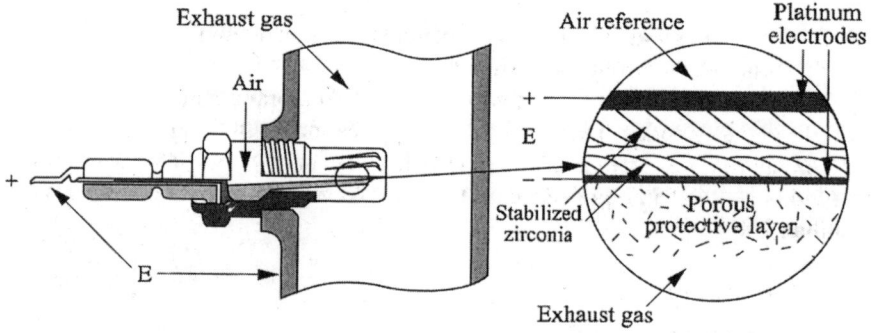

Figure 11.52. *Lambda probe used for controlling combustion in spark ignition engines: probe assembled on the exhaust pipe and section view of the membrane*

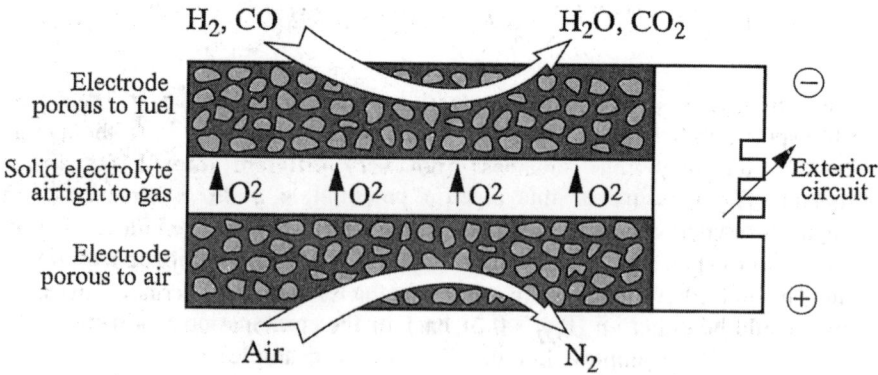

Figure 11.53. *Schematic diagram of a fuel cell with solid electrolyte*

The chemical energy produced by the reaction:

$$\text{fuel} + \text{air} \Rightarrow CO_2, H_2O \quad \Delta H$$

is thus converted into electrical energy. The voltage at the terminals is given by the relationship:

$$U = E^o - RI - \varepsilon_\alpha - \varepsilon_c = E^o - \Delta U$$

R being the resistance of the electrolyte, $\varepsilon_{a,c}$ the potential difference, which is a non-linear function of the current. By connecting a large number of elementary cells in series and in parallel, it is possible to make a generator giving significant power. Thus, a battery of 576 elementary cells in a tubular form operated for a period of 5,582 h with a nominal power of 25 kW without a deterioration in performance (see [SIN 00]).

The cathode must satisfy a certain number of requirements:

– a high electronic conductivity;

– the absence of chemical reactivity with the other materials, at different stages of battery manufacturing as well as during service;

– a coefficient of thermal expansion as similar as possible to the other materials;

– a significant porosity providing easy transport to the gases and a large specific area facilitating oxidation.

The selected material is doped lanthanum manganite ($La_{1-x}Sr_xMnO_3$). This is a mixed conductor, electronic and ionic (oxygen ions), developing a relatively low polarization voltage (100–150 mV for a current of a few 100 mA). Making use of standard technology (injection, pressing), the electrode is developed in the form of a tube of a thickness of approximately 2 mm, which will also serve as a base to other elements, the solid electrode, the anode and the interconnecting elements.

The solid electrolyte, yttrium stabilized zirconia (YSZ10), is deposited by an electrochemical process in the form of a thick dense layer of approximately 40 μm. This small thickness ensures a low resistance.

The anode must be an electronic conductor, but unlike the cathode, the operating atmosphere is reducing. It is therefore possible to consider using a metal like nickel. A composite Ni-YSZ10 has the advantage of minimizing the mismatch of the thermal expansion coefficients and increasing the adhesion of the metal on the solid electrolyte.

The interconnection is ensured by a conductive band, 9 mm wide and -85 μm thick, deposited along the length of the tube by plasma technology. The stresses are particularly severe. The selected material is doped lanthanum chromite $(La_{1-x}Sr_xCrO_3)$, a p-type semi-conductor in a large domain of P_{O2} (air–10^{-18} bar).

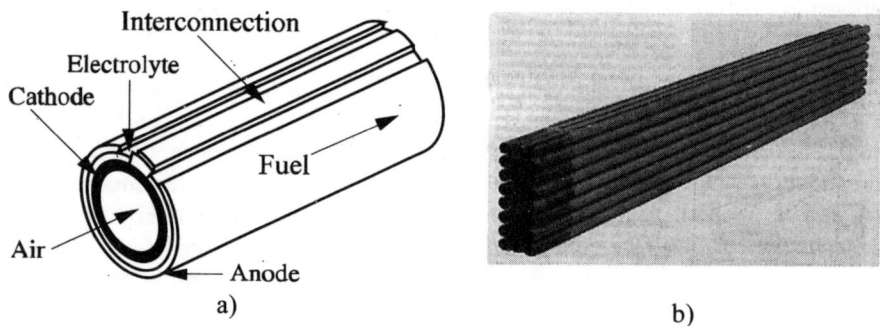

Figure 11.54. *a) Different elements of a cell, b) assembly of 3 × 8 cells*

The tube is closed at one end. The reaction takes place all along the tube and consumes between 50 and 90% of constituents depending on experimental condition. The unreacted gases are burnt at the outlet of the fuel cell, enabling pre-heating of incoming gas. The voltage at the terminals of a cell is 0.5 to 0.8 V and the current of some hundreds of mA. The efficiency is approximately 50%. It increases if the system is under pressure.

There are of course several technological variants and research, both public and private, is now very active in this domain.

11.7. Bibliography

[CHI 90] CHING W.Y., "Theoretical Studies of the Electronic Properties of Ceramic Materials", *J. Am. Ceram. Soc.*, vol. 73, no. 11, p. 3135-3160, 1990.

[CIC 90] *Livre Blanc de la Céramique Technique Française*, publication du Syndicat des Poudres, Fibres, Céramiques Techniques, Confédération des industries ceramics de France, Paris, 1990.

[COX 95] COX P.A., *Transition Metal Oxides. An Introduction to their Electronic Structure and Properties*, Clarendon Press, 1995.

[DES 84] DESPORTES C. *et al.*, *Electrochimie des solides*, Presses Universitaires de Grenoble, Collection Grenoble Sciences, 1984.

[GOL 87] GOLDMAN A., "Magnetic ceramics", in *Electronic Ceramics,* Levinson L.M. (Ed.), Marcel Dekker Inc., p. 147-189, 1987.

[GOO 73] GOODENOUGH J.B., *Les Oxydes des Métaux de Transition,* Editions Gauthier-Villars, 1973.

[GOO 91] GOODMANN G., BUCHANAN R.C. and REYNOLDS T.G., "Ceramic Capacitor Materials", in *Ceramic Materials for Electronics*, Buchanan R.C. (Ed.), Marcel Dekker Inc., p. 69-127, 1991.

[GUR 87] GURURAJA T.R., SAFARI A., NEWNHAM R.E. and CROSS L.E., "Piezoelectric ceramic-polymer composites for transducer applications", in *Electronic Ceramics,* Levinson L.M. (Ed.), p. 92-128, Marcel Dekker Inc., 1987.

[HAR 85] HARBEKE G., "Polycrystalline semiconductors. Physical properties and applications", *Proceedings of the International School of Materials Science and Technology*, Springer Verlag, 1984.

[HIL 91] HILL D.C., TULLER H.L., "NTC Thermistors: Device, Manufacture and Testing", in *Ceramic Materials for Electronics*, Buchanan R.C. (Ed.), p. 272-302, Marcel Dekker Inc., 1991.

[KIT 98] KITTEL C., *Physique de l'état solide*, Editions Dunod, 1998.

[KOF 72] KOFSTAD P., *Nonstoichiometry, Diffusion and Electrical Conductivity in Binary Metal Oxides*, Wiley-Interscience, 1972.

[MAR 79] MARFUNIN A.S., *Physics of Minerals and Inorganic Materials*, Springer Verlag, 1979.

[MOU 90] MOULSON A.J., HERBERT J.M., *Electroceramics. Materials, Properties, Applications*, Chapman and Hall, 1990.

[NEW 86] NEWNHAM R.E., "Composite Electroceramics", *Ferroelectrics*, vol. 68, p. 1-32, 1986.

[NIE 94] NIEPCE J.C., HAUSSONNE J.M. (Eds), *BaTiO₃: Matériau de Base pour les Condensateurs Céramiques*, vol. *1 I- Les condensateurs céramiques à base de BaTiO₃; II- Structures et propriétés de BaTiO₃; vol. 2 III- Les méthodes d'élaboration de BaTiO₃; IV- Réalisation des condensateurs: contraintes liées aux matériaux et aux technologies*, Editions Septima, Collection Forceram, 1994.

[NOW 88] NOWOTNY J. and DUFOUR L.C., *Surface and Near Surface Chemistry of Oxide Materials*, Editions Elsevier, 1988.

[NYE 87] NYE J.F., *Physical Properties of Crystals: Their Representation by Tensors and Matrices*, Oxford University Press, 1987.

[OUC 85] OUCHI H. and KAWASHIMA S., "Dielectric Ceramics for Microwave Application", *Japanese J. Appl. Phys.*, vol. 24, Supplement 24-2, p. 60-64, 1985.

[POH 87] POHANKA R.C., SMITH P.L. and FREIMAN S.W, "Strength fracture and fractography of piezoelectric ceramics", in *Electronic Ceramics,* Levinson L.M. (Ed.), p. 51-91, Marcel Dekker Inc., 1987.

[ROS 87] ROSS MACDONALD J., *Impedance Spectroscopy Emphasizing Solid Materials and Systems*, John Wiley and Sons, 1987.

[ROY 99] ROYER D. and DIEULESAINT E., *Ondes élastiques dans les solides*, vol. 2, Masson, 1999.

[SCH 90] SCHMITT S., *La Microélectronique Hybride. La Couche Epaisse*, Hermès, 1990.

[SHA 84] SHARMA B.L., *Metal-semiconductor Schottky Barrier Junctions and Their Applications*, Plenum Press, 1984.

[SHI 95] SHI D. (Ed.), *High Temperature Superconducting Materials Science and Engineering*, Pergamon, 1995.

[SIN 00] SINGHAL S.C., "Science and Technology of Solid Oxide Fuel Cells", *MRS Bulletin*, vol. 25, no. 3, p.16-21, 2000.

[TAL 74] TALLAN N.M. (Ed.), *Electrical Conductivity in Ceramics and Glass*, 2 vol., Marcel Dekker Inc., 1974.

[TAN 82] TANAKA T., "Piezoelectric devices in Japan", *Ferroelectrics*, vol. 40, p. 167-187, 1982.

[VAN 61] VAN DYKE K.S., "Transducteurs et résonateurs piézoélectriques", in *Les Matériaux Diélectriques et leurs Applications,* Von Hippel A.R. (Ed.), Editions Dunod, p. 252-263, 1961.

[VES 91] VEST R.W., "Materials aspects of thick-film technology", in *Ceramic Materials for Electronics*, R.C. Buchanan (Ed.), p. 435-488, Marcel Dekker Inc., 1991.

[WOL 91] WOLKENSTEIN T., *Electronic Processes on Semiconductor Surfaces during Chemisorption*, Consultants Bureau-Plenum Press, 1991.

[XU 91] XU Y., *Ferroelectric Materials*, Editions North-Holland, 1991.

Chapter 12

Bioceramics

12.1. Introduction and history

Bioceramics are meant to be used as implants in living organisms or, more generally, during prolonged contact with biological fluids or tissues. This broad definition includes ceramics which are or could be used in extracorporeal circulation systems (dialysis for example) or engineered bioreactors. However, this chapter will deal essentially with ceramics used as implants.

The substitution of faulty organs or tissues has always been man's preoccupation. The first attempts at implantation probably date back to prehistoric periods and most probably involved teeth, on account of their easy accessibility, their nutritional role, and due to their social function, of course, and their place in the collective imagination. We find traces of surgical attempts on skeletons, particularly in such civilizations where it was customary to mummify the dead. The most elaborate applications were carried out by Larrey, with the use of plaster of Paris particularly for setting internal bone fractures. However, the extensive use of bioceramics began after World War II, when the development of the medical insurance system and the rapid technological progress led to a widespread use and an improvement of these surgical interventions benefiting a larger number of people. Biomedical ceramics used today essentially come from other fields of application. However, various adaptations were necessary and ceramics for exclusively medical use began to be developed. The place held by ceramics and polymers as well as ceramic-polymer composites in the field of biomaterials will probably increase in

Chapter written by Christèle COMBES and Christian REY.

the coming years at the expense of metals, whose physicochemical characteristics, both mechanical and biological, are generally unsuitable for use as biomaterials, apart from a few exceptions.

In practice, biomaterials are classified into three types depending on the reaction of biological tissues with which these are in contact: biotolerated, bioinert and bioactive materials. Biotolerated materials induce tissue alterations and are isolated by a covering of fibrous tissue. Several metals and certain polymers come under this category. Bioinert materials do not induce any visible tissue reactions; the majority of ceramics belong to this group. Bioactive materials, on the other hand, favor tissue repairs and the integration of associated devices. Some ceramics and various polymers belong to this last category. Bioactive effects are aimed at very specific tissues and the applications of these materials are oriented and limited to specific organs or tissues. As far as ceramics are concerned, bioactivity is essentially used in orthopedics. It favors bone repair and the integration of implants in bone tissues.

12.2. Biomedical ceramics and their field of use

12.2.1. *Usage properties of biomedical ceramics*

Ceramics have numerous uses in the field of biomaterials, mainly because of their physicochemical properties. Their chemical inertness helps to minimize organic reactions of the host organism and their hardness and resistance to abrasion makes them suitable for substitution of hard tissues (bones and teeth). Some ceramics also have excellent tribological properties and are utilized in friction couples intended to replace malfunctioning joints. Finally, other properties (appearance, electrical insulation) also determine certain biomedical applications.

12.2.2. *Multipurpose ceramics*

A number of implanted ceramics have not actually been designed for specific biomedical applications and are used in different implantable systems because of their properties and their good biocompatibility.

12.2.2.1. *Alumina*

Alumina is one of the most widely used multipurpose ceramics. It is essentially used in orthopedics for its good tribological properties and its outstanding chemical inertia. One of the advantages of alumina is that it is a very bad substrate for the crystalline growth of calcium phosphates, which can alter other friction couples [ROY 93]. It constitutes the heads of femoral prostheses and is used also in the development of the acetabulum. Early applications raised some problems of

. mechanical strength, now very rare, and led to the creation of wear debris. These problems were attributed to different causes: too large grain size of sintered piece, loosening at grain boundaries, insufficient density and shaping flaws. Today, the alumina used have evolved and most of these problems have been eliminated. Properties of alumina used have been stringently standardized. In all cases, it refers exclusively to rhombohedral alpha phase, and high purity. The lifespan of alumina heads is now very often longer than the patient's. The main cause for failure is the wear of high density polyethylene used in the acatebulum. Alumina acatebulum may also be used. Their characteristics are identical to those of the heads, the essential problem here being the alumina-bone contact. Alumina is in fact considered to be a bioinert ceramic and it does not directly bind with the bone. There is always a thin layer of fibrous tissue between the bone tissue and alumina which can cause osteolysis, pain and loosening. Hip prosthesis stem made totally in alumina were also developed, but these initiatives seem unlikely to lead to development on an industrial scale, due to inadequate mechanical properties such as a very high brittleness and a Young's modulus which is very different from that of bone tissue. Alumina has also been proposed as an adhesive underlayer for bioactive coating, generally biodegradable, on the stem of metallic prostheses [DEM 98]. The undercoat deposition is generally obtained by plasma spraying. However, this process leads to several phases, whose biological properties are still little known. Nevertheless, these phases are relatively more soluble than alpha alumina and the release of aluminum *in vivo* can induce bone lesions (osteomalacia) [FRA 94]. Other uses of alumina ceramics to be mentioned are: inner ear ossicles, ocular prostheses, electrical insulation for pacemakers, catheter orifices. Finally, alumina is also used in numerous prototypes of implantable systems (cardiac pumps for example).

12.2.2.2. *Alumino-silicates and glasses*

Alumino-silicates are essentially used in dental prostheses, either as massive ceramic, or cermet or in ceramic-polymer composites. Polymers, usually associated with alumino-silicates, are also increasingly used in the filling of cavities replacing amalgams suspected to have toxic effects [MJO 97]. The alumino-silicates used are characterized by a glassy structure, sometimes incorporating crystalline phases (vitroceramic). Particular care is taken by manufacturers in the coloring of materials to ensure a perfect visual integration with natural teeth. Contrary to artificial teeth in resin, the color of tooth ceramic remains stable. However, the enamel of natural teeth has often a tendency to turn yellow with age and differences may then appear. More important problems like bonding with biologic tissue (bone, enamel, dentine, oral epithelium) and mechanical strength arise in the case of crowns and implants. From an esthetic point of view, the ceramic crown is superior to a metallic one or to those made of cermets [BAS 98], but their making entails a high level of precision, on the part of both practitioner and prosthesist. Their brittleness also leads to breakages, notably on molars [FUZ 98]. Regarding mechanical properties, microstructure plays an important role in the resistance of ceramics, especially the size and nature of crystalline phases which are associated with alumino-silicates (mica, alumina, zircon, etc.). Moreover, they are relatively sensitive to fatigue. Cermets always pose problems of metal-ceramic bonding and also appear less resistant to abrasion than ceramics. These different materials introduced in the oral cavity, do not seem to have any incidence on the development of dental plaque and the appearance of dental caries. However, these can cause abrasion on the dental enamel of the opposite teeth during mastication, which can result in losses much more important than those caused by crowns in metal alloys. The reverse can also be observed and some ceramics could be damaged prematurely by friction and pressure on the opposite teeth. Progress is yet to be made in this respect.

Different types of fixation of crowns or implants have been studied. Implants, more often in titanium, are directly fixed in the jawbone and are crowned by the artificial tooth. An osteo-conductive bioceramic coating is sometimes done, as for artificial hips, in order to favor the integration of the implant in bone (see section 12.2.3). Dental crowns are simply attached with organic resins, the positioning is particularly crucial during the setting period and could determine the longevity of the substitute.

Alumino-silicate glasses were also proposed as substitutes for bones; their chemical composition is then adapted for making these materials bioactive and this aspect will be dealt with in section 12.2.3.2.

12.2.2.3. *Zirconia*

Zirconia doped with yttrium oxide has also been proposed as a substitute for alumina in the heads of osteoarticular prostheses. Important developmental studies have been made [CAL 95] and these heads have been commercialized. The main advantages of zirconia compared to alumina are a greater failure strength, mainly bending strength, as well as a good resistance to fatigue [DRO 97]. These properties make the use of prostheses heads of very small dimensions possible, thus reducing the wear debris. Besides, zirconia has a better coefficient of friction and a better wear resistance, despite the quite controversial results. One of the unknowns is the role of zirconia in the nucleation of calcium phosphates from supersaturated body fluids. Moreover, as we have mentioned, alumina has the advantage of a longer use and currently gives satisfactory results.

12.2.2.4. *Vitreous carbon and diamond carbon*

Vitreous carbon has several interesting physicochemical and biological properties: it is light, resistant to wear and haemocompatible. It is essentially used for making cardiac valves and replaces natural valves taken from pigs which have a tendency to calcify and have a more limited life. Essential problems which remain are the formation of thromboses and bleeding due to the degradation of the junction between the prosthesis and the artery. Mechanical valves appear, however, to be tolerated in the long term [PET 99].

Diamond is an interesting coating from a biological point of view: it does not induce any cytotoxic or haemolytic effects and can be used for various vascular applications or in cardiac surgery [DIO 92].

12.2.2.5. Other ceramics

A number of other ceramics have been subjected to biomedical tests for implantation, without currently being developed industrially. Among these ceramics, we can cite silicon carbide, titanium nitrides and carbides, and boron nitride. TiN has been suggested as the friction surface in hip prostheses. While cell culture tests show a good biocompatibility, the analysis of explants shows significant wear, related to a delaminating of the TiN layer [HAR 97]. Silicon carbide is another modern day ceramic which seems to provide good biocompatibility and can be used as bone implant [SAN 98].

12.2.3. Ceramics for specific uses

Ceramics for specific uses, in addition to their traditional properties, have biological activity. We thus refer to bioactive ceramics. These ceramics are essentially used as bone replacements and today form part of the daily practice of orthopedic, maxillofacial and plastic surgeons. These are used in case a loss of bone substance (tumor, important trauma, infection, etc.) requires to be filled. According to the type and shape of the defect to be filled, the location of the implant and mechanical stresses, different types of bioceramics with varying biological properties are available. Biodegradable ceramics will be resorbed and replaced by reconstructed tissue, whereas non-biodegradable ceramics are intended for a permanent implantation. Very often, synthetic ceramics compete with natural materials and it seemed more interesting to start this section with a description of the latter.

12.2.3.1. Natural materials

Different types of natural substitutes for bone tissue are available on the market, generally of animal origin. These are subjected to physical, chemical or biochemical processes before utilization as biomaterials. Calcium carbonates, particularly those produced by marine organisms (coral, mother of pearl, etc.), have been used for many years as substitutes for bone.

The idea that coral can replace defective parts of bone comes from the similarity in the structure of some coral skeletons with cancellous bone matter allowing colonization by cells and blood vessel penetration. The exoskeleton of coral polyps constitutes blocks of calcium carbonates with regular and interconnected porosity, according to a structure specific for each species. After selecting, cleaning and shaping, these materials can be implanted in bone locations, to serve as a framework for the new bone tissue. They are degraded due to a carbonic anhydrase, thus leaving place for newly synthesized bone tissue [GUI 95]. These materials are biocompatible and their advantage essentially lies, on the one hand in their open porosity which facilitates bone colonization and on the other hand, on their rapid resorption due to the good solubility of calcium carbonates and their enzymatic degradation. Despite good mechanical strength, this is not sufficient to allow their use in bones subjected to high mechanical stresses (load bearing bones). Moreover, their structure is fixed with the considered species and their chemical composition is not well controlled, particularly with respect to trace elements.

Mother of pearl, also composed of calcium carbonate and an organic matrix, has been proposed as a substitute for bone. Mother of pearl powder implanted in a bone defect has shown behavior similar to that of a coral [LOP 98].

12.2.3.2. Bioglasses

Since the development of Bioglass® by L. Hench et al. [HEN 71] in the beginning of the 1970s, several types of glasses and vitroceramics, in particular those belonging to the family $Na_2O-CaO-SiO_2-P_2O_5$, have shown a certain capability of adhering to the bone. However, this real chemical bonding with the bone tissue occurs only for a narrow range of SiO_2 content (42–52%). For SiO_2 contents higher than 60% [HEN 99], the bioglass appears isolated from the bone by a non-adhering fibrous capsule leading to failure of the implant.

Fundamental studies bearing on the bioglass-bone bonding mechanism show that the groups Si-OH on the surface of these materials induce the formation of a layer of apatite analogous to bone minerals, guaranteeing the durable integration of biomaterials [HEN 91]. Moreover, the slow dissolution of silicates would improve cell proliferation and the formation of an osteoid matrix. Furthermore, the release of calcium and phosphorus contained in bioglasses encourages the heterogenous nucleation of bone minerals in the osteoid matrix, thus forming the new bone very rapidly.

Bioglasses constitute a large variety of materials depending on their composition and structure. However, the implantation of bioglass in an area of high mechanical stresses cannot be considered on account of their brittleness. To overcome this disadvantage, these glasses can be heat treated so as to obtain a vitroceramic, or even to use this bioglass in the form of a deposit on a metallic substratum [BRI 97].

Thus, considerable research was directed towards the development of multiphase vitroceramic materials, so as to reinforce the mechanical properties of bioglasses. All these materials are available in different forms: massive, deposit, powder or composite. Vitroceramics are first produced as glass and then transformed into crystallized ceramics by heat treatment. The vitreous stage offers possibilities to mould complex forms. The next stage of crystallization allows a fine microstructure with little or no porosity to be obtained, which confers on the material a good mechanical strength against impact, by virtue of the relaxation of stresses around the pores. With the crystallization process still incomplete, the glassy part fills space between grain boundaries, so as to create a structure without pores.

A biphasic material with excellent mechanical properties, called vitroceramic A/W and made up of an apatitic phase $(Ca_{10}(PO_4)_6(OH, F)_2)$, a wollastonite phase $(CaO \ SiO_2)$ and a residual vitreous phase $MgO\text{-}CaO\text{-}SiO_2$ is clinically used, particularly in vertebral reconstructive surgery [KOK 85; YAM 88].

Some bioglasses contain aluminum. However, these are not well developed and the risks of alteration of bone mineralization have been stated [NIZ 99].

12.2.3.3. *Calcium phosphates*

Calcium phosphate-based ceramics constitute, at present, the preferred bone substitute in orthopedic and maxillofacial surgery. They are very similar to the mineral phase of the bone, by their structure and/or their chemical composition. Calcium phosphates usually found in ceramics are:

– hydroxyapatite (HAP): $Ca_{10}(PO_4)_6(OH)_2$;

– tricalcium phosphate β (β TCP): $Ca_3 (PO_4)_2$;

– mixtures of HAP and β TCP.

Name and chemical formula	Abbreviation	Type of materials and utilization
Hydroxyapatite $Ca_{10}(PO_4)_6(OH)_2$	HAP	ceramics, plasma sprayed coatings, composites, drug carrier
Calcium deficient apatites	ns-HAP	Low temperature coating composites, drug carrier
α and β tricalcium phosphates $Ca_3(PO_4)_2$	α and β TCP	ceramics, composites, plasma coating cements, drug carrier
Dicalcium phosphate dihydrate $CaHPO_4, 2H_2O$	DCPD	cements
Anhydrous dicalcium phosphate $CaHPO_4$	DCPA	cements
Octocalcium phosphate $Ca_8(PO_4)_4 \cdot (HPO_4)_2, 5H_2O$	OCP	cements
Tetracalcium phosphate $Ca_4(PO_4)_2O$	TTCP	cements
Amorphous calcium phosphate	ACP	cements, drug carrier low temperature coating

Table 12.1. *Calcium phosphates used as biomaterials*

These ceramics are bioactive and can be degraded to various degrees.

The stoichiometric HAP, characterized by an atomic ratio Ca/P = 1.67 and a hexagonal structure, is the nearest relative of biologic apatite crystals. Moreover, the HAP is the least soluble and the least resorbable calcium phosphate. When an HAP ceramic is implanted in a bone site, the bone tissue formation is observed on its contact (osteoconduction) (see Figure 12.1). Besides, in certain conditions, calcium phosphate ceramics can induce the formation of bone tissue in ectopic sites. HAP implants appear in the form of dense ceramics or with variable porosity or again, in the case of prostheses, as thin coatings deposited by plasma sprayed on a metal.

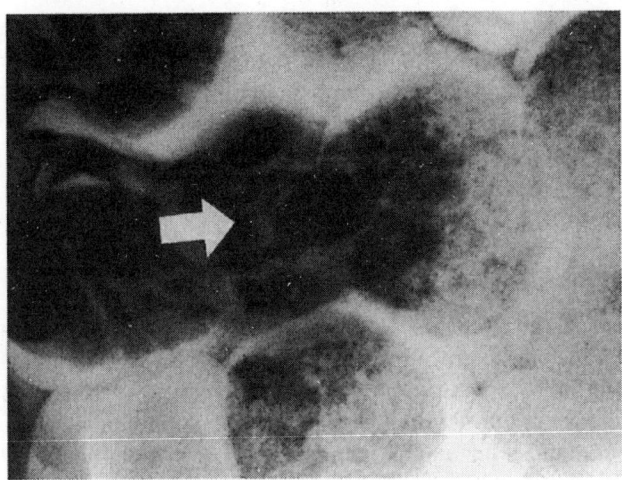

Figure 12.1. *Micro-radiography of a section of an HAP-DCPD composite showing the invasion of DCPD by bone tissue*

The β TCP, characterized by an atomic ratio Ca/P = 1.5, is perfectly biocompatible and bioresorbable. Like HAP, it is capable of developing a chemical bond with the bone and to stimulate its growth, but its resorption is more rapid.

It is difficult to make pure HAP or β TCP and biphasic materials HAP-β TCP have been developed initially by accident and later deliberately; they combine the physicochemical properties of each of the compounds. These can be advantageously used to prepare materials with controlled resorption and bone substitution [DAC 90].

The presence of pores in materials provides anchor points for the bone and improves the mechanical quality of the bone/implant interface; the increase in the specific surface further encourages cell colonization and the revascularization. While calcium phosphate-based bioceramics are excellent materials for bone reconstruction, they have a low mechanical strength (less than that of the bone), not lending themselves to machining. This resistance diminishes while porosity increases, making the utilization of very porous implants very delicate.

Even if the HAP remains the most important calcium phosphate, from an industrial point of view, the development of low temperature processes, particularly those concerning mineral cements and coating on metal, have led to the utilization and emergence of other calcium phosphates which are more reactive. Table 12.1 lists the different calcium phosphates used as biomaterials.

The development of calcium phosphate-based ceramics at high temperature requires taking into account the thermal stability of these compounds. We can distinguish two schemes of decomposition according to the temperature: irreversible decompositions (condensation of hydrogenophosphate ions, decomposition of carbonate ions, of hydroxide ions, etc.) at low temperature (150–1,000°C) and reversible decomposition (decomposition of the apatite into TCP, TTCP and lime) at high temperatures (T > 1,000°C).

12.2.3.4. *Oxides and hydroxides*

As discussed in the previous section, it would seem that the layer of hydrated silica, formed at the surface of implanted bioglasses and vitroceramics, plays a very important role in the formation of the neo-formed apatitic crystals, hence the idea of using pure silica or titanium hydrogels as bioactive compounds. Thus, gels of silica, titanium or zirconia, after being subjected to heat treatment, can induce the formation of the apatite when these are immersed in a metastable solution analogous to biological fluids, while alumina gels are not bioactive. The development through sol-gel processes is an interesting method for the preparation of bioactive materials, since it helps to obtain a product favoring the heterogenous nucleation of the apatite. While the silica gel is heated to a temperature greater than 900°C, the formation of the apatite is delayed [LI 93a]. It therefore appears that the speed of rehydroxilation and the rate of hydroxyl group on the surface of the silica gel control the formation of apatite. In the same way, sufficiently hydrated titanium dioxide, that is with Ti-OH groups on the surface, can join with the bone while it is placed in a bone site. This property offers the possibility to render titanium bioactive by a treatment before implantation, in such a way as to form a gel or a very hydrated layer at the surface of the implant [KOK 99; LI 93b].

12.2.3.5. *Composites*

The development of mineral-organic composite materials offers the possibility of combining the favorable properties of bioceramics such as the HAP, alumina or titanium dioxide with the molding capacity of biocompatible polymers (polymethylmethacrylate): PMMA [KHO 92], poly(L-lactic) acid: PLLA [ROD 95], poly(ethylene): PE [DOW 91]). It is also conceivable to attain a value of the modulus of elasticity near to that of the bone.

We can differentiate composites as bioresorbable or non-bioresorbable. The non-bioresorbable composites are the result of the combination of a non-bioresorbable calcium phosphate (HAP) with a non-bioresorbable polymer (PMMA, PE). In this case, we have to avoid the covering of ceramic grains of the surface, so as to preserve their biological activity.

Bioresorbable composites combine a bioresorbable polymer (PLLA, poly(glycolic) acids and poly(butyric) acids) with HAP particles or of resorptive calcium phosphate. In order to guarantee a successful combination of calcium phosphates with a bioresorbable polymer, it is important to adapt the resorption of the two constituents to avoid inflammatory reactions due to the release of ceramic particles.

These materials should grow in the future on account of the great many combination possibilities and their aptitude at combining a biological activity with mechanical properties similar to those of the bone.

Besides, various combinations of calcium phosphates – with chitosan [VIA 98], with cellulosic composites [OKA 97], with collagen [NIS 95] – have also been well studied, but these combinations cannot be considered as real composite materials. It refers most often to sintered ceramic particles, quite coarse (in order to avoid an inflammatory reaction) and poorly bonded by the macromolecule. Some injectable formulations have been developed.

Certain ceramic-ceramic composites have also been studied, particularly in the fluorohydroxyapatite-alumina and hydroxyapatite-alumina systems [DIM 95; GAU 95].

12.3. Biological properties

Biological properties of ceramics have been studied only recently and essentially from a practical point of view with respect to the envisaged application. There is no possibility at present to predict, from a knowledge of the composition and surfaces properties of ceramic, their biological behavior (hemocompatibility, adhesion proliferation and cell expression, bonding with the tissues, etc.) despite some attempts having been made. We will describe in this section the biological properties of some ceramics and what we can learn from them. Different levels of interaction can be discerned: with biological fluids, with tissues and with cells. Furthermore, the biological behavior gets modified with time, often related to modifications in the surface of the implanted ceramics or their degradation.

12.3.1. Ceramic-tissue interactions

Ceramic-tissue interaction determines the integration of ceramic in its environment (biointegration). Several parameters play an important role in this interaction, and we can distinguish the mechanical anchorage, which ensures the initial stability of the implant, and the chemical anchorage, which actually determines the integration of the implant with its host organism.

12.3.1.1. *Implant stabilization*

The introduction of a chemically inert foreign body in a living organism gives rise to a series of reactions leading to the formation of a fibrous encapsulating tissue. This fibrous tissue isolates the foreign body which could not be destroyed and ensures at the same time the joining between the healthy tissue and the implant. This primary form of biointegration of the implant is considered to be of relatively poor quality, notably for the bone tissue where this layer allows a relative mobility of the implant and the existence of micromovements, eventually associated with the phenomena of abrasion and with an inflammatory process. The existence of micromovements increases the thickness of the encapsulating fibrous layer [HOL 92; PIL 95]. Movements of low amplitude (less than 50 micrometers) do not appear to have any effect. Above 200 micrometers, on the contrary, we observe the formation of a thick fibrous tissue. Other than the surgical skill, various morphological factors especially determine the mechanical anchoring of the implant: a porous or rough surface favors mechanical stabilization of the implant. Likewise, partially biodegradable surfaces resorbed in an irregular manner may be considered as favorable to mechanical integration.

12.3.1.2. *Ceramic-tissue bonding*

The integration of the implant can also be ensured by the interactions of a physicochemical nature with the tissue. These phenomena have been more particularly studied for ceramics used in orthopedics. Interactions with the mineral part have been more particularly described; those with the collagenic part of the bone are not known, despite the fact that they may play an equally important role. In the case of implants of apatitic structure, it has been shown by high resolution electron microscopy that there exists a continuity between apatite crystals of ceramics and those of bone tissues [BON 91]. These interactions of epitactic type justify in some measure the use by apatites as orthopedic biomaterials. With other types of non-apatitic materials, even other amorphous materials such as bioglasses, the formation of a carbonated apatite, analogous to bone mineral, is also implied, even though epitaxy relations do not necessarily exist. Phosphate ions appear to play an important role at the interface between certain materials and the bone mineral. Different authors have clearly shown that the bonding between apatitic type ceramics and the bone is among the strongest, and is generally the line of fracture is situated in the bone tissue and not at the bone-implant interface. An exception that remains generally unexplained is the bioglass-bone bonding. The interactions with organic matrix, as well as the possibility of having locally weaker but greater number of bonds, have been suggested. The type of interaction involved is in fact highly dependent on the number of crystals in direct contact with the biomaterial. But the methods of nucleation of these crystals can be affected by a number of factors associated with the implanted biomaterials (nucleation sites) and its environment (protein adsorption for example).

12.3.1.3. *Mechanical stresses*

Cell activity, of the bone tissue in particular, is closely connected to mechanical stimuli. This effect is now well established, based on observations that in the absence of mechanical stress the remodeling of bones tends to slow down. The implant integration in the bone tissue depends also on biomechanical factors. Besides, since the Young's modulus of ceramics is generally much higher than that of the bone tissue, the implant can cause mechanical stresses at the bone interface. Moreover, it produces a modification of the lines of force (stress shielding) which can result in bone defects close to the implant. This phenomenon is sometimes visible in x-rays and manifests itself by a diminution of the bone density near the implant related to the lack of mechanical stimulation of the tissue.

12.3.2. *Cell-ceramic interactions*

The presence of materials may modify cell activity and affect tissue reconstruction. We have known for many years how to grow eukaryotic cells on different materials and studies have been made on a wide variety of ceramics. Concerning ceramics used in orthopedics, there is a succession of events which ends in the integration of the implant. The sketch in Figure 12.2 details these events in the case of a bioactive ceramic.

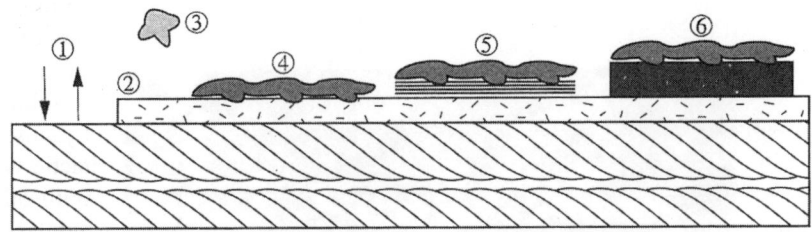

1. Equilibrium of the ceramic surface and the solution. Eventual release of mineral ions, adsorption of ions and/or of proteins.
2. Nucleation and growth of a layer of carbonated apatite analogous to bone mineral from supersaturated biologic fluids. This layer contains a number of bioactive proteins.
3. Cells go towards the modified surface on which these will settle and become differentiated in order to give rise to osteoblasts.
4. The osteoblasts multiply and colonize the surface of the biomaterial.
5. The cell layer synthesizes a collagenic organic matrix.
6. The organic matrix mineralizes and the new bone is deposited.

Figure 12.2. *Events happening at the surface of a bioactive ceramic leading to the formation of a bone tissue*

12.3.2.1. *Surface phenomena*

Biological fluids are generally supersaturated with respect to the apatite of bone tissues. The presence of nucleation and crystal growth inhibitors of calcium phosphates, however, generally prevent the phenomena of uncontrolled mineralization. In this environment, the nucleation and the crystalline growth of apatitic calcium phosphate analogous to bone mineral are relatively easy. It has been shown that all bioactive compounds which facilitate the formation of bone tissues also favor nucleation of calcium phosphates. Several authors in fact now consider this property as a direct measurement of the biological activity of ceramics. We can distinguish ceramics which simply play the role of a nucleator without bringing mineral ions from those which modify in biological environment with release of mineral ions and accelerate the formation of a layer of neo-formed apatite analogous to bone mineral. The first are constituted by calcium phosphates with apatitic structure (hydroxypatite, fluoropatite, carbonated apatite) and certain oxides or hydroxides (titanium oxide, titanium hydroxides, silica gels) (see Figure 12.3). The latter are constituted by hydrolyzable phases with release of calcium ions, phosphate and/or hydroxide or carbonate (non-apatitic calcium phosphates, bioglasses, alkaline surfaces, or calcium carbonates). The latter are capable of generating a neo-form crystalline layer faster than the former and can also create a denser crystalline layer. In certain cases, we can combine composites of the first family with those of the second.

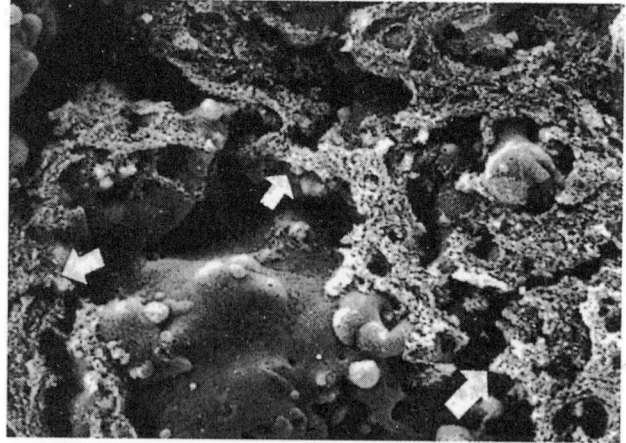

Figure 12.3. *Photograph of the formation of immature bone tissue (after removal of the organic material) at the surface of an HAP layer on a hip prosthesis implanted in a man for several weeks*

Let us note that the formation of an apatitic layer *in vivo* is not desirable for ceramic applications other than for bone substitutes (cardiac valves, friction surfaces of joints). While choosing ceramics for such applications, we have to take care to choose ceramics which do not favor nucleation of these crystals. It has been precisely shown that alpha alumina is a very bad substrate for the nucleation of calcium phosphates. On the contrary, titanium nitrides and carbides also proposed as friction surfaces are quite good nucleators [ROY 93] and yet these can be questioned with respect to their use in friction couples for joint repair.

12.3.2.2. *Recruitment, development and cell expression*

The formation of the neo-formed layer is a determining element of biological activity. The neo-formed nanocrystals of apatite are extremely reactive and have the property to bond specific bone proteins (osteopontin, osteocalcin, etc.) which later promote adhesion, multiplication and cell expression. The formation processes of this layer are, however, quite different from those that take place during primary ossification, which can be described as the mineralization of a pre-formed collagenic matrix. Furthermore, this layer does not have the characteristic organization of a bone tissue with apatite crystals, well oriented with respect to the collagen fibers, and should rather be considered as an uncontrolled mineralization. The events which occur at the surface of bioactive minerals, however, have a certain analogy with those which come into play during the bone remodeling in the cementing zone which limits the bone osteons [MCK 93]. The same phenomena and the same proteins seem to be involved. The neo-formed layer is favorable to the adhesion of osteoblastic cells, which then multiply and form the collagenic matrix at first and mineralize it later (see Figure 12.4). We must mention that the osteoblasts can also attach and develop and form bone on materials which are not necessarily bioactive, for example, cell culture dishes in polymer materials, normally more effective than apatite substrate [HOT 97]. However, the chemical bond between the substratum and the tissue is missing in these systems. To deserve the adjective bioactive, a material should, at the same time, be able to establish a bond with the tissue and help in its renewal.

The cells generally settle on a substratum by means of adhesive proteins characterized by the presence of amino acids sequences (particularly the tripeptide sequence arginine-glycine-aspartic acid (RGD)). On the other hand, the relationship between secretion and/or adsorption of these proteins and the surface characteristics remain unknown. The energy and the surface load appear to play a preponderant role; thus, the proteins fixed on the substratum with different loads exhibit different characteristics and can influence cell recruitment and development [SHE 88]. Generally, the positively charged substrate are more favorable for recruitment and adhesion of negatively charged cells. Similarly, the surfaces which have a high bipolar energy encourage cell development [HOT 97]. However, when over a period of time the protein film develops on the surface of the material, these initial factors no longer have an effect on cell activity. It appears therefore that the kinetic factors of the protein film formation and the competitive adsorption of different proteins, probably determined by the characteristics of the surface, play a decisive role on cell activity [DEW 99].

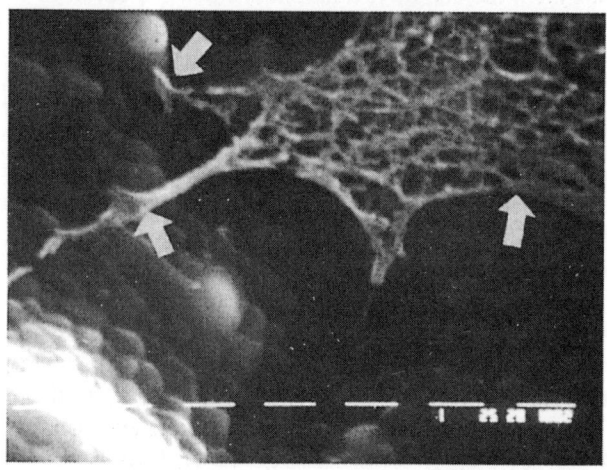

Figure 12.4. *Photograph of cytoskeleton of an osteoblast (after dissolution of the plasmic membrane) fixed on an HAP ceramic*

12.3.2.3. *Cell rehabitation*

Some porous materials may be reoccupied by the cells and invaded by a newly formed bone tissue. These materials should have at the beginning a definite pore size. Very large pores do not encourage bone growth, while the very small ones do not allow an adequate vascularization. The pore diameter should be a minimum of 100 μm, but could go up to 500 μm (see below). Moreover, tissue rehabitation requires interconnected pores.

12.3.3. *Biodegradation*

Biological environments are particularly corrosive: they can generate relatively acidic pH values, associated with the formation of oxide radicals (superoxides, hydroperoxides) and molecules with strong chelating capabilities. These phenomena may lead to a biodegradation of ceramics, at times desired and at times accidental. Overall, biodegradation appears connected to the solubility of these ceramics. Thus, very insoluble calcium phosphates, corresponding to stœchiometric apatites are non-biodegradable, while those that are relatively more soluble are biodegradable. It seems that this general rule is equally valid for other types of ceramics.

We can distinguish between the phenomena of passive biodegradation not involving cells and the phenomena of biodegradation activated by the cells.

12.3.3.1. *Passive degradation by biological fluids*

Modes of passive degradation are not different from what we can observe in other aqueous media. We may, for example, observe a dissolution at grain boundaries associated with the release of particles and hydrolysis phenomena, sometimes limited to the surface of the material. In the case of multiphase ceramics, phenomena connected to the transfer of material between phases of different solubility may appear. The release of particles in tissues or biological fluids may sometimes produce inflammatory reactions which we will analyze later. Certain plasma sprayed coatings of calcium phosphate can be associated with small proportions of calcium oxide, whose hydration is accompanied by a large increase in molar volume, leading to the formation of cracks and to a detachment from the metallic substrate.

12.3.3.2. *Mechanical degradation*

Mechanical degradation is frequently observed in ceramics used in friction couples. The wear debris causes an acceleration of the degradation process, deterioration of the friction couples and above all induces a local inflammation which could be responsible for some loosening of implants. These modes of degradation can be observed *in vitro* and the results obtained confirm the very good performance of the alumina-alumina couple. We must however note that solutions imitating biological fluids which are used during mechanical tests (Ringer solution) are not supersaturated with respect to apatitic calcium phosphates, like biological fluids, and the aspects connected with the nucleation of this phase are often neglected. It is not rare, however, to find calcium phosphate particles, on the explanted friction surfaces of prostheses [FRA 97].

Other phenomena of degradation can be due to micromovements of implants subjected to periodical mechanical forces (stems of hip prostheses for example). These phenomena generally affect the ceramic coatings of some metallic prostheses. Like in the previous case, the production of wear debris can produce inflammations and have devastating effects on the performance of the prostheses, thus accelerating the process.

12.3.3.3. *Cell degradation*

In the first few days after surgery, all implants are subjected to attacks from cells in charge of defending the integrity of the organism (multinucleate giant cells, macrophages, monocytes, etc.). The main cell involved in this degradation is the macrophage. They are cells which can, initially, create detrimental conditions of pH and oxidation for materials. It has thus been shown that ceramics in bioactive calcium phosphates deposited on metallic implants suffered a significant dissolution in the first few days after the implantation. These phenomena however stop naturally when these implants are not toxic in nature and do not generate particle inflammation. However, in the opposite case, the local inflammatory reaction can take place and deteriorate the implant and neighboring biological tissues.

12.3.3.4. *Toxicity of constituting elements*

Several elements present in ceramics, either as impurity or as constitutive elements, may cause toxicity. It is customary to distinguish essential trace elements, oligoelements necessary for the life, from toxic trace elements. It appears, however, that all these elements have different toxic limits. The latter are difficult to determine in an absolute manner and in the case of implanted substances, the local effect on tissues should be distinguished from the remote effect of the implant on other tissues or organs (Table 12.2).

The toxicity of alumina, lead, arsenic, mercury and cadmium are well documented. These studies however often relate to the elements ingested in nutriments. Very few studies have been made on the toxicity of elements associated with implants and in their immediate neighborhood and the standards in vigor are not often based on any experimental data.

The solubility of phases and their reactivity in a biological environment is evidently a determining factor. Thus, for example, alpha alumina implants have never shown toxicity associated with aluminum, despite the large quantities of aluminum introduced into the organism, while degradable glasses containing smaller proportions of aluminum may considerably impair the formation of bone tissue [NIZ 99]. Similarly, fluorapatite, which is very insoluble, does not induce any negative reaction despite the toxicity of fluorine. As far as dental amalgams are concerned, the poisoning effect of mercury appears very controversial. The toxicity of elements depends equally on their degree of oxidation and on their state of complexation. Thus, the arsenates, eventually present as trace elements in calcium phosphates, are much less toxic than arsenites; likewise, certain organic derivatives of phosphorus are highly toxic, while orthophosphates are harmless.

Lastly, high temperature heat treatments (sintering, plasma spraying) tend to reduce the strength of volatile impurities in ceramics, particularly mercury. It can also help the segregation of certain impurities at grain boundaries or on the surface and increase their bioavailability.

Elements	Biological effect
Mg	Essential element, well regulated in healthy substance, may affect locally the cell functioning; inhibits the crystal growth of bone mineral.
Al	Neurological effects of the element are not well established. At low doses, it helps cell multiplication. At high doses, it inhibits cell development, alters mineralization and inhibits crystal growth of the bone mineral.
Si	Essential element, may induce a cell toxicity in the form of isolated silicate ions, seems to help the formation of the neo-formed layer of apatite.
As Pb, Hg, Cd	These trace elements are relatively spread out and can be found in most ceramics. They can have important neurological consequences and seriously disturb cell function. Maximum acceptable concentrations are usually standardized.
Ti	Does not seem to have any important toxic effect. Metal and oxide particles can be toxic.
Zr	Zirconium has a higher toxicity, at equal dose, than aluminum (test on mice). It will set essentially in the ovaries and to a lesser degree in the bones and the lungs.

Table 12.2. *Toxicity of elements encountered in inorganic biomaterials*

12.3.3.5. *Inflammatory reaction connected with particles*

Insoluble particles, irrespective of their nature, can provoke serious inflammatory reactions. These phenomena can be compared to those provoked by crystal coating disease: arthritis (whitlockite, apatite, calcium pyrophosphates), gout (uric acid), calcification of tendons (calcium phosphates). The inflammatory reaction is generally a periodic phenomenon which can however seriously damage the tissues and the biomaterials and eventually lead to the loosening of the prosthetic device. The predominant factor in the inflammatory phenomena connected with particles is the size of the latter. Particles smaller than 50 micrometers are phagocyted by the macrophages for possible degradation in the lysosomes. Phagocytose activate macrophages which in turn synthesize various cytokines, growth factors which are active on the effector cells. Among these are immunizing cells with remote action or cells with local action like the osteoclasts which, once activated, are responsible for massive bone destruction. The inflammatory phenomenon is all the more accentuated the finer the size of the particles. Other factors are, however, involved, such as chemical composition, crystallographic nature, crystal morphology, composition of surface, the adsorption and desorption of proteins. Particles with prominent angles, for example, appear more phlogogenic than those with a globular morphology. However, the majority of studies on inflammatory processes have been carried out on soft tissues and generally the inflammatory responses of bone tissues appear more limited.

12.3.4. *Standards and biological tests*

Biological characteristics, that is the behavior of different tissues and of the organism vis-à-vis the implant, at a distance or at its contact, can be evaluated by different tests defined by standards.

A ceramic will be biocompatible if it does not cause inflammatory reactions or toxicity (no noxious elements brought in: Pb, Cd, As, Hg). This expanded biocompatibility opens up the notion of this very contemporary concept of bioactivity, according to which we aim at a material that is no longer inert but triggers reaction from the living tissue. The standards look essentially to evaluate the quality of the array of materials used for implants and their innocuousness. There is at present no standard concerning the measurement of bioactivity.

The specific uses of bioceramics *in vivo* require their full compatibility with the living tissues. Taking into account the reliability requirements, implantable ceramics, whether massive or in the form of coatings, should be subjected to a number of tests (physical, chemical, mechanical, mineralogical and biological characterization), for ensuring guaranteed performance without failure in future.

These different types of characterization are presently the subject matter of important works of standardization in the AFNOR and ISO organizations. We give below a non-exhaustive overview of tests practiced at present on pieces and ceramic coatings, with the knowledge that some drafts of ISO, AFNOR or ASTM standards are under preparation.

Physical and mechanical characterization:

– on massive parts: the basic appraisal consists of dimensional check and static load test. Other tests can also be done:

- porosity and bulk density (soaking of specimens under vacuum NF EN 623-2: B41-203),

- micro hardness,

- impact resistance,

- wear resistance;

– on coatings and depositions: the main characteristics required for determining the quality of deposition on the metallic medium (alloy TA6V) are thickness (AFNOR S94-069, metallographic sections), porosimetry (ISO/CD 13779), surface condition (AFNOR S94-071, rugosimetry) and the adherence strength in tension (AFNOR, S94-072).

Chemical and mineralogical characterization:

– on massive parts: the general techniques of analysis used for ceramics (fluorescence X and plasma spectrometry) are applicable to bioceramics. For measurement of low contents (< 100 ppm), the inductive coupled plasma mass spectrometry (ICP-MS) seems to be the most suited technique. Mineralogical analysis takes place by x-ray diffraction;

– on coatings and depositions: ceramic depositions for medical applications essentially consist of hydroxyapatite. Various tests check the purity of the HAP (AFNOR S94-066, Ca/P determination and AFNOR S94-067, qualitative and quantitative determination of foreign phases), the crystallinity (AFNOR S94-068) and the presence of impurities (AFNOR S94-065, As, Hg, Cd, Pb assays).

Biological characterization: mechanical behavior in biological medium can be studied from tests in Ringer solution [CAL 94] considered as being representative of the biological environment. These tests apply on cylindrical bars (diameter 5 mm, length 125 mm) subjected, while immersed in Ringer solution, to a static flexural stress [LED 95]. The duration of the test is 10 months and the solution is changed every month. These undergo the tests, including the dimensional check, weighing, measurement of fracture modulus by three point method and the measurement of dynamic modulus of elasticity (ultrasound technique).

12.4. Processing of bioceramics

Technical ceramics are composed of raw materials generally as powder and of natural or synthetic chemical additives, favoring either compaction (hot, cold or isostatic), or setting (hydraulic or chemical) or accelerating sintering processes. According to the formulation of the bioceramic and the shaping process used, we can obtain ceramics, dense or with variable porosity, cements, ceramic depositions or ceramic composites.

12.4.1. *Massive ceramics*

Molding, traditional or under pressure, and pressing, unidirectional or isostatic, is part of the ceramist's range of tools for obtaining the final form of the piece to be made.

The heads of hip prothesis in zirconia or alumina are formed by isostatic pressing (approximately 1,200 bars) from micronized powders. The sintering operation which follows is carefully defined by heating and cooling parameters (temperatures and rates), so as to obtain a finished product with a nil total porosity, especially in the case of prostheses heads [MOC 93]. The choice of a micronized and reactive powder combined with isostatic pressing leads to this result [SED 93]. More recent manufacturing methods borrowed from the electronic industry like strip molding can also be used for making bone substitutes [RIC 98]. Pieces obtained are of small thicknesses (from 100 μm to 2 mm) and of large surface (many tens of cm^2).

Porosity is an important parameter for cell rehabilitation. It can be obtained by different methods. The utilization of a pore forming agent is the method most widely used. The pores generated by the release of a gaseous compound (H_2O_2, CO_2), combined either with slurry or a powder, are generally irregular and we prefer calibrated organic substances (naphthalene, polystyrene, polyethylene), added to the powder, which decompose slowly during heating. These techniques can be employed for different types of ceramics. The use of calibrated polymer foams, impregnated with slurry, has also been proposed and achieved successfully.

In parallel, more specific techniques have been sought for calcium phosphates using smaller temperatures. Thus, porous ceramics in calcium phosphates resembling coral skeletons can be obtained by the "*replamineform*" technique (Interpore) in aqueous media [RIC 98]. Moreover, a hydrothermal process helping to obtain apatitic ceramics at temperatures clearly lower than those used by ceramists has been perfected and developed: this is the way of hydrothermal cements [DON 98]. Other methods using hydraulic cements have also been proposed [DRI 95]. These methods allow easy preparation, molding and machining of the sintered piece, as well as a more flexible check of the composition and the additives. All these methods lead, however, to porous compounds with weak mechanical performance.

Another way of material processing is developing; it is based on the biomimitic processes aiming at imitating natural and biological processes and offers the possibility of making these bioceramics at a temperature lower (ambient temperature) than that of conventional or hydrothermal processes [GRO 96]. The prospect of using these very low forming temperatures opens up possibilities for mineral organic combinations with improved biological properties by the addition of proteins and biologically active molecules (growth factors, antibiotics, anti-tumor agents, etc.). However, these materials have poor mechanical properties which can be improved, partially, by combinations with bonding proteins.

12.4.2. *Thin coatings*

Ceramic coatings are obtained in different ways (chemical vapor deposit (CVD), plasma spraying, sol-gel, lazer assisted deposition, etc.). For materials used in orthopedics, they are made on the surface of metallic parts of the prosthesis which are in direct contact with the bone, so as to enable integration with latter. The ceramic coating should particularly be chosen according to its biological properties, the nature of the substrate to be covered, the required nature of the bonding between the substrate and the deposition of the physical, chemical and mechanical properties.

The example of the hydroxyapatite coatings by plasma spraying on a hip prosthesis in titanium or titanium alloy illustrates all the complexities of processing of this type of bioceramic. HAP powder is introduced in the inside in a carrier gas on the jet of a plasma torch before being sprayed on the substratum prepared mechanically for this operation (sand blasting). This technique poses a number of problems, essentially due to the partial and superficial decomposition of HAP particles at a very high temperature. Consequently, the deposit obtained is composed of different phases heterogenously distributed in the deposit. These different foreign phases (CaO, TCP, TTCP, oxyapatite) will, depending on their quantity, modify the chemical, mechanical and biological behavior of the biomaterial [RAN 96]. For example, the rehydration of CaO as $Ca(OH)_2$ induces a large increase of the molecular volume, thus causing the formation of cracks in the deposit, and even its detachment from the surface. On the other hand, the presence of phases which are more soluble (TCP, TTCP) than the HAP may help nucleation of a layer of neo-formed apatite *in vivo* and the biological activity of biomaterials.

Taking into account the inherent disadvantages of high working temperatures required for plasma spraying of HAP, other techniques of covering metallic substrate at lower temperatures have been developed at the laboratory level: the sol-gel method [MIY 95] and electrodeposition [DUC 90]. However, most of these techniques require some heating to obtain a cohesive layer and its adherence to the substrate. Other techniques, using supersaturated solutions to promote precipitation of apatites closer to biological apatites, have been suggested. These help to obtain coatings on surfaces which were up to now hardly suitable for covering by conventional techniques of depositing [GRO 96]. The electrolysis of a supersaturated metastable solution also yields a coating when the metal to be covered is placed as a cathode [ROY 93]. However, the main disadvantage of these techniques is the slow speed of formation of the deposit and the poor quality of adhesion of the deposit on the metal surface.

12.4.3. *Cements*

Bone defects are often irregular and their filling can conveniently be carried out by hydraulic cements [DRI 95]. Plaster of Paris was one of the early cements to be used in reconstructive surgery, but its rapid degradation has favored the emergence of phosphocalcic cements, which have been the subject matter of considerable development in the last few years. The applications for the filling of these bone defects are very promising, whether in the form of a molded paste placed in the surgery site or in the form of an paste hardening *in situ*. These products are generally biodegradable and are progressively being replaced by bone neo-formed tissue.

Two reactive principles can be used in phosphocalcic cements. The first consists of achieving a reaction between one or more calcium phosphates, basic in nature (rich in calcium) and one or more calcium phosphates, acidic in nature, (rich in phosphate). In aqueous medium, the reaction between these phases leads to less soluble new phases, whose crystallization brings about the setting. Various combinations have been proposed. The second principle of cement formation consists in using only one phase, whose hydrolysis leads to a crystallized phase and the setting. The alpha tricalcium phosphate has been proposed; more recently, an injectable cement with an amorphous phosphate base has been developed and marketed [KNA 98].

The setting time of a cement is generally of the order of 15 to 45 minutes for orthopedic or dental usage. The advantage of ionic cements is that they are easy to use. Besides, they can easily be associated with the active principle (antibiotics, growth factors). Nevertheless, their high porosity is harmful to their mechanical resistance after hardening and the small size of pores does not favor cell rehabitation.

12.4.4. *Composites*

Ceramic-polymer composites are generally obtained by traditional techniques of component mixing. However, original methods have been developed, notably injectable compounds with hydroxymethyl cellulose base capable of hardening *in vivo* [WEI 97]. New avenues for research opened up, aimed at the mineralization of organized organic matrices, rich in sites for calcium phosphate nucleation (biometric processes). In order to help the nucleation of calcium phosphates on existing polymer matrices, various techniques have been successfully used: phosphatation is one of the most effective technique. It is inspired by the different methods of calcification of bone tissues where the phosphoproteins and the phosphopolipides will be involved in the nucleation phases and growth of apatitic phosphates. The phosphatation of cellulose helps the mineral growth of calcium phosphate on fibers which will be without particular affinity for calcium. Other techniques such as silanation would also be effective.

The combinations between polymethylmethacrylate (PMMA), one of the widely used polymer cements for prostheses stabilization and fixation, and apatites can be achieved by using a phosphate monomer combined with apatite crystals. The composites with compound matrices obtained did not show improvement in properties, biological or mechanical, justifying the additional cost involved in the process [DEL 90].

The biomimetic processes can be used in the preparation of apatite-protein composites at low temperature, in aqueous suspension. The proteins used generally have the property of binding strongly with calcium phosphates (casein, albumin etc.). Crystals of calcium phosphates involved are nanocrystalline and of apatitic structure. The evaporation of water from the suspension results in solid materials with generally reduced porosity and having mechanical properties close to those of polymer cements but still relatively different from those of natural tissues [SAR 99].

12.5. Bibliography

[BAS 98] BASSI F., CAROSA S., PERLA P. and PRETI G., "All-ceramic restorations: an overview", *Minerva Stomatol.,* vol. 47, no. 9, 1998.

[BON 91] BONFIELD W. and LUBLINSKA Z.B., "High resolution microscopy of bone implant surface", *The Bone-Biomaterial Interface*, ed. J.E. Davies, Toronto University Press, 1991.

[BRI 97] BRINK M., Bioactive glasses with a large working range, PhD Thesis, Abo (Finland), 1997.

[CAL 94] CALES B., STEFANI Y. and LILLEY E., "Long-term *in vivo* and *in vitro* aging of a zirconia ceramic used orthopaedy", *J. Biomed. Mater. Res.,* vol. 28, 1994.

[CAL 95] CALES B. and STEFANI Y., "Yttria-stabilized Zirconia for improved orthopedic prostheses", In *Encyclopedic Handbook of Biomaterial and Bioengineering*, vol. 1B: Applications, Ed. Marcel Dekkert, MIT Boston, 1995.

[DAC 90] DACULSI G., PASSUTI N., MARTIN S., DEUDON C., LEGEROS R.Z. and RAHER S., "Macroporous calcium phosphate ceramic for long bone surgery in humans and dogs. Clinical and histological study", *J. Biomed. Mater. Res.*, vol. 24, 1990.

[DEL 90] DELPECH V., Etude de l'influence d'une liaison apatite-polymère dans un ciment orthopédique, Thesis, INP Toulouse, 1990.

[DEM 98] DEMONET N., Etude physico-chimique de dépôts plasma duplex alumine/ hydroxyapatite pour applications médicales relations élaboration/structure/propriétés (dissolution, adhérence, contraintes résiduelles), Thesis, Ecole des Mines de Saint-Etienne, 1998.

[DEW 99] DEWEZ J.L., DOREN A., SCHNEIDER Y.J. and ROUXHET P.G., "Competitive adsorption of proteins: key of the relationship between substratum surface properties and adhesion of epithelial cells", *Biomaterials*, vol. 20, no. 6, 1999.

[DIM 95] DIMITROVA-LUKACS M., SUBA Z.S., MIKLOS L., MINK J. and LUKACS P., "Bone ash based composite bioceramics of good mechanical properties and high bioactivity", *Proceedings of the 8th International Symposium on Ceramics in Medicine,* Bioceramics, vol. 8, p. 403-408, 1995.

[DIO 92] DION I., LAHAYE M., SALMON R., BAQUEY Ch., MONTIÈS J.R. and HAVLIK P. "Blood hemolysis by ceramics", *Biomaterials,* 1992.

[DON 98] DONAZZON B., Céramiques apatitiques basse température. Elaboration-Propriétés, Thesis, INP Toulouse, 1998.

[DOW 91] DOWNES R.N., VARDY S., TANNER K.E. and BONFIELD W., "Hydroxyapatite-polyethylene composite in orbital surgery", *Proceedings of the 4th International Symposium on Ceramics in Medicine,* Bioceramics, vol. 4, p. 239-246, 1991.

[DRI 95] DRIESSENS F.C.M., "Chemistry of calcium phosphate cements", *Proceedings of the 4th Euro Ceramics,* vol. 8, p. 77-83,1995.

[DRO 97] DROUIN J.M., CALES B., CHEVALIER J. and FANTOZZI G., "Fatigue behavior of zirconia hip joint heads: experimental results and finite element analysis", *J. Biomed. Mater. Res.,* vol. 34, no. 2, 1997.

[DUC 90] DUCHEYNE P., RADIN S., HEUGHEBAERT M. and HEUGHEBAERT J.C., "Calcium phosphate ceramic coatings on porous titanium: effect of structure and composition on electrophoretic deposition, vacuum sintering and *in vitro* dissolution", *Biomaterials,* vol. 11, 1990.

[FRA 94] FRAYSSINET P., TOURENNE F., ROUQUET N., CONTE P. and BONEL G., "Biological effects of aluminum diffusion from plasma-strayed alumina coatings", *J. Mat. Sci. Mat. Med.,* vol. 5, 1994.

[FRA 97] FRAYSSINET P., GINESTE L., BONEL G. and ROUQUET N., "Calcium phosphate particles are found at the polyethylene insert surface wether implanted with HA-coated devices or not", *Proceedings of the 10th International Symposium on Ceramics in Medicine,* ed. Sedel L., Rey C., Elsevier, Bioceramics 10, p. 143-146, 1997.

[FUZ 98] FUZZI M. and RAPELLI G., "Survival rate of ceramic inlays", *J. Dent.,* vol. 7, 1998.

[GAU 95] GAUTIER S., CHAMPION E. and BERNACHE-ASSOLANT D., "Microstructure and mechanical properties of hot pressed hydroxyapatite-alumina platelets composites", *Proceedings of the 4th Euro Ceramics,* vol. 8, p. 201-208, 1995.

[GRO 96] DE GROOT K. and LEITAO E., "New biomimetic HA coatings", *Proceedings of the 2nd International Symposium on Inorganic Phosphate Materials,* Phosphorus Research Bulletin, vol. 6, p. 71-74, 1996.

[GUI 95] GUILLEMIN G., PATAT J.L. and MEUNIER A., "Natural corals used as bone graft substitutes", *Proceedings of 7th International Symposium on Biomineralization,* Bulletin de l'Institut Océanographique, no. 14-3, p. 67-77, 1995.

[HAR 97] HARMAN M.K., BANKS S.A. and HODGE W.A., "Wear analysis of a retrieved hip implant with titanium nitride coating", *J. Arthroplasty,* vol. 12, 1997.

[HEN 71] HENCH L.L., SPLINTER R.J., ALLEN W.C. and GREENLEE T.K., "Bonding mechanisms at the interface of ceramic prosthetic materials", *J. Biomed. Mater. Res. Symp.,* vol. 2, 1971.

[HEN 91] HENCH L.L., ANDERSSON O.H. and LATORRE G.P., "The kinetics of bioactive ceramics part III: surface reactions for bioactive glasses compared with inactive glass", *Proceedings of the 4th International Symposium on Ceramics in Medicine*, Bioceramics, vol. 4, p. 155-162, 1991.

[HEN 99] HENCH L.L., "Bioactive glasses and glass-ceramics", *Materials Science Forum*, vol. 293, 1999.

[HOL 92] HOLLIS J.M., HOFMANN O.E., STEWART C.L., FLAHIFF C.M. and NELSON C., "Effect of micromotion on ingrowth into porous coated implants using a transcortical model", *Fourth World Biomaterials Congress*, Berlin, p. 258, 1992.

[HOT 97] HOTT M., NOEL B., BERNACHE D., REY C. and MARIE P.J., "Proliferation and differentiation of human trabecular osteoblastic cells on hydroxyapatite", *J. Biomed. Mater. Res.*, vol. 37, 1997.

[KHO 92] KHORASANI S.N., DEB S., BEHIRI J.C., BRADEN M. and BONFIELD W., "Modified hydroxyapatite reinforced PMMA bone cement", *Proceedings of the 5th International Symposium on Ceramics in Medicine*, Bioceramics, vol. 5, p. 225-232, 1992.

[KNA 98] KNAACK D., GOAD M.E.P., AILOVA M., REY C., TOFIGHI A., CHAKRAVARTHY P. and LEE D., "Resorbable calcium phosphate bone substitute", *J. Biomed. Mat. Res.*, vol. 43, 1998.

[KOK 85] KOKUBO T., ITO S., SHIGEMATSU M., SAKKA S. and YAMAMURO T., "Mechanical property of a new type of apatite-containing glass-ceramic for prosthetic application", *J. Mater. Sci.*, vol. 20, 1985.

[KOK 99] KOKUBO T., "Novel biomedical materials based on glasses", *Materials Science Forum*, vol. 293, 1999.

[LED 95] LE DOUSSAL H., "Characterization of ceramics for bio-medical uses", *Proceedings of the 4th Euro Ceramics*, vol. 8, p. 273-283, 1995.

[LI 93a] LI P., DE GROOT K. and KOKUBO T., "Correlation of physical-chemistry of gel derived silica and its apatite induction", *Proceedings of the 3rd Euro Ceramics*, vol. 3, p. 101-106, 1993.

[LI 93b] LI P., *In vitro* and *in vivo* calcium phosphate induction on gel oxides, PhD Thesis, Leiden (Netherlands), 1993.

[LOP 98] LOPEZ E., ATLAN G., BERLAND S., VIDAL B. and BALMAIN N., "Utilization de la nacre dans la réparation des pertes osseuses de l'os alvéolaire maxillaire humain", *Actualités en Biomatériaux*, vol. 4, p. 197-202, 1998.

[MCK 93] MCKEE M.D. and NANCI A., "Ultrastructural cytochemical and immunocytochemical studies on bone and its interfaces", *Cells and Materials,* vol. 3, 1993.

[MIY 95] MIYAJI F., KIM H.M., KOKUBO T. and NAKAMURA T., "Bioactive titanium alloys by chemical surface modification", *Proceedings of the 8th International Symposium on Ceramics in Medicine*, Bioceramics, vol. 8, p. 323-329, 1995.

[MJO 97] MJÖR I.A., "Selection of restorative materials in general dental practice in Sweden", *Acta Odont. Scand.*, vol. 55, no. 1, 1997.

[MOC 93] MOCELLIN A., "Microstructures de quelques céramiques polycristallines à vocation médicale", *Actualités en Biomatériaux*, vol. 2, p. 99-104, 1993.

[NIS 95] NISHIHARA K. and HIROTA K., "Successful pressure sintering of hydroxyapatite-collagen composite", *Advances in Science and Technology – Materials in Clinical Applications*, vol. 12, p. 297-304, 1995.

[NIZ 99] NIZARD R., Recouvrement d'un substrat d'alumine par bioverres. Study *in vitro* and *in vivo*, PhD Thesis, Paris-13, 1999.

[OKA 97] OKADA K., YOKOGAWA Y., KAMEYAMA T., KATO K., KAWAMOTO Y., NISHIZAWA K., NAGATA F. and OKUYAMA M., "Antibacterial property of Ag-doped calcium phosphate compound-cellulose composites", *Proceedings of the 10th International Symposium on Ceramics in Medicine*, Bioceramics, vol. 10, p. 329-332, 1997.

[PET 99] PETERSEIM D.S., CEN Y.Y., CHERUVU S., LANDOLFO K., BASHORE T.M., LOWE J.E., WOLFE W.G. and GLOWER D.D., "Long-term outcome after biologic versus mechanical aortic valve replacement in 841 patients", *J. Thorac. Cardiovasc. Surg.*, vol. 117, 1999.

[PIL 95] PILLAR R., DEPORTER D. and WATSON P., "Tissue-implant interface: micromovements effects", in *Advances in Science and Technology, Materials in Clinical Applications*, vol. 12, ed. P. Vincenzini, Faenza Techna, 1995.

[RAN 96] RANZ X., Développement and caractérisation de dépôts d'apatite obtenus par projection plasma sur prothèses orthopédiques, Thesis, INP Toulouse, 1996.

[RIC 98] RICHART O., GALLUR A., DESCAMPS M., SZARZYNSKY S., THIERRY B., ANSELME K., LU J., HARDOUIN P. and FLAUTRE B., "Techniques d'élaboration des substituts osseux en hydroxyapatite. Nouvelle voie d'élaboration: le coulage en bande", *Actualités en Biomatériaux*, vol. 4, p. 285-293, 1998.

[ROD 95] RODRIGUEZ LORENZO L.M., SALINAS A.J., VALLET-REGI M. and SAN ROMAN J., "Degradative behaviour of biomaterials based on alumina/PLLA/PMMA composites", *Proceedings of the 4th Euro Ceramics*, vol. 8, p. 61-68, 1995.

[ROY 93] ROYER P., Etude du recouvrement à basse température de matériaux orthopédiques par des phosphates de calcium, Thesis, INP Toulouse, 1993.

[SAN 98] SANTAVIRTA S., TAKAGI M., NORDSLETTEN L., ANTTILA A., LAPPALAINEN R. and KONTTINEN Y.T., "Biocompatibility of silicon carbide in colony formation test *in vitro*. A promising new ceramic THR implant coating material", *Arch. Orthop. Trauma. Surg.*, vol. 118, 1998.

[SAR 99] SARDA S., TOFIGHI A., HOBATHO M.C., LEE D. and REY C., "Associations of low temperature apatites ceramics and proteins", *Phosphorus Res. Bull.*, vol. 10, 1999.

[SED 93] SEDEL L., "Les céramiques en chirurgie orthopédique", *Actualités en Biomatériaux*, vol. 2, p. 81-88, 1993.

[SHE 88] SHELTON R.M., RASMUSSEN A.C. and DAVIES J.E., "Protein adsorption at the interface between charged polymer substrate and migrating osteoblasts", *Biomaterials*, vol. 9, 1988.

[VIA 98] VIALA S., FRÈCHE M. and LACOUT J.L., "Preparation of a new organic-mineral composite: chitosan-hydroxyapatite", *Ann. Chim. Sci. Mat.*, vol. 23, 1998.

[WEI 97] WEISS P.-J., Caractérisation and mise au point d'un biomatériau de substitution osseuse and dentinogénique, PhD Thesis, Nantes, 1997.

[YAM 88] YAMAMURO T., SHIKATA J., KAKUTANI Y., YOSHII S., KITSUGI T. and ONO K., "Novel methods for clinical applications of bioactive ceramics", *Annals of the New York Academy of Sciences*, vol. 253, p. 107-114, 1988.

Chapter 13

Nuclear Ceramics: Fuels, Absorbers and Inert Matrices

13.1. Introduction

Fuel is generally the fundamental component of a nuclear power plant. It is in fact by the fission of either uranium or plutonium that it is possible to release recoverable thermal energy, according to one of the two standard methods used:

$$^{235}U \text{ (or } ^{239}Pu) + n => FP1 + FP2 + (2 \text{ to } 4) \, ^1n + E_f \qquad [13.1]$$

Without fuel, there is no fission and hence no nuclear energy. This leads to two corollaries:

– fuel progressively undergoes a change in its composition because of the destruction of fissile atoms and the creation of the fission products FP_1, FP_2 and also plutonium (and other actinides) during fertile capture. Thus, at the end of its life, the fuel contains more than 10% of foreign atoms, covering half of the Mendeleyev table (Figure 13.1);

– fuel is subjected to an intense radiation. Most of the E_f energy liberated during the fission E_f is in the form of kinetic energy of the fission products which suffer significant damage in their journey range (from 6 to 10 µm). During the life of the fuel, we can evaluate this radiation damage at several ten-thousands of displacements per atom (dpa).

Chapter written by Clément LEMAIGNAN and Jean-Claude NIEPCE.

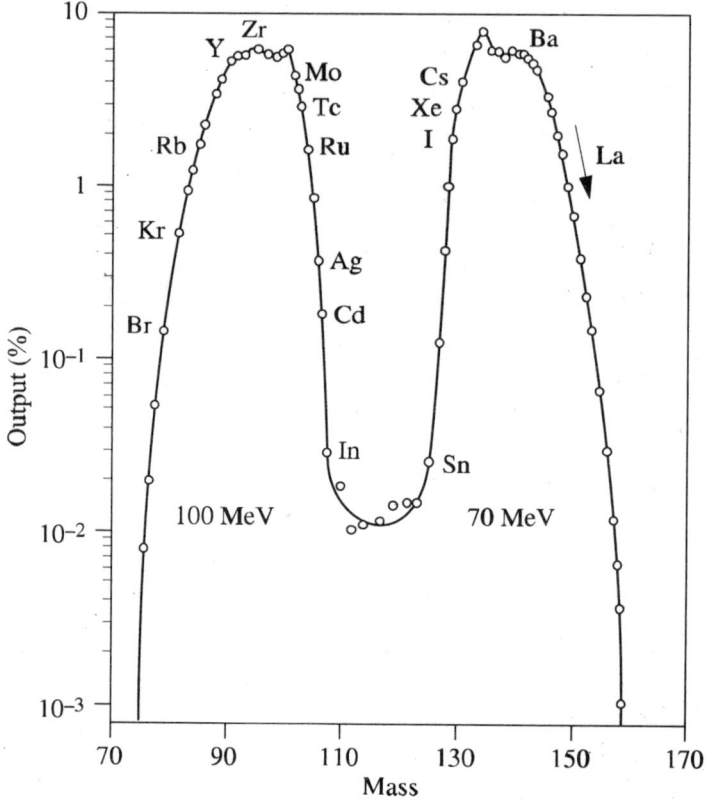

Figure 13.1. *Distribution of atomic masses of fission products of uranium*

Contrary to other elements of a nuclear reactor, the fuel gets exhausted under radiation: it is therefore designed to be regularly replaced. Done in a partial manner (one-third or one-quarter of the core), the annual reloading of the new fuel provides important feedback. This also helps in establishing, without causing much disturbance to the overall design of the nuclear plant, new fuel designs with improved performances or capable of satisfying new requirements, like those arising at the end of the cycle (storing, stocking).

13.2. Fuel element

The fuel element is the simplest part of the fuel. It consists of a stack of oxide UO_2 (or MOX) as ceramic pellets, which are filled in a zirconium alloy tube, welded at the ends and filled with helium for improved internal heat exchange. These elements, 4 to 5 meters long, are grouped in handling units: the fuel assembly (Figure 13.2).

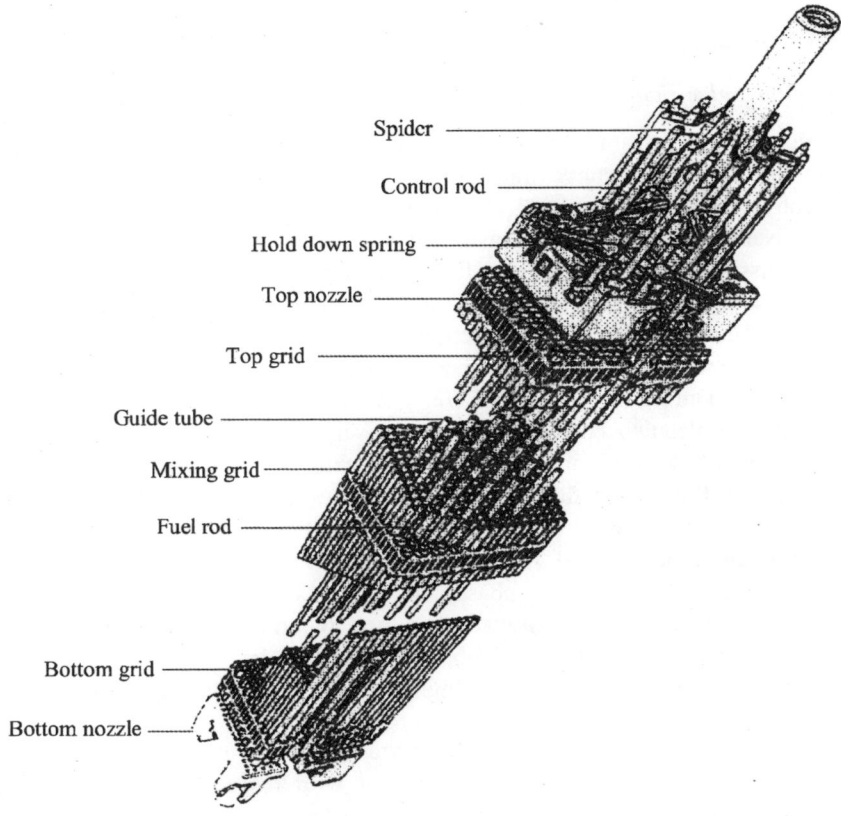

Spider

Control rod

Hold down spring

Top nozzle

Top grid

Guide tube

Mixing grid

Fuel rod

Bottom grid

Bottom nozzle

Figure 13.2. *Diagram of a Framatome fuel assembly*

For a standard pressurized water reactor (PWR), the assembly contains 264 rods (geometry 17 × 17 with 24 guide tubes to accommodate control rods containing absorbent materials, and another central one, for nuclear detectors). A PWR contains 157 to 193 of such assemblies: these are regularly renewed (by a third or a quarter) as and when the fuel is exhausted.

13.2.1. *Fuel fabrication (UO$_2$ or MOX)*

Even though it is protected from the coolant by the fuel cladding, which acts as the first barrier, the fuel should still be compatible with the coolant. This is the first criteria for selection. The fissile material (fuel) is generally in an oxide form UO$_2$

(or MOX, a mixture of UO_2 + (U, PuO_2)), which is quite inert in the presence of water, even at high temperature.

13.2.1.1. Preparation of UO_2

Extraction and enrichment of uranium isotope

After extraction from mines, rocks containing uranium are chemically treated to obtain an intermediate concentrate, rich in uranium, known as "yellow cake". It is then converted into an oxide U_3O_8, standard compound of uranium in the global market. Natural uranium being constituted of two principal isotopes (0.71% of ^{235}U the only fissile and the rest as ^{238}U, fertile), it is necessary to achieve an enrichment in fissile isotopes to enable fission reaction in the reactors. This is done in France by gaseous diffusion: by successive reactions with hydrofluoric acid HF and then with fluorine F_2, uranium hexafluoride UF_6 is obtained as a compound used in the gaseous state. During the passage of UF_6 gas molecules across a porous membrane, the lightest molecules corresponding to the lightest isotope ^{235}U travel faster. Downstream of the diffusion membrane, we thus obtain a fraction very slightly higher in hexafluoride of this isotope. The process is repeated a large number of times to attain the required degree of enrichment. For industrial reactors, the usual maximum enrichment is at 4.5% for water reactors and at 15-30% for fast neutron reactors. After the enrichment phase, the UF_6 gas is converted into UO_2 oxide powder form by hydrolysis with water and then by reducing with hydrogen at about 850°C. This makes it possible to obtain a fine powder, very pure, with a large specific area (a few m^2/g), thus being highly suitable for sintering.

It may be relevant to point out that research was done to obtain other techniques for this stage, since the cost for enrichment with respect to the fuel cost was significant. Thus, isotopic enrichment by lazer helped to obtain high energy yield, by specifically ionizing the fissile atoms in a vapor of uranium, a process known as atomic vapor laser isotope separation of uranium (AVLIS). Though the principles and attempts made are actually very promising, the technological difficulties for industrial development remain considerable. These arise mainly from the materials: thermal behavior and interaction between liquid uranium and refractory metals with which it is in contact.

Fabrication of UO_2 pellets

The oxide powder UO_2 is at first shaped by cold pressing at about 400 MPa, in order to obtain a casting with a density of about 6 gm/cm^3. The pellets obtained are then sintered in dry hydrogen inside continuous furnaces with two zones: preheating until 700°C, necessary for the removal of organic additives (pore forming material and lubricant) and sintering at 1,700°C. Their density then corresponds to 95% of the theoretical density (10.96 gm/cm^3). It is necessary to state that the large remaining porosity is not due to an incomplete sintering: this porosity is needed in order to balance the great strain increase due to the point defect formation and gas release during irradiation. It is obtained by the introduction of polymer grains in the green ceramic paste. After sintering, the grain size of the UO_2 is generally between 7 and 10 micrometers (Figure 13.3).

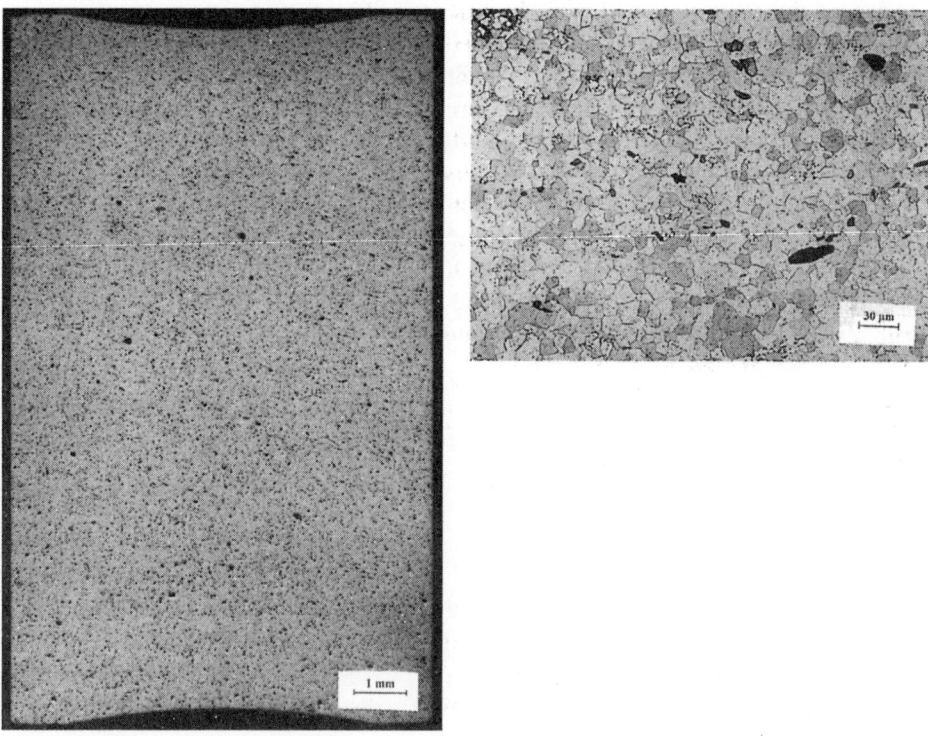

Figure 13.3. *Fuel pellet after sintering (left) and microstructure of UO_2 (right)*

The design criteria for fuel elements impose a very precise definition for the geometry of the pellet, especially its diameter (8.19 mm ± 12 µm), which determines the thermal behavior of the fuel during irradiation. The two faces of the cylinder are partially hollowed out in its central part so as to reduce the thermal expansion of the fuel column. Considering the non-homogenous compaction during sintering, grinding is necessary to obtain the final desired diameter. This grinding is a process not very easily achieved and which requires the recycling of the abrasive products (grinding mud). Already handicapped in the case of UO_2 oxide, this stage strongly hampers the fabrication of MOX. Significant progresses in understanding the mechanisms of decompression and sintering is required. They would help to bypass this stage, while respecting the final geometrical constraints and would thereby translate into considerable economic gains.

13.2.1.2. Specific problems of MOX fuel

Plutonium being an α emitter, the fabrication of the plutonium-based fuel should be carried out fully in glove boxes. In order to reduce the operations of mixing and grinding of powders of UO_2 and PuO_2 oxides, the process known as MIMAS has been developed in two stages: first a parent mixture is obtained, rich in PuO_2 oxide, (about 30%). It is then added in the form of aggregates of significant sizes (an average of 30 μm, but with a large proportion higher than 100 μm) within an oxide matrix of UO_2, normal or weakened (Figure 13.4). A total strength variable between 2% and 8% of plutonium can then be obtained. A significant effect of this mode of preparation, and therefore of the final microstructure, exists in the location of the fission essentially in this mass, leading to a very high rate of local combustion (about three times the average rate of the pellet). We will discuss this point later with respect to the release of fission gas and the oxidizing power of the fuel.

13.2.2. Behavior of nuclear fuel under irradiation

13.2.2.1. Impact of fission

Taking into account the fission densities of reactor fuels, 10^{13} fission cm3/sec, the values of irradiation damages are very high, particularly when compared to the values obtained in the tubing alloys and all the more with respect to the tank. For a rate of combustion of the order of 45 GWd t^{-1}, the irradiation damage is 2 to 5,000 dpa, at a speed corresponding to a few dpa per day. Such a rate of creation of defects evidently calls for a continuous process of timely restoration of these point defects.

In addition to the elastic recombinations, with activity increasing with the loading of each defect in this matrix with ionic bonds, the temperature of the fuel allows a rapid diffusion of oxygen, whose mobility will be the principal process of continuous restructuring of the UO_2 oxide network.

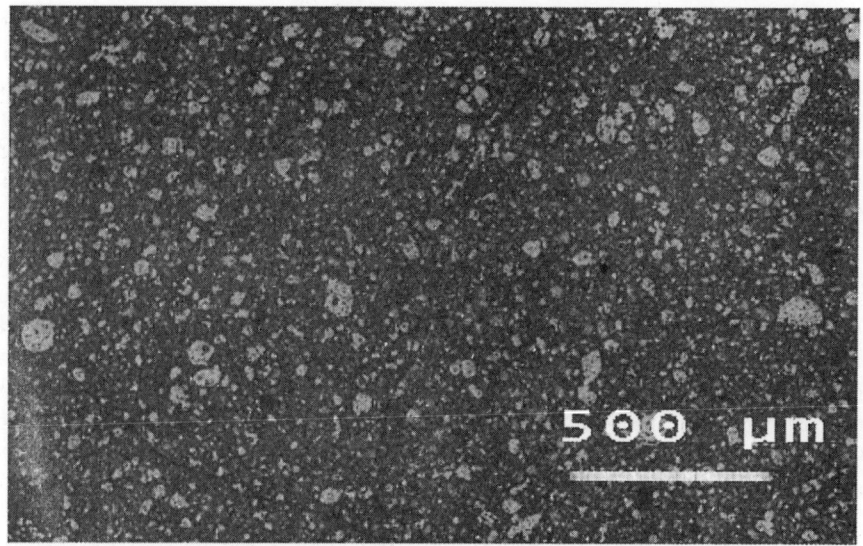

Figure 13.4. *Micrograph of an MOX ceramic showing the presence of Pu rich aggregates in the UO$_2$ matrix*

The damages due to radiation in ionic crystals are more complex than in metals. In fact, crystal cohesion is ensured by Columbian attraction between anions and cations. With both these types being charged, all point defects (vacancy, interstitial or substitution element) can correspond to an electric charge defect, which considerably increases its energy. Thus, the energies of formation of oxygen and uranium vacancies have recently been calculated in quantum chemistry and have been found equal to 10 and 19 eV respectively, when compared with a few eV for metals. For reducing the internal energy due to the presence of vacancies, a combination of vacancy defects is easily carried out in the form of a large neutral defect: two oxygen vacancies and one uranium vacancy combining to form a Schotky defect.

The interstitials have a complex behavior: from a volume point of view, the fluorite crystalline structure of oxide UO$_2$ is sufficiently open for allowing the steric insertion of these ions in non-occupied octahedral uranium sites. However, the electrostatic repulsions limit the possibilities of indifferently placing oxygen or uranium interstitials in the link. Also, the interstitials of uranium created by irradiation are hardly stable and recombine extremely fast, because of the great mobility of O^{2-} ions, in order to be wholly transformed in an oxygen vacancy.

However, during irradiation, one effect cannot be restored. This is the doping by new chemical elements, notably transmuted atoms (^{238}U => ^{239}Pu) and fission products (FP). In the first case, the plutonium can be substituted, without difficulty, for uranium in the oxide UO_2. On the other hand, in the case of a doping in fission products which do not have the same state of charge as the ions which they replace, it is necessity to consider the formation of complexes where the differences of ionic charges are compensated by the additional formation of other defects, like oxygen vacancies or U^{5+} ions. This point will be examined in detail during the description of the behavior of fission gas.

13.2.2.2. Peripheral zone ("RIM")

A particular location corresponds to the peripheral zone of the pellet (called "RIM") (see Figure 13.5), where the capture of epithermal neutrons create a crown rich in Pu fissile isotopes. When the average rate of combustion is quite high, we obtain a very high rate of local combustion due to plutonium fissions (about 150–200 GWd t^{-1} locally, for an average BU pellet of 50 to 60 GWd t^{-1}). Moreover, this zone, which remains at low temperature, will be completely restructured.

The structure of the RIM is presently the subject matter of wide research, particularly because of the general tendency to increase the rate of combustion of discharge of assemblies. For a long time already, we have noticed the appearance of a significant and fine local porosity and a loss of a clear definition of the grain size in this zone of 100 to 150 micrometers wide. The initial grains ($\varnothing$ 7 to 10 μm) subdivide into sub-grains of the order of a tenth of a micrometer. This type of microstructure can also be found in stacks rich in plutonium on the outer cold parts of the MOX fuel, with the rate of local combustion being two or three times more.

13.2.2.3. Thermal behavior of fuel element

The thermal behavior of the fuel element during irradiation depends on the details of power evolution. Taking into account the low thermal conductivity of the UO_2 oxide, which gets degraded progressively under the accumulation of fission products and the defects of radiation, the average central temperatures of PWR pellets are slightly below 950°C, attaining 1,100°C during maximum power. This results in a thermal gradient in the pellet of the order of 500 to 600K between the center and the periphery, i.e. 100–120 K mm^{-1} on average, and almost double in the periphery. A consequence of this gradient is the creation of thermal stresses and pellet fracture. The fracture density depends on the maximum power attained and the microstructure of the fuel. It is an average of a few tenths of the fragments over a section (see Figure 13.6).

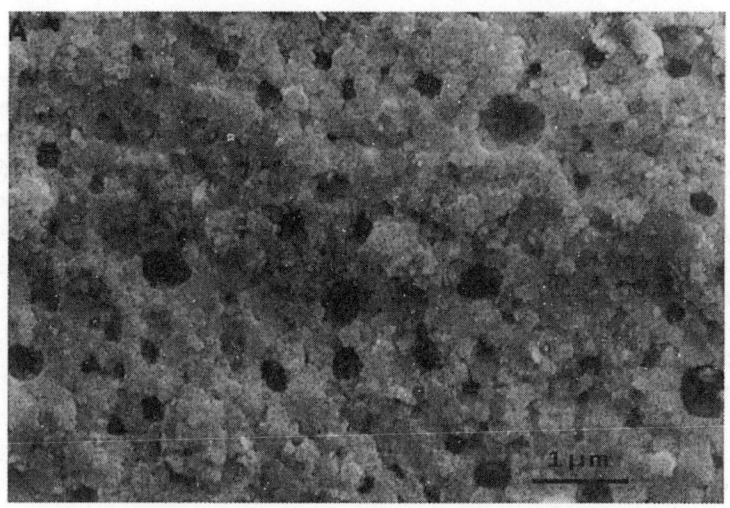

Figure 13.5. *Micrograph of the peripheral zone of a pellet showing the very fine and high porosity structure of the RIM*

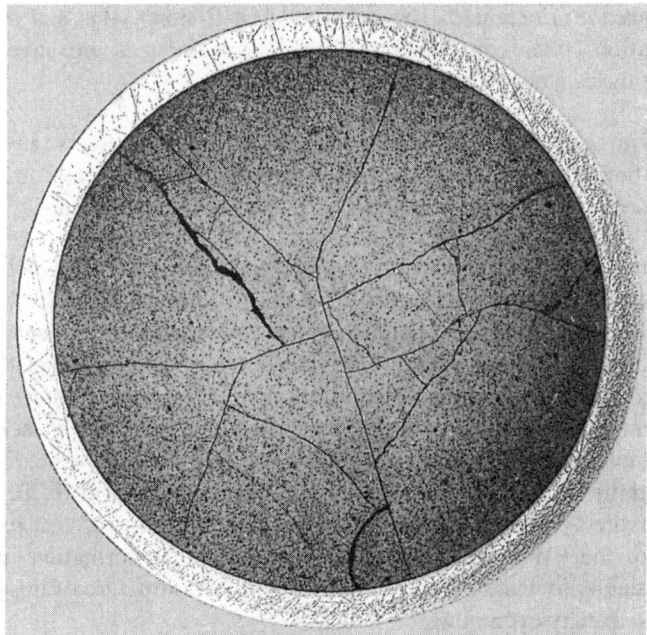

Figure 13.6. *Macrograph of a diametric section of a UO$_2$ pellet embedded in its tube, after irradiation, showing fragmentation of the material*

13.2.2.4. *Creation and evolution of fission products.*

The fission of a fissile atom results in the creation of two new atoms: the fission products (FP). The balance of the creation of the FP depends on the type of atom fissured, the irradiation procedure, the energy of incident neutrons and any eventual disintegration or other nuclear reactions. The fission yields are sensibly different for uranium and plutonium, particularly for the different isotopes of zirconium, platinoids, iodines and gases. For example, in the case of MOX, the fission yields relative to the iodine are almost two times higher for the fission of plutonium than for uranium.

A very large number of experiments have been carried out with the object of defining the thermochemical transformations induced in the oxide by the fission and creation of FP. This transformation is mainly due to the fact that for each fission of U or Pu atoms, two oxygen atoms are released. Of the fission products, some will interact with these oxygen ions, zirconium atoms for example, which will combine as an oxide, others will be neutral like xenon atoms, and some others will finally precipitate in the metallic form or in an oxide phase.

The first stage is to ascertain the material balance, that is the number of atoms which disappear or are created. The knowledge of fission yields and the utilization of transformation codes can solve this problem as long as we have satisfactory knowledge of the power evolution.

Emitted with a high level of kinetic energy, the fission products may move a few micrometers before entering a UO_2 oxide lattice site. This leads to two distinct effects: the doping by the fission products and the irradiation damage.

During irradiation, the oxide is thus doped little by little with foreign elements. These extraneous atoms rapidly go beyond the maximum solubility level and the deviation from thermodynamic equilibrium increases markedly. There is, therefore, a thermochemical driving force for developing a transformation in the microstructure towards a return to equilibrium by the precipitation of secondary phases. Thermodynamically, we can expect a precipitation of rare gases (Xe, Kr) in the form of bubbles, a dissolution of certain soluble FPs in the oxide, like the group Zr, Nb, Sr and the lanthanides, or lastly a metallic precipitation (Mo, Tc, Ru, Rh, Pd, etc.) or of oxides (Sr, Ba, etc.). During normal functioning, the relatively low temperature of the PWR fuels hardly allows such a transformation in the central portion. Detailed knowledge, at the atomic scale, of the structure of this oxide doped in FP remains sketchy even now.

On the contrary, during power transmission where the average temperature increases sharply or when the rate of combustion is significant, which considerably increases the equilibrium deviation, certain thermochemical processes can be activated. We thus notice a significant release by thermochemical processes. There are also other exit mechanisms for the FP from the fuel. These are impervious mechanisms essentially activated by ballistic processes.

13.2.2.5. *Release of fission products outside the fuel*

As long as the fuel remains quite cold, that is, in normal working conditions, the fission products remain fixed at a few micrometers of the site where they have been emitted, without any notable displacement. Thus, the analysis by gamma spectrometry of a fuel element after irradiation shows a remarkable uniformity in the distribution of isotope ^{137}Cs, although this element, a tracer of the fission, should be quite mobile as soon as the temperature conditions allow it. In the absence of long displacements, which are seen only for gases, there may be small structural reorganizations between different chemical substances present, under the effect of the electrochemical driving force induced by the appearance in the matrix of fission products.

The group of fission products constituted by the gases Kr and Xe, forms about 15% of the total FP (see Figure 13.1). These inert gases are practically insoluble in the oxide and tend to escape towards the exterior of the pellet or to precipitate in the form of bubbles, according to the different mechanisms which vary depending on the temperature, structure and history of the oxide. The impact of this release on the composition, pressure and thermal interplay of gases leads to the detailed study of the kinetics and the associated processes.

In normal and static conditions of PWR functioning, the rate of exit of the fission products is relatively slow. Investigations carried out on the elements after irradiation show a low release rate. This exit level corresponds to what one should expect from impervious processes which are reversal (where the FP ion exits the fuel by means of its own kinetic energy) and ejection (when the reversal fragment transfers a part of its energy to atoms present on the surface of the oxide, pulverizing them).

For powers of irradiation, temporarily larger and therefore with higher temperature, the precipitation and the coalescence of bubbles of fission gases can be activated. Since a heterogenous nucleation is always easier than a homogenous nucleation, gas bubbles are formed preferentially at grain boundaries (see Figure 13.7). However, the gas is not the only one to precipitate and the metallic insoluble species, like platinoids, also precipitate at grain boundaries and in the grains. The relative speeds of nucleation of these different phases are actually far from quantifiable, even though the physical principles that govern these phenomena are quite well known.

For the MOX, the rates of exit of gaseous fission products seem higher, without it being presently possible to determine whether this observation is only due to a larger power with the third irradiation cycle, or whether it is mainly a phenomenon allied to the microstructure of the cluster. Of the specific mechanisms of MOX, we can observe that wherever clusters are rich in Pu, there is a mechanism of precipitation of bubbles of fission gas and therefore release. This is partially due to the very high value of the rate of local combustion in the clusters, since these are concentrations of almost the whole of the fissile material; moreover, the higher oxidizing nature of the Pu fission generates an increase in the movement at the same rate as the local combustion. Crack formation of thermal origin in the pellet, passing through these clusters, will facilitate the release of the FP which were locally created. On the other hand, when these cracks cross the matrix, corresponding to a slow rate of combustion, they do not cause any specific release. The critical point is therefore the existence, during cracking, of a percolation of porosities within these clusters. This particular point may lead to attaining a much more homogenous microstructure for MOX. Though technically possible, this option would strongly reduce the present level of productivity of units manufacturing this type of fuel.

13.3. Absorptive ceramics

The starting and stopping of reactors, as well as the variations in power are ensured by virtue of materials containing nuclides absorbing neutrons, unlike fuels which are fissile and therefore multipliers of neutrons. In addition to the absorber dissolved in the water of PWR (boric acid), the absorbent materials used in the control rods of nuclear reactors enable the regulation of the neutron flow, in which the fuel rod assembly is immersed.

Absorbents are materials containing chemical elements of which one or more isotopes capture the neutrons according to different nuclear reactions. For example:

$$^{10}B + {}^1n \rightarrow {}^4He + {}^7Li + 2.6 \text{ MeV or } {}^{177}Hf + {}^1n \rightarrow {}^{178}Hf + \gamma \qquad [13.2]$$

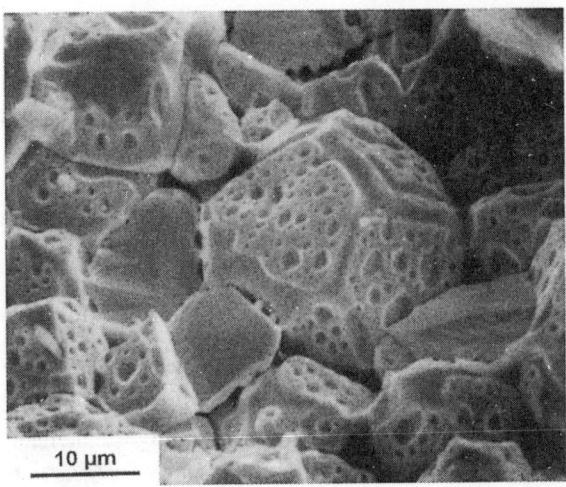

Figure 13.7. *Precipitation of the fission gas in UO₂ after irradiation and formation of a network of intergranular channels making their release outside the fuel possible*

Generally, isotopes possess a neutron effectiveness well adapted to a field of the neutron spectrum. Thus, hafnium is interesting for controlling a reactor burning MOX fuel, where the neutron spectrum is less thermalized. On the other hand, the isotope ^{10}B has, compared to other absorbents, an effectiveness in a very large spectrum, hence its role in the power control of rapid neutron reactors (RNR) and in the stoppage of pressurized water reactors (PWR).

The PWRs, which account for almost the entire electronuclear installations in France, are controlled by two main absorbent materials:

– ceramic materials, boron carbide B_4C;

– metallic materials, ternary alloy of silver-indium-cadmium, commonly known as SIC. This second material is not therefore a ceramic and will not be dealt with here, even though transmutations induced by capture are at the root of problems under radiation.

These materials appear in the form of a pellet columns of B_4C or of a SIC rod, sealed in a stainless steel tube, water tight, thus constituting an "absorber rod". For evident reasons of safety, the length of rods are such that, even in raised position, they always reside inserted in the guide tubes.

13.3.1. *Fabrication of ceramic absorber materials*

Pellets of B_4C carbides are prepared by the sintering of powders. The B_4C powders are synthesized from boron oxide B_2O_3 powder according to two main processes which are the magnesiothermy and the carbothermy. In the first process, the boron oxide powder is mixed with magnesium and carbon powders. The reaction is highly exothermic and results in a fine powder of boron carbide which can directly be sintered. On the other hand, the second process, in which only the carbon reacts directly with boron oxide, the reaction is endothermic. The carbide produced is thus purer, but it is in a highly agglomerated form; it has to be ground in order to obtain a powder form fit for sintering.

Pellets for the PWR are lightly compressed, at 70%, so as to leave a residual open porosity enabling the release of helium formed by the nuclear capture reaction. The boron used is natural boron which contains only 19.8% of the isotope ^{10}B, a very effective absorber. The pellets are sintered by natural sintering at around 2,000°C. On the other hand, the absorbent pellets in B_4C for the RNR Super-phoenix should have a high density of nuclides pick up. Hence, they are compacted much more densely, 96%, and the sintering should be assisted by the application of a very high pressure. The boron should be strongly enriched in ^{10}B.

13.3.2. *Behavior of B_4C in reactor*

This section essentially concerns the behavior of B_4C in the RNR, because in the PWR , considering their position in the rods, it is subjected only to a very weak neutron flow. On the other hand, in an RNR, B_4C is subjected to very severe stresses. First of all, while it has a very poor thermal conductivity, it is the seat of a very high radial thermal gradient, of about 1,400°C at the core, and 600°C at the surface. Also, the energy of the reaction, 2.6 MeV, per neutron captured, is released within the material; this results in a significant local heating and in high thermal stresses. The pellets are therefore subjected to fractures as soon as they are used. Moreover, by ensuring their role, pellets become the seat of an important formation of gaseous helium (see the reaction given above), which is of the order of 380 cm^3 (under normal conditions of temperature and pressure) for 1 cm^3 of B_4C. A part of this gas tends to get retained in the material, causing it to expand noticeably; this expansion is smaller for a less strongly compacted B_4C material. The gaseous nanobubbles which are formed in the ceramic B_4C cause a progressive disintegration of this. The expansion and the disintegration progressively cause a mechanical interaction between the ceramic and the tube. This strong interaction, which acts on a tube embrittled by neutron irradiation and chemical diffusion emanating from B_4C, may cause a tube fracture. Evidently, in normal functioning, it is not possible to attain this situation. In order to increase its life in a reactor, various improvements have been brought about on absorptive B_4C rods. These mainly consist of diminishing the rate of enrichment of the boron subjected to the most severe conditions of irradiation and enrobing the pellet column for preventing the interaction ceramic-tube.

13.4. "Inert matrix" ceramics of the fuel and other nuclear ceramics

During the irradiation of a standard fuel like UO_2 oxides or MOX, a significant part of nuclear reactions are not fission reactions, but capture reactions or more complex nuclear transmutations. Thus, in the course of time, a part of the isotopes ^{238}U transforms into plutonium ^{239}Pu, a fissile material which will increasingly contribute to the supply of power. However, some of these reactions will lead to the formation of isotopes difficult to recover (isotope pairs of Pu, Am and other minor actinides, etc.), whose accumulation would cause major concern. The fission of these isotopes can be achieved by irradiation in certain reactors whose neutron characteristics will be adapted to this function. We must however avoid the recreation of similar isotopes as and when they are "burnt". This could lead to striving for fuel designs where the fissile material is dispersed not in the UO_2 oxide, but in a neutronically inert matrix.

The "inert matrices" can be considered as standard ceramics, in which globules of actinide oxides will be integrally dispersed. These ceramics will have to ensure the mechanical support role of accommodating the deformations arising out of the fission and the removal of heat. On the other hand, these ceramics will be subjected to a very high degree of radiation induced by fission products. They therefore should be capable of supporting a few 10,000 dpa of damages without major degradation and thus get restored during irradiation.

The research work on such materials is still in the early stages. The "inert matrices" on which research has been carried out are from the group generally known for other applications: zirconia, spinels, etc. We must however emphasize that the materials have been little studied only under intense irradiation, the corresponding problems being relatively new.

In the same manner as in the release mechanisms of fission gas, the knowledge of physical mechanisms responsible for the behavior of these "inert matrices" ceramics during irradiation in the reactor can profitably be adapted for ceramics envisaged as host medium matrices for storage. In fact, they will be submitted to the same types of radiations and damages, induced by helium ions during the α disintegration of transuranic elements.

13.5. Bibliography

[BAL 96] BAILLY H., MENESSIER D. and PRUNIER C. (Ed.), *Le combustible nucléaire des réacteurs à eau sous pression et des réacteurs à neutrons rapides,* Eyrolles, Paris, 1996.

[BEL 61] BELLE J. (Ed.), *Uranium Dioxide*, US Atomic Energy Commission, 1961.

[MAT 99] MATZKE H.J. *et al.*, "Material research on inert matrix", *J. of Nuclear Materials*, 274, p. 47-53, 1999.

[USDC 85] Fundamental aspects of nuclear reactor fuel elements, TID 26711-P1, National Technical Information Service, US Department of Commerce, Springfield, Va 22161 USA, 1985.

[WEB 98] WEBBER *et al.*, "Radiation effects in crystalline ceramics", *J. Material Research*, 13, p. 1434-1484, 1998.

Chapter 14

Sol-gel Methods and Optical Properties

14.1. Physico-chemistry of gels

In standard methods of preparation of oxide-based materials (glasses and ceramics), the raw materials are crystalline powders (size of a few μm). Ceramics are prepared through solid state reactions of powders and by high temperature of a granular agglomerate. Single crystals and glasses are generally obtained starting from a molten bath. In this case, the homogenity in liquid state facilitates the formation of the material, which does not involve long-distance atomic diffusion.

In sol-gel procedures [YOS 00], the basic idea is simple: starting from a mixture of liquid molecular precursors, thus homogenous at the molecular scale, followed by the transformation of the liquid into a solid by a chemical reaction of inorganic polymerization at ambient temperature. The homogenous solid obtained is porous, amorphous and hardens at a low temperature, thus making glass or ceramic preparation possible, without going through fusion.

14.1.1. *Sols and silica gels*

Synthesis by the sol-gel method is generally done by using alkoxides having the formula $M(OR)_m$, where M is an atom, very often metallic, and R an organic alkaline group C_nH_{2n+1}. In alcoholic solution and in the presence of water, these alkoxide precursors undergo hydrolysis and condensation resulting in gelation. We will describe the reactions taking into account the specific case of tetraethoxysilane,

---

Chapter written by Jean-Pierre BOILOT and Jacques MUGNIER.

$Si(OC_2H_5)_4$ or $Si(OEt)_4$, which constitutes a model system for inorganic polymerization [BRI 90].

Hydrolysis is a nucleophilic substitution reaction. This reaction, which can get repeated in each of the OEt groups of the molecule, leads to the formation of the silanol (Si-OH) groups and to the release of alcohol molecules.

$$\text{RO} \diagdown \underset{\text{RO}}{\overset{}{}} Si \overset{\text{OR}}{\underset{\text{OR}}{}} \; + \; 4\,H_2O \longrightarrow \text{HO} \diagdown \underset{\text{HO}}{\overset{}{}} Si \overset{\text{OH}}{\underset{\text{OH}}{}} \; + \; 4\,R\,OH$$

Condensation is a nucleophilic substitution which is reflected either by the expulsion of water through reaction between two silanol groups or by the loss of alcohol due to reaction between a silanol group and an alkoxy group. This stage of condensation results in the formation of siloxane bridges which represent the basic unit of the inorganic polymer.

$$\text{HO} \diagdown \underset{\text{HO}}{} Si \diagup^{\text{OH}}_{\text{OH}} \;\; + \;\; \text{HO} \diagdown \underset{\text{HO}}{} Si^{\text{OH}}_{\text{OH}} \longrightarrow \underset{\text{HO}}{\overset{\text{HO}}{}} Si \diagup^{O} \diagdown Si \underset{\text{OH}}{\overset{\text{OH}}{}} \;\; + \; H_2O$$

These reactions ensure the growth of the molecular architecture, with a tetrahedron as a basic brick and a silicon atom at the center, also a constituent brick of massive silica. Competition between hydrolysis and condensation determines the equilibrium between the coarsening of dense particles and their aggregation (see Figure 14.1); it thus determines, at the nanometric scale, the geometry of the formed structure. This competition is controllable chemically by the pH and the salinity of solutions, which modify the reaction speed and the superficial charge of formed particles [ILE 79].

In an acidic medium (pH > 1), hydrolysis is rapid before condensation, which releases all the monomers for the rapid formation of small nanometric particles (0.5 to 2 nm) by chain cyclization. These particles subsequently aggregate to form ramified polymer clusters. The clusters progressively occupy a larger volume fraction, approaching a unit. The viscosity of the medium then increases and the liquid coagulates: this is gelation. Macroscopically, this assembly ends in the appearance of a rigidity and an elasticity, similar to solids, arising out of the gel. Solid and transparent, the gel obtained is thus made up of a polymeric network of silica binding the solvent and eventually any clusters still in solution.

In neutral or moderately basic medium (Figure 14.1), the condensation of silicates is faster than hydrolysis, and the polymer is therefore progressively fed with monomers. The stage of formation of the elementary units is a monomer-cluster aggregation with its kinetics limited by chemistry. This mechanism leads to the formation of dense particles of silica. These, the size of which may attain several hundreds of nanometers, are negatively charged. The resulting electrostatic repulsions impede a new aggregation between particles which remain in suspension in the solvent. This particle-solvent system constitutes a sol. By the addition of ionic salts or by tilting the pH (towards an acidic pH), it is possible to screen these Coulombian interactions and to allow the destabilization of the sol. The aggregation between particles thus leads to a gelation like for the acidic system. Finally, in a very basic medium (pH > 11), the depolymerization (reaction of siloxanes with OH⁻ ions) wins and the silica is transformed into soluble silicate.

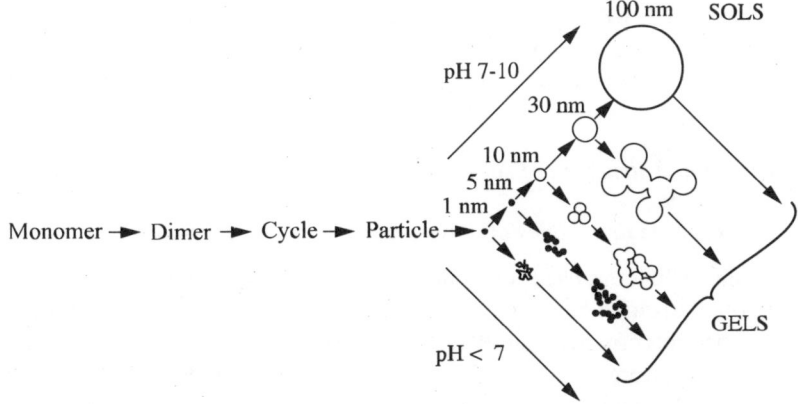

Figure 14.1. *Polymerization of aqueous silica. In neutral or basic medium (pH 7-10), the size of the particles increases and their number decreases. In acidic solution (pH < 7), the particles aggregate to give a three-dimensional network and form a gel*

A variation is the hydrolysis-condensation of alkoxides of transition metals [LIV 89]. These precursors show a strong reactivity with respect to water, resulting in the rapid precipitation of a hydrated oxide. For these alkoxides, gelation requires the preliminary modification of alcoxides by complex chemicals, known as ligands, whose role is to contaminate hydrolysable groups and thus slow down the kinetics of hydrolysis-condensation. In the modified precursors, the OR groups are easily hydrolyzed, while complex groups resist an excess of water. The condensation between hydrolyzed groups then results in the gelation of the system.

In the gel, the polymer network extends to the full volume of the initial liquid and represents only a small portion of the total volume, with the liquid occupying the pores of the gel. The elimination of the solvent without polymer deterioration is therefore a delicate operation. The solid cannot endure the capillary stresses induced by normal drying through evaporation and the porous structure is thus partially crushed (xerogel). On the other hand, the hypercritical evacuation of the solvent helps to preserve the structure of the gel. This procedure consists of bringing the liquid to a temperature and pressure higher than those defined by its critical point. The surface tension of the solvent is thus nil. Carried out under optimum conditions, the hypercritical drying results in obtaining aerogels (polymer-air composite), whose structure is almost identical to the structure of the solid part of the humid gel. Aerogels are ultra-light solids, whose porosity may attain 99% and whose density (10 to 200 kg/m^3) is between that of air and that of water. Their internal surface is considerable: it can attain 1,000 m^2/g. Another remarkable property of aerogels is their very high thermal insulating capacity. An aerogel conducts heat 100 times less than standard glass and even five times less than air. The thermal insulating capacity, associated with the transparency in visibility and opaqueness against infrared, make aerogels remarkable thermal bulkheads. The field of applications is very vast: it ranges from double glazing to its utilization in space rockets [FRI 86].

14.1.2. Growth and structure of sol-gel polymers

14.1.2.1. Study of the first stages of polymerization by nuclear magnetic resonance

Among the techniques used for evaluating the progress of the system undergoing inorganic polymerization, nuclear magnetic resonance (NMR) occupies an important place. This spectroscopy helps in identifying in a very precise manner the chemical bonds involving the precursor atoms under observation. This identification is based on the fact that the deformation of the electron cloud induced by a chemical bond creates a small variation in the magnetic field seen by the nuclear spin and, consequently, a modification in its magnetic resonance frequency. This shift in the frequency, known as chemical shift, is usually small (it is expressed in relative terms of ppm or parts per million), but is fully representative of the considered chemical bond. The atom nucleus under study must of course possess a nuclear spin! This is the case of the isotope 29 of silicon for studying inorganic polymerization which is somewhat equivalent to the isotope 13 of carbon for studying organic macromolecules. In the NMR of silicon 29, the SiO$_4$ tetrahedrons are usually represented by the letter Q with the number of oxygen bridges (Si-O-Si bond) as exponent. Their chemical shift ranges between −70 and −120 ppm (the reference being the TMS: Si(CH$_3$)$_4$).

Thus, Figure 14.2 shows the monitoring by NMR of silicon 29 at the beginning of the hydrolysis-condensation in tetraethoxysilane acidic medium Si(OC$_2$H$_5$)$_4$ [DEV 90]. The reactions take place *in situ*, i.e. in the specimen carrier of the NMR spectrometer and the spectra are recorded continuously. It reveals the appearance of different chemical species, first the products of hydrolysis, Si(OC$_2$H$_5$)$_x$(OH)$_{4-x}$, followed by the condense species where a silicon possesses one or more siloxanes Si-O-Si bonds.

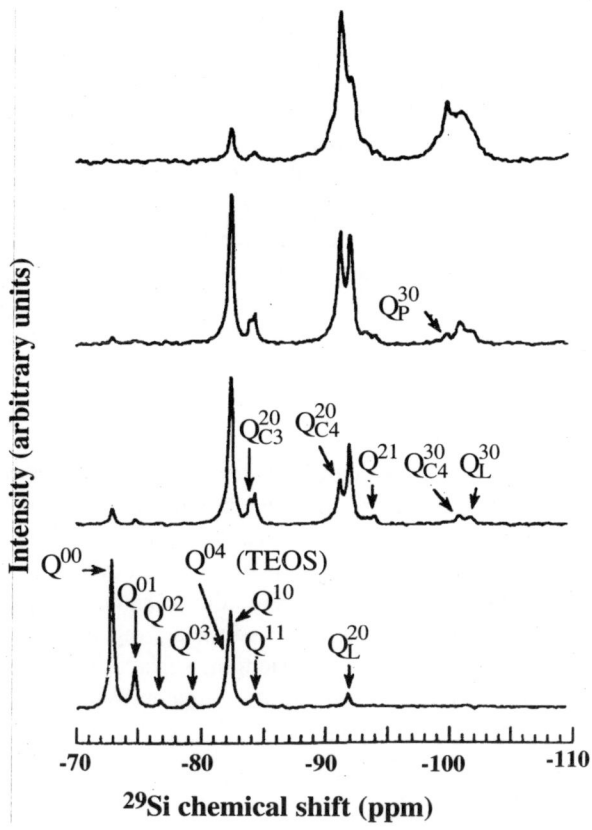

Figure 14.2. *NMR ^{29}Si spectra in the first stages of hydrolysis-condensation of TEOS (pH = 2.5) (left-hand side). The notation Q^{ij} corresponds to a silicon with i siloxane bonds and j alkyl groups. The indices L, C3, C4 and P indicate respectively the presence of linear groups, of cycles at 3Si, 4Si and of polyhedrons. The right side shows the development of the spectrum during the entire process of gelation ($t_g/2$ to $8t_g$). The peaks correspond to silicon atoms resonating with n = 0, 1, 2, 3 and 4 siloxane bonds (Si-O-Si) [DEV 90]*

The NMR is sensitive to a change in environment due to chemical bonds affecting not only the atom under observation and its immediate neighbors, but also its distant neighbors. It is for example possible to distinguish, among twice condensed silicon atoms of the type $Si(OH)_2(OSi\equiv)_2$, between those which belong to the chains and those which belong to the cyclic structure with three or four atoms of silicon. It appears in Figure 14.2 that the polyhedral structure with 8-10 atoms of silicon becomes rapidly predominant. These particles are elementary units for the formation of disordered aggregates.

As polymerization progresses, the possible number of environments for a silicon atom gets larger and larger and the resonance lines of each of the chemical species become difficult to differentiate. Also, while the solution transforms itself as gel, the resonance lines expand on account of the appearance of anisotropic interactions. By artificial techniques, it is possible to partially restore the resolution of the spectra. This is what is shown in Figure 14.2 (right side), where we can distinguish rather wide lines which correspond to all the atoms of silicon with n siloxane bonds $Si(OH)_{4-n}(OSi\equiv)_n$ (n = 0 to 4). We can thereby follow depending on time the transformation of the less condensed species into more condensed species and study the kinetics of polymerization. We note in particular that a long time after gelation (or time of gel, t_g), the number of non-condensed bonds is not negligible (n = 2 and n = 3). This shows the open polymeric structure of the gel.

14.1.2.2. Structure of aggregates

On the other hand, the NMR hardly gives any indications on the stage of aggregation. It is difficult to describe these aggregates, as the geometries displayed are of irregular structure, making a quantitative description a problem. For example, Figure 14.3 shows the negatives of a transmission electron microscopy done on aggregates of gold. On this scale, where the elementary particles are not distinguishable, it is clear that the description of the aggregate in terms of Euclidean objects is impossible, because the results (length, surface, volume) depend on the scale of observation. The concept of fractality is generally used for describing the structure of these aggregates [MAN 82].

14.1.2.2.1. Growth of fractal objects

The notion of fractality has been introduced by B. Mandelbrodt. From a qualitative point of view, a fractal object possesses irregularities in all dimensions. Whether we look at it with the naked eye or through a microscope, it never displays a smooth contour. In other words, its surface or its contour does not lend itself to any definition by a tangential plane or by a tangential line. The theory of fractality introduces a number, the fractal dimension, in general a non-integer, which helps to quantify the generally twisted appearance of the object. Let us take the example of an aggregate which can be constructed by an iterative method. It is made up of spherical particles of the size of a. Around a ball placed at the origin, six balls are stuck along the positive and negative directions of the three axes of a right-handed Cartesian system. Subsequently, we consider the assembly as a single ball, and stick the same pattern at the end of the six branches and so on. We redo this operation many times. Figure 14.4 represents such objects fabricated in two and three dimensions.

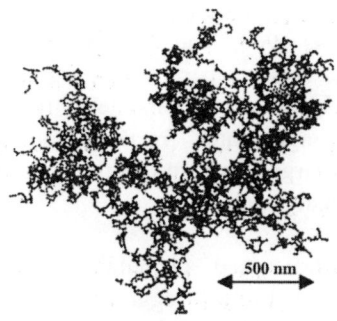

Figure 14.3. *Aggregates of gold observed by electron microscopy [WEI 84]*

Figure 14.4. *Regular fractal aggregate constructed in two dimensions (d = 2) and in three dimensions (3d). The aggregate d = 3 (b) is shown along its principal diagonal. The fractal dimension of such an object is D = Log (2d + 1)/Log 3 [JUL 87]*

This mathematical object possesses a remarkable property: self-similarity, that is, invariance when scaled up. If we enlarge a portion of the object, the enlarged portion is identical to the initial object. Another property is as follows: if we calculate the mass content in a sphere with radius R centered on a point in the object, then this mass varies as R^D: $M(R) \propto R^D$, where D is the fractal dimension in volume which can be exactly calculated for our mathematical object (D = ln7/ln3) and which, more generally, is between 1 and 3. The fractal object considered above has properties of regularity which usually, coarse objects found in the nature do not possess. For an aggregate, self-similarity is valid only as an average and the property $M(R) \propto R^D$ is true only when we take an average over a large number of statistically independent samples.

By using the law $M(R) \propto R^D$, we can see that the material contained in the sphere of radius R varies as R^{D-3}. The density of a fractal $(D < 3)$ therefore decreases when we consider a larger part. By enlarging the sphere, we should get larger holes. This therefore gives another image of fractals: that of a sponge with holes in all dimensions. For aggregates, fractality does not exist except in certain ranges of length, which are large compared to the size of elementary particles and small compared to the aggregate itself.

The aggregation of silica-based elementary particles is limited by the condensation reaction (the probability of aggregation between the clusters depends on the site reactivity), which signifies on the one hand that a number of cluster-to-cluster collisions are necessary for forming the intercluster chemical bond and, on the other hand, that diffusion plays only a minor role. Different modes of aggregation have been studied by digital simulation, by imposing on the particles a random trajectory and by varying the probability of aggregation [JUL 87]. In the model corresponding to a kinetics limited by chemistry and to the aggregation between clusters (RLCA or reaction-limited cluster aggregation), the screening of clusters leads to an open structure whose average density decreases while size increases $(D = 2.1)$, which is a characteristic of a fractal object. The fractal structure of aggregates makes it possible to understand how the final gel, whose solid skeleton occupies only a few percentage points of volume, has a rigid appearance and can enclose so much solvent. The process of growth generating larger and larger holes creates a large distribution of pores, ranging from one nanometer to hundreds of nanometers. Gelation occurs at the time when the clusters which occupy a larger and larger volume, but with also more and more gaps, start to penetrate each other and combine together by a mechanism similar to percolation. We then observe a spectacular modification of rheological properties: the system, which until then appeared in the form of a liquid, ceases to flow when we shake or even when we invert the container. This is the transition of gelation or transition sol-gel.

14.1.2.2.2. Structural determination by small-angle diffusion

The left side of Figure 14.5 is a schematic representation of an aggregate structure on different scales [SAP 90]. For the lowest magnification (top of the figure), we observe an irregular object, greatly branched, with a size R (typically 100 nm). If we enlarge a portion of the aggregate, the structural information then concerns the geometry and the topology of the skeleton. In this regime, the structure of aggregates is often self-similar. The fractal domain is limited to a small scale by the elementary particles. For this magnification, the structural information concerns the size (typically 1-5 nm) of elementary units. Structural details about the surface of these micro-aggregates are then obtained, in particular about its roughness. The maximum magnification pertains to the crystalline nature of the elementary brick from which the aggregate is built.

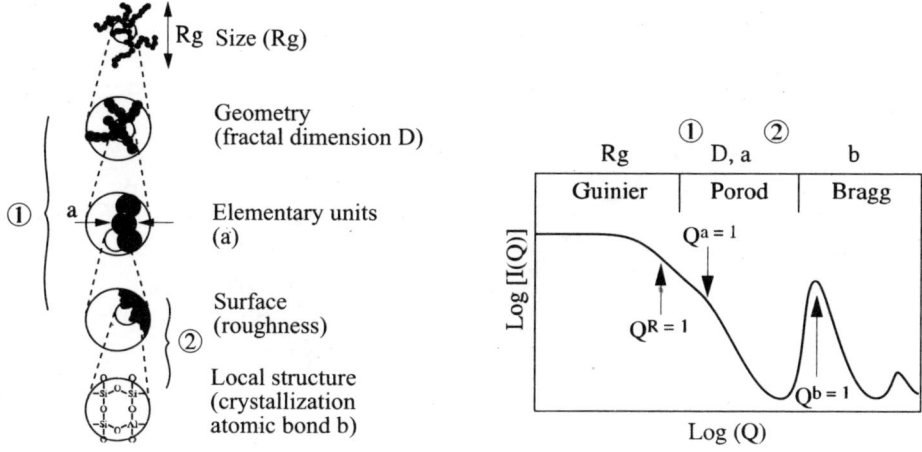

Figure 14.5. *Structure of a colloidal aggregate and corresponding diffusion curve*

The techniques of small-angle diffusion (RX, neutrons) provide the whole range of structural information in the range 1-100 nm. In a diffusion experiment, we measure the diffusion intensity $I(Q)$ according to the diffusion vector $Q = 4\pi\sin\theta/\lambda$, where θ is the Bragg angle. For low values of Q, we observe a plateau and the intensity is proportional to the square of the object's mass. While we increase the value of Q, a first break in the slope is seen (Guinier regime) which directly indicates the size of the aggregate. The laws of power ($I \propto Q^{-D}$) observed in the Porod regime contain structural information, namely the fractal dimension in volume D, the size of elementary units and their roughness.

Figure 14.6 shows for example the diffusion spectra of neutrons on the aerogels of alumino-silicates obtained by hypercritical drying [CHA 90]. It shows clearly that the aerogels display a fractal structure with mutual self-similarity in a large range of density ($\rho = 40$–160 kg/m³). An aerogel of density ρ appears like a homogenous assembly of fractal aggregates in volume, of average size ξ (10 to 40 nanometers). Each fractal aggregate is formed of dense elementary bricks (density ρ_0), of size a approaching 1 nanometer. In the investigated range of density, the rule for construction of aggregates is identical for all the aerogels. The variation of size of the fractal domain is in fact only linked with the properties of a fractal in volume, which requires that the average density of material around a point decreases while moving away from this point. The size of fractal field ξ decreases as the apparent density increases, following a law of type $\rho/\rho_0 = (\xi/a)^{D-3}$. The fractal dimension D is around 2.1 for each sample and seems to be in accordance with a model of cluster-to-cluster aggregation limited by chemical reactivity.

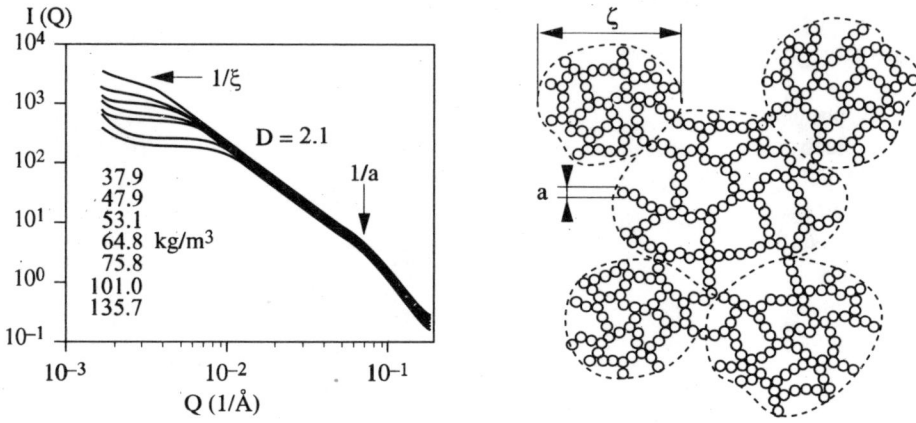

Figure 14.6. *(On the left) diffusion curves for different aerogels of alumino-silicate. The order of density in kg/m³ corresponds to the order of diffusion curves of neutrons. b) (On the right) schematic representation of an aerogel showing the two characteristics scales a and ξ which limit the fractal domain in the material [CHA 90]*

The excellent capacity of thermal insulation of aerogels directly originates from the fractal structure. The propagation of heat is linked to the vibrations of atoms of the structure and to the propagation of these vibrations. The low thermal conductivity of aerogels can be understood in the following manner: in an ordinary solid, the modes of excitation (phonons) extend to the entire system, these are delocalized. In a fractal medium, the existence of modes of localized atomic vibration (fractons) does not allow the propagation of heat for a long distance. The underlying reason for this difficulty in the propagation resides in the property of the average density of the material of a fractal, decreasing further away from that point. The difficulty for propagation by contiguity the excitations of atoms therefore increases since there are fewer atoms as we move away [SAP 90].

14.1.3. *From gel to materials*

An interesting aspect of the sol-gel technology is the possibility of directly making gels in very different forms: films, fibers, monoliths and submicronic powders (see Figure 14.7). During the sol-gel transformation, the control of viscosity enables the drawing of fibers by simple extrusion, using a technology comparable to the technology used for synthetic fibers. We can also cover a substrate with a film a few microns thick. In the dip coating technique the substrate is soaked in a solution of alkoxides and removed at controlled speeds, when it gets coated with a film which polymerizes spontaneously by reaction with ambient water vapor. The firm Schott, in Germany, has carried out for the last several years lining by dip coating on flat plates with surface areas of several square meters. Glazing is thus fabricated industrially with anti-reflection coatings or heat absorbing filters

A number of other applications are possible: protective coatings, ferroelectric film, electrochrome film, etc. For example, the possibility of making organomineral hybrid gels starting from a non-hydrolyzable ligand precursor (for example from modified precursors with the formula $RSi(OEt)_3$, where R is an organic group that remains stable during polymerization) helps to combine the properties of organic polymers and ceramics. Abrasion resistant coatings on acrylic glasses, soft abrasives for dermatology and corneal lenses were thus developed. Lastly, the preparation and applications of ceramic liners for purely optical use are described in section 14.3.

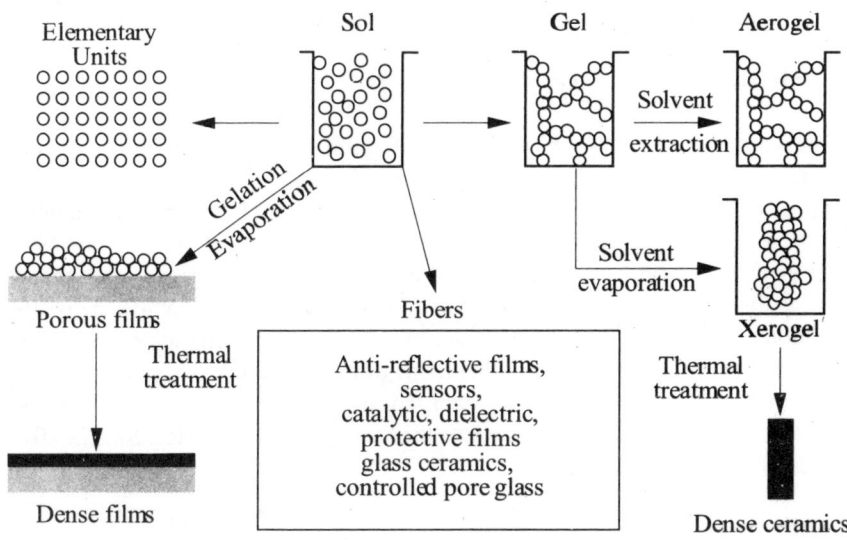

Figure 14.7. *Sol-gel materials*

Besides, the temperature at which xerogels are made is compatible with the stability domain of many molecular groups, in particular with those of organic compounds [AVN 92]. The wealth of molecular chemistry has helped in the preparation of a range of materials by a simple insertion of molecules or aggregates in transparent xerogels. Numerous applications have been suggested for these solids, in particular in the field of optics (sensors, lazers, memories, non-linear optics, etc.) [BOI 96 and 98].

Alkoxide hydrolysis-condensation in stoichiometric quantities of water yields monolithic gels, while submicronic powders are obtained by using a large excess of water. In all cases, after drying, xerogels are endowed with a high level of surface energy (surface area of 100 to 500 m^2/g) which helps in the preparation of dense materials at lower temperatures (200 to 500°C less than standard techniques). We can thus prepare glasses and vitroceramics without resorting to fusion, ceramics by direct heat treatment (without calcination) of a granular agglomerate with controlled porosity (dense or porous objects for the membranes), stoicheometry (no grinding of powders) and microstructure. The major drawbacks of this method of preparation of materials are the high cost of raw materials and the high shrinkage of parts during heat treatment (due to the release of residual solvents).

14.2. Sol-gel process for multi-elemental oxides

14.2.1. *Metallo-organic molecular precursors*

Speaking very broadly, the synthesis of multi-elemental structures requires the combination of oxides, whose behavior in solution is often opposed. This is for example in the case of perovskite structures which form the majority of electroceramic systems and non-linear optics. Their chemical composition is ABO_3. As far as their behavior in solution is concerned, A is less charged and coarser than B, and is thus less polarizing. The oxide AO is generally basic while the oxide BO_n is amphoteric or acidic [JOL 95]. We should note that this separation is arbitrary and helps mostly to broadly outline the behavior of each cation. For example, the cations Cu^{2+}, Mn^{2+}, Co^{2+}, Ni^{2+} are lightly charged and belong to the category A for their behavior in solution (basic oxides). But they oxidize while heated and usually enter the site B of the perovskite.

14.2.1.1. *Precursors of type B cations*

These are the ionocovalent oxide systems, such as Ti, Zr, Nb, Ta, etc., whose majority of alkoxides are commercial and soluble. These elements form oxides which are much more ionic than the silicon. Their reactivity with respect to hydrolysis and condensation is therefore much higher, but this can be easily controlled to obtain condensed systems of controlled rheology [LIV 89]:

– by the choice of the alkoxide: in the case of $Ti(OR)_4$, reactivity with respect to hydrolysis varies according to the alkoyl R group, as follows: the reactivity decreases in the sense R = OEt > OPr > OBu. This is linked to the increase in the steric strength and the hydrophobic character of the complex. So, the reactivity of isopropoxide is also generally much higher than that of n-propoxide, because the former is a monomer where Ti is in coordination 4 highly unsaturated, while the latter is a more stable dimer of coordination 5;

– by modifying agents, i.e. by multi-pronged systems with a tendency to increase the coordination of the cation. The most frequently used are the carboxylates (for example, acetic acid) and the β-dicetones (for example, the acetyl acetone). Their behavior is however different in as much as the acetates are often bridging between cations, playing a role in the oligomerization and tend to remain trapped in the solid phase after condensation. As far as β-dicetones are concerned, they form the chelating enolates. These are less strongly fixed but remain as final condensation products ending on the surface of colloidal particles helping them to control their size but not their density;

Acetic acid

Acetylacetone

– by the pH of the hydrolysis water: the hydrolysis reaction being a substitution reaction, the nature of the attacker depends on the pH. Thus, the hydrolysis of alkoxides of Ti and Zr by acidulated water can, for a given concentration, result in obtaining sols, gels or precipitates depending on the acid/metal ratio. This therefore helps to finely control the size of oxide particles by playing with their surface charge. However, this method has however been very little used in the case of multi-element systems.

14.2.1.2. Precursors of type A cations

These are usually more difficult to handle than the type B cations.

Li, Na, K: these are obtained very easily by the reaction between the metal and the alcohol with which they form an ionic salt. They do not give any condensation and by virtue of their RO⁻ group basic and nucleophilic, they can modify the reactivity of other cations. These ions remain in solution or get adsorbed in the colloids formed by the condensation of other cations. But this adsorption has to be limited for avoiding the flocculation of these colloids by stabilizing them with polymers or by adjusting the surface charge with an appropriate pH.

Ba, Sr, Ca, Mg: these are also obtained very easily, but give poorly soluble alkoxides. Because of the small charge of metals of type A and their unsatisfactory coordination, we unfortunately obtain alkoxides which are often oligomeric or polymeric and are difficult to dissolve. Their acetates are also in the organic solvents.

Y, La, Pb, Bi: these are obtained by the reaction of alcohol with the metal or by exchange of chlorides with alcohol in basic medium. Very hygroscopic, they give rise to precipitations which are difficult to control.

To overcome these problems of reactivity or solubility, several methods have been proposed [CHA 93]. Probably the most interesting solution is the formation of hetero-alkoxides for increasing the solubility of certain cations [CAU 90].

14.2.2. *Different methods of synthesis*

The process variations may be classified in different methods, such as the organo-metallic route, the polymeric route and the sol-gel process, a term often wrongly used to refer to the other two techniques. It is a question of adapting in each case the rheology of the solution for helping the creation of a film and to hinder the separation of different elements during drying and calcination. The methods are differentiated by the manner of controlling solution viscosity, but their boundaries remain subjective and insignificant.

In the polymeric route [LES 89], it is an organic polymer which, when added to the solution, helps to modify its viscosity. The effectiveness of the technique increases with the polymer randomly combining the cations such that they remain dispersed all through the process. The best known method is based on the polymerization of polyesters by reaction of polycarboxylic acids (citric acid) with polyalcohol (ethylene glycol) [PEC 67]. Other syntheses use chelating polymers like polymethyl metacrylates. The deposition of films is easy, and the films can also be easily formed either by chemical treatment or by photochemical methods. They can be engraved like photosensitive polymers by UV, by lazer pyrolysis, and then washing (solid) or by lazer ablation (hollow). Their flaw is the decomposition of a large organic residual quantity which can provoke glazing and hinder the densification of films.

For the organo-metallic route (MOD) [VES 90], the elements are provided in the form of chelates with a long carbon chain like the neodecanoates or 2-ethylhexanoates. We can even bypass such precursors by merely using more ionic salts, chlorides, acetates, nitrates or trifluoroacetates, on condition that these are hot deposited or by nebulization (pyrosol@ method). The method is simple, but also generates a stage of exothermic calcination with a sizeable porosity of the film. This does not always lead to a good densification of the film. Various PZT films have however been manufactured by using 2-ethylhexanoate of lead as precursor.

Lastly, the proper sol-gel method is based on the control, which is not always simple, of the viscosity by the formation of an inorganic polymer. This is dependent on the condensation of the oxides of cations which compose the final ceramic film. The principal interest resides in the fact that the condensation of the oxide network, already well advanced, induces a very low crystallization temperature, while the absence of organic residues makes a good densification during heat treatment possible.

14.2.3. *Crystallization problem*

A priori, the best method to obtain a reactive mixture is the homogenity in the smallest scale possible when precursors have formed stable bonds by means of oxygen bonds. But it is not obvious that such bonds can resist the heating necessary for the establishment of a long-distance bond and for the crystallization of the desired phase. After drying, the mixture of precursors is homogenous but thermodynamically unstable. At ambient temperature, the low coefficients of diffusion hinder a phase separation. During heat treatment, there is therefore a competition between a separation of phase controlled by the speed of diffusion of elements and the crystallization of the desired phase.

Phase separation has been analyzed in the more general framework of glasses and metallic alloys [CAH 68]. The miscibility of elements is less at low temperature, but the kinetics are higher at high temperature. There is therefore a compromise between thermodynamics and kinetics so that we can observe a temperature T_{max} corresponding to the highest speed of phase separation. We have of course to avoid this temperature zone during heat treatment. The methods which are currently employed for fighting these often inevitable phase separations are the following:

– ultra-fast heating: starting from films deposited by MOD (neodecanoates) and heated in an ultra-fast thermal heat treatment furnace (rise in temperature 100°C/s), the phase $YBa_2Cu_3O_7$ can be formed in a few seconds at 900°C (compared to 24 hours by standard methods or by chemistry of solutions with a heating at normal speed);

– confinement: in the case where the separation of phases will be done according to a mechanism of nucleation-growth, the grain size of the precipitate may play an important role. In fact, if it is less than the critical nucleation radius, the separation of phases will not be possible;

– guided heterogenity: in the case of ternary systems, an additional problem linked to phase separation is that the recombinations occur according to the lowest energy routes by forming first the most stable binary compound, which is therefore the least reactive for forming the multi-element phase. On the other hand, we can then encourage heterogenity by forming a binary containing the least reactive oxide and then obtaining the phase by reacting the binary with the most reactive oxide. Besides, this method has been known for a long time, because the ceramists often prepare the columbite $MgNb_2O_6$ before making it react with PbO for obtaining a perovskite in the system Pb-Mg-Nb [SHR 87].

14.3. Optical properties of sol-gel thin layers

14.3.1. *Sol-gel thin layers*

The sol-gel procedure is particularly well adapted for making thin layers (essentially of oxides), because this method uses a liquid (the sol) for obtaining a thin solid film (the gel), thus offering a very low level of roughness. Moreover, the sol can be easily doped in a homogenous way and at a molecular scale, thus constituting an undeniable advantage of the process. These layers are deposited over substrates of different nature (glass, metal, plastic, etc.) with varying properties (chemical durability, corrosion resistance, protection, etc.) [SAK 82]. By virtue of the excellence in their quality, they are of real interest for optical applications. Some examples are given in Table 14.1.

A first classification is to categorize coatings intended for optical use according to two configurations:

– where the luminous beam is perpendicular to the surface [*] (Figure 14.8a);

– where it propagates in the layer itself [**], which corresponds to the configuration of a planar waveguide (Figure 14.8b).

In both these geometries used, the relevant parameter is the optic wavelength used which implies rigorous criteria for the film quality.

Optical applications	Example of materials
Passive optical wave guides [**]	SiO_2, TiO_2, ZrO_2
Electro-opticals [**][*]	PLZT, $KnbO_3$, $LiNbO_3$
Magneto-opticals [**]	Fe_2O_3
Film with index gradient [*][**]	SiO_2, TiO_2, $PbTiO_3$
Lazer [**]	SiO_2 or Al_2O_3 doped with R6G
Non-linear optics [**]	SiO_2 doped with CdS, ZnS,PbS,CdTe
Transparent coatings and conductors [*]	ITO
Electrochrome glazing [*]	WO_3, V_2O_5
Anti-reflection coatings [*]	SiO_2, TiO_2, Ta_2O_5, ZrO_2
Mirrors [*]	TiO_2, Ta_2O_5, Al_2O_3
Filters [*]	TiO_2: Fe^{3+},TiO_2: Cr^{3+}, Fe_2O_3, VO_3

Table 14.1. *Examples of materials deposited in thin optical layers by sol-gel route, used either perpendicularly to the film plane [*], or as waveguide [**]*

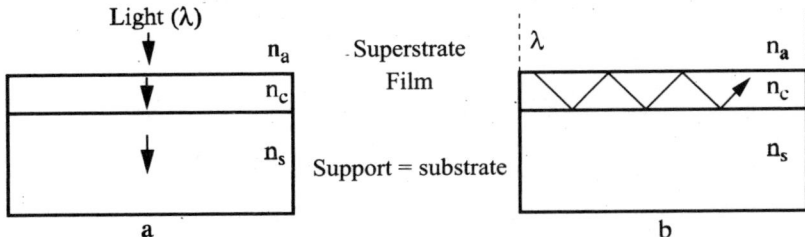

Figure 14.8. *Configurations used for optical layers: perpendicularly to the layer (a) and in guided mode (b). n_a, n_c, n_s are the indices of refraction of the top strata of the layer and of the substrate respectively, at the considered wavelength*

14.3.2. *Planar optical waveguides*

The development of integrated optics necessitates planar waveguides of high quality and low cost. The sol-gel process is attractive with regard to these requirements. A planar waveguide is a planar optical fiber which has a core constituted by the coating and with the tube as the support (substrate) and surrounding medium (top strata, often air) (see Figure 14.8b). These three media should be of good optical quality and transparent for the wavelength used. If the refraction index of the layer n_c is higher than that of the surrounding media, we have shown [TIE 77] that it is possible to inject a luminous beam and to propagate it in a layer of adequate thickness.

The sol-gel process is particularly well adapted for the making and the functionalization of similar flat structures on account of its flexibility (ease of depositing, adjustment of porosity, doping, etc.). We refer to planar passive waveguides, if the function of the coating is limited to the transportation of light and to active waveguides, if the coating possesses in addition another function like, for example, the emission of luminescence (development of optical effect lazer amplifier, etc.).

Many methods are used for injecting the lazer beam in the sol-gel guide [TIE 77]. The easiest way of doing this is by prism coupling (Figure 14.9). This method consists of pressing a prism, whose refractive index is higher than the refractive index of the coating, against this by maintaining an air gap (of the order of a tenth of a micron) adjustable by mechanical pressure. This coupling zone is lighted by a lazer beam which is subjected to total reflection on the base of the prism. A part of the energy is transferred into the layer by optical tunnel effect across the air gap (evanescent wave).

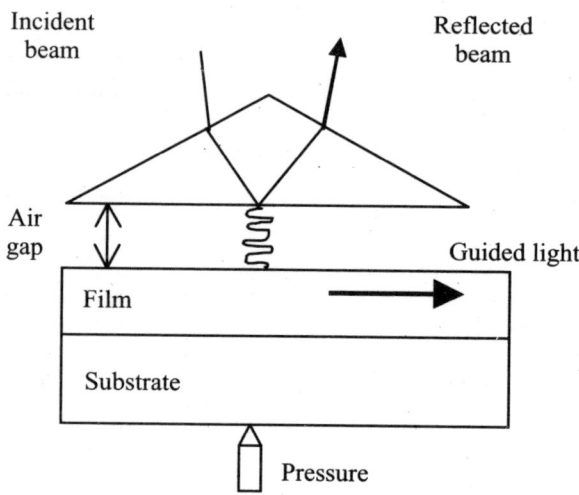

Figure 14.9. *Injection of beam by prism coupling in a planar waveguide*

14.3.3. Development and method of analysis of planar sol-gel wave guides

We can identify two principal techniques for making sol-gel waveguides [SCH 69]:

– dipping the substrate in sol and removing it at controlled speed;

– spin coating which enables spreading the sol by rotating the substrate.

Coatings of about a tenth of a micron, free from cracks (which can be supplemented by successive deposits) are thus obtained. To enable light propagation in the sol-gel layer, this should be of high quality so as to maintain the confinement of light energy in the film. Therefore, we have to take extra care at each stage of the process of depositing the guide, by using:

– a sol stable for long duration, filtered, of suitable viscosity;

– a transparent and clean substrate, of minimum roughness possible (we usually use silica wafers for optical polish or silica on silicon or cleaved substrate);

– a dust-free environment (clean room, glove box).

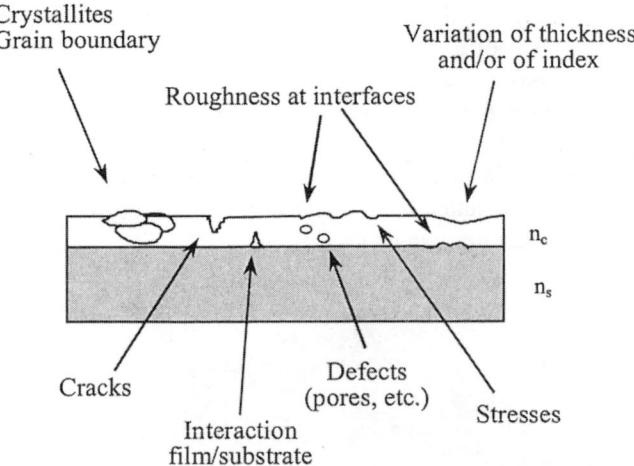

Figure 14.10. *Example of the origin of losses by diffusion in planar waveguides achieved by the method of sol-gel. n_s and n_c are the optical indices of the substrate and the coating respectively, at the considered wave length*

After deposition, the film is heat treated at a low temperature ($t \leq 100°C$) in order to eliminate the more volatile solvents and then annealed for densification at a higher temperature t_r according to the structure desired ($t_r \leq 400°C$: amorphous structure, $t_r \geq 400°C$: crystalline structure). These heat treatments, which cause a contraction of the coating, necessitate a satisfactory balance between the respective coefficients of expansion of the sol-gel layer and the substrate for limiting the stresses which could generate cracks.

Defects may arise at each stage of processing, (see Figure 14.10) and disturb light propagation. We qualify the coating by the coefficient of attenuation of the film α (dB cm^{-1}). α is measured by taking a digital image of the path of the beam in the waveguide layer. We should note that in an ideal theoretical guide a luminous path does not appear, but in practice, the diffusion due to flaws allows the viewing of the light propagation. The analysis of this image (decrease of beam intensity) makes an assessment of α possible and gives an excellent estimation of the quality of the coating achieved by the sol-gel process. α is very dependent on the heat treatment and therefore on the structure of the coating. If the coating is amorphous, the attenuation is generally smaller than 1 dB cm^{-1} and the more the guide is crystalline, the more α increases and therefore the more the propagation decreases. The value of α can be correlated to the roughness (Ra) of the sol-gel film determined by atomic force microscopy (see Figure 14.11) [URL 97].

Moreover, the image of the light propagation in the guide has been optically focused on the entry slit of a monochromator equipped with a detection system. A highly sensitive analysis of the film by a Raman diffusion spectroscopy is possible because the sol-gel layer behaves like a massif sample whose thickness is equal to the length of propagation of the light in the guide [URL 96]. This helps to determine, through a non-destructive method, whether the guide is amorphous and/or crystalline. In the latter case, the nature of the crystalline phase and the average size of the crystallites can be ascertained.

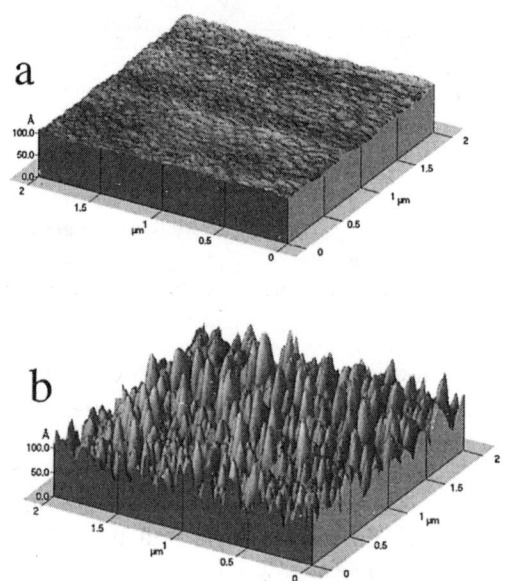

Figure 14.11. *Images of atomic force microscopy of a guide ZrO₂ amorphous (a): roughness (Ra) = 0.2 nm and attenuation (α) ≈ 0.8dB/cm, and (b) crystalline at 600°C: Ra = 1 nm and α ≈ 2.5dB/cm*

An analysis of the luminescence of active guides is carried out with the same setting.

We can also follow the variations in the index and in the thickness of the coatings, with an accuracy of 10^{-3}, according to the treatment given by a fine analysis of the reflected beam (see Figure 14.9): that is, black line spectroscopy [TIE 77].

Active or passive, single-layer or multi-layer guides have been developed by the sol-gel method and classified. Different oxides such as TiO_2, ZrO_2, Al_2O_3, etc. and their mixtures are used, in such a way as to obtain the desired refractive index. Doping by rare earth ions or by dye lazers for optical amplification or for lazer effects have been perfected. The state of the art concerning sol-gel optical amplifiers has just been presented by Originac [ORI 99]. The porosity of SiO_2 or of TiO_2 coatings is used for designing waveguides containing CdS nanocrystallites; such films have attractive non-linear optical properties [JUO 98]. Some active guides using chosen materials (PZT < PLZT, etc.) [UHL 94] have also been developed by the sol-gel method.

Let us mention that planar waveguides can be structured laterally by annealing under CO_2 lazer or by lithography and are thereby transformed into guides of limited width, so as to improve the confinement of light [FAB 94].

Keeping the applications in mind, it is indispensable to correlate the useful properties of active or passive guides to their structural properties. In order to obtain the best performance during guiding, the analysis of the guided configuration is completed by more conventional methods, like transmission or scanning electron microscopy and the diffraction of x-rays, in addition to the specific methods of guided optics described earlier.

Even if the sol-gel coating is, as we will see later, used perpendicularly, its suitability for optical guidance is a guarantor of optical quality (absence of cracks and defects, homogenity, etc.).

14.3.4. Other applications of thin sol-gel coatings

Beyond the applications where light propagates in the layer, sol-gel coatings act optically perpendicular to the coating. In this configuration, the sol-gel coating should also be of high quality. We briefly describe below some of the functions mentioned in Table 14.1.

14.3.4.1. Anti-reflection coatings

The anti-glare effect consists of reducing the natural reflection of a surface. This effect is generally obtained by depositing one or more layers, so as to generate an interferential system. When the considered substrate has a refractive index of around 1.5 (glass, Plexiglas, etc.), the desired effect is obtained by depositing a single layer with an index of 1.23. While there are no materials with such a low index, sol-gel layers with such an index can be obtained by depositing silica in colloidal form [FLO 92]. Since this anti-glare treatment is restricted to a well defined spectral range, the anti-glare reflection is enlarged to a visible range (400 to 700 nm) by usually depositing three coats with controlled index and thickness (1.42–

88 nm/1.88–118 nm/1.65–58 nm/substrate). Such a build-up is possible by the sol-gel method, having a coefficient of reflection of only 1% [BIS 96].

These sol-gel anti-glare treatments find applications, as much for ophthalmologic glasses (Kelar procedure) as for the coating of very large diameter lenses (80 cm) used for focusing the powerful lazer of the Phebus program of CEA, for example [FLO 92]. In the latter case, the sol-gel coatings have a remarkable resistance to lazer flux.

14.3.4.2. *Mirrors*

We can also make optical mirrors by the sol-gel route, starting by stacking dielectric coatings having alternately a high index (TiO_2, Ta_2O_5, Al_2O_3, etc.) and a lower one (essentially SiO_2). A reflectivity of 99% has been obtained by a complex stack of four layers [FLO 92].

14.3.4.3. *Transparent and conductor coatings*

This relates to tin oxides doped with antimony or indium doped with tin (ITO) which serve as transparent electrodes. These thin coatings obtained by the sol-gel route [DIS 81] are transparent to visible solar radiations, but are reflectors of infrared rays; therefore these can be used as heat reflectors.

14.3.4.4. *Filters*

We develop, by the sol-gel route, high-pass filters which suppress ultraviolet rays. These deposits on transparent substrate have cut-off wave lengths adjustable by judicious mixing of different doped oxides of the type TiO_2, Fe_2O_3, VO_3, etc., in variable concentrations [SCH 69].

14.4. Conclusions and future prospects

The flexibility of the sol-gel procedure should not conceal the difficulty to obtain coatings of high optical quality as flaw-free as possible. In the optical applications mentioned above, the attraction of the procedure is the ease of processing, the property of lending itself to doping and its low cost. Also, these coatings offer a good adherence to substrate while OH radicals are present on the latter's surface.

Other specific applications are presented in the book published by L.C.Klein [KLE 94] like micro lenses, composite layers (organic-inorganic) etc., as well as optical applications of massif sol-gel materials.

The future prospects for thin coatings obtained by this process are vast and extend from performance enhancement of guides of limited width to the development of coatings for medical imagery, to mention only a few examples.

14.5. Bibliography

[AVN 92] AVNIR D., BRAUN S., OTTOLENGHI M., "Encapsulation of molecules and enzymes in Sol-Gel glasses", in: *Supramolecular Architecture ACS 499*, T. Bein (Ed.), American Chemical Society, Washington DC, 1992.

[BIS 96] BISSUEL F., Réalisation de couches minces antireflets sur des substrats organiques par le procédé sol-gel et par spin coating, PhD Thesis, University of Montpellier-II, 1996.

[BOI 96] BOILOT J.-P., CANVA M., SANCHEZ C., LEBEAU B., CHAPUT F., LEVY Y., ZYSS J., "Applications des matériaux hybrides dans le domaine de l'optique", *Matériaux hybrides*, série Arago, Observatoire français des techniques avancées, Masson, 17, p. 181, 1996.

[BOI 98] BOILOT J.-P., CHAPUT F., MALIER L., BRUN A., LEVY Y., GALAUP J.-P., "Hybrides organo-minéraux et applications en optique", *Images de la Physique*, p. 44-51, 1998.

[BRI 90] BRINCKER J., SCHERER G.W., *Sol-Gel Science, the Physics and Chemistry of Sol-Gel Processing*, Academic Press, 1990.

[CAH 68] CAHN J.W., *Trans. Met. Soc. of AIME*, 242, p. 166, 1968.

[CAU 90] CAULTON K.G., HUBERT-PFALZGRAF L.G., "Synthesis, structural principles, and reactivity of heterometallic alkoxides", *Chem. Rev.*, 90, p. 969, 1990.

[CHA 93] CHANDLER C.D., ROGER C., HAMPDEN-SMITH M.J., "Chemical aspects of solution routes to perovskite-phase mixed-metal oxides from metal-organic precursors", *Chem. Rev.*, 93, p. 1205, 1993.

[CHA 90] CHAPUT F., BOILOT J.-P., DAUGER A., DEVREUX F., DE GEYER A., "Self-similarity of alumino-silicate aerogels", *J. of Non-Crystalline Solids*, 116, p. 133, 1990.

[DEV 90] DEVREUX F., BOILOT J.-P., CHAPUT F., LECOMTE A., "Sol-gel condensation of silicon alkoxides", *Phys. Rev.*, 41, p. 6901, 1990.

[DIS 81] DISLICH H. and HUSSMAN E., "Amorphous and crystalline dip coatings obtained from organometallic solutions: procedures, chemical processes and products", *Thin Solid Film*, 77, p. 129, 1981.

[FAB 94] FABES B.D., "Lazer processing of sol-gel coatings" in Klein L.C., *Sol-Gel optics: processing and applications*, Kluwer Academic Publishing, Norwell, Massachusetts, p. 483-510, 1994.

[FLO 92] FLOCH H.G. and BELLEVILLE P.F., "Couches minces par le procédé sol-gel", *Revue scientifique et technique de la DAM*, 5, p. 91, 1992.

[FRI 86] FRICKE J., "Aerogels", *Springer Proc. in Physics*, 6, Springer, Berlin, 1986.

[ILE 79] ILER R., *The Chemistry of Silica*, John Wiley & Sons, New York, 1979.

[JOL 95] JOLIVET J.P., *De la solution à l'oxyde,* Savoirs actuels, CNRS Editions, 1995.

[JUL 87] JULLIEN R. and BOTET R., *Aggregation and Fractal Aggregates*, World Scientific Publishing Co, Singapore, 1987.

[JUO 98] JUODKAZIS S., BERNSTEIN E., PLENET J.C., BOVIER C., DUMAS J., MUGNIER J., VAITKUS J.V., "Optical properties of CdS nanocrystallites embedded in $(Si_{0.2}Ti_{0.8})O_2$ sol-gel waveguide", *Optics Com.*, 143, p. 242, 1998.

[KLE 94] KLEIN L.C., *Sol-gel optics: processing and applications*, Kluwer Academic Publishing, Norwell, Massachusetts, 1994.

[LES 89] LESSING P.A., "Mixed-cation oxide powders via polymeric precursors", *Ceramic Bulletin*, 68, p. 1002, 1989.

[LIV 89] LIVAGE J., "Sol-gel Chemistry of Transition Metal Alkoxides", *Proceedings of the Winter School on Glasses and Ceramics from Gels: Sol-Gel Science and Technology*, edited by M.A. Aegerter, M. Jafelicci Jr., D.F. Souza, E.D. Zanotto, World Scientific Publishing Co. Pte. Ltd., p. 103-152, 1989.

[MAN 82] MANDELBRODT B., *The Fractal Geometry of Nature*, Freeman, San Francisco, 1982.

[ORI 99] ORIGNAC X., BARBIER D., DU X.M., ALMEIDA R.M., MCCARTHY O., YEATMAN E., "Sol gel silica/titania on silicon Er/Yb doped waveguides for optical amplification at 1.5 μm", *Opt. Mat.*, 12, p. 1, 1999.

[PEC 67] PECHINI, US Patent No. 3 330 697, July 11, 1967.

[SAK 82] SAKKA S., "Gel method for making glass", *Treatise on Materials Sciences and Technology*, edited by M. Tomazawa and R.H. Doremus, 22, p. 129-167, 1982.

[SAP 90] SAPOVAL B., *Les Fractales*, Aditech, 1990.

[SCH 69] SCHROEDER H., "Oxide layers deposited from organic solution", *Physics of Thin Films*, edited by Haas G., Thun R.E., 5, Academic Press, New York, p. 87-141, 1969.

[SHR 87] SHROUT T.R., HALLIYAL A., "Preparation of lead-based ferroelectric relaxors for capacitors", *Am. Ceram. Soc. Bull.*, 66, p. 704, 1987.

[TIE 77] TIEN P.K., "Integrated optics and new wave phenomena in optical waveguides", *Rev. Mod. Phys.* 49, p. 361, 1977.

[UHL 94] UHLMANN D.R., MOTAKEJ S., SURATWALA T., WADE R., TEOWEE G., BOULTON J.M., "Ceramic films for optical applications", *J. of Sol-Gel Science and Technology,* 2, p. 335, 1994.

[URL 96] URLACHER C., MUGNIER J., "Waveguide Raman Spectroscopy used for structural investigation of ZrO_2 sol-gel waveguiding layers", *J. of Raman Spectroscopy,* 27, p. 785, 1996.

[URL 97] URLACHER C., DUMAS J., SERUGHETTI J., MUNOZ M., MUGNIER J., "Planar ZrO_2 waveguide prepared by the sol gel process: structural and optical properties", *J. of Sol-Gel Science and Technology,* 8, p. 999, 1997.

[VES 90] VEST R.V., "Metallo-organic Decomposition (MOD) Processing of Ferroelectric and Electro-optic Films: A Review", *Ferroelectrics,* 102, p. 53, 1990.

[YOS 00] YOSHIMURA M., LIVAGE J., "Soft processing for advanced inorganic materials", *MRS Bulletin,* 25(9), p. 12-13, 2000.

[WEI 84] WEITZ D.A., HUANG J.S., "Self-similar structures and the kinetics of aggregation of gold colloids", *Kinetics of aggregation and gelation,* edited by Family F., Landau D.P., Elsevier Science Publishers B.V., 1984.

List of Authors

Pierre ABÉLARD
ENSCI, Limoges

Jean-François BAUMARD
ENSCI, Limoges

Philippe BOCH
Formerly Pierre and Marie Curie University, Paris

Jean-Pierre BOILOT
Ecole Polytechnique, Palaiseau

Jean-Pierre BONNET
ENSCI, Limoges

Anne BOUQUILLON
C2R des musées de France, CNRS, Paris

Thierry CHARTIER
CNRS, ENSCI, Limoges

Christele COMBES
ENSIACET, Toulouse

Sylvie FOUCAUD
University of Limoges

Jean-Marie GAILLARD
ENSCI, Limoges

Paul GOURSAT
University of Limoges

Clément LEMAIGNAN
Atomic Energy Commission, Grenoble

Anne LERICHE
University of Valenciennes

Jacques MUGNIER
Formerly Claude Bernard University, Lyon

Jean-Claude NIEPCE,
University of Burgundy, Dijon

Henri PASTOR
Retired from CERMEP, Grenoble

Jacques POIRIER
CRMHT-CNRS and University of Orléans

Christian REY
ENSIACET, Toulouse

Tanguy ROUXEL
University of Rennes 1

Index

A

abrasion 283, 286, 287, 342, 348, 496
aerogel 542, 547, 548
agglomeration 126, 127
aggregation 540, 541, 546
aging 157, 430
alumina 10, 11, 16, 21, 45, 58, 59,
82, 86, 99, 116, 134, 158, 199-216,
246, 313, 337, 362, 363, 368, 460,
463, 494-496, 510
alumino-silicates 495, 496, 547
apatite 498, 502, 504, 506, 507, 515
    carbonated 504, 506
apparent conductivity 447, 448, 451,
452, 455

B

B$_4$C 287, 535-537
barium titanate 17, 59, 227, 428, 477
Bayer 202-204
binder 5, 7, 40, 61, 119, 152, 155-
157, 160-163, 171, 172, 177, 179,
185, 188-190, 210, 287, 326, 335,
336, 348, 364, 366
biocompatibility 494, 497, 512
biodegradation 508, 509
bioglasses 498, 499, 502
biointegration 503, 504
Biot number 314

blast furnaces 359, 368
boehmite 202
bolometer 481
bond 5, 9, 13, 21, 25, 26, 133, 186,
205, 236, 241, 243, 288, 341, 364,
366, 372, 374, 391, 501, 504, 542-
544, 553
    ionic 9, 216, 288, 304, 528
bone china 115, 116
brittleness 4, 6, 26, 123, 216, 263,
264, 285, 495

C

calcination 59, 203, 552
calcium 40, 47, 49, 96, 99, 101, 115,
116, 216, 364, 494, 496-503, 506-
510, 514, 516
carbide 10, 17, 56, 86, 193, 207, 232,
233, 247, 326, 333, 335, 341-344,
347, 360, 393
    silicon 15, 17, 21, 61, 232, 235,
    238, 243, 246, 247, 254, 255, 338,
    348, 460, 497
    tungsten 5, 15, 17, 313, 334, 336
carbon 26, 234, 238, 246, 247, 257,
332, 341, 342, 360, 366, 370, 380,
456, 536, 552
    diamond 496
    vitreous 496

castables 207, 360, 364

casting 96, 108, 114, 117, 119, 147-158, 328, 361

cement 8, 209, 218, 227, 360, 364, 501, 513-516

cement works 359, 368, 370

ceramic processes 9, 124, 125, 143, 184

cermets 325, 326, 329, 336, 340, 496

clay 6, 11, 13, 29, 30, 38, 44, 96, 98, 99, 109, 111, 114, 170, 172, 368

coarsening 57, 61, 64, 65, 83, 540

coating 132, 154, 164, 192, 194, 223, 239, 242, 243, 248, 332-335, 495-497, 509, 511-515, 548, 549, 554-560
    anti-reflection 549, 559

Coble 74, 90, 212, 213

collagen 503, 507

compression 279, 280, 293, 338, 341, 347, 385

condensation 539-542, 546, 550, 551

connectivity 454, 455

coral 497, 498

corrosion 308, 346, 358, 361, 362, 366, 363, 368, 376, 379, 385, 386

corundum 361, 363, 368

covalent 9, 17, 21, 25, 26, 201, 232, 233, 243

crack 74, 104, 132, 183-185, 208, 221-225, 264-268, 271-275, 280, 310, 311, 317, 337, 338

creep 74, 87, 90, 203, 213, 270, 297-303, 310, 311, 375

cristobalite 23, 81, 101, 103, 111, 250

crown 495, 496, 530

crystal 8, 17-19, 23, 24, 40, 65, 73, 81, 201, 203, 244-246

crystal field 403

crystalline structure 18, 99, 217, 243, 244, 257

Curie temperature 423, 465, 466, 469, 470, 477

cutting 15, 86, 207, 208, 226, 287, 326, 327, 329, 332, 333, 336-340, 347

D

damage 296, 298, 303, 306, 307, 310, 311, 315, 316, 345, 381, 384, 528, 529, 532

debinding 158, 184-190

Debye 419

decoration 35-40, 48, 103, 105, 108, 112, 243

densification 57, 60, 61, 70-74, 82, 83, 85, 90, 91, 127, 245

deposition technique 191, 193

diamagnetic 406, 433

diamond 286, 335, 339, 342, 344, 345, 347, 496, 497

dielectric loss 420, 465, 474

diffuse transition 466

diffusion 40, 58, 64-70, 74, 76, 82, 83, 85, 87, 90, 213, 242, 251

dip coating 548, 549

dispersion 4, 133, 138, 151, 156, 172, 216, 222, 226

dissolution 59, 69, 76, 236, 246, 253

Dolni Vestonice 30

dolomite 360, 370

domain 300, 310, 413, 420, 426-428, 430, 432, 438, 546
    of ionic conductivity 413

drilling 340-345

ductile 277, 278, 312, 332, 358

ductility 264, 277, 279, 305, 333

E

earthenware 12, 32, 41-43, 47, 106, 108-110

effective mass 392, 398, 400

effective medium 450, 451, 454

electrochemical potential of the electrons 440

electrofused 361-363, 372

electron holes 392, 393, 397
electronic affinity 444
electrostriction 421, 433
enamel 12, 97, 103-106, 108-110, 114-116, 119
energy 72, 74-79, 85, 87, 133, 139, 141, 167, 223, 245
energy approach 265, 275, 315
exchange energy 402, 434
extrusion 96, 110, 145, 160, 170-172, 176, 177, 182, 236

F

faience fine 48, 49
feldspar 11, 13, 47, 49, 81, 82, 96, 101-103, 108, 110-112, 116, 117
Fermi level 390, 392, 395, 402, 405, 439, 440, 443-445, 478
ferrimagnetism 435, 438
ferrites 17, 218, 436, 485, 486
ferroelasticity 416
fibers 31, 73, 130, 209, 210, 235, 236, 238, 255, 272, 455
filters 549, 560
firing 4, 6, 13, 30, 31, 34-38, 44, 45, 97, 105-108, 110, 111, 114-117, 387
fission 523, 526, 528-530, 532-534, 537
fission products 523, 529, 530, 532-534
flux 13, 44, 45, 82, 96, 101-103, 110, 111, 117, 303, 382
forbidden band 392-394, 404, 405, 475
fractality 544, 546
fracton 548
friction 164, 166, 177, 207, 257, 272, 283-285, 348, 473, 494, 507, 509
fuel 487, 488, 523-525, 527, 528, 530, 532, 533, 537
    MOX 528, 530, 535
fuel
    assembly 525

cells 220, 488
element 524, 527, 530, 533

G

gel 502, 515, 539-542, 546, 548, 550, 554-556, 559, 560
    silica 502
gelation 540, 541, 546
gibbsite 202
glasses 5, 7, 11, 23, 211, 294, 495
glaze 39-41, 47, 97, 109
    lead- 39
    stanniferous 41
grain 15, 25, 57, 61, 65, 68, 71-78, 80, 82, 83, 85, 87, 111, 126, 127, 146, 150, 203, 207, 234, 245, 247, 301, 303, 449, 454, 466
grain boundary 56, 66, 72, 75, 82, 87, 213, 247, 302, 306, 449, 458, 478
granulation 132, 160, 162, 334
graphite 329, 340, 344, 380
Griffith 265, 266
grinding 112, 119, 132, 157, 232, 286, 287, 332, 527
growth 61, 71-78, 85, 191, 208, 212, 225, 234, 238, 245, 247, 250, 271, 307, 311, 338, 459, 467, 542, 544
    crystalline 81, 238, 245, 494, 506

H

halides 11, 13, 18
hardness 26, 46, 123, 166, 201, 205, 215, 243, 257, 280, 285, 294, 299, 327, 329, 333, 336, 343
Hubbard 403, 404
hydroxide 502, 506
hydroxyapatite 116, 304, 499, 513, 515
hysteresis loop 431, 438

I

impedance spectroscopy 459
inflammatory reactions 503, 510-512

injection 145, 170-172, 176-179, 182, 183, 190, 328, 334
inorganic compound 9
interaction spin-lattice 438
interface 57, 61, 62, 66, 68, 72, 133, 135, 210, 250, 251, 296, 379, 439, 444, 445, 478
ionic superconductor 413, 414
iono-covalent 9, 10, 13, 25, 26, 201

K

kaolin 8, 45, 46, 49, 99, 103, 109, 111, 115
kaolinite 8, 11, 81, 98, 99, 101
kiln 31, 35, 38, 39, 45, 46, 115, 118
kinetics 56, 61, 78, 90, 149, 150, 184, 186, 213, 241, 250-252, 279, 307, 396, 431, 544, 553
Kröger-Vink 394

L

law of scale 69
Lichtenecker 450-452, 465
luster 40, 41, 47

M

machining 255, 286, 287, 326, 334-336, 338-340, 501
magnesia 10, 13, 17, 82, 199, 200, 216, 217, 304, 360, 370, 385
    carbon 380
materials
    inorganic 5, 8
    refractory 8, 12, 13, 15, 17, 26, 74, 126, 170, 203
metallic elements 9, 10
metals 6, 10, 86, 205, 226, 232, 277, 286, 294, 302, 326, 359, 475
metastable 23, 74, 200, 202, 205, 214, 221-223, 225, 337, 430, 445, 502, 515
microcracks 225, 265, 310, 315, 337

microstructure 18, 25, 55, 56, 68, 72, 73, 75, 82, 113, 125, 128, 129, 146, 147, 150-152, 172, 192, 204, 208, 225, 243, 244, 246, 258, 374, 447, 449, 466
milling 132, 160, 194, 204, 333, 334
mirrors 560
mobility
    defect 410
    electronic 399, 400
modeling 106, 118, 242, 334, 382, 447, 449
modulus
    of elasticity 10, 288, 502, 513
    shear 279, 293, 197, 300
    Young's 205, 206, 215, 224, 265, 280, 293, 294, 299, 331, 495, 505
monoclinical 74, 219
monocrystal 8, 16, 64, 200, 210, 214, 219, 238, 250, 299
morphotropic 469
mother of pearl 497, 498
MOX 525, 527, 532, 534, 537
mullite 13, 46, 60, 102, 111, 113, 199, 200, 213-216, 361, 368
multilayer technology 462

N

Nabarro-Herring 74, 87, 213
Neolithic 31-33, 35, 45
nitride 8, 17, 125, 191, 193, 233, 234, 243, 257
    aluminum 210, 243, 246, 259
    silicon 60, 61, 82, 192, 234, 242, 244-246, 258
non-linear optics 549, 550
nucleation 250, 428, 498, 502, 504, 506, 515, 516, 533, 553

O

optical properties 16, 18, 104, 219, 243, 554, 559

organic additives 120, 124, 158, 160, 162, 170, 173, 184-187, 189
Orowan 265
osteoblasts 507
output work 444
oxidation 26, 37, 211, 217, 234, 236, 243, 249-252, 348, 376, 489, 510
oxide 8, 10-12, 16-18, 39, 60, 96-98, 103-105, 113, 116, 133-135, 202, 205, 216, 219, 226, 227, 232, 233, 244, 248, 301, 339, 358, 363, 368, 377, 379, 391, 395, 402, 403, 406, 413, 435, 437, 462, 475, 484, 485, 502, 526-530, 550, 560
    calcium 509
    multi-elemental 550
oxygen probe 487

P

Paleolithic 30
paraelectric 416, 423-425, 465, 469, 478
paramagnetic 402, 405, 434
particles 30, 32, 33, 56-59, 62, 64, 69-72, 81, 87, 90, 118, 125-134, 137-139, 142, 143, 146, 155, 156, 160, 171, 223, 224, 234, 245, 363, 368, 503, 511, 540, 546
paste 31, 34, 46-49, 96, 98, 101, 109, 112, 114-118, 142, 170, 176, 177, 183
Pauling 18, 21
PCD 339, 340, 344, 345, 347, 348
Peierls 299, 301, 405, 406
percolation 432, 451, 452, 454, 456, 534
phase
    inorganic 9
    liquid 58, 59, 61, 65, 66, 70, 76, 78, 80-82, 114, 173, 204, 245, 338, 454, 484
    solid 59, 66, 70, 71, 81-83
phosphate

calcium 499-501, 503, 506, 511, 514, 516
    tricalcium 499, 516
piezoelectric 417, 418, 420-422, 429, 433, 456, 457, 469, 472, 474
piezoelectric motors 472
pitch 18, 32, 366
plasticizer 156, 157, 163, 167, 172
PMMA 502, 516
point defects 213, 217, 395, 399, 404, 407, 427, 432, 527-529
Poisson's ratio 205, 224, 265, 294, 300, 313, 314
polaron 400, 401, 405, 413
polycrystal 56, 73, 201, 211, 301
polymorphism 23, 25, 219
porcelain 12, 43, 45, 46, 49, 73, 102, 112, 114-117
    soft-paste 47, 48, 115
porosity 43, 57, 59, 61, 71, 85, 90, 110, 112, 114, 150, 164, 211, 244, 264, 329, 337, 363, 374, 376, 453, 489, 512, 514, 516, 526, 536, 542, 559
powder 6, 7, 16, 55-59, 61, 64, 86, 90, 119, 125-127, 131-133, 146, 150, 155, 159, 160, 164, 166, 182, 194, 203, 210, 232-235, 250
pressing 85, 86, 88, 96, 108, 110, 117, 159, 160, 164, 166, 168, 169, 176, 183, 244, 247, 329, 334, 337
process-zone 273
prostheses 16, 494-497, 509, 513
PSZ 220-222, 226
pyroelectric 415, 416, 423, 431, 432, 481
pyrotechnology 31

Q

quartz 8, 11, 22, 23, 47, 81, 103, 108-111, 115, 202, 391, 474

R

R-curve 271-273, 275
relative permeability 435
relaxors 421, 431, 466
resonance 421, 422, 470, 473, 474
    nuclear magnetic 542
rheology 98, 107, 131, 142, 183
RIM 530
rock 5, 9, 11, 30, 102, 200, 202, 341-343
route
    organo-metallic 552
    polymeric 552
    sol-gel 560
ruby 200, 202, 210

S

Schottky barrier 441, 443, 445
screen printing 460, 461, 467, 470
self-similarity 545, 547
semi-conductivity
    extrinsic 394, 395, 404
    intrinsic 393
shard 30, 33, 96-99, 103-111, 113-117
shear 142-145, 151, 156, 172, 173, 179, 219, 279, 293, 297, 300, 461
shock 208, 327, 342-344, 382
    thermal 4, 48-50, 206, 209, 215, 257, 258, 313-316, 327, 331, 337-339, 361, 368, 370, 487
silica 10, 12, 16, 23, 47, 79, 98, 101-103, 109-111, 134, 148, 199, 214, 238, 239, 249-252, 338, 368, 475, 502, 539-541, 546
silica-alumina 360, 361, 368
silicates 10, 11, 18, 21, 22, 250, 380, 390, 392, 495, 540
silicon 10, 11, 21, 26, 98, 192, 233, 244, 248, 300, 339, 372, 391, 542-544

silicon carbide 21, 232, 235, 238, 243, 246, 247, 254, 255, 338, 348, 460, 497
sintered hard metals 326, 329, 333-336, 340-342, 346, 348
sintering
    reactive 59, 60, 216
slag 44, 217, 358, 370, 375-377, 379, 380
small-angle diffusion 546, 547
sol 541, 554, 556
solid electrolyte 407, 413, 487-489
solubility 81, 133, 148, 156, 189, 214, 245, 498, 508-510, 552
spin coating 556
spinel 17, 101, 199, 216-218, 360, 363
stabilization 133, 137, 138, 140, 141, 220, 226, 504
stacking of particles 128
state density 390, 392, 402, 443
steel industry 17, 209, 258, 368, 370
steel
    high-speed 329, 332, 333, 336, 340
stoneware 43, 44, 102, 106, 110-114
strength
    mechanical 4, 74, 86, 96, 98, 109, 117, 148, 152, 167, 172, 215, 216, 224, 336-339, 495, 501
stress 25, 66, 69, 104, 142, 144, 145, 166, 172, 182, 210, 216, 222-225, 265-268, 273-277, 289, 296, 297, 307, 310, 331, 361, 368, 381, 382, 384, 385, 417, 418, 420, 430, 432, 438, 505, 536
substrate 154, 191-194, 209, 210, 242, 249, 335, 459, 460, 494, 507, 555, 556
superconductors 4, 16, 17, 406, 414
super-exchange 404
superplasticity 264, 303, 306

suspension 96, 131, 139, 147-158, 160
SZ 219-221

T

temper 33, 102
tension 268, 276, 279, 281, 305, 313, 338, 377, 385, 542
thermistors
    NTC 475, 479
    PTC 456, 477, 479
thermodynamics 56, 141, 376, 553
thermoelastic analysis 314
thermomechanics 358, 381, 387
terra cotta 3, 4, 8, 12, 31, 41, 99, 106, 107
tetragonal 23, 74, 219, 221, 423
thin layers 191, 554
tile 107, 108
TiN 332, 334, 335, 337, 497
tool 286, 326, 327, 329, 332, 334-339, 341, 342, 344, 346, 347, 376, 381
toughness 4, 25, 74, 206, 222-225, 253, 266, 270, 271, 280, 281, 327, 343
toxicity 338, 510, 512
tridymite 23, 102
type
    I 465
    II 465
    III 468
TZP 222, 226

U

UO_2 524-527, 529, 530, 537

V

vitreous 7, 39, 47, 66, 69, 73, 97, 105, 106, 117, 204, 303, 462, 496, 499
vitroceramics 7, 498, 499, 550

W

Wagner 452, 454
wave guides 554, 556
wear 15, 132, 207, 283-286, 327, 333, 341-343, 346, 348, 375, 512
wear debris 285, 495, 496, 509
Weibull 264, 273, 275, 277, 306, 309
wetting 63, 75, 79, 133, 156, 182
wheel 34, 38, 61, 286

X

xerogel 542, 549, 550

Z

zirconia 17, 23, 74, 199, 219-221, 226, 337, 360, 363, 412, 489, 496, 513